Birkhäuser

Classic Texts in the Sciences

Classic Texts in the Sciences offers essential readings for anyone interested in the origin and roots of our present-day culture. Considering the fact that the sciences have significantly shaped our contemporary world view, this series not only provides the original texts but also extensive historical as well as scientific commentary, linking the classical texts to current developments. Classic Texts in the Sciences presents classic texts and their authors not only for specialists but for anyone interested in the background and the various facets of our civilization.

David Coleman
Editor

Thomas Robert Malthus – An Essay on the Principle of Population

A New Look for the 21st Century

Birkhäuser

Editor
David Coleman
Department of Social Policy and Intervention
University of Oxford
Oxford, UK

ISSN 2365-9963 ISSN 2365-9971 (electronic)
ISBN 978-3-031-94502-1 ISBN 978-3-031-94503-8 (eBook)
https://doi.org/10.1007/978-3-031-94503-8

Mathematics Subject Classification: 00-01; 01-01; 01-02; 00Axx

Classic Texts in the Sciences

This book is published under the imprint Birkhäuser, www.birkhauser-science.com by the registered company Springer Nature Switzerland AG
The registered company address is: Gewerbestrasse 11, 6330 Cham, Switzerland

SACRED TO THE MEMORY
OF THE REV. **THOMAS ROBERT MALTHUS**
LONG KNOWN TO THE LETTERED WORLD
BY HIS ADMIRABLE WRITINGS ON THE SOCIAL BRANCHES OF
POLITICAL ECONOMY

PARTICULARLY BY HIS ESSAY ON POPULATION

ONE OF THE BEST MEN AND TRUEST PHILOSOPHERS
OF ANY AGE OR COUNTRY
RAISED BY NATIVE DIGNITY OF MIND
ABOVE THE MISREPRESENTATION OF THE IGNORANT
AND THE NEGLECT OF THE GREAT
HE LIVED A SERENE AND HAPPY LIFE
DEVOTED TO THE PURSUIT AND COMMUNICATION
OF TRUTH
SUPPORTED BY A CALM BUT FIRM CONVICTION OF THE
USEFULNESS OF HIS LABORS
CONTENT WITH THE APPROBATION OF THE WISE AND GOOD

HIS WRITINGS WILL BE A LASTING MONUMENT
TO THE EXTENT AND CORRECTNESS OF HIS UNDERSTANDING

THE SPOTLESS INTEGRITY OF HIS PRINCIPLES
THE EQUITY AND CANDOUR OF HIS NATURE
HIS SWEETNESS OF TEMPER URBANITY OF MANNERS
AND TENDERNESS OF HEART
HIS BENEVOLENCE AND HIS PIETY
ARE THE STILL DEARER RECOLLECTIONS OF HIS FAMILY
AND FRIENDS.

BORN FEB:14.1766. DIED 29.DEC:1834.

Malthus's memorial in Bath Abbey

Contents

Malthus: The Last Words. Comments and Reflections of the 6th and last Edition of Malthus's 'Principle of Population' (1826).

1

David Coleman

> 'Towards the extinction of controversies about Malthus, no observable progress whatever has hitherto been made.[1]' (anon, after Malthus).

> 'The passion between the sexes has appeared in every age to be so nearly the same that it may always be considered, in algebraical language, as a given quantity'. (Malthus 1826, p. 293).

> 'Malthus was a clear-headed theoretician of traditional society, but he was a prophet of the past. He was born too late in a world too new. (William Hazlitt, 1825).

Preface

The Rev[nd] Thomas Robert Malthus[2] FRS was a clergyman of benign temperament but divisive views on the way the world works. Writing initially in 1798 he has continued to inspire and infuriate[3] ever since, as a shelf of recent books on his work testifies. Setting

The original version of the chapter has been revised. A correction to this chapter can be found at https://doi.org/10.1007/978-3-031-94503-8_5

[1] See Flew edition p. 148

[2] Usually known as 'Robert'. The surname 'Malthus' is often assumed to be of foreign origin, but it is likely to be a contraction of 'Malthouse'.

[3] Far from reactionary as often accused, he was in favour of most progressive causes (Dupâquier 1983, p x). But see Shermer (2016) and Stone (1986, 2021).

D. Coleman (✉)
Department of Social Policy and Intervention, University of Oxford, Oxford, UK
e-mail: david.coleman@spi.ox.ac.uk

D. Coleman (ed.), *Thomas Robert Malthus – An Essay on the Principle of Population*, Classic Texts in the Sciences,
https://doi.org/10.1007/978-3-031-94503-8_1

aside the demographic detail of his argument, among analysts and prophets of progress Malthus is in the pessimistic corner. Provoked into print by the endorsement of perpetual progress promoted by philosophic writers of the late 18th century Enlightenment, Malthus damped those dreams by reminding the world that resources were limited but the capacity of man to increase in numbers is great. That inevitable and perpetual collision, he insisted, would defer the golden age forever ensuring that progress, if any, was slow, painful and finite. Although situated in the mostly agrarian circumstances of the late 18th century his prescription of the power of population to (prevent) progress has obvious resonance for today's concerns about the limits of growth - now a further source of controversy. And so the debate is given new life. This introduction, published on the eve of its 200th anniversary, presents his last[4] word; the final fully worked out edition of his 'principle of population', and comments on its possible implications for contemporary issues.

1.1 Malthus and His Times

What were the circumstances in which Malthus wrote, which provoked him to such pessimism – apparently perverse, in the face of much obvious advance? He was born, in 1766 at a time of accelerating change in economy, society, politics, opinion and science. Rational Enlightenment views about the prospect of progress had challenged traditional authority and continuity with the past. In England, since the 'Glorious Revolution' of 1688 monarchical authority had been dethroned, with parliament in the ascendant. Modern party politics was emerging; and acquired a head in the first' Prime Minister': Robert Walpole 1721-42. Dissenting churches, persistent recusant Catholicism and atheist and freethinking ideas challenged the authority of the established Church. Malthus himself was in the thick of it. Born into an intellectual household; his progressive father a correspondent of Rousseau and his home a salon for intellectuals and thinkers. Malthus was by no means a follower of his father's ideals. Home was a hotbed of intense discussion and argument. In fact Malthus's polemic was provoked by a disagreement with his father about the views of the contemporary *bien pensant* on the perfectibility of man and his estate (Charbit 2000 p. 9).

In the world outside this domestic intellectual ferment, change of every kind was under way. When Malthus was born Britain was still primarily an agrarian country although economic growth was picking up as manufactures gained ground. Land comprised half the national capital, rents 20% of national income. Britain and the Netherlands were in the forefront of economic change, industry initially riding on the back of the products of the 17th century agricultural revolution and from overseas slavery. Adam Smith's 'Wealth of Nations' (1776) showed a way to economic growth. From the 16th century, coal became the main source of energy, displacing firewood and animal power (Table 1.1, Warde 2007 Table 4 p. 69.). Already in the year of Malthus's birth, coal provided 64% of the nation's

[4]The 'Summary View...' was in fact published later, in 1830. But this work really dates from 1824; it consists of the larger part of Malthus's article on 'Population' published in the 1824 Supplement to the *Encyclopaedia Britannica.*

Table 1.1 Energy production in England and Wales 1560-1900.

	Energy consumption England and Wales 1560 = 1900 (percent of total). Source: Warde 2007 Appendix Table 1.						
	Draught animals	Population	Firewood	Wind	Water	Coal	All
1560	33.3	22.0	35.8	0.3	0.8	12.8	100
1700	19.9	16.4	13.7	0.8	0.6	46.6	100
1760	13.4	12.2	8.9	1.2	0.5	63.7	100
1834	4.5	6.0	0.7	1.5	0.2	87.1	100
1900	1.7	2.0	-	0.3		95.2	100
	oil made a contribution by 1900 of 0.8%						

energy. At his death in 1834 that proportion had reached 87%. Britain led the world into the industrial revolution, becoming the first country - from about 1850 = where most of the population lived in cities. Towns especially London were growing fast.

Scientific and technical knowledge had picked up speed over the previous century, encouraging a rational measure of the world and its workings. Although science and religion started to part company over the period, initial progress in 'natural philosophy' owed much to Puritan millennarial thought, as Webster explains (2002, Wallis 2003). Clergymen were no stranger to the debates of the Royal Society. Malthus apart, the Revnd John Mitchell FRS first proposed the concept of a black hole in 1783. William Buckland FRS, Reader in Mineralogy, Dean of Westminster, announced and described the first dinosaur in 1824[5], The Revnd James Bradley FRS, Savilian Professor of Astronomy in the University of Oxford, established the first accurate estimate of the speed of light and its aberration in 1729. In the provinces, the Revnd' Gilbert White pioneered the study of nature in action; his 'Natural History and Antiquities of Selbourne' (1798)', never out of print, created a new genre. One wonders if these divines ever had any time for the cure of souls[6]. Telescopes and microscopes, differential calculus, logarithms, astronomical developments underlined the scale and complexity of the universe and opened doors to its understanding. Buckland's and Hutton's revelations of the antiquity of the world and the variety of its past life expanded intellectual horizons and challenged biblical accounts. The measurement of longitude with Harrison' chronometer expanded international trade; financial and banking innovations made possible investment in canals, a near-invincible navy and the subsidy of other nations to fight Britain's continental wars. Science and practical knowledge flourished in the intellectual independence encouraged by the Royal Society (1660) and by the Academies of Science established around that time in a number of European countries. Markets, a mobile labour force, canals and empirical innovations in agriculture had long put famine in the past and made modest inroads into mortality. Watt's improved

[5] *Megalosaurus* welcomes visitors to his display cabinet in Oxford's University Museum of Natural History

[6] Malthus, however, remained actively involved in theological matters (Waterman 1983).

steam engine further freed the economy from the constraints of an organic economy subject to a severe limit to growth. Hitherto domestic food supplies and power had been limited to human and animal muscle, wind and river, as in all previous centuries. Now man had begun to control his environment, not merely accept it.

There was little echo of these contemporary developments in the Second Essay; a reflection of Malthus' reluctance to acknowledge the transformation into an industrial society and economy. The colossal European war then in progress when he was writing the later editions, goes unremarked except for a brief period of demographic tourism in 1803. Although mentioned as a check on population, war only receives extensive attention in the ethnography. As it is, we hear little more about that conflict than in 'Pride and Prejudice' (published 1813). However, in his other writings on political economy Malthus the celebrity public intellectual was fully engaged in the debates of the time on the economy, poverty, theology and other disputatious topics (Winch 1996). This included the problem of theodicy[7] arising from the evils which he acknowledged followed from his Principle of Population. In the Second Essay he devoted much attention to that problem, although the theological chapters in the First Essay had been dropped. In a glimpse of the future he was 'cancelled' by opponents, banned from publishing anything more in 'The Edinburgh Review'.

1.1.1 The Social and Economic Context of the Essay on Population

Debates on poverty in the past have surged back and forth from 'optimistic' e.g. Clark (2001) to 'pessimistic' interpretations about the condition of the poor. Overall, economic growth then was moderate, never dramatic. Flatlining during the Napoleonic war (Crafts and Mills 2009), it took off from the 1820s. Then gross domestic product (gdp) per head was £1937 and had reached £2717 by 1850[8]. In that evolving economy there were losers and gainers. Most economic historians agree that the early stages of the industrial revolution benefited the poor little (e.g. Feinstein 1998). But not all (Griffin 2018, Clayton and Rowbotham 2008). However their conditions seem to have improved slowly in the first quarter of the 19th century (Gallardo-Albaran and de Jong (2020) and then at a faster pace. Expectation of life at birth was about 37 years in the 1780s, rising to 41 years (both sexes) in the first English Life Table of 1841. This heavy mortality was concentrated in infancy and childhood, which skewed the average. About one in seven babies did not pass their first birthday. Once the hazards of infancy were passed, expectation of life for males increased from 40 years at birth to 47 at age 10. A young man of 20 could expect to live to about 60 years and if still alive then, a further 14 years. Men aged 60 were not rare and were not

[7] An argument that attempts to explain the co-existence of evil with a benevolent and omnipotent God. As in other matters, Malthus changed his mind on this in the course of his career.

[8] All data in 2019 pounds Sterling. Source Measuring Worth

old. Women enjoyed an advantage of about two years at birth. Malthus was not that exceptional in living to 68[9].

By Malthus's time the rural population could not be described as 'peasants' in the classical sense of the word (Macfarlane 1978). Most were small proprietors ('yeoman farmers'] or customary tenants or wage labourers with modest holding of land if any, in a relatively mobile monetised economy of rents, wages and markets. In lowland England, mobility was such that few villagers had local descendants after a couple of generations (Coleman and Salt 1991 table 1.5). Famine had been consigned to the past; the last regional famine was in 1623 (Healey 2011). But the condition of the poor was marginal and wages stagnant. The situation of agricultural labourers (wage labour, 'farm servants' and 'husbandry servants') changed around the turn of the century, mostly for the worst. Hitherto many of the young and unmarried had boarded with the farmer (Kussmaul 1979, Laslett 2004) and by some accounts fared well (Burnett 1968). This custom of accommodating non-relatives was one of a number of distinctive features of the English household (Hajnal 1982). Standards fell when farm servants began to live out and became day labourers, following the increase of prices during the Napoleonic wars. After the end of that war wheat prices fell to less than half the wartime value. The labourer's income was now in cash but they had to fend for themselves. Prices fell but his wages more so.

Engels[10] (1845) blamed the advent of the factory system for immiserating workers. Nonetheless many moved from the land to the growing towns, attracted by the higher, regular industrial wages on offer. In a new world of factories some traditional trades did badly. Handloom weavers and framework knitters, formerly the aristocracy of labour, fell from grace, wages dropping from 30/- week to 5/- by the 1830s. Others; carpenters, bricklayers, compositors, boilermakers did well. Incomes, accordingly diet, were correspondingly highly varied between trades and over time. An estimate of 1867 by Dudley Baxter put 50,000 into an 'upper class' with £1000 or more per year, 1 million into a middle class with £300 or more per year, and 2.8 million unskilled or agricultural labourers with under £100 per year. (Burnett 1968 p. 124). According to Rowntree and Booth at the end of the 19th century, between a quarter and a third of the population lived in 'poverty',

1.1.2 Welfare and the Poor Law

The Poor Law is a central feature in Malthus's story. From the Middle Ages at least, in England and some other parts of NW Europe the nuclear family was the usual form of household. It often included relatives, lodgers and among the better sort, husbandry servants, but not the extended family typical of continental peasant society. Because its

[9]By this stage, higher social status conveyed longer life. It was not necessarily so in earlier periods.

[10]Friedrich Engels, 1820-1895. Businessman, revolutionary socialist, author, philosopher. Marx's closest collaborator.

composition and self-sufficiency varied over its lifespan, it often needed external support, from church, manor and neighbours. The Poor Law (1597, 1601) formalised these arrangements, established during a crisis of welfare to prevent social unrest by protecting the poor from want. In the 16th century the abolition of the monasteries with their (erratic and unsystematic) charitable activities (Scott 2015), and new rural conditions, had increased destitution. With changes in profitability pasture replaced arable, sheep replaced crops, a few shepherds displaced numerous ploughmen. Groups of unemployed 'sturdy beggars' disturbed the peace. The Poor Law was a 'welfare state in miniature' (Blaug 1963, 1964). It obliged each of the 15,000 parishes of England to support their local poor (paupers unable to support themselves). This through 'outdoor relief 'in payments in money or kind, or residential 'indoor relief' in poorhouses. Recipients included the aged and infirm, the unemployed and workers too poor to support their family. Poor Law expenditure was very high and increasing, from about two million pounds in around 1784 to six million in 1815 (Marshall 1968). Over 20% of the population received relief in the difficult years between 1817 and 1821. Malthus considered the alarming increase in cost to be a proof of his principle of population.

The system was made more malign, in Malthus' view, by the 'Speenhamland' reforms (Taylor 1991), which extended the scope of poor relief, subsidising the wages of labourers in proportion to the number of children of the recipients and the price of bread, thus (according to Malthus) exacerbating the very process of excess family building which was responsible for the distress in the first place. Introduced by the magistrates of Speenhamland (Berkshire) in 1795 it extended relief to the employed poor in a crisis, increasingly unable to support themselves and more than a small family on their wages. Thus if a loaf cost 1/- a married man received 4/6d per week if 2/-, 7/8d. On the higher bread scale, with two children 12/6d, with seven 23/- (Burnett p. 33)[11]. However little the famer paid, the unemployed received the same as the employed. 'High prices first transformed the farm servant into a day-labourer; low prices and the poor law had turned the day-labourer into a pauper' (Burnett 1966 p. 33). This increasing burden on local ratepayers (over 8% of gdp) led many to agree with Malthus that the system encouraged idleness, subsidised farmers enabling them to offer lower wages, and encouraged imprudent early marriage and more dependent children. These concerns led to the replacement of this system by the New Poor Law in 1834, a reform in which Malthus played a decisive role. Intended to make recourse to benefit less attractive, this measure (finally abolished by the National Assistance Act 1948) confined support to indoor relief in workhouses, where menial work was exchanged for welfare.

The British Government's Two-Child Policy: The Poor Law Revisited

Despite the general triumph of welfare, an unintended natural experiment in restricting welfare has been in progress since 2017. It might throw light on Malthus's argument about

[11] Until 1971 there were 240 pence (d, for denarius) and 20 shillings in one pound sterling.

the effect of income supplements on fertility. The chief welfare support for the family in the UK is the means-tested Universal Credit. On top of that families can claim 'child tax credit' for each child, up to £3455 per year each. In 2017 this benefit was restricted to the first two children only, born after 6 April 2017. That has affected almost 1.6 million children in 440,000 families, saving the exchequer £1.7 billion per year according to the Child Poverty Action Group (CPAG). The authoritative Institute for Fiscal Studies, however, estimate that its abolition would cost £3.4 billion per year but reduce the number of children in relative poverty by about 500,000. This is not the benefit cap, which separately restricts the amount of support a working-age household can receive from the social security system. That affects 123 000 families according to the CPAG.

This rule, unique in Europe, is claimed seriously to increase child and family poverty among some of the poorest families. Despite strong feelings the new (2024) Labour government in the UK has so far been unable to promise to abolish the rule. It was not introduced for demographic reasons or - heaven forbid - eugenic ones, but to save money. However some anonymous ministerial comments[12] have welcomed its supposed demographic effects, removing an incentive for family growth in poor households. HMRC (tax) statistics show that in 2023, 50% of families affected by the two-child limit were single parents and 57% had at least one adult in paid work (retrieved 17 Jun 2024). According to the CPAG the measure seems only to have reduced the chance of having a third child by five percent. This is attributed to lack of foresight, lack of knowledge and preference for larger families especially among Muslims (the rule particularly affects Pakistani families). So this potential, unintended reinforcement of the preventive check seems to have been ineffectual. Malthus would have been disappointed. Had it been effective, it would of course have exacerbated to the current birth dearth which has taken total fertility in England and Wales to its lowest level ever of 1.44 (retrieved 2024).[13]

1.2 Aims of this New Edition of the Second Essay

This is a new edition of the Second Essay by Malthus on population. What new can possibly be said, to justify this additional burden on our bookshelves? Hundreds of journal articles offer comment, support and refutation of his views. Many distinguished books address all aspects of the whole Malthusian scene (Appendix Table 1). Old controversies still rage, whether Malthus was essentially right in his analysis of the relationship between population and resources in the past times, never mind any relevance his model may have for contemporary problems. Dismissed by Hazlitt (1825) as a 'prophet of the past, born too

[12] In the previous administration.

[13] It is generally thought that a sustainable level of population ageing cannot be maintained without a total fertility of at least 1.6. Replacement fertility (about 2.1 in a modern population) requires at least 30 percent of women to have at least three children.

late in a world too new', Malthus nonetheless remains popular both as demon and as prophet. This introduction does not revisit the controversies of the past, about which much has already been written by a glittering array of experts attracted – or repelled – by his ideas. The arguments about the Poor Law and welfare, which occupy much of the latter part of the Second Essay, are considered only briefly. That issue has been resolved by the triumph of welfare, in the English context by the Welfare State of 1948 and the National Health Service 1948, in an agnostic age 'the nearest thing the English have to religion'.

Instead it has two aims: First, by republishing the long 'Second Essay', to bring to the interested public the Full Malthus, his extensively worked-out theory with supporting data and historical examples, which are thin on the ground was both the First Essay (FE, 1798, see Flew 1970) and the Summary View (1830). Of the First Essay of 1798, there are numerous excellent editions. The Second Essay is less easily available. Second, to explore critically the extent to which the Malthusian framework of population regulation works in the contemporary world including issues such as global population growth and climate change. To some these appear to be Malthusian problems, but they are highly contentious. 'It was Malthus's fate to frame an analysis of the relationship between population, economy and society during the last generation to which it was applicable'. (Wrigley 1983, p.112). Without doing too much violence to the ideas of Malthus, let us see how far that remains true.

1.2.1 A Stumbling Block

Those who condemn Malthus without reading much of him do have a slight excuse. The Second Essay is very long. As the editor of an 1890 one-volume 612-page reprint, G.T. Bettany, remarked 'The author purposely abandoned the higher graces of style when he enlarged his book... and the book lost somewhat of balance and proportion in its different parts.'[14] (Introduction p. xxxiv). Malthus himself admitted, 'I am fearful that I shall appear to have treated some parts of it too much in detail, and to have been guilty of unnecessary repetitions.' (Preface to the second edition p xxxvi). The philosopher John Stuart Mill commented unkindly 'Malthus made what was obvious seem astonishing, what was simple complicated, what was clear obscure and what was easy difficult' (cited in Charbit 2000 p. 11.) The focus on the First Edition has encouraged much misunderstanding. Many critics (e.g. Foster (1998, p. 2) insist that Malthus' work was 'an attempt to prove that future improvements in the condition of society, and more fundamentally in the condition of the poor, were impossible'. In his Second Edition Malthus says otherwise.

[14]Reprinted from the 6th edition London, Ward Lock 1890.

1.2.2 What Is New in the Second Essay?

The first edition of the Principle of Population was a polemic written in irritation and exasperation provoked by the 'utopian optimism' about the perfectibility of man and society of the brilliant mathematician the Marquis de Condorcet (1743-1794), the political philosopher William Godwin (1756-1836) and others. It was a riposte to the ideologies and enthusiasms unleashed by the French Revolution, in sharp contrast to its rapturous welcome by Wordsworth and other British romantics[15]. Their views, fuelled by the huge popularity of Godwin's master work 'Enquiry Concerning Political Justice (1797) and the inspirational 'Sketch for an Historical Picture of the Progress of the Human Mind' (1794) by Condorcet, enjoyed a tremendous vogue at the time, chiming with the hopeful sentiments underwritten by the early days of the French Revolution.

Malthus felt that this optimism would be tripped up, as it always had been, by collision between two inexorable forces: the antithesis of which was then already well known; the tendency of human populations to grow faster than the slower growth of the means of their subsistence. The former able, without check, to double every twenty five years driven by the constant urgings of the 'perpetual passion' between the sexes; the latter, essential for life, limited by the modest availability of new land for agriculture and the slow progress of increasing productivity, stifled by the law of diminishing returns. If actively in conflict these forces could only be resolved by premature mortality; Malthus's 'misery'. If averted by 'improper arts' of contraception or abortion[16] then by 'vice'. Those dismal checks would see off any utopian fantasies of progressive perfectibility. These conclusions led him to oppose policies to ameliorate poverty such as the Poor Relief Act ('The Poor Law', McIntosh 2011) which he thought would exacerbate the situation as discussed earlier.

The Second Essay presents a thorough account of Malthus' mature thought, supported by abundant data and examples from diverse places and times. Some of this evidence persuaded him to change his mind on some topics or at least undermined his confidence in previously firm assertions, for example on industry and the Poor Law. So although they are the 'last words', ambiguities remain. Aware of the limitations of the evidence that he could present in the First Essay, responding to comments on it, he wrote what is in effect a new book. In it he implicitly recognises the significant shortcomings of his Principle of Population and attempts to revise and qualify it. It marshals as much demographic, social and ethnographic data as he could find from home and abroad to support his thesis and develop its theory and solutions. In the First Essay there are just a handful of tables. Now there are many, detailed and meticulous calculations in the text, and formulae relating to growth. A different book which should have had a different title. The 'Second Essay' went

[15]'Abhorrence of 'An Essay on the Principle of Population' was almost the only matter on which the major poets were firmly united'. Woodring 1970, cited in Winch 1996 p 289

[16]As matters *inter Christianos non nominanda,* neither ever explicitly mentioned by Parson Malthus. Until the 1817 edition he appears not to have understood exactly what 'contraception' involved.

through a number of revisions; the sixth of 1826 being the last. That differs little from the 1817 edition. But it very long (614 pages in the 1890 single-volume edition), and little read; hence the source of much misunderstanding.

1.2.3 The Structure of the Second Essay

The work is organised into four 'Books' I to IV, subdivided into 'Chapters'. Beginning by reprising the general principles of his population model, Malthus then addresses the lack of statistical evidence which he deplored in the First Essay. Fully one third of the book attempts to demonstrate how checks to population operate across the full spectrum of human society. In Books I and II Malthus gives us an extensive ethnography and history, possibly the first of its kind, in search of the checks to population in every possible culture. Professional, academic ethnography had not then been invented; Malthus is a pioneer. He scoured travellers' and explorers' accounts, the reports of ambassadors, the reports of the ubiquitous, inquisitive Jesuits, colonial officials and missionaries, local officials and classical authors, and extensive correspondence, with exemplary thoroughness. That's all there was, and usable 'data' were usually completely absent. He made many errors, especially about hunting and gathering societies leading an essentially Palaeolithic life. Revelations about the relatively leasured life led by many of those societies were not widely known before the publication of the seminal 'Man the Hunter' (Lee and DeVore 1968). While vital rates of different hunter-gatherer societies have been quite varied, their population sizes were typically well below the carrying capacity of their environment (Read and LeBlanc 2003, Guzmán et al. 2018).[17]

Malthus cites Hindu scriptures (105-6) and Chinese censuses and classical authors. Some comments (e,g, on India) stand up reasonably to modern knowledge, although much more is known today (Dyson 2018). Lee and Feng (1999) show that at least among the Manchu elite in China, a different kind of preventive check kept fertility low, through late marriage and frequent bachelorhood, while abortion and infanticide were routine in wider society. The extensive use of abortion and infanticide in Tokugawa Japan escapes attention (Woods (2000) 386-396, Eng and Smith (1976)). As Patricia James (1979, p. 93) remarks 'The Material is enthralling. I wonder nobody has ever produced an annotated edition of the second version of the essay simply as an excuse for the pleasure of thoroughly enjoying the sources'. In fact this early achievement has been recognised in a remarkable book by Bashford and Chaplin (2016). Malthus interprets with sensitivity the check to population from Brahminical avoidance of self -indulgence but reacts with aunt-like horror to the diet of large nutritious grubs favoured in Australia (pp. 16-17).

[17] This comment is in the past tense because so few hunting and gathering societies still pursue a fully traditional way of life.

Malthus emphasised the persistent diversity of moral principles between societies (Kreager 1986). 'While societies everywhere and at all times confront the same dilemmas posed by the Principle of Population their moral and political economies give rise to distinctive solutions'.(pp. 145-49). The protection of native peoples is a repeated concern. He expressed a strong and unfashionable opposition to the displacement of native people by colonists, which he thought would eventually lead to their extermination (Bashford and Chaplin 2016). He deplored slavery and rejected at length the abuse of his works by employers in favour of slavery (Appendix p 568-570) adding this double-edged footnote to the edition of 1806: 'As long as the nations of Europe continue barbarous enough to purchase slaves in Africa, we may be quite sure that Africa will continue barbarous enough to supply them'. However Malthus does not discuss the considerable effect that slavery had on the populations of Africa and the Americas. He believed that despite slavery and other checks 'the population is constantly pressing against the means of subsistence' (First Essay p. 44) a view which has had mixed notices (Godelier 1983 p 127, 145-146).

Book II presents with abundant data the checks to population in modern Europe. The Napoleonic War prevented travel to much of Europe but Malthus was able to visit neutral Norway and Finland (then part of Sweden) in 1799 where he gathered valuable data from that country's excellent registers, for example on late marriage. He was able to talk and correspond with local experts and see local conditions. But he underestimated the severe positive check inflicted by epidemics of the time. Recurrent crises crippled northern agriculture (Drake 1966, Dribe et al. 2015, Pitkänen 1992) made marginal by high latitudes. Later, he could visit France and Switzerland and collect much data, profiting from the travel permitted by the short-lived tourist interlude offered by the Treaty of Amiens (1803). He made use of the analyses of his fellow parson Johann Peter Süssmilch[18] and other statistical gurus, with the new data collected by church and state from early censuses and the older parish registers, including some highly focussed data from individual parishes. The discussions of checks to population in England, France and Switzerland in Chapter V are intense. Further chapters illustrate mortality, the fruitfulness of marriage, public understanding of the Causes of Poverty and Civil Liberty, the effects of Charity, and Plans to improve the condition of the poor, concluding in Chapter XIII where, emphasising the need for general principles, he extracts generalisations from the exhaustive, and exhausting, mass of data which he has assembled.

1.2.4 Books III and IV

The second half of the Essay considers problems and remedies. Book III 'Of the different systems...proposed as they affect the evils arising from the principle pf population' (p. 297) evaluates once again of the views of Godwin, Condorcet et al. He sets out his arguments

[18] 'The father of German demography' 1707-1767. Hecht 1987.

with considerable clarity but also at considerable length. Emigration (Chapter IV). is one of the few proposals which passes muster with Malthus, as a temporary expedient in response to the surge of population born in the boom years of the Napoleonic Wars (Pullen 2009). Malthus became much involved presenting evidence to the House of Commons Select Committee on Emigration (1827) from the United Kingdom to encourage emigration; to relieve poverty at home and strengthen the settlements of the empire. That impetus survived, as the Oversea Migration Board (1960) until 1960. England lost population to emigration for most of its modern history, in net terms until about 1990.

Chapter VIII and following on agriculture, commerce and international trade do not reveal Malthus at his best, showing his initial reluctance to accept the industrialisation that was already in progress. The section on trade, and the following Chapter XIII on increasing wealth and its effect on the poor connects with current controversies and the radical views of Thomas Piketty (2014) (pp. 415-425). In exchanges with Ricardo, Malthus was forced to step back somewhat from his Principle of Population (Kreager 2022). Chapter XIV on 'General Observations' makes wide-ranging generalisations on development anticipates modern analyses of the role of bad governance, and of culture for economic backwardness and failure by David Landes (1998) and others,

Book IV 'Of Future Prospects...of Mitigation of Evils arising from the Principle of Population' is devoted to improvement. It has quite a lot about sex, its attractions, and the essential benefits to future society of moral restraint (delayed marriage) in improving the condition of the poor. Malthus' discussion of these arguments (Chapter IV 'Objections....'), especially of possible consequent shortage of labour, looks forward to present problems. Chapter V begins with an apparently obnoxious but misunderstood paragraph, widely cited, on the need to maximise mortality among the poor. The final paragraphs of the preceding Chapter IV, however, show that Malthus is being sarcastic, responding to proposals to encourage marriage (and consequently population) among the poorest. Avenues for improvement include egalitarian reforms, Poor Laws, Emigration, the terms of commerce, reform of Corn Laws,[19] but for Malthus prudential restraint is sovereign (pp. 455-506). Public relations is not ignored in a long and rather tedious essay on public understanding, and changing it. The need to reform (gradually abolish) the Poor Laws is emphasised (Chapter VIII).

The book ends with a chapter of qualified optimism (Chapter XIV) on expectations for the future. Mortality has diminished (plague, smallpox) and the prevalence of the preventive check increased. Fertility[20] has moderated most where prudential restraint is most marked: Norway, Switzerland, Scotland and England (p. 539). Progress, apparent in the past, continues today (i.e. 1826) and into the future. This upbeat approach is in marked

[19]Here, 'Corn' refers to the food grains wheat, barley and oats. Maize was not widely cultivated in England until 1970.

[20]In demography 'fertility' refers to actual reproductive performance' that is the number of live born babies, not the physiological capacity to produce them, which is termed 'fecundity'.

contrast to the pervasive gloom of the First Essay. Malthus concludes that: ... if it (prudential restraint) were generally adopted, by narrowing the supply of labour in the market, would in the natural course of things, soon raise its price. The period of delayed gratification would be passed in saving... which would enable (the labourer) to in a few years to enter into the matrimonial contracts without fear of its consequences. The operation of the preventive check would give a real value to the rise of wages and the sums saved by labourers.... As the wages of labour would thus be sufficient to maintain with dignity a large family...all abject poverty would be removed' (p. 435).

1.2.5 A Step Backwards: The 'Satanic Verses' of the Principle of Population

Although the Second Essay is more optimistic and genial than the First, the most offensive of Malthus' observations on the responsibility of the poor for their own poverty are to be found in the 'allegory of the banquet' in the 1803 edition (Malthus 1803, page 531). Removed in all later editions from 1806, these remarks still did lasting damage to his reputation. These are the notorious passages where Malthus comments:

"With regard to illegitimate children, after the proper notice has been given, they should on no account whatever be allowed to have any claim to parish allowance.... The infant is, comparatively speaking, of no value to the society, as others will immediately supply its place." In the same vein he wrote: 'A man who is born into a world already possessed, if he cannot get subsistence from his parents on whom he has a just demand, and if the society do not want his labour, has no claim of right to the smallest portion of food, and, in fact, has no business to be where he is. At nature's mighty feast there is no vacant cover for him. She tells him to be gone, and will quickly execute her own orders, if he do not work on the compassion of some of her guests. If these guests get up and make room for him other intruders immediately appear demanding the same favour.... The order and harmony is disturbed, the plenty that before reigned is changed into scarcity.... The guests learn too late their error, in counteracting those strict orders to all intruders, issued by the great mistress of the feast (i.e. natural law), who, wishing that all her guests should have plenty, and knowing that she could not provide for unlimited numbers, humanely refused to admit fresh comers when her table was already full".

"We cannot, in the nature of things," Malthus further wrote, "assist the poor, in any way, without enabling them to rear up to manhood a greater number of their children." The essence of the Malthusian doctrine, Marx observed in 1844, was that "charity...itself fostered social evils." The very poverty that "formerly was attributed to a deficiency of charity was now ascribed to the superabundance of charity." (Wiltgen 1998). According to Malthus's Marxist critics, notably Foster (1998. p. 78), the author of this quote.

"...the claim that the poor were not entitled to the smallest portion of relief, and that any attempt to invite them to the "mighty feast" against the will of its "mistress" (who represented natural law in the allegory) would only come to grief—remained the central ideological thrust of the Second Essay throughout its numerous editions". (Foster 1998).

It seems extraordinary that views so contrary to conventional Christian charity should come from the pen of a clergyman, and from one reputed to have so amiable a disposition. It is as though Malthus had completely forgotten the Sermon on the Mount (Matthew 5.7) or the unqualified duty of charity implied by Matthew 25.35-40, and in many other passages in a book with which he should have had more than a passing acquaintance. But Malthus was much more attracted to natural rather than doctrinal theology. While divine providence and divine purpose is often referenced, neither Essay refers to scripture and only once (p. 477) Christ. Malthus's theology is both heterodox and confused (Waterman 1983). It is possible, of course, that these brutal comments are but early signs of his transubstantiation from parson into economist.

1.2.6 Malthus Rebuffed.

These remarks epitomise the moral grounds for the hostility towards Malthusian views, or what are assumed to be Malthusian views, antithetical not just to ordinary charity but more fundamentally hostile to alternative understandings of poverty and its persistence. A wide range of libertarian writers criticise what they see as attempts to restrict choice in reproduction and the coercion of population policies which (Follett 2020) wrongly believed to have been encouraged by Malthus. He is seen as an enemy of equality (Kallis 2019) or just plain wrong, ignorant about food and its production, and racist (Valenze 2023). Malthus obstructs Marx's view of poverty as a consequence of unjust distribution of wealth and entitlement (Marx 1867). Marx's philippic against Malthus is memorably vitriolic and amusing, nevertheless his hostility on economic grounds is ambivalent (Charbit 2000 Ch.5). Generally speaking, with Marx, left wing and progressive opinion condemned Malthus's model for plagiarising the work of previous authors, blaming the poor for their own misfortunes and being dishonestly motivated by a desire to defend the interests of the rich and privileged. In doing so, Marx claims, he wrongly ignored the arguments for more equal ownership of the means of production and exchange in the explanation, and solution, of poverty. Marx's intellectual leadership has inspired hostility to policies suspected of Malthusian origin (Meek 1953 and Chaplin 2018). But Marxian hostility is not absolute (Charbit 2000 p.1 *passim*). In 1845 Engels recognised the problems posed by population growth to the working class and that 'Malthus's theory had a good deal of truth in it'. In 1880 Kautsky[21] found a middle way between Marx and Malthus, arguing in favour of birth control (all cited in Petersen 1989). According to Wiltgen (1998) those unqualified criticisms have powerfully influenced thinking about population growth and poverty in

[21] Karl Kautsky 1834-1938. Theorist of Marxist social democracy, promoted a democratic road to socialism.

the developing world, delaying the implementation of family planning reform and thereby perpetuating and magnifying the poverty and want which exercised Malthus Bongaarts and Sinding (2009). Of course for other, religious reasons, Parson Malthus might have joined the godless Marx in criticising such programmes.

As Ross (2000) put it (1528-1531 '(Malthus's) greatest achievement, has been to provide an enduring argument for the prevention of social and economic change and to obscure, in both academic and popular thinking, the real roots of poverty, inequality and environmental deterioration.' The Malthusian argument, he claims, has consistently overwhelmed other explanations of poverty. Malthusian famine scenarios have systematically distracted attention from the fact that it is not people's reproductive habits that are the principal source of most of the misuse or waste of the world's resources, but the contradictions and motives of capitalist development (Shermer 2016). In view of the partial eclipse of family planning programmes since the 1994 Cairo International Conference on Population and Development (Wheeler 1999), that opinion is not entirely convincing.

1.2.7 A Counterpoise to the Harsher Elements in the Second Essay - Malthus and Love –

The second essay offers a mild antidote to these hard comments, because Malthus also speaks softly of love. Malthus is firm on the importance of sexual continence before (delayed) marriage, otherwise the whole point of the sacrifice is lost. And he understands, as anyone would, that it is a sacrifice. Later, Malthus softens those moralising strictures, admitting that such restraint is hard and lapses understandable. For Malthus there are worse evils than failure of chastity e.g. temptations arising from continual distress (e.g. p. 444). He even admits that there may have been 'some irregular connections with women, which have added to the happiness of both parties, and have injured none'. While Malthus was intellectually committed to 'moral restraint', he seems to have been rather pessimistic about it in the 1803 edition: "whatever hopes we may entertain of its prevalence in the future, has undoubtedly in past ages operated with inconsiderable force" (1989a I, pp. 329-30). But the parish registers show that delayed marriage was without doubt a major phenomenon in England and elsewhere in NW Europe at least from the 16th century, as Malthus seems to have been first to recognise.

He is often portrayed as harsh and unfeeling. Some of his policies may seems so. But the strength and constancy of the passion between the sexes is a cornerstone of his theory. He is clearly not indifferent to Cupid's darts and he goes as far as an early 19th century clergyman can go to allude, if elliptically, to the importance of sexual pleasure. Specifically, *eros* as well as *philia*. It throws some light on his expectations concerning moral restraint. Here is one of a number of his endorsements of the importance of love:

'...the most powerful and general of our desires is the passion between the sexes, taken in an enlarged sense. Of the happiness spread over human life by this passion very few are unconscious. Virtuous love, exalted by friendship, seems to be that sort of sensual and intellectual enjoyment particularly suited to the nature of man and most powerfully calculated to waken the sympathies of the soul and produce the most exquisite gratifications. Perhaps there is scarcely a man, who has once experienced the genuine delight of virtuous love, however great his intellectual pleasures have been, who does not look back to the sunny spot in his whole life, where his imagination most loves to bask, which he contemplates with the fondest regret and which he would wish to live over again'… .It is 'the symmetry of person, the vivacity, the voluptuous softness of temper, the affectionate kindness of feeling, the imagination and the wit' of a woman, which excites the passion of love and not the mere distinction of being a female' (Book IV, Ch.ii p. 431). And in pressing for later marriage, he states: The passion of love is a powerful stimulus to the formation of character and often prompts to the most noble and generous exertions. Late marriages taking place in this way would be very different from those of the same name at present, where the union is too frequently prompted solely by interested views'.

1.3 New Elements in the Second Essay

1.3.1 Prudential Restraint, the Preventive Check and Homoeostatic Systems

The most significant new aspect of the Second Essay is the amelioration of Malthus' pessimism concerning the way that population adjusts to its means of subsistence. Malthus' basic assumption that populations are homoeostatic ('density dependent') systems, in which reproduction and survival levels adjust the numbers of people in relation to resources, is regarded as his most important contribution. (Malthus did not use those words). That is how the checks are supposed to work: as numbers rise, further growth meets increasing resistance from limited resources; mortality or emigration increase and fertility declines. Growth stops.[22]As population growth eases surplus can accumulate, conditions improve and numbers rise again. This rational empirical approach of Malthus to demographic truth, with its emphasis on laws and principles[23], was often described in the spirit of Enlightenment rationalism as 'Newtonian'. Reflecting his preference for a natural theology, devoid of specific divine interventions, he assigned the checks to population to 'laws of nature'. In contrast Godwin argued (Hazlitt 1807) that they arose from remediable defects in society and the constitution, Some felt that Malthus's somewhat rigid approach

[22]In the mistaken understanding that Malthus believed that growth could be unending, Pierre François Verhulst (1804-1849) modelled these density-dependent constraints on growth, producing the very useful logistic growth model. This embodies built-in density dependent constraints on growth which eventually produce a stationary population. Leach (1981), Schtickzelle (1981).

[23]The formula for exponential growth $P_t = P_0.e^{rt}$ is sometimes called the 'Malthusian growth model... 'r' the growth rate is also Lotka's 'intrinsic rate of natural increase' and named by R.A. Fisher (1930) the 'Malthusian parameter' in his seminal 'Genetical Theory of Natural Selection' (1930).

was at variance with religious concepts of nature (Winch p. 229-230). That caused Malthus some trouble. In the Second Essay, inflexible 'laws' become more adaptable, 'tendencies' able to adjust to circumstances.

The First Essay dismally insisted that 'the catalogue of disagreeable circumstances, positive and preventive checks 'may be fairly resolved into misery and vice' (FE p. 102, 103), the true cause of the slow increase of population in all the states of modern Europe' so that 'the principal argument of this essay tends to place in a strong point of view the improbability of that the lower classes of people of any country should ever be sufficiently free from want and labour to achieve any high degree of intellectual improvement (Flew 1970 pp. 148, 149.). In the Second Essay Malthus is more optimistic (Winch 1987 ch. 4): 'Finding, therefore, that from the laws of our nature we could not proportion the food to the population, our next attempt should naturally be, to proportion the population to the food. If we can persuade the hare to go to sleep, the tortoise may have some chance of overtaking her. We are not however to relax our efforts in increasing the quantity of provisions but to combine another effort with it; that of keeping the population, when once it has been overtaken, at such a distance behind, as to effect the relative proportion which we desire; and thus to unite the two grand desiderata, a great actual population, and a state of society, in which abject poverty and dependence are comparatively but little known; two objects which are far from being incompatible' (Malthus 1826 p. 443).

1.3.2 The Preventive Check and the Prospect of Improvement.

The Second Essay moderates constraints on growth with a more flexible system which opens the possibility for improvement, because potentially under human control, by placing more emphasis on the preventive check. 'These checks to population, may be classed under two general heads – the preventive, and the positive checks......The preventive check, as far as it is voluntary, is peculiar to man and arises from that superiority in his reasoning faculties, which enables him to calculate distant consequences.' (Book 1, Chapter II, p. 84). The ultimate check to population appears to be a want of food. .... But the ultimate check is never the immediate check except in case of actual famine.' Population can adjust to resources, one way or another, in a homoeostatic, 'density dependent' relationship. 'Of the preventive checks the restraint from marriage which is not followed by irregular gratification, may properly be called moral restraint' (p. 85). That way Malthus opens the door to an easier future and the hope of progress. That, he felt, was typical of more 'civilised' societies and opened the way to general improvement in society by taking the pressure off the economy – and families - through later marriage and therefore fewer children.

1.3.3 Evidence for a Homoeostatic System.

What is the evidence for this process, so essential to his model? Direct evidence of homoeostatic feedbacks among simpler societies; hunter gatherers and pastoralists is lacking, but it is possible to make a powerful indirect inference. That is, that over the million years or more of the Palaeolithic, human populations remained tiny and net reproduction must have been on average close to 1.0 and growth therefore, nearly zero (Binford 1968, Wrigley 1986). Only a system of regulation could account for that (Lee 1986). In human populations, demonstrating homoeostatic density dependent responses in human behaviour is difficult. As we will see later, animal populations provide easier models. Human generation length is long so cohort effects, anyway, may take a long time to unfold, and culture change, altering the rules of the game, may proceed at a faster pace. Substantial long-term studies are rare. This model of population regulation received strong support by a remarkable study.

The reconstruction of English population by Wrigley and Schofield (1981) from 1541-1871 concluded that Malthus had essentially got it right, at least up to the beginning of the 19th century (Fig. 1.1) (Wrigley 1986). The outer circuit in Fig. 1.1 represents the positive check; the inner circle the preventive check (Woods 2000; pp. 10-15). Completed family size of married women varied with circumstances but on average was about five children. Taking all women together the average was about four, - up to 20% of women never married. For example, in England national total fertility in 1600-1649 was 4.4, completed family size of married women 5.2. (Wrigley and Schofield 1983). The potential power of the preventive check can be seen from the fact that women in England marrying at ages 15-19 would have 6.2 children, at age 20-24, 5.0 children and at age 25-29, 3.6 children. (Table 1.2)

In that huge reconstruction they were able to relate time-series of marriages, fertility, population and real wages. Increasing wages promoted earlier and more popular marriage which in the general absence of parity-specific birth control led to higher fertility and population growth. That in turn eventually depressed wages, and hence depressed marriage and fertility. Systems of this kind cannot respond instantly; they develop lags as feedback and response cannot be instantaneous. Malthus realised (Book I Ch. Ii p. 88) that these 'oscillations' were an inevitable feature of the connection between economic and demographic variables. They offered opportunities for improving conditions when population pressure relaxed, but distress when population had moved ahead of resources (Waterman 1987). It was thought that such a great work could never be repeated, but others have followed, mostly confirming that broad verdict (Macfarlane 2004) that in England from the 16th century, population size and wages were indeed related by a Malthusian equilibrating mechanism (Lee 1986), and in the Malthusian trap[24] in which all societies were caught up

[24] A Malthusian trap is a situation where populations are confined to a certain population maximum by limited resources, so that population exceeding the maximum is pulled down by increased mortality or reduced fertility.

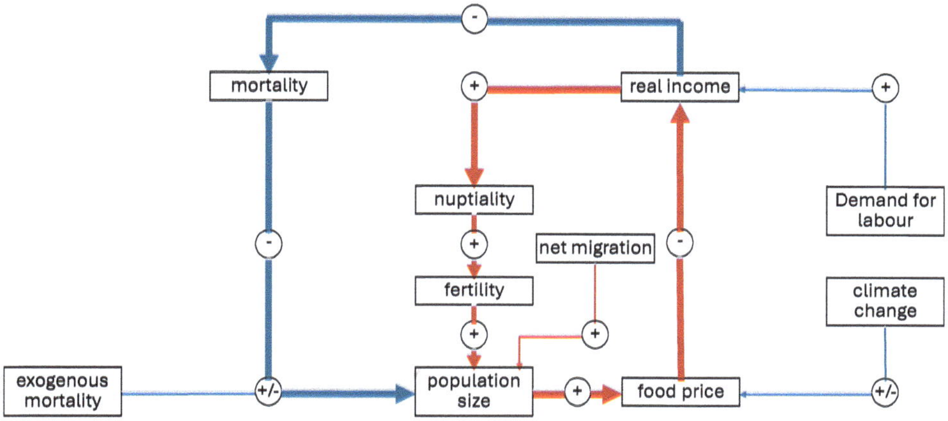

Fig. 1.1 The English population system in the late 16th century. Drawn by the author, inspired by: Wrigley and Schofield (1981) fig. 11.6

Table 1.2 Trends in marital behaviour, England 1576 – 1849

England: variation in marital behaviour						
England. Proportion of people never marrying.						
Year in which cohort was aged 20-24	% Never Marrying	England 1600-1849. Mean age at first marriage and percent never married.				
1576-60	8.4	Marriage cohort	Men	Women	Never maried %	Marriage cohort
1601-05	14.7					
1626-30	24.2	1600-1649	28.0	26.0	24.1	1601-6
1651-55	18.1	1650-1699	27.8	26.5	27.8	1651-6
1676-80	26.7	1700-1749	27.5	26.2	11.2	1701-6
1701-05	17.6	1750-1799	26.4	24.9	4.9	1751-6
1726-30	9.6	1800-1849	25.3	23.4	9.6	1801-6
1751-55	7.7					
1776-80	5.2	Wrigley and Schofield 1981 mean of 12 reconsitution studies t 7.26				
1801-05	7.1	Wrigley and Schofield 1981 men and women together t 7.28				
1826-30	10.2					

to that time (Roser 2020). Le Roy Ladurie (1966) traced a Malthusian system back to earlier times among the peasants of the Languedoc. Data from seventeen countries over the millennium from 900AD supports this view, that rises in wages and consequent population increases were snuffed out by rising mortality (Madsen et al. 2019). Improved economic data (Clark 2005), however, persuaded Crafts and Mills (2007) that Malthusian stagnation in England persisted until the end of the 18th century but that Malthusian checks ceased in the mid 17th century.

Wrigley and Schofield's data show that variation in fertility predominated over mortality, as the solid lines in Fig. 1.1 indicate. The long-run upward drift in population size was due to a secular trend in the demand for labour. Wages, mortality and fertility had little long-term trend but all showed two or three deep swings which are correlated with each other, in the two centuries covered by the data. And three periods of decline.

1.3.4 Malthusian Stagnation and the Demographic Transition.

Most scholars agree that a Malthusian regime obtained in Europe until the 19th century (Ehrlich and Kim 2005, Galor and Weil 2002, Macfarlane 2003, Kelly 2005, Lee 1986). That is understood to mean that population was contained by positive or preventive checks in a regime with high birth and death rates, growing very slowly in a cyclical fashion. Any increases in output simply led to increases in population, not improvements in living standards (Fernandes-Villaverde 2001, Roser 2020). That is 'Malthusian stagnation' or a 'Malthusian trap'. Since the 16th century the parish registers show that the preventive check of late, variable and avoidable marriage had established itself in England (Table 1.2) and some other NW European countries[25], a feature first recognised by Malthus and revived in modern times by Hajnal (1965). By making the schedule of reproduction dependent on changeable economic circumstances not the fixed timetable of biological maturity, variable marriage not only moderated fertility but kept reproduction in line with production.

Before 1538 we have no parish registers to guide us. But other evidence indicates a much greater antiquity of the system: it had been established at least by the end of the 13th century (Kelly and Ó Gráda 2012). That implies lower fertility and therefore lower mortality in earlier centuries than had been supposed. In these and other respects the middle ages were not as 'mediaeval' as popularly imagined. Others, however, dismiss the whole idea of Malthusian stagnation and demographic transition (Nielsen aka Nurzynski 2013), and some biologists reject the parallel idea that natural systems are kept in balance, and protected from extinction by homoeostatic feedbacks (Andrewartha and Birch 1954, Boer 1990). Murray (1999) insists that at least in animal populations that density dependence cannot adequately be defined and in any case is contradicted in some animal cases.

---

[25] Some revisionist historians have extended this NW European marriage pattern to parts of Eastern Europe (Szoltysek and Zuber-Goldstein 2009).

1.3.5 Malthus Among the Animals

Malthus paid considerable attention to the regulation of reproduction in animals, making a strong distinction between the instinctive behaviour of animals, whose numbers were limited only by positive checks, and the foresightful behaviour of humans. He believed that 'the preventive check, insofar as it is voluntary, is peculiar to man and arises from that distinctive superiority in his reasoning faculties (Book 1, Chapter II). Are animal numbers then solely limited by 'misery and vice'? Innocent animals know no vice. But plenty of misery, from disease, starvation and predation. But do animals show any form of preventive check to mitigate the direct effects of famine and pestilence? If so then the contrast between the natural regulation of brute creation, and that of civilised man, might not be as great as Malthus assumed (Lee 1987). He comments thus: (Lee 1987, Chapter I Statement of the Subject p. 79). :'The constant tendency in all animated life (is) to increase beyond the nourishment prepared for it. ...There is no bound to the prolific nature of plants or animals, but what is made by their crowding and interfering with each other's means of subsistence. Through the animal and vegetable kingdoms Nature has scattered the seeds of life abroad with the most profuse and liberal hand. But has been comparatively sparing in the room and nourished' 'The germs of existence contained in this earth, if they could freely develop themselves, would fill millions of worlds in the course of a few thousand years.'. (Chapter II Of the General Checks to Population... pp. 7-8.). Darwin later noted (1859, p.47) that 'There is no exception to the rule that every organic being naturally reproduces at so high a rate that, if not destroyed, the earth would soon be covered by the progeny of a single pair'. Even elephants, whose population unchecked could reach 19 million after 750 years. No wonder, after reading Malthus in 1838, that the penny dropped for Darwin (1887, p. 118) about the consequences of this inevitable struggle for existence on the selection of those who survived or managed to reproduce.

The operation of the positive check through mortality is abundantly obvious in the animal kingdom. And in every animal species each individual behaves as though to maximise its own reproductive success. Nonetheless it has been realised since the 1950s at least that density-dependent regulation of animal populations could follow from the depression of reproduction as well as from from lethal collisions with the environment (McClaren 1971. P.10, Murdoch 1994). Of course it is not suggested that there would be any deliberate act on the part of the animal to limit its own reproduction by modifying its own behaviour in some way. Instead what might seem to be a 'preventive check' could follow from physiological restriction of reproduction in response to environmental constraints or the concentration of reproduction in particular individuals through the variable filters of territory and hierarchy (Lack 1954), Lack's work has stood the test of time, a whole range of species from invertebrates to mammals show population growth depressed by higher population density; (Tanner 1966). Food-supply, nest-sites, predators, parasites and competitors limit vital rates in a density dependent fashion (Newton 2003, Sibly, Hone and Clutton-Brock 2002, p. 1149). In some mammals, shortage of food, and stressful interactions at higher population density, suppress female fecundity especially in gregarious species (lemmings and marmots) prone to population surges in simple environments. In social animals such as wolves; the breeding of subordinates is suppressed at higher densities, being confined to animals of higher status (Cubaynes et al. 2014). So reproduction is adjusted to circumstances (Clark and Cummings 2015) and some stability in numbers is maintained (Henderson and Magurran 2014). In chimpanzees and generally in primates male rank typically predicts male reproductive success. Through different (socio-economic) mechanisms, human reproduction too was traditionally concentrated among higher status individuals, although that requires no psychosocial explanation, Reversed during the demographic transition (Skirbekk 2008. Dribe et al. 2017.) in some populations, that traditional pattern seems to be re-asserting itself.

1.3.6 The Unimportance of Population Density

Despite the emphasis on what we call 'density-dependence', density per se does not feature much in Malthus' comments. It is more a question of the pressure of numbers on resources. But he cites 'great urban places', unwholesome occupations, crowding and interference' as species of vice likely to suppress population, without being specific about the nature of the harmful effects of urban life. Does density itself actually matter? Cramped housing in great cites require no analysis to see its obvious depressing effect upon family size, but that is readily understood in terms of living space without any psychosocial or physiological complications. However a large multi-national study of 145 countries, adjusted for confounding factors, showed that fertility was clearly lower in high density areas (Lutz et al. 2006). Limited living space is thought to be an important factor limiting family size in modern East Asia.

But could there also be more subtle effects of density on human behaviour and reproduction, akin to the responses observed in the animal kingdom? Some claim that this is so, and in the 1960s a small industry of such studies arose, focussing on territory, dominance, hierarchy and individual distance (Boyden 1970). Negative effects were expected because human psychology had evolved during the millions of years of the Palaeolithic at very low population density (Tiger and Fox 1971), Marmot and Sapolsky (2014) and would be maladapted to high density urban living. There is recurrent interest in the old work by Calhoun (1962). who confined rats at different population densities although always with abundant food. with rather nightmarish consequences. Stress triggers the production of cortisol, activating the body but with harmful side effects (Christian and Davis 1964). At very high densities rodent infant mortality, blood pressure and stress hormones rose to disastrous levels, nest building ceased. All was resolved with emigration. These extreme responses would not occur in wild populations and are highly implausible in humans. And in fact humans live successfully in diverse societies whose densities differ by orders of magnitude (Freedman 1980). Many people enjoy high density living; young families, old people, (Mitrany 2005) and seek them out. High rise, high density living does nor automatically foster community spirit despite the hopes of planners (Walter 1978). Depending on design, the reverse may be likely (Lawson 2010, Mitchell 1971).

Nonetheless a Royal Society project did look into possible harmful effects of urban density. But a comparison of stress levels of urban compared with rural populations in Oxfordshire revealed only mild differences (Harrison 1995). Joyful feelings awaken in all of us upon arrival in the country. But in developed societies, rural populations are in many ways already highly 'urbanised'. And despite living amidst very large populations, the social circles of individuals are manageably small, necessarily so by the limits of human cognition. The number of stable social relationships has been estimated (Dunbar 1992) on average to be about 150 individuals; the 'Dunbar number' (consult the number of names in your address book). But not all are impressed, the idea of defining such a number being dismissed as 'futile' by Lindenfors et al. (2021).

1.3.7 The Limit to Births.

Except for those more advanced societies where the prudential preventive check prevailed, Malthus seemed to assume that reproduction proceeded at a necessarily high rate, to offset the depredations of inevitably high mortality and thus guarantee survival, even if actual population growth was minimal or zero. (Book II ch, xiii p. 204 – 206). Here Malthus oversimplified; Few human societies have marital and sexual arrangements which maximise fertility to the achievable limit. That limit on average is generally reckoned to be about sixteen live births per woman, although the individual record is much higher. The (disputed) concept of 'natural fertility' refers to average fertility levels which arise when no target of family size exists; 'controlled' fertility when sexual behaviour changes when some target family size is achieved (Wilson et al. 1988). But even without any 'target', habits of sexual behaviour, marriage patterns, inadequate food, breast feeding, separation of partners, always keep average fertility below the achievable maximum, to an extent determined ultimately, by prevailing levels of mortality (Gagnon 2015). Many 'natural fertility' populations, even if not concerned about number of children, have practices which space births or end reproduction. Thus the Kung San ('bushmen') of Botswana, whose expectation of life under premodern conditions was reckoned to be about a (very creditable) 40 years, had long birth intervals and a completed fertility of about four. Some other traditional populations had similarly low fertility (Hilevych and Kreager 2025) although most hunter/gatherer populations (e.g. the Hadza of Tanzania, Dyson 1977) had larger families. The pattern of natural fertility of the Hutterites, an Anabaptist religious sect in North America, where completed fertility was 8.1 in the 1950s, is used as a standard of an achievable average (Leridon and Mencken 1982, Smith 1988) for comparing different societies. Since the 1950s standards have lapsed somewhat among the Hutterites but not, it seems, among the Amish. Such high fertility in a modern low-mortality environment is only sustainable in the long run with infinite room to expand. Even in the United States the Hutterites and other philoprogenitive sects risk running out of cheap land; in Israel public opinion is running out of patience with the prolific Haridim.

1.3.8 Breakout – The Eclipse of the Organic Economy, Escape from the Malthusian Trap

Before the 18th century, technological progress was slow everywhere and population growth prevented any sustained rise in income per capita. Up until that date, the economy was 'organic' that is, not only food but also almost all energy and motive power depended upon plants and animals; upon the area of the land and its productivity. That inevitably limited the scope for population growth and ensured that high densities of population could only be sustained at low levels of individual consumption. By the end of the 18th century, however, a new world of growth was emerging. In the absence of suitable statistics, knowledge of population growth was then uncertain, the source of fierce contemporary controversy (Glass 1973[26]). But now we know better. A new kind of evidence came from the first census of England and Wales in 1801. In conjunction with the census, the director John Rickman (1771-1840) required parishes to submit abstracts of their registers of baptisms and burials from 1700. In rural areas at least, these data were almost equivalent to births and deaths, and the difference between them to natural increase. Those made it clear that the population was growing, and at an accelerating rate (Fig. 1.2): population had changed course at around mid-century (Wrigley 1983).

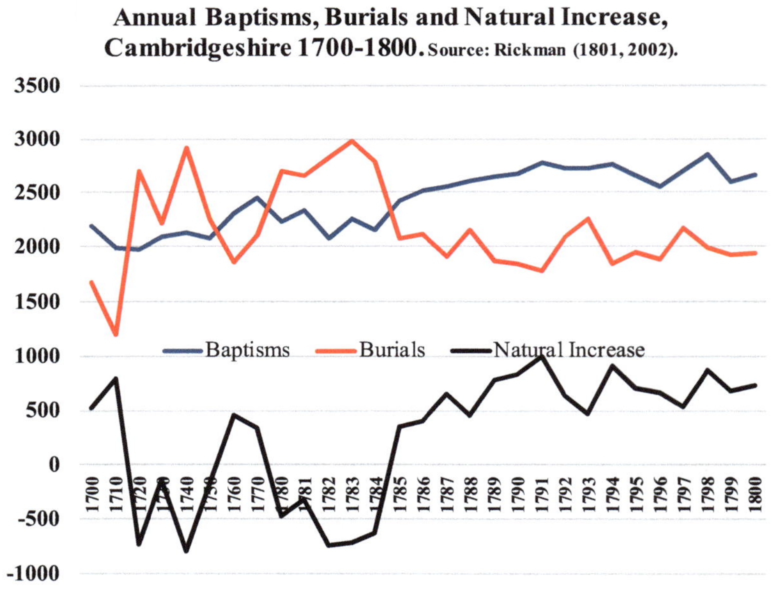

Fig. 1.2 Baptisms, Burials and Natural Increase, Cambridgeshire

[26]Glass, David Victor FRS, FBA, 1911-1978. My late revered supervisor.

Table 1.3 England and Wales 1801 – 1901. Trend of population (millions) and growth (percent)

		Intercensal growth		
Year	England and Wales	millions	percent	annual
1801	8,893	1,271	14,3	1,34
1811	10,164	1,836	18,1	1,66
1821	12,000	1,897	15,8	1,47
1831	13,897	2,017	14,5	1,36
1841	15,914	2,014	12,7	1,19
1851	17,928	2,138	11,9	1,13
1861	20,066	2,646	13,2	1,24
1871	22,712	3,262	14,4	1,34
1881	25,974	3,029	11,7	1,10
1891	29,003	3,525	12,2	1,15
1901	32,528			
	Source: decennial census			

Leaving behind the epidemics of the earlier 18th century, population had at last shrugged off the constraints of the past and was headed for a dramatic new increase evident in the 19th century (Table 1.1). Figure 1.2 shows the baptisms, burials and natural increase for the County of Cambridge. Cambridge in the Fens is a damp unhealthy marshy sort of place much depressed by malaria in the past (the last epidemic was in 1857-60; Nicholls 2000); an unlikely place for a revival of population and therefore all the more convincing. By the 1780s it is clear that, in Fenland, the birth rate was winning against the death rate. A new era had begun; 'organic' demographic regimes were on the way out. These changes were widespread in Western Europe, the positive check having diminished over time, the preventive check remaining steady until around the end of the 18th century. In Germany, for example, the transition from a Malthusian economy has been dated to about 1810 (Pfister and Fertig 2020).

Successive censuses in Malthus's lifetime showed that population continued to increase at an unprecedented pace. (Table 1.3). That was not meant to happen under the Principle of Population. The rise of technological competence meant that the 'organic' economy was being supplemented, eventually dominated, by one based on external sources of energy, specifically coal. Harnessed by increasingly efficient steam engines, coal transformed locomotion through railways (from 1825) and the cheapness of manufactures through factories (from about 1790, Wrigley 2016). The area of land to provide food, previously fixed, was effectively greatly enhanced because manufactures could be traded for produce overseas. Aided by the abolition of the Corn Laws in 1846 (whose abolition Malthus had opposed) and the enormous production from the United States, Canada and Australia, imports of grain exceeded domestic production by 1867.

While Malthus the theorist did not abandon his model, Malthus the empiricist had to recognise the new reality. He did so, but a little grudgingly. He presents (Book III. Chapter VIII p 362) in detail the attributes and advantages of three kinds of economic system; agricultural, commercial and a mixture of both, preferring the last: 'a country with an extensive and rich territory, the cultivation of which is stimulated by improvements in agriculture, manufactures and foreign commerce, has such varied and abundant resources that it is extremely difficult to say when they will reach their limits.' (p. 409. 887 my copy). This is really a new departure for him, an admission of the possibility of more substantial and continued growth than could be possible with the limitations of an organic economy dependent, ultimately, on a fixed area of land and the fixed productivity of photosynthesis. The replacement of human and animal power by new resources ended the Malthusian phase of historical restraint and opened the door to a new era of growth in production, welfare and population freed from previous limitations (Wrigley 2016).

Malthus was not a prophet of doom and disaster. He is able, eventually, to embrace the machine. Overcoming earlier anxieties, he remarked 'the presumption always is that (the increase of machinery) will lead to a great extension both of wealth and value' (Principles of Political Economy, Vol. 1, p. 409). And managed a cheer in favour of steam engines (Winch 1996 p. 392). In thinking about economic development, he develops two major new concepts in the 'Principles'. The critical role of elite investment as a motor of growth, and his concept of effective demand. The wider economic balance of capital promotes high working class fertility, not just lack of foresight (Kreager 2022). It planted the seed for the development of his mature ideas on political economy. Malthus did not foretell the industrial revolution only then beginning. In the post-Malthusian regime population growth now only absorbed part of the growth of output (Galor and Weil 2020) and may have promoted it (Boserup 1965, Wood 1998). The rest, eventually, increased the standard of living.

Malthus' reputation diminished as the 19th century progressed. Despite persuading the government to pass the radical revision of the Poor Laws through the Poor Law Amendment Act of 1834 his later economic writings were not well received. Even he had doubts as to whether the old Poor Laws had really served to encourage improvident marriage and undesirable aspects of population growth. The humane 'Parson Malthus' prevailed against the hard logic of 'Population Malthus' (Digby 1983, Flinn 1983). Experience and facts were against it. Such evidence as exists has showed only a small effect, if any, or even the reverse (Huzel 1968); Wrigley and Smith (2019) conclude that he was simply wrong. It was evident that the population growth of the early part of the 19th century had not had deleterious consequences. On the contrary despite some persistent poverty, prosperity and standards of living were clearly improving.

Poor Malthus could not let go of his assumption that a large agricultural sector – equivalent to the industrial – was necessary for national prosperity. He could not foresee the astonishing acceleration of grain yields to come and therefore the possibility of supporting the people through the agricultural labour of just a few percent of the population. By 1867 agriculture contributed just 16% of national income and 0.76% in 2022. In Malthus's lifetime grain yields had increased only marginally, as he supposed, in roughly an 'arithmetical' fashion. Later they increased beyond his imagination (Table 1.4), albeit at a cost in fertiliser and in the nature of the soil. And important for our story, that stellar increase appears at the moment to have peaked.

Table 1.4 England. Wheat yield in tonnes per hectare 1600-2020. Source: Ritchie, H., et al. (2022) Crop Yields, tonnes per hectare. Our World in Data

1600	0.74
1700	0.95
1800	1.26
1900	2.12
2020	6.96

1.3.9 The Progress of Population into the 20th Century: How Did Malthus Fare?

Following the end of this period of so-called 'Malthusian stagnation' in an organic economy, we find ourselves in more turbulent waters. Malthus and all his works were left behind in an era of growth of population and economy. Both were emancipated from the constraints of population until the rise of modern partial demographic equilibrium through the demographic transition. That led us out of the land of Malthus into the promised sunlit uplands of constant economic improvement without population constraints. For now. But in the developed world, the period of post-Malthusian population growth has been relatively brief. And increasingly so in the developing world outside Africa, Whether the parallel growth of the economy can be sustained is an issue which brings Malthus out of the cupboard again.

1.3.10 Prosperity Without Check – Malthus Marginalised

Here we telescope 150 years and emerge in the 21st century. While material prosperity has continued to grow, indeed accelerated, brushing aside any resistance from population, the population growth which slowly emerged in the 18th century could not endure. Not just because of pressure of population on resources in the large scale, but from the impossibility for families to cope with the new numbers of children who survived. Depicting changes with a very broad brush, epidemics moderated, survival increased, birth rates went up as marriage became more popular. Surviving family size expanding beyond experience, drove population size upwards. Only in recent decades has population begun to reverse as fertility ebbed. Now in Europe natural increase has stopped and even reversed as fertility fell, after a considerable lag. Family size, now with all children surviving, is at a manageable level for parents but throughout the Western world too low to sustain population size. In most of Europe only migration now drives population, a development unimaginable to Malthus (Table 1.5).

All this made no sense in Malthus' scheme of things. Before the mid 19th century his influence waned as events – population and prosperity advancing together - contradicted him (Mayhew 2015). His ideas, demographic and economic, had been absorbed into general knowledge. But to professional economists – Marshall, Toynbee, Bonar - a new

Table 1.5 Population trends 1750-2050

	Population of selected countries 1750-2050, millions									
	France	Germany	UK	England & Wales	Ireland	Belgium	Netherlands	USA	Russia	Japan
1750	24,0	15,0		5,4	3,0	2,3	2,0	1,2	18,0	29,0
1800	29,0	18,0		8,9	5,3	3,3	2,0	5,3	30,0	28,0
1850	36,0	27,0	22,3	17,9	6,5	4,5	3,0	23,2	60,0	32,0
1900	38,9	56,4	38,2	32,5	4,7	6,7	*5,9*	76,9	73,0	43,8
1925	40,3	63,2	45,0	40,6	3,9	*7,8*	*7,4*	115,8	*92,0*	59,7
1950	41,7	68,4	50,2	43,8	3,0	*8,6*	10,0	152,3	103,3	83,2
1975	52,6	78,5	56,2	49,5	3,2	9,8	13,6	213,1	134,2	111,3
2000	59,5	82,2	59,1	52,0	3,8	10,3	16,0	281,5	144,8	126,9
2025	66,6	84,1	69,6	59,0	5,3	11,8	18,3	347,3	144,0	123,1
2050	67,4	89,2	75,6		6,2		18,8	420,0	137,8	97,0

species in late Victorian Britain – he had become an historical figure, not a contemporary icon (Mayhew 2015). His first biography by Bonar (1885) marked his ascension from current controversialist into intellectual history. The pace of publications on the central Malthusian controversy, lively in the 1820s and 1830s, flagged thereafter (Banks and Glass 1953 Table 3, pp 85-92) until its revival in recent times.

Malthusian interest was stimulated again at the end of the 19th century by concern not for the poor of the non-European world – that was a cloud on a distant horizon – but by the distress of Europe's poor themselves. Growth arising from the relaxation of mortality, both of national populations and of the size of families, was already under way. Children survived as never before. increasing the burden of the poor. In response infanticide and (always illegal) abortion correspondingly rose; the need for an alternative was pressing.

1.3.11 Prudential Restraint by Other Means. The Rise of Family Planning

By the later 19th century the more educated classes, and some working people – had learned to moderate their family size effectively without the rigours or frustrations of moral restraint. Indeed, Wrigley and Schofield (1983) showed further that some families were moderating their family size by adjusting birth spacing for centuries before on a parity-specific basis (i.e. according to the number of children) although not stopping reproduction early (Cinnirella et al. 2017). Prominent among the 19th century pioneers were doctors and clergymen, whose professional bodies resolutely refused to sanction contraception until well into the 20th century. The Lambeth Conference of the Church of England did not give approval to family planning within marriage until 1930. The Church of Rome still has not, whatever the faithful may be up to. Malthus was prayed in aid of contraception (called Malthusianism) despite his own views,

Anxiety about the welfare of the poor in the midst of obvious signs of national prosperity brought Malthus' name, if not his principles, back to the front line. Pressure groups such as the Malthusian League, (1877) tried to persuade the public, especially workers, of the need to limit family size. It had influence far beyond its numbers (McLaren (1978). Strong support for Malthus thus came from where he least sought it, from the advocates of artificial birth control., Francis Place[27] proposed in his 'Illustrations and Proofs of the Principles of Population' (1822) that the solution to the 'Malthusian' difficulty was the very un-Malthusian expedient of contraception. His so-called "Diabolical Handbills," offered explicit advice on contraception 'To the Married of Both Sexes' and 'To the Married of Both Sexes in the Genteel Life'. A flurry of publications on similar lines followed; Richard Carlisle's (1828) 'What is Love ?[28], Dale Owen's 'Moral Physiology; (1831) and Charles Knowlton's 'Fruits of Philosophy' (1832) Caldwell p 479.).

Malthus would not have welcomed support from such quarters, whose propaganda appropriated his name. He would not want perfect control over reproduction. For him family size and (moderate) population growth arising from the 'passion between the sexes' was an essential God-given spur to industry and improvement. Malthus knew Place, and helped him, but there seems to be no note of any comment or contact in the Essay. In the nature of things those prepared to raise their voices in favour of so risky and radical a promotion of contraception were often radicals in the political realm as well, unorthodox and secular, radical and revolutionary. Towards the end of the century there was a wave of prophets and martyrs to contraception, with their attributes of sponge, pessary and condom, braving prosecution for their 'obscene' publications. The radical reformers Charles Knowlton and Annie Besant sought publicity for their views by courting prosecution. They had their wish. For re-publishing Knowlton's tasteful old pamphlet 'The Fruits of Philosophy' they were prosecuted for obscenity-in the end unsuccessfully. But newspapers now circulated millions of copies to a population 80% literate. 'Birth control' was brought onto every Englishman's breakfast table'; Knowlton's pamphlets sold by the millions, not the tens of thousands as before (Teitelbaum 1984). The real shocker was the heroic Marie Stopes, whose books 'Married Love' (1918) and 'Wise Parenthood' (1921) spoke of sex and sexuality with a new frankness. She opened the first birth control clinic in England in 1921 and provided the first education for doctors in sexual medicine (Soloway 1997). Thanks to her approach, independent of Malthus, motivated more by her interest in sexuality and eugenics, 'birth control' became called as such, no longer the highly inappropriate label 'Malthusian'. But that is another story.

[27] The 'father of family planning' and of ten children.

[28] Subtitled 'Containing much important information for the prudent regulation of the Principle of Love and the Number of the Family

1.3.12 The 20th Century Reproduction Roller-Coaster: Fear of Fertility Decline, and Baby Boom

As death rates fell before birth rates and in some cases birth rates increased; population grew without precedent until birth rates themselves declined in Europe and later in the developing world. This 'demographic transition', from high birth and death rates to low ones, ushered in decades of population growth thanks to the prior decline of the death rate, Even when fertility has fallen to the level of replacement, the inertia built into the age-structure drives further growth for up to fifty years. Eventually in the developed world contraception has become nearly universal. Its label is unimportant; the central point is that couples are taking prudential steps to minimise the chances of early conception, through delayed marriage or by blocking the biological process. The prudential intent is the same, to protect family standards from the impact of too many children. As Malthus had feared that they might, 'improper arts' eventually reduced fertility rates in the 1930s in several European countries to levels inadequate to replace population, Malthus' 'Devil U of unemployed resources' and the failure of effective demand had elbowed aside Malthus' 'Devil P of Population', as Keynes put it (1937) and Toye (2000). The birth rate differentials between the classes excited eugenic concerns (of which Malthus had been oblivious). Across Europe, the failure of reproduction and the economy brought insecurity in its train, a 'Malthusian' worry despite Malthus, in the general sense of the nexus of population and economy (Mayhew p199 *passim*). Concern in Europe was sharpened by the advent of modern methods of cohort-component population projection which gave numerically precise but often misleadingly gloomy future forecasts of decline. (Charles 1934). Pronatalist polices were developed to counter the 'twilight of parenthood' (Teitelbaum and Winter 1985)'. But post-war recovery lifted the birth rate, bringing a baby boom peaking in the 1960s and with it transient fears of overpopulation (Fig. 1.3, Coleman and Salt 1992 Ch. 4, 13). Thirty years of relative demographic stability followed. In the British context at least, Hollingsworth (1983) could claim (p. 220) that 'The very word 'Malthusian' has almost dropped out of the language'...Malthus is either forgotten or his views are so well known that his name can be forgotten' "Thus it would seem wise not to take Malthus very seriously...' Now population is on the move again, with a severe fertility decline which vexes demographers and worries politicians throughout the developed world. Frets about rapid growth in the 1950s and 60s have given way from the 1980s to concern about population ageing and stagnation with decline deferred only by transformational levels of migration, demanded by employers but usually unwelcome to the people. Western families are too small and too long delayed; in the developed world fertility is everywhere below the level required to replace population size (Fig. 1.3).[29]

[29]Replacement fertility is the total fertility required to replace the population in the long run. Conventionality taken to be about 2.07 children, in fact it is variable, depending entirely on the death rate. In high mortality populations the replacement level can be 3 or even 4.

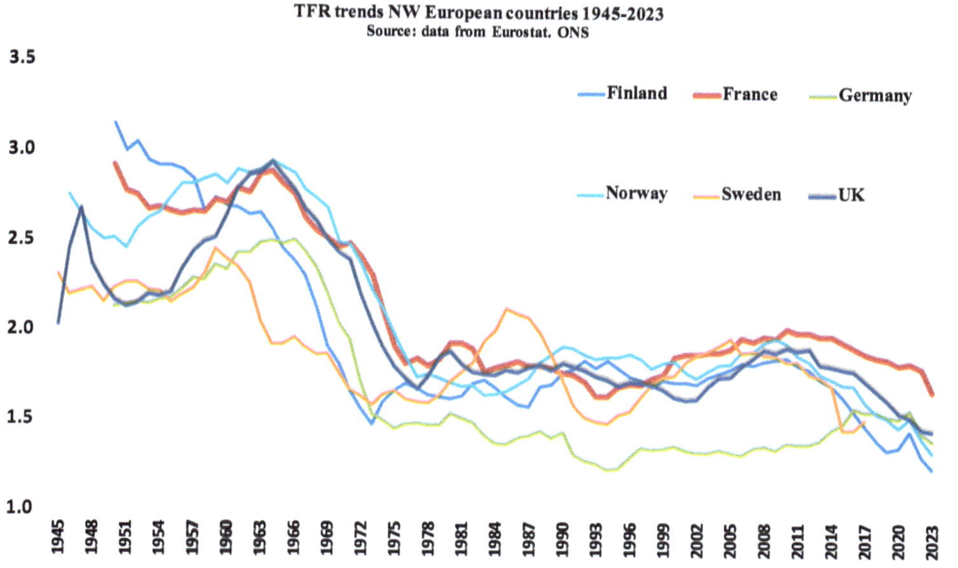

Fig. 1.3 Total fertility, selected. European countries 1945-2022

In the 'Third World' Malthusianism was rescued from its ashes after the Second World War by the so-called 'population explosion'[30] (Charbit p 165). There, mortality had retreated from a reduction in warfare, economic and agricultural progress, effective public health and vaccination programmes while for decades fertility continued traditionally high or even increased. For a while population growth even exceeded the maximum that Malthus had imagined. Development and population policy has brought fertility back into alignment, Outside Africa, population growth in the developing world increasingly runs on population momentum[31] and that tank is running dry.

Only parts of sub-Saharan Africa and a few countries in Asia have not yet joined this transition. There, population grows apace, promising to add a further three billion to the world total, for whom we must find places at 'nature's feast'. The UN projects that even with falls in fertility, sub-Saharan Africa will increase from 1.2 billion now to four billion by 2100. In some African countries, family size real and desired has moved little from historical levels (Madsen 2015, Bongaarts 2020); In parts of NW Africa and the Sahel Burkina Faso, Mauretania, Ivory Coast, Senegal, total fertility ranges between 5 and

[30] 'population explosions' only occur in the minds of journalists. The reality is a little slower.

[31] Population momentum. The capacity of a population to keep on increasing when total fertility is equal only to the level of replacement or less. This often arises when past higher birth rates have created a transient age structure with large numbers of young people of reproductive age. When these move up to older ages with cohort progression, and the population becomes stable, the momentum dissipates. Such growth can continue for half a century and the population double.

7 children. (Garenne 2014), and in East Africa Uganda and Ethiopia. Not so mortality, which has fallen so that most children survive, creating for the first time in history a large, living family of children (Lam and Marteleto 2008) at once private workforce and private dependent poor (Merchant 2022). 'In Western Africa, progress towards adoption of contraception has been dismally slow' (Cleland et al. 2011, Cohen 1996, Cohen et al. 1980) but there are signs of progress (*Cleland 2016).

1.3.13 Population Growth in the Developing World and Concerns About Food

The fizzing of this 'population bomb' began to alert the world that a new demographic era had arrived, rejuvenating awareness of Malthus and waking it from its slumbers. And Neomalthusianism with it. Vogt (1948) was the pioneer advocate of family planning to protect the global environment from the new population growth. 'Demographic time-bombs' became fashionable in the middle of the 20th century, from the workshops of Paul Ehrlich (1968) and others, posing a threat of global calamity and imminent starvation. In this Malthus was much (mis) quoted Merchant (2022). He seemed to be vindicated by the coming crisis that he was (wrongly) believed to have foretold, Malthus had gone green and embraced environment; for a while 'Friends of the Earth' and the UK Green Party adopted policies to restrain population to what were regard as sustainable levels, 30 million in the case of the UK. defined by calculations of ecological footprint[32]. That implausible romance has not endured. For most on the left Malthus has resumed his traditional pantomime role as villain.

Concern grew about the adequacy of food production in those fast-growing poorer countries. How to bring the rejuvenated geometric growth and the arithmetic growth closer together, and to rescue the teeming poor of Asia and Africa from poverty and from the prospect of starvation which began to be forecast. Malthus was appropriated with forecasts of doom of a kind he had never made. Forecasts aside, the obvious pressure of population on resources, the poverty and malnutrition of the people, the propensity to local famines, had led to Malthusian ideas being exported to the countries of the 'third world,' then in the early stages of the demographic transition. Most began under colonial rule. India alone of large agrarian countries had a modern census (1871) and other statistics well developed (Caldwell 1998). But when the immense glacier of Indian population had started to move, in the 1930s (Table 1.6), the colonial authorities felt unable to respond. The new independent government of India had no such inhibitions, initiating a family planning programme in 1954. It was even of a kind that Malthus might have approved, being based on Ghandian

[32]Ecological footprint is a measure of the resources required to support sustainably a specified population living a specified way of life. It is usually expressed in terms of the land area required to provide those resources. This concept is far from universally accepted.

Table 1.6 Population of British India 1871 – 1941. Source: Census of India

	Population of India 1871-1941		
	British Raj only.		
	Millions	Population growth (%)	
		Decennial	Annual
1871	238.8		
1881	253.9	6.3	0.61
1891	287.2	13.1	1.23
1901	293.5	2.2	0.22
1911	316.0	7.4	0.74
1921	318.9	1.2	0.09
1931	352.9	10.6	1.01
1941	388.9	10.2	0.97
	Source: Census of India		

notions of restraint – and correspondingly ineffective. As population growth accelerated worldwide that concern grew into a massive family planning industry, promoted both by many newly independent ex-colonial countries and international organisations both official such as the UNFPA, and non-governmental such as the Carnegie Foundation. In all of this Malthus was patron saint, even in China.

There, in an extraordinary *volte face*, a Communist government then as now theoretically committed to anti-Malthusian Marxist doctrine, almost overnight had an epiphany, a conversion experience to Malthusian analysis It translated analysis into powerful policies only possible for a dictatorship with every means of coercion at its disposal and, to begin with, the enthusiastic loyalty of most of the population (Jiang 2011, Wang et al. 2016). The nagging, but non-coercive, later-longer-fewer policy (1962-66) helped to bring total fertility down to nearly replacement,[33] The one-child policy (1971-2016), only partly enforced and partly successful, brought fertility little lower (Gietel-Basten et al. 2019) and seems to have cemented it into a sub-replacement trajectory (Malthus 1820, Minzner et al. 2023). All this accompanied by (rather misguided) enthusiastic Malthusian quotes which would make most Neo-Malthusians blush.

1.3.14 A Pause for Thought. Is Our Krafft Ebbing?

Before moving onto contemporary Malthusian themes it would be appropriate to review the security of the fundamentals behind Malthus's argumen*t*, which underlies both the first and the second essays. That concerns his *postulata*. We have already considered the essential need for food, which none denies. Is all well with the other? That is: 'Towards

[33]How far fertility decline can be attributed to population policies, and how far to development, remains contentious.

the extinction of the passion between the sexes, no observable progress whatever has hitherto been made.[34]'

Could the passion between the sexes be failing? There is startling evidence that it is so in some societies. If so, then the Principle of Population may start to deflate. Malthus' almost lyrical invocation of the power of love has already been cited (p. 443, 444). But it need not go with passion. A minority of people, drawn to the religious life, have for millennia set aside the joy of sex for the consolations of philosophy, religious devotion and the prospect of salvation (or oblivion, according to taste - other salvations are available). This propensity varies with culture. In Islam there is no monasticism (Bashir 2007), all the faithful being under an obligation to marry and raise a family. Only the Sufi orders offer an exception, usually temporary. Judaism shares the same strong injunction, although a few exceptions are allowed (Diamond 2007). In Ch'ing China, as well as Buddhist monasticism, many men remained bachelors without joining any society (Lee and Feng 1999). In both societies numerous eunuchs had been made offers they couldn't refuse. Following the example of its founder, there is a strong tradition of professional celibate monastic clergy in Buddhism, entered in youth or later in life and sometimes as a temporary sabbatical. This clashed with traditional Chinese culture when Buddhism was introduced; sex being regarded as essential for good health and necessary to continue the obligation to revere ancestors (in Japan today most monks are married. Kieschnick 2007).

In some Buddhist societies the great popularity of monasticism must have had negative demographic consequences. Malthus, commenting on the frequency and social significance of Buddhist monks in the harsh environment of Tibet, and on the unique polyandrous custom of the brothers of a household marrying the wife of the senior, notes 'Combined with the celibacy of such a numerous body of ecclesiastics must operate in the most decided manner as a preventive check to population' (Book 1 Ch XII 171)., Following the example of its founder and his precepts and those of his disciples, virginity and chastity were elevated over the married life from the earliest times in the Orthodox and Roman Christian tradition both in private life in the world (Biller 2000, Cheatham 2007) and in the monastic orders which had their beginnings in the 4th century and formalised in the 6th under Benedict. Religious orders and their adherents proliferated; numbers varying with the conditions of the time until, in NW Europe, the Reformation. On the dissolution of the 900 religious houses in England there were 12,000 religious (Heale 2013), probably not enough to make much difference to the birth rate. There was no suggestion that the historical ups and downs of the celibate life owed anything to changes in sex drive. Until now.

Now we have an interesting challenge to Malthus' postulate; a story for which the world is not yet prepared. In the developed world at least, not a few people are losing interest in sex, and others are becoming incapable of fathering or bearing children. This scarcely believable development has its beginning in the countries with the lowest birth rate in the

[34] See Flew edition p.148

world including Japan (Moriki 2012, Sato and Iwasawa 2013) and Korea (Matsuda et al. 2024). Survey after survey in Japan has shown that the trend is real (Atoh 1998, Konishi et al. 2022). Among adults in their 30s about one in ten still had no sexual experience. (Ghaznavi et al. 2019). A survey in 2011 found that 61% of unmarried men and 49% of women aged 18-34 were not in any kind of romantic relationship, a rise of almost 10% from five years earlier. One in three under 30 had never dated at all.

Although there has long been a pragmatic separation of love and sex in Japan, sex fares no better. In 2023 the Japan Family Planning Association (JFPA) found that 45% of women aged 16-24 "were not interested in or despised sexual contact". More than a quarter of men felt the same way. This restrained hurly-burly on the chaise longue apparently translates seamlessly into the peace and quiet of the marital bed for that shrinking proportion who do get married. In 2012 the JFPA found that 41% of couples had not had sex in the previous month.

These dry statistics hide interesting sociology. Japan and Korea are strongly familist with vertically organised social connections – between generations – rather than horizontal ones such as between spouses and partners. Informal social customs for linking sideways are weak and formal dating a difficulty, too informal now that go-betweens are unfashionable. Some Japanese youth, mostly males, without sexual contacts are developing some extreme forms of self isolation; *sekkusu shinai shokogun*, or "celibacy syndrome" is the name given to this flight from human intimacy, and the related severe self -isolation at home (Travers 2024). *hikikomori* are individuals who refuse to leave their parents' house, do not work or go to school, and isolate themselves in a single room for months; one step beyond the 'parasite singles' the *parasaito shinguru* who hang around the parental home sometimes until their 30s;. Furlong (2008) suggests that this anomic behaviour follows the failure of the ordinary labour market for young people and its replacement by a precarious secondary sector, eroding traditional expectations and norms in a rigid and formal society where appropriate behaviour is no longer obvious. It is difficult to be Japanese. This isolation was estimated by the Japanese Cabinet Office in 2022 to affect 1,46 million between ages 15-64. Korea, with its rigorous expectations of career progression and academic success, might be expected to be fertile ground for the *hikikomori* syndrome and so it has proved, with an estimated 244,000 people aged 19-32 in 2022 in 'reclusive isolation' according to the Korean Institute for Health and Social Affairs. There are no reports of parasite singles *a la japonaise*. However the birth rate is the lowest in the world and marriage is falling through the floor.

1.3.15 No Sex Please, We're British

Looking elsewhere, in the UK also, passion appears to have gone off the boil somewhat. The decennial National Survey of Sexual Attitudes and Lifestyles (15,000 people, Mercer et al. 2013) found those aged 16-44 were having sex fewer than five times a month, compared with more than six times a month in 1990-91 and 1999-2001. Money worries and the distractions of social media are responsible, researchers say. However a key insight from the survey is the paradox that people are having sex earlier but less often and having children later.[35] Similar trends have been reported from the US, where between 2000 and 2018, the proportion of 18- to 24-year-old individuals who reported having had no sexual activity in the past year increased among men from one in five to one in three with similar declines in coital frequency. (Ueda et al. 2020). In 33 European countries a survey of adolescents revealed that the proportion of 15 year olds who were sexually experienced had fallen between 2010 and 2018, from 25% to 18% (de Graaf et al. 2004). These delays in all aspects of reproduction except initiation is the key to understanding Europe's low fertility (Skirbekk et al. 2024).

1.3.16 Fecundity and Sperm Count

Another element in this reproductive problem has nothing to do with 'passion' although it is highly important. That is, that the human species, or at least some of its populations, is becoming less fecund. Measured declines in sperm count first came to light in 1974 (Nelson and Bunge 1974). By 1973, sperm count had halved from 107 million per millilitre in 1951 to 48 million by the 1970's. Hardly any populations measured have not shown a decline. Sperm counts globally fell on average by over one percent per year between 1973 to 2018, from 104 to 49 million/ml (Levine et al. 2022, a finding confirmed by others (e.g. Mann et al. 2020). Poor semen quality is thought to contribute to the decline of the birth rate (Jensen et al. 2002). Women's fecundity declines with age. These trends interact most unhelpfully when the timetable of reproduction is so delayed. More couples are having trouble conceiving. One in six couples in Denmark is infertile. These are now important factors in the failure of couples to realise their fertility plans. (Lazzari and Beaujouan 2024).

1.4 Malthus in the 21st Century: Modern Controversies

In recent years a number of profound questions have arisen which Malthus did not address himself but which his votaries have tried to illuminate with his ideas (or anti-Malthusians damn them), Thus Malthus finds himself summoned to controversies on global warming, low fertility, population decline, international migration, most of them beyond his experience. Malthusian responses to them are often inferred ('what would Malthus have said') as a kind of exegesis of a near-theological nature, like speculating what the common law says about climate change, or the American constitution about abortion. Can we usefully think

[35] All this has nothing to do with the violent and misogynistic 'incel' culture where the celibacy is involuntary and resented. (Brace 2021)

of the demographic questions of today's world, or their solution, in Malthusian terms? How much stretching old text and ideas beyond breaking point to encompass concepts not then imagined or intended? (Mayhew 2014 Ch.9 p.229. Epilogue p. 233). There are a number of connections to explore.

Rapid population growth in poor countries, as noted above, became a dominant NeoMalthusian concern in the 20th century. Caught in a Malthusian trap, food shortage would follow, economic growth stifled. Family limitation policies adopted by many developing countries, pushed hard by international organisations, were to solve it. Argument continues to the present. But the world does not starve and growth continues; the analysis is not universally accepted. As described earlier, critics objected on ideological and practical grounds, observing reasonably that food production (thanks to the green revolution) had latterly at least kept pace with population growth. But that was possible only because fertility had fallen, undermining population growth.

Disaster avoided, the misery of chronic malnutrition continues, especially of children. Some big states have solved this problem; now there is no food shortage in China (and fewer people). In most regions health and nutrition has improved. In others the people remain chronically underfed and subject to periodic famine (Ó Gráda 2009). The FAO (2024) reports that food security remains constant globally but is on the edge in parts of Africa, About 30 percent of the global population – 2.4 billion people – did not have access to adequate food in 2022; about nine percent actually going hungry; between 700 and 800 million people. In space-age India child malnutrition, hunger, stunting and wasting of children may be getting worse (Chatterjee 2021), India ranks 97th out of 104 in the 2020 World Hunger Index – and this in the context of declining total fertility[36] and population growth.

Economic literature pays little attention to population growth, climate change or food adequacy, although high fertility is identifed as a handicap (Akintunde et al. 2013). Population growth in Africa in 2023 was estimated to be 2.5%, down from 2.8% in 2010. That still implies a dooubling of population every 28 years. Economists' view have been mixed on this topic; indeed rather relaxed. Weil and Wilde (2010) conclude "The Malthusian channel by which a high level of population reduces income per capita is still relevant in poor developing countries that have large rural populations dependent on agriculture, as well as in countries that are heavily reliant on mineral or energy exports. … ., such a country would be 26 percent richer in per capita terms if it had half as many people. This increment is significant, but it is pretty small in comparison to the differences in income between rich and poor countries, which exceed a factor of 20.'

[36]'total fertility' This rather pedantic usage is preferred to the usual 'total fertility rate'; or TFR. The TFR is not a 'rate'; it is a quantity; namely the number of children on average which women would have by age 49 if current birth rates by age were to persist'. It is a 'what if…' snapshot specific to the year in question and sensitive to changes in the timing and quantity of births. Total fertility in South India is now below replacement.

In the future ten billion population will only be fed if there is a radical change in the pattern of nutrition, particularly a reduction of livestock, claims Smith (2015). Malthus made a parallel point. In the *First Essay* 1798 he laments the reduction in food production caused by using land to produce high quality meat and to maintain a great number of horses for pleasure (FE p. 319). Most governments of developing countries have introduced policies to moderate fertility. Whether for that reason or not, fertility rates almost everywhere outside Africa are falling. World population is projected to pass its peak before the end of the century. Up to then, the growth of economies will depend on technological change. Trewavas (2002) puts his faith in continuous technical change, to keep Malthus foiled. Keeping production growing faster than population, warn Montano and García-Lopez (2020), will be essential to keep Malthus at bay, Naso et al. (2020) also insist on the crucial importance of technical progress. The waning of global population growth means the rise of global population ageing, the problem then being a deficit, not a surplus, of working population. Malthus waits in the wings. Eventually this contest will evaporate. But in the meantime global climate change will keep up the pressure, diminishing cultivatable land in a new race with population.

1.4.1 We Need to Talk About Africa

This politicised argument now centres on Africa. To some, Africa seems to be the epitome of a Malthusian situation. Despite being a continent of farmers, it is a net importer of food. Food production in the early 2000s exceeded population growth by just 0.1 percentage points. The usual suspects are indicted: population growth, stagnating agricultural productivity, policy distortions, weak institutions and poor infrastructure (Rakotoarisoa et al. 2012).

But while many high fertility populations in Africa are growing rapidly their numbers are obviously not yet limited by any check, although standards of living of some are constrained. While the economies of some African countries are growing fast, the overall picture in the last decade if not encouraging. (Fig. 1.4). In a nuanced analysis, van Wissen and van der Weijden (2024) show that while Africa is not 'overpopulated', its rapid population growth is a problem. Much of the impressive-looking gdp growth is swallowed up in population growth. Those redictions are not unanimous. The World Bank expects gdp growth in Sub-Saharan Africa to fall from 3.6% to 2.5% from 2022 to 2023,[37].while the African Development Bank group (2024) expects 3.8% in 2024, higher than the global 2.9% [38] (Fig. 1.4). In sub-Saharan Africa 280 million people (twenty percent– 57 million

[37] https://www.worldbank.org/en/region/afr/overview

[38] https://www.afdb.org/en/news-and-events/press-releases/africa-dominates-list-worlds-20-fastest-growing-economies-2024-african-development-bank-says-macroeconomic-report-68751#:~:text=Overall%2C%20real%20gross%20domestic%20product,fastest%2Dgrowing%20region%20after%20Asia.

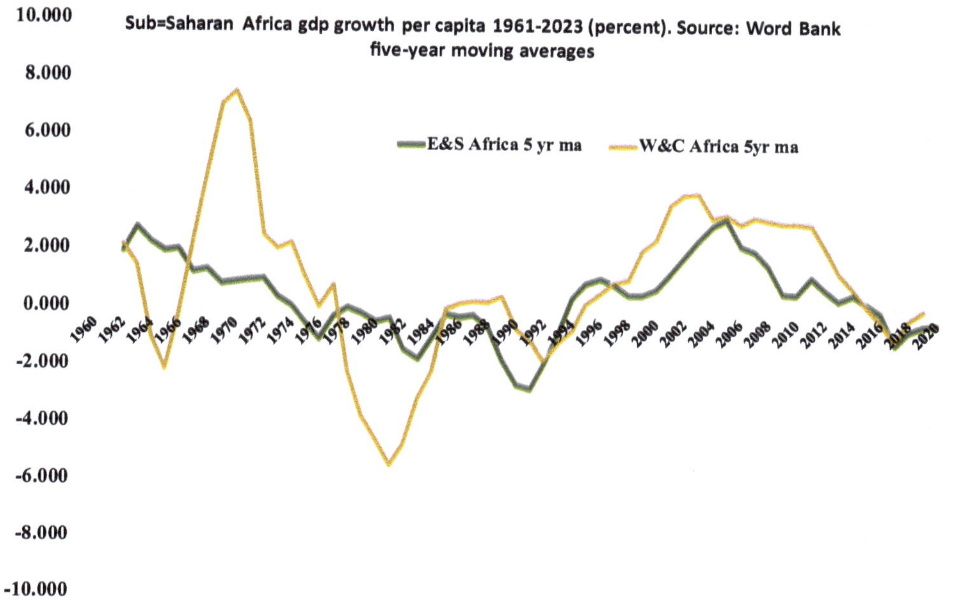

Fig. 1.4 Sub Saharan Africa gdp per capita trend 1961-2023. Source: World Bank. Source: https://data.worldbank.org/indicator/NY.GDP.PCAP.KD.ZG?end=2023&locations=ZG&start=1961

-more than in 2019), did not have access to adequate food. The rapid reproduction noted earlier makes food security in African populations vulnerable to droughts and civil war, currently (2024) at disastrous levels in Sudan, Congo and eleven other African countries.,

'Africa rising' is a popular optimistic interpretation of events, pointing to the demographic bonus of a youthful workforce (Beresford 2016) and examples of rapid economic growth. But each year 12 million young Africans arrive at working age to find only 3 million new jobs in the formal sector. That is not a demographic bonus. Economic growth must become general if Africa is to accommodate a population rising from 1.2 billion to about 4 billion by 2100; almost 40 percent of the Earth's projected population. Pessimists worry about 'the tragedy of Africa's dismal economic growth in the 20th century' (Artadi and Sala-i-Martin 2003) and its excessive reliance on primary resource extraction (Taylor 2016). Grimm et al. (2016) claim that dealing with Malthusian forces is the main challenge for countries such as Burkina Faso in the decade to come. African values, expensive investment goods, low levels of education, poor health, adverse geography, closed economies, high public expenditure and frequent military conflicts, keep the region on the edge of entrapment.

Is the problem too many mouths, or deficiencies of equality, supply and entitlement, as Amartya Sen (1983) and others insist, citing especially the Bengal famine of 1943 and the Irish Potato famine of 1841-5. NGOs such as OXFAM claim that people in Africa do not 'have too many children'. Problems are attributed to unequal land law, unfair restrictions on markets, bad administration and official indifference to suffering. Unwise strategy on agriculture and settlement can have disastrous results (Diamond 2005). Believing such factors are always the root cause of famine implies belief in the possibility of permanent perfection in human and political arrangements; like the aspirations of Godwin and Condorçet, unlikely ever to be attained. The history of famines (Ó Gráda 2009) provides many examples of natural events indifferent to human control. In the fourteenth century violent climatic change and epidemics halved the population of Europe; in the enduring cold and epidemics of the seventeenth, one third of the world's population may have died (Parker 2013). Even in the 20th century, when we ought to have known better, 70 million people are believed to have died in famines (Devereux 2000). The combination of vulnerable population, drought and civil war in sub-Saharan Africa has made famine endemic. This seemingly perpetual argument goes to the heart of the dispute between Malthus and his critics (The Economist 2008).

Are we (nearly) all Malthusians now? Is the near-universal adoption of artificial means of contraception, together with other measures, essentially a continuation of Malthus's 'prudential restraint', or something quite new because Malthus ruled out contraception? Malthus was quite specific about contraception. Contraception was vice; perverting natural law[39]' Through its precision, sidesteping the providential need for exertion, without which industriousness and innovation would not flourish and society would not develop. Positive or otherwise the mismatch of domestic food production and the increase of people is a potential often actual, source of trouble. Malthus' unequivocal prohibition forbids contraceptive use being labelled "Malthusian' in any strict sense. But a reasonable view would be that prudential restraint is using foresight to avert adverse consequences for welfare of the family and by extension the population. Motive and

[39] Malthus makes no allusion to the biblical injunction to 'be fruitful and multiply, and replenish the Earth' (Genesis 1.28). That is all very Old Testament, an injunction to demographic inflation after the Big Bang of Genesis. The command is repeated later: 'Now be fruitful and multiply, and repopulate the earth.' (Genesis 9:7); a response to a near –extinction event; negative feedback as punishment for sin (there really was a near-extinction event. But a little earlier than Noah; Ashton and Stringer 2023);. Mediaeval theologians Jerome, Augustine and others took the view that the divine injunction had been obeyed and that the world was already quite full enough (Brown 1989); one of the reasons why chastity and celibacy, monkish or otherwise, was often esteemed over marriage in the pre-reformation church (Biller 2000). In line with that tradition of restraint the Church of Rome was not, historically, devoted to large families or large population; Initially, Malthus was quite well received. Attitudes changed after 1870 (Vallin 1983). The founder of Christianity, unlike the founder of Islam, made no recommendation of marriage. Indeed Paul, His chief promoter, grudgingly concedes 'Now to the unmarried and the widows I say: 'it is good for them to stay unmarried, as I am. But if they cannot control themselves, they should marry, for it is better to marry than to burn with passion' (1 Corinthians 7). Malthus makes no recommendation of lifetime celibacy, conventual or otherwise, as a form of prudential restraint. Too Catholic, perhaps. But in this context he notes the unhappiness arising from marrying without the means to support children and the impropriety of doing so (Book iv, Ch.ii. p 462).

outcome are the same in each case. To that limited extent we are (nearly) all Malthusians now.

1.4.2 Malthus' 'Dire Warnings' About Future Problems. Will They Come to Pass?

A complaint often made against Malthus is that his 'dire warnings' of starvation and catastrophe from population growth, attributed to him by many authors, have not [yet] come to pass (Lam 2011; Allendorf and Allendorf 2012; Nekola and others 2013). Many disasters have been made worse by their population being too close to the sustainable level. But Malthus never spoke of 'catastrophe' or forthcoming disaster. He made no projections, forecasts of doom or global famine. Misery and vice were enough for him. Qualified pessimism was his line, that his Principle of Population would impede the progress of individuals and society. In a number of places in the Essay (e.g. p. 314) he regards numerical forecasts with great caution '…no estimates of future population or depopulation, formed from any existing rate of increase or decrease, can be depended upon'. So true. Malthus ' pre-occupations were on a smaller scale, micro-demographic, domestic not large scale; essentially the issues of the here and now, the contrast between population and resource growth keeping the former in check at the cost of discomfort, distress or even death, Apocalyptic prophesies were the enthusiastic projections of his followers; logical enough, to be sure.

Malthus was more definite about the ill effects of a declining population and its indications about the state of society, as noted above: But in focussing on food as always the ultimate cause of decline (p. 429) Malthus is surely wrong; history is full of the destruction of population by war, even if the other horsemen of the apocalypse also lent a hand. Despite the horrors of the 20th century, German historians think that the Thirty Years' War (1618-1648), reducing the population by 40% was the most destructive event in German history. The invasion of North China by the Mongols in the 13th century reduced its population by over 40 million. And yet another depopulation of Ukraine is in progress today (Coleman 2023).

In general Malthus favoured population growth, welcoming larger populations, as indicating social success if accompanied by higher living standards but fearing too rapid an increase, noting rather indecisively that 'Scarcity and extreme poverty may, or may not accompany an increasing population, according to circumstances' (Book III Ch xiv p. 419). Hedging his bets, he made a distinction between a large population and population that was 'redundant' (p. 126). He opposed population growth when economic circumstances were adverse, or when growth was too rapid; 'It is almost impossible, in theory, to overestimate the mischief attendant on a redundant, a growing and a unemployed population, converting that which ought to be the strength into the peril of a nation……leading to the boundless multiplication of human beings satisfied with the lowest condition of existence'. Malthus might think that in his time, populations did not need to be limited.

Density dependent regulation inevitably has a crucial time-lag, making it incapable of reacting precisely or effectively to population change. Malthus considered this latency ('oscillations') explicitly in Chapter II of the first essay (Hofmann 2012, Waterman 1987). These time lags may mean that human populations are in the process of exceeding comfortable limits, ensuring that the delayed response will be severe. But if all goes well he is sanguine about population growth. 'That an increase in population, when it follows in its natural order, is a great positive good in itself, and absolutely necessary to a further increase in the natural produce of the land or labour in any country I would be the last to deny.'. ... (p. 422). 'I can easily conceive that this country, with a proper direction of the national industry, might, in the course of some centuries, contain two or three times its present population and yet every man in the country be better fed and clothed than he is at present' (p. 461). Well, we can only hope. In 1821, Great Britain had 14.4 million, Ireland 6.8 million inhabitants so UK 21.2 million. Three times that is 63.4 million, not far short of the present 67.7 million but of course two centuries later, not one!

1.4.3 Overpopulation and the Limits to Growth – Our Concern, but Less So for Malthus.

One of the notable features of the Second Essay was an emphasis on the hope of progress, and equivocation on population growth. Population growth could act not so much as a road to distress and misery but as a spur to innovation, invention, the improvement of society and in the spread of civilisation. In this regard Malthus shared some of the opinions of the economist Ester Boserup (1965), generally supposed to be at odds with his views (Lee 1986, Lee and Anderson 2002) and certainly rather rude about them. He expressed few concerns about 'overpopulation', although in fact his allusions to 'redundant population', noted above, come to the same thing. He is explicit about Scotland, observing that it is 'certainly still over-peopled' and that the situation in Ireland is even more grave, the population growing such as to double every 45 years, more than any country in Europe (Book II, Ch. xi p. 199-200). Scarcity and extreme poverty, therefore, may or may not accompany an increasing population, according to circumstances. (Book III Ch xiv p. 420). In a society with markets for food and labour, population was not constrained by shortage of food directly except under extreme circumstances, but by the effect on wages. Agriculture could support a bigger population than those engaged in it, 'the whole population of the country may continue for ages increasing with the improving state of agriculture and yet be able to export corn.' (Book III Ch. XIV p. 418).

So students will search the editions of the 'Principle' in vain for any mention of maximum or sustainable population totals, and the word 'overpopulation' is used but twice. His comments on the US come closest to 'sustainable population': 'The United States is an instance of the agricultural system in which the condition of the labouring classes is the best of any we know' (Book III Ch viii pp. 374). 'America is perhaps the only modern instance of the fair operation of the agricultural system'. In arguing for the merits of combined systems of agriculture and commerce in an ideal state free of the disadvantages of being a purely agricultural or manufacturing country (Ch X p. 384 *passim*) 'the countries which such unite great landed resources with a prosperous state of commerce and manufactures and in which the commercial part of the population never exceeds the agricultural part are eminently secure from sudden reverses. Their increasing wealth seems to be out of reach of all common accidents and there is no reason to say that they might not go on increasing in riches and population for hundreds, nay almost thousands of years' (Tupy et al. 2021). Malthus the mathematician must have known that this was nonsense, even with a very low growth rate. And in fact he wrote:

'We must not, however, say that there is no limit to this progress though it is distant, and has certainly not been attained by any large landed nation yet known'. We have already seen that the limit to the population of commercial nations is the period when, from the actual state of foreign markets, they are unable regularly to import an increasing quantity of food. And the limit to the population of a nation which raises the whole of its food on its own territory is, when the land has been fully occupied and worked, that the employment of another labourer on it will not, on an average, raise an additional quantity of food sufficient to support a family of such a size as will admit of an increase in population'. (Ch X p. 882) he argues. Book 2 p. 260. Rutherford (2007) reviewing the options open to Malthus, considered the mixed approach to be transient and that Malthus missed the future in not seeing the possibility of a commercial/industrial economy partly self-sustaining but partly relying on imports. Which was the real future. As noted above this marked an important concession by Malthus. However, with increasing rumblings about food security it might just turn out that Malthus's warnings were right. 'Malthus the merchant of doom' is an avatar created both by friends and foe; a vision of what he might have thought, on the logic of his Principle of Population, if confronted with the extraordinary demographic developments of the 20th century, the scale of which he could not have imagined.

The prospect of decline: a Malthusian density-dependent feedback?

In most countries, especially Western countries birth rates are at an all-time low. Is it possible that we see here a correction of 'overshoot'? The lags inherent in feedback systems may have left populations approaching the end of the demographic transition systematically over-populated because their growth was inevitably corrected late (Rees 2023). That might be particularly applicable to the developing world. There, high fertility has fallen fast leaving behind substantial demographic momentum which will fuel population growth for decades in circumstances no longer appropriate for it. Transition from the previous regime implies a temporary demographic bonus. But then developing world faces the novelty of population ageing and decline.

In the developed world fertility has fallen far below replacement level, way beyond anything explicable by the possible retreat of fecundity discussed earlier. These reductions in birth rates (Fig. 1.3) have spread alarm and despondency about the prospect of severe population decline, and of population ageing accelerating to levels difficult to sustain (Coleman 2000, Coleman and Basten 2015), Fears of the damaging consequences– even extinction in the case of Korea - have revived the fears of the 1930s. Pronatalist policies, although seldom effective in the past, have been dusted off and brought up to date (The Economist 2024). Some see salvation in even higher rates of immigration. Others claim that the optimum for welfare is in any case a fertility rate somewhat below replacement level (Striessnig and Lutz 2013), that modest decline is manageable and would resolve some problems (Coleman and Rowthorn 2011, Lutz 2023), Furthermore a retreat of population in the face of climate change may be inevitable and desirable (Sconfienza 2020). But recent falls in fertility could take us into very steep decline and ageing, and clearly out of the world of Malthus.

Demographers are puzzled why this recent fertility decline is so substantial and so general and why previous champions of relatively high fertility; especially the Nordic countries, should decline so far (fertility in France has fallen least of all in Europe). In a few cases, in the Anglosphere and some Nordic countries, population still grows thanks to residual demographic momentum and immigration. In the Neo-Europes space is abundant and the population less hostile to migration, although increasingly restive. Not so in Britain, constrained by limited geography and pressed by inflating population.

Conventional explanations may take us part of the way. The modern world has not properly recovered from the 2008 financial crash. On top of that has come Covid, heightened awareness of a future of climate change which threatens the future of our children and grandchildren, an unsettled, dangerous international situation. Employment in all countries has become less secure. That is known to discourage family building (Clark and Lepinteur 2022). The Ukraine war adds to uncertainty,. With ageing of populations, parents are squeezed between the demands of their children and the care of their own frail or demented parents (Lazzari and Zurla 2024). In England and Wales family building is discouraged additionally by the most costly child care in Europe and very expensive housing. Average house-price has risen from four times to nine times average annual income. Many factors, including migration, have contributed to this increase in price.

A speculative suggestion may help to explain the very depressed birth rate in England and Wales, down to 1.44 in 2023. Immigration usually has a positive effect on the birth rate. But if it puts pressure on housing and its price then it might, most unusually, have the opposite effect. What might look like a Malthusian situation may not be one. Since the late 1990s population growth in the UK has risen considerably, on average to about 0.6% per year between 2001 and 2021 – the highest since 1930. Over that time the proportion of additional households whose head was born outside the UK has risen from 46% to 94%[40]. In 2018 natural increase fell below zero. Almost all population increase now, and all projected, is due to migration. That population growth, despite the fall of fertility, generates Malthusian pathology (population growth and pressure on resources) without Malthusian causation (a high birth-rate). (Malthus paid no attention to inward migration. Effectively,

[40]According to the annual Labour Force Survey. The census 2011-2021 shows 79% overall, from 2001 = 2021 74% (2.3 million out of 3.1 million) of additional households in England and Wales.

there was none). In the UK, thanks to a combination of very slow housebuilding and that relatively rapid population growth (Fig. 1.5) housing has become very expensive – average house price is about nine times annual income. Traditionally it was three or fourfold. In England the average age of first time house buyers has increased from 24 in the 1970s to 30 in 2021. Home ownership has declined.

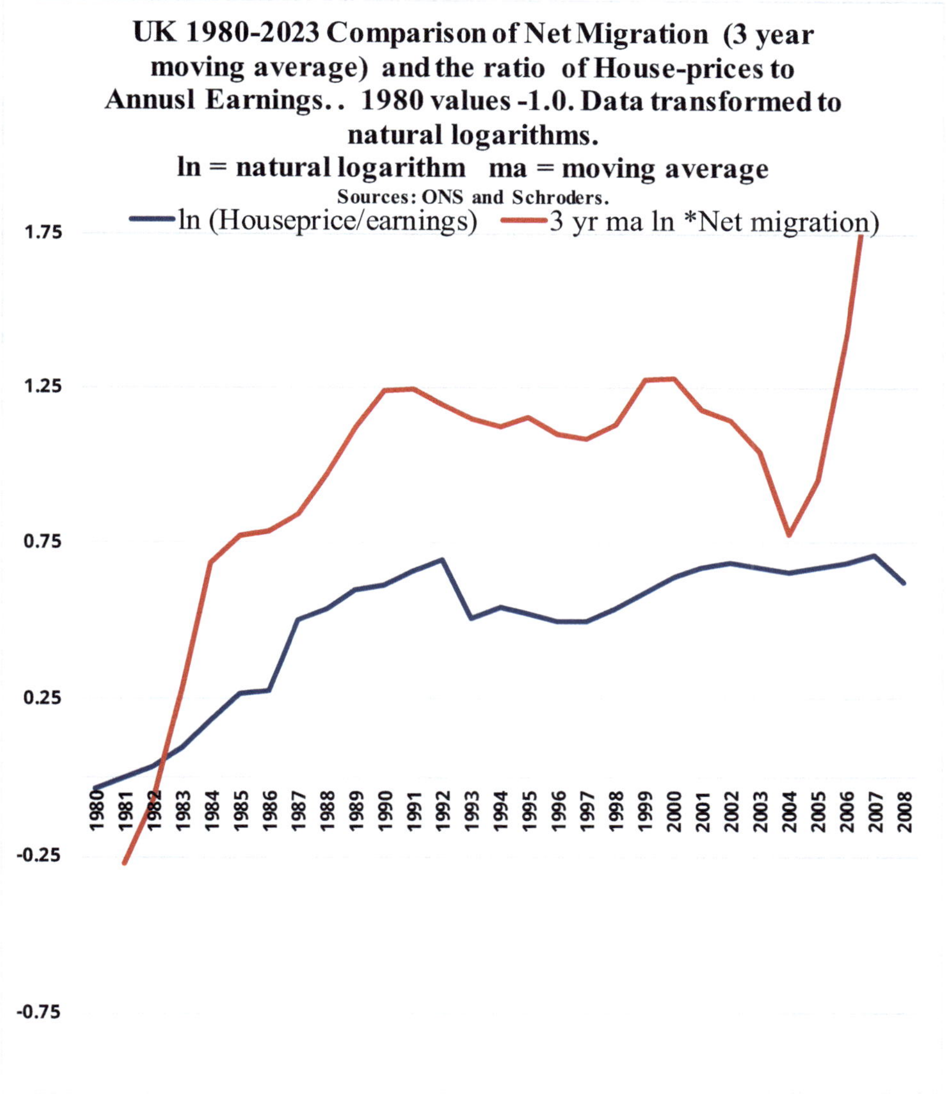

Fig. 1.5 Trend of house-prices and rents in relation to net migration

While results are ambiguous, research by DLP Associates (2023) showed that a 10% increase in house price was associated with a 1.3% fall in births. Since 1995 house prices have increased by 230%. Total fertility of those born in 2020 is projected to fall to 1.35. And in the US, living in an expensive area delays first birth by 3-4 years (Clark 2012). In a conventional Malthusian feedback, fertility would be depressed, eventually easing pressure of demand on housing and encourage recovery of the birth rate. Fertility is likely to be depressed but population growth will continue, pressing on housing. That is because it is not the birth rate that drives population growth but the unrelated, powerful factor of immigration, which continues unaffected (Fig. 1.6) at high levels, projected to be net 340,000 per year. No demographic response by the domestic population can bring this down. Instead the domestic population must adapt, possibly by minimising or delaying their consumption of increasingly unaffordable housing, with consequences for the birth rate. This possible effect of immigation on fertility via pressure on housing remains a speculation, beyond anything known to Malthus. This speculative hypothests, resting on a chain of correlations, remains to be tested with comparable situations oversea, if any exist. It fails if no such effect can be found in other high migration countries. Malthus said litle about immigration. But in their formalisation of Malthus's model of the economy, Wrigley and Schofield (Fig. 1.1) have net migration contributing to population size and hence to higher prices and less marriage, in that case through food price food not housing.

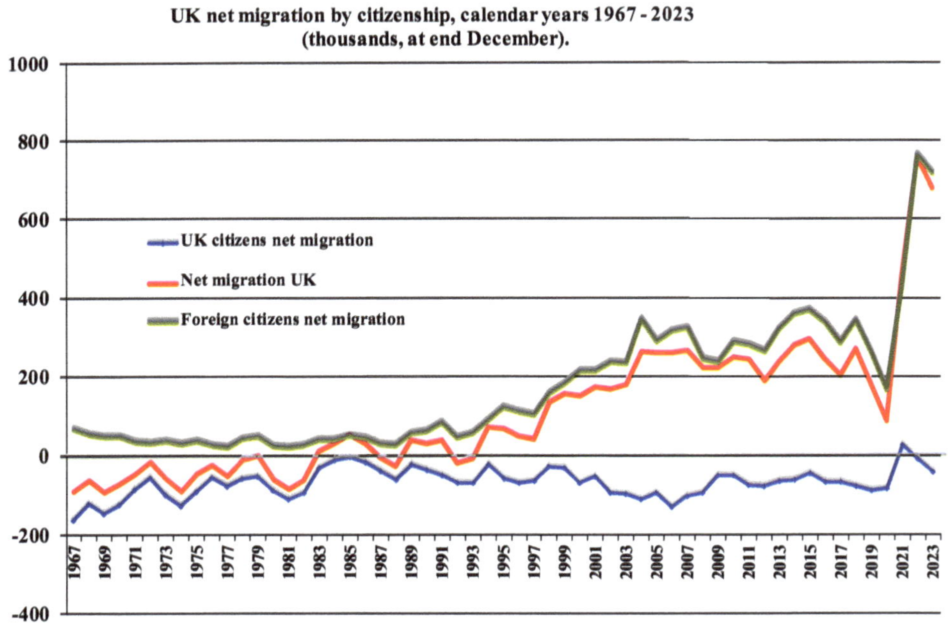

Fig. 1.6 Trend of net migration, United Kingdom 1964 – 2023. Source: ONS

Beyond such conventional factors is the suspicion that attitudes and preferences are moving against family life and the production of children, in a new and alarming psychological transition. Surveys in Britain, Finland, Sweden the US and elsewhere reveal that private interests, freedom of action, pursuit of career, have priority for many over family and children, no longer seen as the natural progression in life (Kearney et al. 2021, Cantalini et al. 2024). Self-realisation is not to be gained through children, but through their absence. Unprecedented proportions of young people state their intention to have no children. All stages of family building and reproduction are postponed without precedent throughout the Western world – the chief cause of low birth rates (Rotkirch 2021). Couples form through cohabitation but are slow to move on to children.

Malthus strongly deplored population decline. In his view it was always the consequence of a failure of society: 'the causes may always be traced to the want of industry or the ill direction of that industry, arising from violence, bad government ignorance etc. (Book III Ch xiv p 420), We are all familiar with inept government, but we must look beyond Malthus for understanding of today's failure of fertility. This new situation has surely moved beyond his ken.

Global Climate Change– Malthus Justified?

Finally, can climate change through human action properly be regarded as a Malthusian negative feedback on population? 'Population dynamics have not been integrated systematically into climate change science' (Stevenson and Sabater 2010). Insofar as population growth is a factor in the production of greenhouse gases and other factors of climate change, it must contribute to climate change. But not on a suitable timescale. Climate warming gases are increasing rapidly in proportion to carbon burning. By comparison, population changes slowly, certainly too slowly to be able to avert short term rapid increases of temperature (Kuruc et al. 2023). Some models of current climate change show no relation to population growth (Bretschger 2020). Instead many e.g. Toth and Szigeti (2016) claim that the main driver of growth and environmental degradation is no longer population per se, but consumption patterns and levels multiplied by the number of consumers, especially in developed economies. Supposedly the issue is not one of numbers but of high levels of consumption, output and emission in the richer countries which little or no population growth (Deuster et al. 2021). But climate change is a legacy of past population and industrial growth as well as current outputs. It would not have happened without the eightfold growth in world population since 1800, roughly coinciding with the industrial revolution. And the inertia of 'committed warming' ensures that change would continue for decades even if emissions stabilised.

Poorer countries had contributed less but have caught up. Their population size, collectively, is much bigger. Their growing populations aspiring to higher consumption, and have energy systems more difficult to decarbonise. World population is projected to increase by a further three billion people, almost all in the developing world. Furthermore estimates must take into account complex global value chains (GVC) which involve developing countries. Developing countries' GVC-based responsibility for global CO2 emissions surpassed that of developed countries since 2012 and is increasing quickly (Meng et al. 2023).

Climate change will do great damage to all life on earth, and by turning land into sea and desert effectively reduce the size of the habitable Earth. The negative effect on human populations is likely be large; some of the world is becoming unproductive. That is certainly negative feedback. That Malthus could not have imagined it is less relevant than the nature of the system.

Afterword

'There was no one element within Thomas Robert Malthus's Principle of Population that was wholly new, and yet he managed to make everything about it seem new.' (Bashford and Chaplin, 2016, p.7). In placing a barricade on the path to perfectibility Malthus provoked an enduring hullabaloo. As the Second Essay makes clear, his demographic work is not just a question of contrasted mathematical ratios and difficult delayed gratification. Instead, in the nature of demography, he presents on a broad scientific and philosophical canvas an eclectic body of scientific and social knowledge to support his argument. He made mistakes. And the Second Essay introduces an element of uncertainty and indecision in what had been strong assertions in the First: on the Poor Law, population trends, the future of industry, while the basic message remains. Foresight failed him on some basic questions; the future prominence of agriculture; the scale of prosperity and population to come, grown together eightfold after his time. That welfare encourages improvident reproduction received little factual support then or now.

Nonetheless poverty persists affecting four million children (Arnold 2022, Henry and Wernham 2024). Few today would publicly blame the poor for their problems; private opinion is a different matter. Some still take up Malthus's cudgels to confront imprudent behaviour. 'Undeserving poverty' endures, but empathy is now more prevalent. At any rate family size, and diversity of family size by status, are modest everywhere or becoming so. For most, 'prudential restraint' is taken for granted, for reproduction but not for sex, and not through means that Malthus would approve. Despite two centuries of progress, perfectibility in human affairs is not in sight.

The relevance of his 19th century thought to his own time is little disputed. Their insights into 21st century problems remain controversial'. The Malthusian trap still yawns for some regions of rapid population growth. Food insecurity, never absent, will very likely grow more acute as we approach ten billion people. The momentous demographic events of the last century; the demographic transition, the population 'explosion' and its persistent afterglow have Malthusian implications. But contrary to Malthus, eventual global population decline seems to be inevitable and necessary given the threat of global climate change and in response to the shrinking of the *oikoumene*, It could be the final homoeostatic response (Seidl and Tisdell 1999). In the past we fretted about the Earth's carrying capacity; none can now see the lower limit of population.

Two fundamental questions remain unanswered. The decline of passion between the sexes (a minority habit so far) may just reflect a temporary mismatch of values with new realities. The decline of biological fecundity is possibly the result of (potentially reversible) insults by our population to the environment. If so it is in our hands to halt. A decline of the desire to reproduce is a different matter, Malthus never considered it, probably assuming it was axiomatic, as an inevitable by-product of the passion between the sexes. But in modern populations free to choose; explaining why people should choose children given the horrendous cost, and twenty years of partial house arrest, awaits a new Malthus.

The value of Malthus' model in a coherent picture of the limits to growth and regulation of populations over the greatest part of human history seems well established. In England, three times since the beginning of the common era the population has approached the five million barrier and three times rebuffed, once in Roman times about AD 200, once in the high middle ages about 1300, once in the 18th century about 1650. Radical technical change was the key to the breakout. His model illuminates human affairs with notions of equilibrium, control, negative feedback and homoeostasis (Coleman 1986) linking human behaviour and its environments. After breaking free from Malthus, world population grew an order of magnitude greater, from one billion heading to ten billion. There will be no more billions after that, but likely decline, where Malthus is no guide.

Feedback is the essence of the Malthusian process. It provides a continuously applicable theory, not just the one-off theory of demographic transition, often thought to be the key demographic theory in a notoriously thin field (Schofield and Coleman 1986, Charbit 2000 p.9). Malthus's 'Principle' exemplifies dialectical confrontation and contradiction. The eternal needs of living beings destined to grow and reproduce in order to survive – the capacity to reproduce beyond current numbers being necessary for survival, attrition being continual and variable. Living beings are confronted by resources of a finite Earth, limited and unstable, with resources that may shrink or be destroyed by its own inhabitants.

Grateful thanks are due to Tony Champion, Rob Colls, Tim Coulson, Tim Dyson, Philip Kreager, Andrew Lack, Robert Mayhew, Anna Rotkirch and Richard Smith, with special thanks to my editors Armin Stock and Frida Trotter. All errors are mine. David Coleman October 3rd 2024.

Appendix 1

James's monumental biography (1979) covers the man as well as his works; Petersen (1979) gives one of the first modern general appreciations, Dupâquier et al. (1983) edited the first collection of important papers on a broad canvas, addressing some popular myths about Malthus. Coleman and Schofield (1986) celebrated the 150th anniversary of Malthus's death with a set of papers on theory from the wise and the good, in which Stone (1986) offers a short friendly sketch. Pullen (1998, p.344) takes on 'The unenviable task of surveying the last sixty-five years of Malthus scholarship (which) presents those brave enough or masochistic enough to undertake it with prodigious problems of research, appraisal, selection, and commentary'. Charbit (2000) uniquely combines a demographically informed analysis with historical and philosophical insights into the controversy in Britain and in France. Mayhew (2014) present a (mostly) favourable and notably readable interpretations of his works. Flew (1970), Glass (1953), Winch (2003, 2013, Macfarlane (2019) offer erudite extensive commentaries in edited editions, Bashford and Chaplin (2016) open a completely new window on Malthus' international awareness. Chaplin's (2018) offers insightful relevant extracts from the works of others, both critical and supportive. Valenze (2023) offers a strong non-statistical critique of Malthus's whole philosophical, logical and moral position. Lee (1986) reconciles Malthus' view with the apparently contradictory notions of Boserup (1965), who promoted the beneficial role of population growth in forcing improvements in productivity. Lee, among others, placed Malthus views in a formal mathematical economic framework. Tony Wrigley has been the most important analyst and friendly critic in numerous seminal articles and in his indispensable analysis of English historical population with Roger Schofield (1981, 1983). He and David Souden (1988) have put together all Mathus's writings, demographic and economic. Other editions of Malthus are available; apologies to those not mentioned.

David Coleman
Crick Road, Oxford and Bryfold Cottage, Lingen.

References

African Development Bank. 2024. *Africa's macroeconomic performance and outlook - January 2024*. African Development Bank.

Akintunde, T. S., et al. 2013. Population dynamics and economic growth in sub-Saharan Africa. *Journal of Economics and Sustainable Development* 4 (13).

Andrewartha, H. G., and L. C. Birch. 1954. *The distribution and abundance of animals*. Chicago: Univ. Chicago Press.

Arnold, H. 2022. *Exploring new narratives on child poverty*. London: Save the Children.

Artadi, E., and X. Sala-i-Martin. 2003. *The economic tragedy of the XXth century: Growth in Africa*. Cambridge, MA: National Bureau of Economic Research.

Ashton, N., and C. Stringer. 2023. Did our ancestors nearly die out? *Science* 381 (6661): 947–948.

Atoh, M. 1998. *Traditional family values of unmarried women: In relation to the phenomenal rise in the proportion never married. Summary of the Twenty-fourth National Survey on Family Planning*. Tokyo: The Mainichi Shimbun.

Banks, J. A., and D. V. Glass. 1953. A list of books, pamphlets and articles on the population question, published in Britain in the period 1793–1880. In *Introduction to Malthus*, ed. D. V. Glass. London: Watts and Co.

Bashford, A., and J. E. Chaplin. 2016. *The new worlds of Thomas Robert Malthus. Rereading the principle of population*. Princeton: Princeton University Press.

Bashir, S. 2007. Islamic tradition and celibacy. In *Celibacy and religious traditions*. C. Olson: Oxford University Press.

Beresford, A. 2016. Africa rising? *Review of African Political Economy*, Taylor & Francis 43:1–7.

Biller, Peter. 2000. *The measure of multitude: Population in medieval thought*. Oxford: Oxford University Press.

Binford, L. R. 1968. Post-Pleistocene adaptations. In *New perspectives in archaeology*, ed. S. R. Binford and L. R. Binford. Chicago: Aldine.

Blaug, Mark. 1963. The myth of the old poor law and the making of the new. *The Journal of Economic History* 23 (2): 151–184.

Blaug, Mark. 1964. The poor law report reexamined. *The Journal of Economic History* 24 (2): 229–245.

Boer, Pieter J. den. 1990. On the stabilization of animal numbers. Problems of testing. *Oecologia* 83: 38–46.

Bonar, J. 1885. *Malthus and his work*. London: Macmillan.

Bongaarts, John. 2020. Trends in fertility and fertility preferences in sub-Saharan Africa: The roles of education and family planning programs. *Genus* 76 (1): 32.

Bongaarts, John, and Stephen W. Sinding. 2009. A response to critics of family planning programs. *International Perspectives on Sexual and Reproductive Health* 35 (1): 39–44.

Boserup, Esther. 1965. *The conditions of agricultural growth the economics of agrarian change under population pressure*. 1st ed. London: Routledge.

Boyden, Stephen Ed, ed. 1970. *The impact of civilisation on the biology of man*. Toronto: Toronto University Press.

Brace, L. 2021. *A short introduction to the involuntarily celibate sub-culture*. Centre for Research and Evidence on Security Threats.

Bretschger, L. 2020. Malthus in the light of climate change. *European Economic Review* 127:103477.

Brown, Peter. 1989.*The body and society: Men, women and sexual renunciation in early Christianity*. London: Faber and Faber.

Burnett, John. 1968. *Plenty and want. A social history of diet from 1815 to the present day*. Harmondsworth: Penguin.

Caldwell, J. C. 1998. Malthus and the less developed world: The pivotal role of India. *Population and Development Review* 24 (4): 675–696.

Calhoun, J. B. 1962. Population density and social pathology. *Scientific American* 206:139–146.

Cantalini, S., et al. 2024. Cohabitation and marriage formation in times of fertility decline: The case of Sweden in the twenty-first century. *European Journal of Population* 40 (15).

Carlisle, Richard. 1828. *Every woman's book, or, what is love?: Containing most important instructions for the prudent regulation of the principle of love and the number of a family*. London: Richard Carlisle.

Chaplin, Joyce E., ed. 2018. *An essay on the principle of population: Influences on Malthus, selections from Malthus's work, economics, population, and ethics after Malthus, Malthus and global challenges, Malthusianism in fiction*. W.W. Norton.

Charbit, Yves. 2000. *Economic, social and demographic thought in the XIXth century. The population debate from Malthus to Marx*. Springer.

Charles, Enid. 1934. *The twilight of parenthood*. London: Watts and Co.

Chatterjee, P. 2021. India's child malnutrition story worsens. *The Lancet Child & Adolescent Health* 5 (5): 319–320.

Cheatham, K. 2007. "Let anyone accept this who can": Medieval Christian virginity, chastity, and celibacy in the Latin West. In *Celibacy and religious traditions*. C. Olson: Oxford University Press.

Cinnirella, F., et al. 2017. Malthus in the bedroom: Birth spacing as birth control in pre-transition England. *Demography* 54 (2): 413–436.

Clark, A. E., and A. Lepinteur. 2022. A natural experiment on job insecurity and fertility in France. *Review of Economics and Statistics* 104:386–398.

Clark, G. 2001. Farm wages and living standards in the industrial revolution: England, 1670–1869. *Economic History Review* liv (496).

Clark, G. 2005. The condition of the working class in England, 1209. *Journal of Political Economy* 113 (6): 1307–1340.

Clark, G., and N. Cummings. 2015. Malthus to modernity: Wealth, status, and fertility in England, 1500–1879. *Journal of Population Economics* 28:3–29.

Clark, P., and D. Souden. 1988. Migration and society in early modern England.

Clayton, P., and J. Rowbotham. 2008. An unsuitable and degraded diet? Part two: Realities of the mid-Victorian diet. *Journal of the Royal Society of Medicine* 101 (7): 350–357.

Cleland John, G., et al. 2011. Family planning in sub-Saharan Africa: Progress or stagnation? *Bulletin of the World Health Organisation* 89 (2): 137–143.

Cleland, John. 2016. Prospects for accelerated fertility decline in Africa. *The Journal of Population and Sustainability* 1 (2): 17–52.

Cobbett, William. 1830 (2001). *Rural rides in the counties of Surrey Kent Sussex and Hampshire*. London: Penguin Classics.

Cohen, Joel E. 1996. *How many people can the Earth support?* New York: WW Morton.

Cohen, M. N., et al., eds. 1980. *Biosocial mechanisms of population regulation*. New Haven: Yale University Press.

Coleman, D. A. 1986. Population regulation: A long-range view. In *The state of population theory: Forward from Malthus*, ed. D. A. Coleman and R. S. Schofield, 14–41. Oxford: Basil Blackwell.

Coleman, David. 2022. The depopulation of Ukraine: A recurrent disaster revisited. *Finnish Yearbook of Population Research* 56:115–136.

Coleman, David A., and Roger Schofield. 1986. *The state of population theory: Forward from Malthus*. Oxford: Blackwell.

Coleman, David, and Stuart Basten (2015). The Death of the West: An alternative view. *Population Studies: A Journal of Demography* 69 (Supplement 1 'Population: The Long View'): S107–S118.

Coleman, David A., and John Salt. 1992. *The British population: Trends, patterns and processes*. Oxford: Oxford University Press.

Coleman, David, and Robert Rowthorn. 2011. Who's afraid of population decline? A critical examination of its consequences. *Population and Development Review* 37 (s1): 217–248.

Crafts, N., and T. C. Mills. 2009. From Malthus to Solow: How did the Malthusian economy really evolve? *Journal of Macroeconomics* 31 (1): 68–93.

Cubaynes, S., D. R. MacNulty, D. R. Stahler, K. A. Quimby, et al. 2014. Density-dependent intraspecific aggression regulates survival in northern Yellowstone wolves (Canis lupus). *Journal of Animal Ecology* 83 (6): 1344–1356.

Darwin, Charles. 1859. *The origin of species*. London: John Murray.

Darwin, Charles. 2003 (1887; Posthumous, revised). *The autobiography of Charles Darwin*. Cambridge: Icon.

de Graaf, H., et al. 2004. Trends and the gender gap in the reporting of sexual initiation among 15-year-olds: A comparison of 33 European countries. *The Journal of Sex Research*:1–10.

Devereux, Stephen. 2000. *Famine in the twentieth century*, IDS working paper 105. Brighton: Institute of Development Studies.

Diamond, E. 2007. 41 "And Jacob remained alone": The Jewish struggle with celibacy. In *Celibacy and religious traditions*. C. Olson: Oxford University Press.

Diamond, Jared. 2005. *Collapse: How societies choose to fail or succeed*. New York: Viking.

Digby, A. 1983. Malthus and the reform if the poor law. Malthus past and present. In *International conference on historical demography*, ed. J. Dupâquier, A. Fauve-Chamoux, and E. Grebenik. London: Academic Press.

DLP Associates. 2023. Housing: The unintended contraceptive. *DLP Briefing Notes* (Note 10.01.2023 Ref No: 379).

Drake, Michael. 1966. Malthus on Norway. *Population Studies* 20:175–196.

Dribe, M., et al. 2017. Socioeconomic status and fertility decline: Insights from historical transitions in Europe and North America. *Population Studies* 71 (1): 3–21.

Dunbar, Robin I. 1992. Neocortex size as a constraint on group size in primates. *Journal of Human Evolution* 22:469–493.

Dupâquier, Jacques, A. Fauve-Chamoux, and E. Grebenik, eds. 1983. *Malthus past and present. International conference in historical demography*. London: Academic Press.

Dyson, Tim. 1977. The demography of the Hadza - In historical perspective. In *Conference on African historical demography*, 19. Edinburgh: Centre for African Studies.

Dyson, Tim. 2018. *A population history of India. From the first modern people to the present day*. Oxford: Oxford University Press.

Ehrlich, Paul R. 1968. *The population bomb. Population control or race to oblivion?* Ballantine Books.

Ehrlich, I., and J. Kim. 2005. Endogenous fertility, mortality and economic growth: Can a Malthusian framework account for the conflicting historical trends in population? *Journal of Asian Economics* 16:789–806.

Eng, R. Y., and T. C. Smith. 1976. Peasant families and population control in eighteenth century Japan. *Journal of Interdisciplinary History* 6:417–445.

Engels, Friedrich. 1845. *The condition of the working class in England*. Leipzig.

Feinstein, Charles H. 1998. Wage-earnings in Great Britain during the industrial revolution. In *Applied economics and public policy*, ed. I. Begg and B. Henry, 181–208. Cambridge: Cambridge University Press.

Fernandes-Villaverde, J. 2001. Was Malthus right? Economic growth and population dynamics *. *Pennsylvania Institute for Economic Research Working Paper 01-046*: 45.

Fisher, Ronald A. 1930. *The genetical theory of natural selection*. Oxford: Oxford University Press.

Flew, Antony. 1970. Introduction. In *Malthus an essay on the principle of population*, ed. A. Flew, 7–56. Harmondsworth: Penguin Pelican Classics.

Flinn, Michael W. 1983. Malthus and his time. Malthus past and present. In *International conference on historical demography*, ed. J. Dupâquier, A. Fauve-Chamoux, and E. Grebenik, 85–95. London: Academic Press.

Follett, C. 2020. *Neo-Malthusianism and coercive population control in China and India: Overpopulation concerns often result in coercion*. Washington, DC: Cato Institute.

Foster, John B. 1998. Malthus' essay on population at age 200. A Marxian view. *Monthly Review* 50 (07): 1–17. https://monthlyreview.org/1998/12/01/malthus-essay-on-population-at-age-200/.

Freedman, J. C. 1980. Human reactions to population density. In *Biosocial mechanisms of population regulation*, ed. M. N. Cohen, R. S. Malpass, and H. G. Klein, 189–208. New Haven: Yale University Press.

Furlong, A. 2008. The Japanese Hikikomori phenomenon: Acute social withdrawal among young people. *The Sociological Review* 56 (2): 309–325.

Gagnon, Alain. 2015. Natural fertility and longevity. *Fertility and Sterility* 103 (5): 1109–1116.

Gallardo-Albarrán, D., and H. de Jong. 2020. Optimism or pessimism? A composite view on English living standards during the Industrial Revolution. *European Review of Economic History* 25 (1): 1–19.

Galor, Oded, and D. N. Weil. 2002. Population, technology, and growth: From Malthusian stagnation to the demographic transition and beyond. *The American Economic Review* 90 (4): 806–828.

Garenne, M. 2014. *Trends in marriage and contraception in sub-Saharan Africa: A longitudinal perspective on factors of fertility decline*, DHS analytical studies no. 42. Rockville: ICF International.

Ghaznavi, C., et al. 2019. Trends in heterosexual inexperience among young adults in Japan: Analysis of national surveys, 1987–2015. *BMC Public Health* 19 (355).

Glass, David V. 1953. *Introduction to Malthus*. London: Watts.
Glass, David V. 1973. *Numbering the people: The eighteenth century population controversy and the development of census and vital registration statistics in Britain*. Farnborough.
Godelier, M. 1983. Malthus and ethnography. In *Malthus past and present. International conference in historical demography* , ed. J. Dupâquier, A. Fauve-Chamoux, and E. Grebenik. London: Academic Press.
Griffin, E. 2018. Diets, hunger and living standards during the British industrial revolution*. *Past & Present* 239 (1): 71–111.
Grimm, M., et al. 2016. Burkina Faso: Shipping around the Malthusian Trap. In *Growth and poverty in sub-Saharan Africa*, ed. C. Arndt, A. McKay, and F. Tarp. Oxford University Press.
Gietel-Basten, Stuart, et al. 2019. Assessing the impact of the "one-child policy" in China: A synthetic control approach. *PLoS One* 14 (11).
Guzmán, R., et al. 2018. The ecosystems of simple and complex societies: Social and geographical dynamics. *Journal of Artificial Societies and Social Simulation* 21.
Hajnal, John. 1965. European marriage patterns in perspective. In *Population in history*, ed. D. V. Glass and D. E. C. Eversley. London: Edward Arnold.
Hajnal, John. 1982. Two kinds of pre-industrial household formation system. *Population and Development Review* 8:449–494.
Harrison, Geoffrey A. 1995. *The human biology of the English Village*. Oxford: Oxford University Press.
Hazlitt, William. 1807. In 'reply to Malthus. In *Malthus and the liberties of the poor*, ed. W. Hazlitt. London: Longman.
Hazlitt, William. 1825. *The spirit of the age or, contemporary portraits*. London.
Heale, M. 2013. *Monasticism in late medieval England, c. 1300–1535*. Manchester University Press.
Healey, J. 2011. Land, population and famine in the English uplands: A Westmorland case study, c.1370–1650*. *Agricultural History Review* 59 (II): 151–175.
Hecht, J. 1987. Johann Peter Süssmilch: A German prophet in foreign countries. *Population Studies* 41 (1): 31–58.
Henderson, P. A., and A. E. Magurran. 2014. Direct evidence that density dependent regulation underpins the stability of abundant species in a diverse animal community. *Proceedings of the Royal Society B* 281:1791.
Henry, A., and T. Wernham. 2024. *Child poverty: Trends and policy options*. London: Institute for Fiscal Studies.
Hilevych, Y., and P. Kreager, eds. 2025. *Low fertilities in the past and present: Studies in compositional demography*. Oxford: British Academy and Oxford University Press.
Hofmann, K. 2012. Beyond the principle of population: Malthus's essay. *European Journal of The History of Economic Thought* 20:1–27. https://doi.org/10.1371/journal.pone.0013206.
Hollingsworth, Thomas H. (1983). The influence of Malthus on British thought. Malthus past and present. J. Dupâcquier, Fauve-Chanoux, A. and Grebenik, Eugene. London, Academic Press.
Huzel, J. P. 1968. Malthus, the poor law and population in early 19th century England. *Economic History Review* 22 (3): 430–452.
James, Patricia. 1979. *Population Malthus. His life and times*. London: Routledge and Kegan Paul.
Jensen, T. K., et al. 2002. Poor semen quality may contribute to recent decline in fertility rates. *Human Reproduction* 17 (6): 1437–1440.
Jiang, L. 2011. *China's one-child policy*. Arizona State University. School of Life Sciences. Center for Biology and Society.
Kallis, G. 2019. *Limits. Why Malthus was wrong and why environmentalists should care*. Stanford: Stanford University Press.

Kearney, M. S., et al. 2021. The puzzle of falling US birth rates since the great recession. National Bureau of Economic Research (Working paper 29286).

Kelly, M. 2005. *Living standards and population growth: Malthus was right.* Department of Economics Working Papers. Dublin, Department of Economics, Trinity College Dublin.

Kelly, Morgan, and Cormac Ó Gráda. 2012. The preventive check in medieval and preindustrial England. *The Journal of Economic History* 72 (4): 1015–1035.

Keynes, John M. 1937. Some economic consequences of a declining population. The Galton Lecture 1937, Eugenics Review XXIX, 13-17 and Introduction by David Coleman. In *Twelve Galton lectures. A centenary selection with commentaries*, ed. Steve Jones and Milo Keynes, 61–74. London: The Galton Institute.

Kieschnick, J. 2007. 225 Celibacy in East Asian Buddhism. In *Celibacy and religious traditions.* C. Olson: Oxford University Press.

Knowlton, Charles. 1832. *Fruits of philosophy, or the private companion of young married people.* New York.

Konishi, S., et al. 2022. Casual sex and sexlessness in Japan: A cross-sectional study. *Sexes* 3 (2): 254–266.

Kreager, Philip. 1986. Demographic regimes as cultural systems. In *The state of population theory. Forward from Malthus*, ed. D. A. Coleman and R. Schofield, 131–155. Blackwell.

Kreager, P. 2022. Smith or Malthus? A sea-change in the concept of a population. *Population and Development Review* 48 (3): 645–688.

Kuruc, K., et al. 2023. Population decline: Too small and too slow to influence climate change. *SSRN.* https://doi.org/10.2139/ssrn.4535022.

Kussmaul, Ann S. 1979. Servants in husbandry in early modern England. *The Journal of Economic History* 39 (1): 329–331.

Lack, David. 1954. *The natural regulation of animal numbers.* Oxford: Clarendon Press.

Lack, David. 1966. *Population studies of birds.* Oxford: Oxford University Press.

Lam, D., and L. Marteleto. 2008. Stages of the demographic transition from a child's perspective: Family size, cohort size, and children's resources. *Population and Development Review* 34 (2): 225–252.

Landes, David. 1998. *The wealth and poverty of nations.* London: Little, Brown.

Laslett, Peter. 2004. *The world we have lost: Further explored.* London: Routledge.

Lawson, Bryan. 2010. The social and psychological issues of high-density city space. In *Designing high-density cities: For social and environmental sustainability*, ed. E. Ng, 285–292. Earthscan.

Lazzari, E., and E. Beaujouan. 2024. *The gap between desired and achieved family size in Europe: The role of infertility. Delayed reproduction: Challenges and prospects.* Vienna: Wittgenstein Centre, Vienna Institute of Demography.

Lazzari, E., and V. Zurla. 2024. The effect of parental caregiving on the fertility expectations of adult children. *European Journal of Population* 40 (1): 35.

Leach, D. 1981. Re-evaluation of the logistic curve for human populations. *Journal of the Royal Statistical Society A* 144 (1): 94–103.

Lee, J. 2020. The impact of sexual initiation timing and sexual experience on depressive symptoms in South Korean adolescents: Gender differences in a nationwide cross-sectional study. *Journal of Pediatric Nursing* 52:e15–e20.

Lee, Ronald D. 1986. Malthus and Boserup: A dynamic synthesis. In *The state of population theory—Forward from Malthus* , ed. D. A. Coleman and R. Schofield, 96–130. Oxford: Blackwell.

Lee, Ronald D. 1987. Population dynamics of humans and other animals. *Demography* 24 (4): 443–465.

Lee, Ronald, and M. Anderson. 2002. Malthus in state space: Macro economic-demographic relations in English history 1540-1870. *Journal of Population Economics* 15:195–220.

Lee, Ronald B., and Thomas DeVore, eds. 1968. *Man the hunter. Proceedings of a symposium held at the University of Chicago April 6–9, 1966*. Chicago: University of Chicago Press.

Lee, J., and W. Feng. 1999. Malthusian models and Chinese realities: The Chinese demographic system 1700–2000. *Population and Development Review* 25 (1): 33–65.

Le Roy Ladurie, Emmanuel. 1966, 1976. *The Peasants of Languedoc*. Tr. John Day. Urbana: University of Illinois Press.

Leridon, Henry, and Jane Menken. 1982. *Natural fertility: Patterns and determinants*. Liege: Editions Ordina.

Lutz, Wolfgang, M. R. Testa, and D. J. Penn. 2006. Population density is a key factor in declining human fertility. *Population and Environment* 28:69–81.

Lutz, Wolfgang. 2023. Population decline will likely become a global trend and benefit long-term human wellbeing. In *Vienna yearbook of population research 2023. Special issue on the causes and consequences of depopulation*, ed. Z. Brzozowska, E. Zhelenkova, and S. Gietel-Basten, vol. 21, 41–55. Vienna.

Macfarlane, Alan. 1978. *The origins of English individualism*. Oxford: Basil Blackwell.

Macfarlane, Alan. 2003. *The savage wars of peace: England, Japan and the Malthusian Trap*. Houndmills: Palgrave Macmillan.

Macfarlane, Alan. 2018. *Thomas Malthus and the making of the modern world*. Cambridge: Cam Rivers Publishing.

McClaren, I. A. 1971. Introduction. In *Natural regulation of animal populations*, ed. I. A. McClaren, 1–21. London: Aldine Transaction.

McLaren, Angus. 1978. *Birth control in nineteenth century England*. London: Croom Helm.

McIntosh, M. K. 2011. The poor laws of 1598 and 1601. In *Poor Relief in England, 1350–1600*, ed. M. K. McIntosh, 273–293. Cambridge: Cambridge University Press.

Madsen, J. B., et al. 2019. Malthus was right: Explaining a millennium of stagnation. *European Economic Review* 118:51–68.

Malthus, T. R. 1803. *An essay on the principle of population,* 2nd ed. London. Printed for J. Johnson. xiii and 610 pp. (1996, London, Routledge).

Malthus, T. R. 1820 (1st edition) 1836, (2nd edition) (1989). *Principles of political economy*. London (Cambridge, ed. Pullen), William Pickering.

Malthus, T. R. 1830. *A summary view of the principle of population*. London: John Murray. Also reprinted in Flew (1970) pp. 219–272 and Glass (1953) pp. 119–181.

Malthus, T. R. ed. Joyce, E. 1798 (2018). *An essay on the principle of population*. New York/London: W.W. Norton.

Malthus, Thomas Robert 1766–1834. 1996. *An essay on the principle of population*. 6th ed. London: Routledge/Thoemmes Press.

Marmot, Michael G., and R. Sapolsky. 2014. Of baboons and men: Social circumstances, biology, and the social gradient in health. In *Sociality, hierarchy, health: Comparative biodemography: A collection of papers*, ed. M. Weinstein and M. A. Lane. Washington DC: Committee on Population; Division of Behavioral and Social Sciences and Education; National Research Council National Academies Press.

Marshall, J. D. 1968. *The old poor law, 1795–1834*. London: Macmillan.

Marx, Karl. 1867. *Capital. Note 6 to Volume 1, Part 7, Chapter 25*. Ware: Wordsworth.

Matsuda, S., et al. 2024. Deterioration in youth employment, social contexts, and marriage decline in Japan and South Korea. *Asian Population Studies* 20 (2): 203–221.

Mayhew, R. J. 2014. *Malthus: The life and legacies of an untimely prophet*. Cambridge, MA: Harvard University Press.

Meek, Ronald L. 1953. *Marx and Engels on Malthus. Selections from the writings of Marx and Engels dealing with the theories of Thomas Robert Malthus*. London: Lawrence and Wishart.

Meng, B., et al. 2023. Developing countries' responsibilities for CO2 emissions in value chains are larger and growing faster than those of developed countries. *One Earth* 6 (2): 167–181.

Mercer, C. H., et al. 2013. Changes in sexual attitudes and lifestyles in Britain through the life course and over time: Findings from the National Surveys of Sexual Attitudes and Lifestyles (Natsal). *The Lancet* 382 (9907): 1781–1717.

Merchant, E. K. 2022. Environmental Malthusianism and demography. *Social Studies of Science* 52 (4): 536–560.

Minzner, C., et al. 2023. China's demographic challenges - Interpret China. *CSIS Interpret China.*

Mitchell, R. E. 1971. Some social implications of high density housing. *American Sociological Review* 36 (1): 18–29.

Mitrany, M. 2005. High density neighborhoods: Who enjoys them? *GeoJournal* 64 (2): 131–140.

Montano, B., and M. García-Lopez. 2020. Malthusianism of the 21st century. *Environmental and Sustainability Indicators* 6:6.

Moriki, Y. 2012. Mothering, co-sleeping, and sexless marriages: Implication for the Japanese population structure. *International Christian University publications. II-B, The Journal of Social Science* 74:27–45.

Murdoch, W. W. 1994. Population regulation in theory and practice. *Ecology* 75 (2): 271–287.

Murray, B. G. J. 1999. Can the population regulation controversy be buried and forgotten? *Oikos* 84 (1): 148–152.

Naso, P., et al. 2020. The return of Malthus? Resource constraints in an era of declining population growth. *European Economic Review* 128:2. https://doi.org/10.1016/j.euroecorev.2020.103499.

Nielsen, Ron W. aka Jan Nurzynski. 1997. No stagnation in the growth of population. *arXiv: Populations and Evolution (q-bio.PE)* (1311.3997 [q-bio.PE])

Nicholls, Alice. 2000. Fenland ague in the nineteenth century. *Medical History* 44:515–530.

Ó Gráda, Cormac. 2009. *Famine: A short history*. Princeton: Princeton University Press.

Owen, Richard D. 1831. *Moral physiology: A brief and plain treatise on the population question*. New York: Leopold Classic Library 2015.

Oversea Migration Board. 1960. *5th Report of the Overseas Migration Board, Mar 1960*. London: HMSO.

Parker, Geoffrey. 2013. *Global crisis. War, climate change and catastrophe in the seventeenth century*. New Haven: Yale University Press.

Petersen, William. 1979. *Malthus*. London: Heinemann.

Petersen, William. 1989. Marxism and the population question. Theory and practice. In *Population and Development Review*, Supplement to Volume 14, ed. Michael S. Teitelbaum and Jay M. Winter, 77–101.

Pfister, U., and G. Fertig. 2020. From Malthusian disequilibrium to the post-Malthusian era: The evolution of the preventive and positive checks in Germany, 1730–1870. *Demography* 57 (3): 1145–1170.

Piketty, Thomas. 2014. *Capital in the 21st century*. Cambridge, MA: Harvard University Press.

Pitkänen, K. 1992. *The patterns of mortality during the Great Finnish Famine in the 1860s*, Acta Demographica 1992. Heidelberg: Physica-Verlag HD.

Pullen, John. 1998. The last sixty-five years of Malthus scholarship. *History of Political Economy* 30 (2): 344–352.

Pullen, J. M. 2009. Malthus on colonization and economic development: A comparison with Adam Smith. *Utilitas* 6 (2): 243–266.

Rakotoarisoa, M. A., et al. 2012. *Why has Africa become a net food importer? Explaining Africa agricultural and food trade deficits*. Rome: Food and Agriculture Organization of the United Nations, trade and markets division.

Read, D., and S. LeBlanc. 2003. Population growth, carrying capacity, and conflict. *Current Anthropology* 44 (1): 59–85.

Rees, William E. 2023. The human eco-predicament: Overshoot and the population conundrum. In *Vienna yearbook of population research 2023. Special issue on the causes and consequences of depopulation*, ed. Z. Brzozowska, E. Zhelenkova, and S. Gietel-Basten, vol. 21, 21–39. Vienna.

Rickman, John. 1801–2. *Abstract of the answers and returns made pursuant to an act, passed in the forty-first year of His Majesty King George III. Intituled, "An Act for taking an account of the population of Great Britain, and the increase or diminution thereof"*. London: HMSO and republished by Routlege, London, 2002.

Ross, E. B. 2000. *The Malthus factor. Poverty, politics and population in capitalist development. Corner House Briefing 20.* The Corner House.

Roser, M. 2020. *Breaking out of the Malthusian trap: How pandemics allow us to understand why our ancestors were stuck in poverty.* Published online at OurWorldInData.org. Retrieved from: [Online Resource].

Rotkirch, Anna. 2021. Birth cohort changes in fertility ideals: Evidence from repeated cross-sectional surveys in Finland. https://doi.org/10.31235/OSF.IO/7VTQM.

Rutherford, Donald. 2007. Malthus and three approaches to solving the population problem. *Population* 62 (2): 213–237.

Sato, R., and M. Iwasawa. 2013. Single, sexless and infertile: Sexuality aspects of very low fertility in Japan. *BSPS Annual Conference 2013.* Swansea.

Schofield, Roger, and David Coleman. 1986. Introduction. The state of population theory. In *The state of population theory: Forward from Malthus D*, ed. A. Coleman and R. S. Schofield, 1–14. Oxford: Basil Blackwell.

Schtickzelle, M. 1981. Pierre-François Verhulst (1804–1849). La première decouverte de la fonction logistique. *Population* 36 (3): 541–556.

Sconfienza, U. M. 2020. Limits. Why Malthus was wrong and why environmentalists should care by Giorgos Kallis. *Environmental Politics* 29 (2): 360–361.

Scott, Anne M., ed. 2015. *Experiences of charity, 1250-1650.* London: Routledge.

Seidl, I., and C. A. Tisdell. 1999. Carrying capacity reconsidered: From Malthus' population theory to cultural carrying capacity. *Ecological Economics* 31 (3): 395–408.

Sen, Amartya. 1983. *Poverty and famines: An essay on entitlement and deprivation.* Oxford University Press.

Shermer, M. 2016. Why Malthus is still wrong, why Malthus makes for bad science policy. *Scientific American* (May 1 2016).

Skirbekk, Vegard. 2008. Fertility trends by social status. *Demographic Research* 18 (5): 145–180.

Skirbekk, Vegard, et al. 2024. *Sustained Postponement of Reproduction Theory (SPORT). Delayed reproduction: Challenges and prospects.* Vienna: Wittgenstein Centre, Vienna Institute of Demography.

Smith, Adam. 1776. *An enquiry into the nature and causes of the wealth of nations.* Edinburgh/London: J.M Dent.

Smith, P. 2015. Malthus is still wrong: We can feed a world of 9–10 billion, but only by reducing food demand. *The Proceedings of the Nutrition Society* 74 (3): 187–190.

Smith, Richard M. 1988. Natural fertility in pre-industrial Europe. In *Natural human fertility: Social and biological determinants*, ed. P. Diggory, M. Potts, and S. Teper, 70–88. London: Palgrave Macmillan UK.

Soloway, R. A. 1997. Marie Stopes, Eugenics and the Birth Control Movement. In *Proceedings of a conference organised by the Galton Institute, London 1996*, ed. R. A. Peel, 49–76. London: The Galton Institute.

Stephenson, J., et al. 2010. Population dynamics and climate change: What are the links? *Journal of Public Health* 32 (2): 150–156.

Stevenson, R. J., and S. Sabater. 2010. Understanding effects of global change on river ecosystems: Science to support policy in a changing world. *Hydrobiologia* 657 (1): 3–18.

Stone, Glenn D. 2021. *Malthusian thought, Oxford research encyclopaedias: Anthropology (online.).* Oxford: Oxford University Press. https://doi.org/10.1093/acrefore/9780190854584.013.312.

Stone, Robert. 1986. Robert Malthus: An appreciation. In *The state of population theory. Forward from Malthus*, ed. D. A. Coleman and R. Schofield, 42–45. Oxford: Blackwell.

Striessnig, Erik, and Wolfgang Lutz. 2013. Can below-replacement fertility be desirable? *Empirica* 40:409–425.

Szoltysek, M., and B. Zuber-Goldstein. 2009. Historical family systems and the great European divide: The invention of the Slavic East. *Demografia* 52 (5, English Edition): 5–47.

Tanner, J. T. 1966. Effects of population density on growth rates of animal populations. *Ecology* 47: 734–740.

Taylor, J. S. 1991. A different kind of Speenhamland: Nonresident relief in the industrial revolution. *Journal of British Studies* 30 (2): 183–208.

Teitelbaum, Michael S. 1984. *The British fertility decline*. Princeton: Princeton University Press.

Teitelbaum, M. S., and J. Winter. 1985. *Fear of population decline. Supplement to population and development review*. New York: Population Council.

The Economist. 2008. Malthus, the false prophet. The pessimistic parson and early political economist remains as wrong as ever. *The Economist*, May 15.

The Economist. 2024. *Fertility policy and practice: A toolkit for Europe*. London: The Economist 1 October.

Taylor, A. M., and J. G. Williamson. 1994. *Convergence in the age of mass migration*, Working paper no 4711, NBER. Cambridge, MA: National Bureau of Economic Research.

Tiger, Lionel, and Robin Fox. 1971. *The imperial animal*. New York: Holt, Rinehart & Winston.

Toth, G., and C. Szigeti. 2016. The historical ecological footprint: From over-population to over-consumption. *Ecological Indicators* 60:283–291.

Toye, John. 2000. *Keynes on population*. Oxford: Oxford University Press.

Travers, M. 2024. The concerning global rise of Hikikomori syndrome. *Psychology Today*.

Trewavas, A. 2002. Malthus foiled again and again. *Nature* 418 (6898): 668–670.

Tupy, M., et al., eds. 2021. *Resource abundance and the limits to growth*. Cato Unbound, Cato Institute.

van Wissen, Leo, and W. van der Weijden. 2024. Five misconceptions about population growth in Africa (2). Agrifood Networks. https://agrifoodnetworks.org/article/five-misconceptions-about-population-growth-in-africa-2

Ueda, P., et al. 2020. Trends in frequency of sexual activity and number of sexual partners among adults aged 18 to 44 years in the US, 2000–2018. *JAMA Network Open* 3 (6): e203833.

Vallin, F. 1983. French Roman Catholics and Malthusianism before 1870. In *Malthus past and present*, ed. J. Dupâquier, A. Fauve-Chamoux, and E. Grebenik. London: Academic Press.

Walter, M. A. H. B. 1978. The territorial and the social: Perspectives on the lack of community in high-rise/high density living in Singapore. *Ekistics* 270:236–242.

Wang, Z., et al. 2016. Ending an era of population control in China: Was the one-child policy ever needed? *The American Journal of Economics and Sociology* 75 (4): 929–979.

Warde, Paul. 2007. *Energy consumption in England & Wales, 1560–2004*. Naples: Consiglio Nazionale della Ricerche.

Waterman, Anthony M. C. 1983. Malthus as a theologian: The 'First Essay' and the relation between political economy and Christian theology. In *Malthus past and present. International conference on historical demography*, ed. J. Dupâquier, A. Fauve-Chamoux, and E. Grebenik. London: Academic Press.

Waterman, Anthony M. C. 1987. On the Malthusian theory of long swings. *The Canadian Journal of Economics/Revue canadienne d'Economique* 20 (2): 257–270.

Weil, N. D., and J. Wilde. 2010. How relevant is Malthus for economic development today? *American Economic Review* 100 (2): 378–382.

Wheeler, M. 1999. ICPD and its aftermath: Throwing out the baby? *Bulletin of the World Health Organisation* 77 (9): 778–779.

Wilson, Chris, et al. 1988. What is natural fertility? The modelling of a concept. *Population Index* 54 (1): 4–20.

Wiltgen, R. J. 1998. Marx's and Engels' views on Malthus: The heritage of a critique. *Organization & Environment* 11 (4): 451–460.

Winch, Donald. 1987. *Malthus*. Oxford: Oxford University Press.

Winch, Donald. 1996. *Riches and poverty. An intellectual history of political economy in Britain 1750–1834*. Cambridge: Cambridge University Press.

Winch, Donald. 2003. Malthus, Thomas Robert. In *The encyclopedia of population*, ed. Paul Demeny and Geoffrey McNicol, vol. 2, 619–621. New York: Macmillan Reference.

Winch, Donald. 2013. *Malthus: A very short introduction*. Oxford: Oxford University Press.

Wood, J. 1998. A theory of pre-industrial population dynamics: Demography, economy and well-being in Malthusian systems. *Current Anthropology* 39 (1): 99–121.

Woods, Robert. 2000. *The demography of Victorian England and Wales*. Cambridge: Cambridge University Press.

Wrigley, E. A. 1986. Malthus's model of a preindustrial economy. In *Malthus and his time*, ed. M. Turner, 3–18. London: Palgrave Macmillan UK.

Wrigley, E. A. 1983. The growth of population in eighteenth-century England; a conundrum resolved. *Past and Present* xcviii:121–150.

Wrigley, E. A. 2016. Malthus: A man of his time, Unpublished paper. Cambridge University, p. 23.

Wrigley, E. A., and R. Smith. 2019. Malthus and the poor law. *The Historical Journal* 63 (1): 33–62.

Wrigley, E. A., and R. S. Schofield. 1981. *The population of England 1541–1871: A reconstruction*. London: Edward Arnold.

Wrigley, E. A., and R. S. Schofield. 1983. English population history from family reconstitution: Summary results 1600–1799. *Population Studies* 37 (2): 157–184.

Vogt, William W. 1948. *The road to survival*. New York: Sloane Associates. (Kessinger Publishing 2007 reprint).

Valenze, Deborah. 2023. *The invention of scarcity. Malthus and the margins of history*. Yale University Press.

An Essay on the Principle of Population, Original Commented Table of Contents

2

Thomas Robert Malthus

T. R. Malthus (✉)
University of Oxford, Oxford, UK

D. Coleman (ed.), *Thomas Robert Malthus – An Essay on the Principle of Population*, Classic Texts in the Sciences,
https://doi.org/10.1007/978-3-031-94503-8_2

Chap. XIII: Of the Checks to Population among the Greeks.
Chap. XIV: Of the Checks to Population among the Romans.
Book II: OF THE CHECKS TO POPULATION IN THE DIFFERENT STATES OF MODERN EUROPE.
Chap. I: Of the Checks to Population in Norway.
Chap. II: Of the Checks to Population in Sweden.
Chap. III: Of the Checks to Population in Russia.
Chap. IV: Of the Checks to Population in the Middle Parts of Europe.
Chap. V: Of the Checks to Population in Switzerland.
Chap. VI: Of the Checks to Population in France.
Chap. VII: Of the Checks to Population in France (continued).
Chap. VIII: Of the Checks to Population in England.
Chap. IX: Of the Checks to Population in England (continued).
Chap. X: Of the Checks to Population in Scotland and Ireland.
Chap. XI: On the Fruitfulness of Marriages.
Chap. XII: Effects of Epidemics on Registers of Births, Deaths, and Marriages.
Chap. XIII: General Deductions from the preceding View of Society.
An Essay on the Principle of Population, vol. 2 [1826, 6th ed.]
Book III: OF THE DIFFERENT SYSTEMS OR EXPEDIENTS WHICH HAVE BEEN PROPOSED OR HAVE PREVAILED IN SOCIETY, AS THEY AFFECT THE EVILS ARISING FROM THE PRINCIPLE OF POPULATION
Chap. I: Of Systems of Equality. Wallace, Condorcet.
Chap. II: Of Systems of Equality. Godwin.
Chap. III: Of Systems of Equality (continued).

It was suggested to me same years since by persons for whose judgment I have a high respect, that it might be advisable, in a new edition, to throw out the matter relative to systems of equality, to Wallace, Condorcet and Godwin, as having in a considerable degree lost its interest, and as not being strictly connected with the main subject of the Essay, which is an explanation and illustration of the theory of population. But independently of its being natural for me to have some little partiality for that part of the work which led to those inquiries on which the main subject rests; I really think that there should be somewhere on record an answer to systems of equality founded on the principle of population; and perhaps such an answer is as appropriately placed, and is likely to have as much effect, among the illustrations and applications of the principle of population, as in any other situation to which it could be assigned.

The appearances in all human societies, particularly in all those which are the furthest advanced in civilization and improvement, will ever be such, as to inspire superficial observers with a belief that a prodigiousChange for the better might be effected by the introduction of a system of equality and of common property. They see abundance in some quarters, and want in others; and the natural and obvious remedy seems to be an equal division of the produce. They see a prodigious quantity of human exertion wasted upon trivial, useless, and sometimes pernicious objects, which might either be wholly saved or more effectively employed. They see invention after invention in machinery brought forward, which is seemingly calculated, in the most marked manner, to abate the sum of human toil. Yet with these apparent means of giving plenty, leisure and happiness to all, they still see the labours of the great mass of society undiminished, and their condition, if not deteriorated, in no very striking and palpable manner improved.

Under these circumstances, it cannot be a matter of wonder that proposals for systems of equality should be continually reviving. After periods when the subject has undergone a thorough discussion, or when some great experiment in improvement has failed, it is likely that the question should be dormant for a time, and that the opinions of the advocates of equality should be ranked among those errors which had passed away to be heard of no more. But it is probable that if the world were to last for any number of thousand years, systems of equality would be among those errors, which like the tunes of a barrel organ, to use the illustration of Dugald Stewart,[9] will never cease to return at certain intervals.

I am induced to make these remarks, and to add a little to what I have already said on systems of equality, instead of leaving out the whole discussion, by a tendency to a revival of this kind at the present moment.[10]

A gentleman, for whom I have a very sincere respect, Mr. Owen, of Lanark, has lately published a work entitled *A New View of Society,* which is intended to prepare the public mind for the introduction of a system involving a community of labour and of goods. It is also generally known that an idea has lately prevailed among some of the lower classes of society, that the land is the people's farm, the rent of which ought to be equally divided among them; and that they have been deprived of the benefits which belong to them from this their natural inheritance, by the injustice and oppression of their stewards, the landlords.

Mr. Owen is, I believe, a man of real benevolence, who has done much good; and every friend of humanity must heartily wish him success in his endeavours to procure an Act of Parliament for limiting the hours of working among the children in the cotton manufactories, and preventing them from being employed at too early an age. He is further entitled to great attention on all subjects relating to education, from the experience and knowledge which he must have gained in an intercourse of many years with two thousand manufacturers, and from the success which is said to have resulted from his modes of management. A theory professed to be founded on such experience is, no doubt, worthy of much more consideration than one formed in a closet.

The claims to attention possessed by the author of the new doctrines relating to land are certainly very slender; and the doctrines themselves indicate a very great degree of ignorance; but the errors of the labouring classes of society are always entitled to great indulgence and consideration. They are the natural and pardonable result of their liability to be deceived by first appearances, and by the arts of designing men, owing to the nature of their situation, and the scanty knowledge which in general falls to their share. And, except in extreme cases, it must always be the wish of those who are better informed, that they should be brought to a sense of the truth, rather by patience and the gradual diffusion of education and knowledge, than by any harsher methods.

After what I have already said on systems of equality in the precedingChap.s, I shall not think it necessary to enter into a long and elaborate refutation of these doctrines. I merely mean to give an additional reason for leaving on record an answer to systems of equality, founded on the principle of population, together with a concise restatement of this answer for practical application.

Of the two decisive arguments against such systems, one is, the unsuitableness of a state of equality, both according to experience and theory, to the production of those stimulants to exertion which can alone overcome the natural indolence of man, and prompt him to the proper cultivation of the earth and the fabrication of those conveniences and comforts which are necessary to his happiness.

And the other, the inevitable and necessary poverty and misery in which every system of equality must shortly terminate from the acknowledged tendency of the human race to increase faster than the means of subsistence, unless such increase be prevented by means infinitely more cruel than those which result from the laws of private property, and the moral obligation imposed on every man by the commands of God and nature to support his own children.

The first of these arguments has, I confess, always appeared to my own mind sufficiently conclusive. A state, in which an inequality of conditions offers the natural rewards of good conduct, and inspires widely and generally the hopes of rising and the fears of falling in society, is unquestionably the best calculated to develope the energies and faculties of man, and the best suited to the exercise and improvement of human virtue;[11] and history, in every case of equality that has yet occurred, has uniformly borne witness to the depressing and deadening effects which arise from the want of this stimulus. But still, perhaps, it may be true that neither experience nor theory on this subject is quite so decisive as to preclude all plausible arguments on the other side. It may be said that the instances which history records of systems of equality really carried into execution are so few, and those in societies so little advanced from a state of barbarism, as to afford no fair conclusions relative to periods of great civilization and improvement; that, in other instances, in ancient times, where approaches were made toward a tolerable equality of conditions, examples of considerable energy ofCharacter in some lines of exertion are not unfrequent; and that in modern times some societies, particularly of Moravians, are known to have had much of their property in common without occasioning the destruction of their industry. It may be said that, allowing the stimulus of inequality of conditions to have been necessary, in order to raise man from the indolence and apathy of the savage to the activity and intelligence of civilized life, it does not follow that the continuance of the same stimulus should be necessary when this activity and energy of mind has been once gained. It may *then* be allowable quietly to enjoy the benefit of a regimen which, like many other stimulants, having produced its proper effect at a certain point, must be left off, or exhaustion, disease, and death will follow.

These observations are certainly not of a nature to produce conviction in those who have studied the humanCharacter; but they are, to a certain degree, plausible, and do not admit of so definite and decisive an answer as to make the proposal for an experiment in modern times utterly absurd and unreasonable.

The peculiar advantage of the other argument against systems of equality, that which is founded on the principle of population, is, that it is not only still more generally and uniformly confirmed by experience, in every age and in every part of the world, but it is so pre-eminently clear in theory, that no tolerably plausible answer can be given to it; and, consequently, no decent pretence can be brought forward for an experiment. The affair is a matter of the most simple calculation applied to the known properties of land, and the proportion of births to deaths which takes place in almost every country village. There are many parishes in England, where, notwithstanding the actual difficulties attending the support of a family which must *necessarily* occur in every well-peopled country, and making no allowances for omissions in the registers, the births are to the deaths in the proportion of 2 to 1. This proportion, with the usual rate of mortality in country places, of about 1 in 50, would continue doubling the population in 41 years, if there were no emigrations from the parish. But in any system of equality, either such as that proposed by Mr. Owen, or in parochial partnerships in land, not only would there be no means of emigration to other parishes with any prospect of relief; but the rate of increase at first would, of course, be much greater than in the present state of society. What then, I would ask, is to prevent the division of the produce of the soil to each individual from becoming every year less and less, till the whole society and every individual member of it are pressed down by want and misery?[12]

This is a very simple and intelligible question. And surely no man ought to propose or support a system of equality, who is not able to give a rational answer to it, at least in theory. But, even in theory, I have never yet heard any thing approaching to a rational answer to it.

It is a very superficial observation which has sometimes been made, that it is a contradiction to lay great stress upon the efficacy of moral restraint in an improved and improving state of society, according to the present structure of it, and yet to suppose that it would not act with sufficient force in a system of equality, which almost always presupposes a great diffusion of information, and a great improvement of the human mind. Those who have made this observation do not see that the encouragement and motive to moral restraint are at once destroyed in a system of equality, and community of goods.

Let us suppose that in a system of equality, in spite of the best exertions to procure more food, the population is pressing hard against the limits of subsistence, and all are becoming very poor. It is evidently necessary under these circumstances, in order to prevent the society from starving, that the rate at which the population increases should be retarded. But who are the persons that are to exercise the restraint thus called for, and either to marry late or not at all? It does not seem to be a necessary consequence of a system of equality that all the human passions should be at once extinguished by it; but if not, those who might wish to marry would feel it hard that they should be among the number forced to restrain their inclinations. As all would be equal, and in similar circumstances, there would be no reason whatever why one individual should think himself obliged to practise the duty of restraint more than another. The thing however must be done, with any hope of avoiding universal misery; and in a state of equality, the necessary restraint could only be effected by some general law. But how is this law to be supported, and how are the violations of it to be punished? Is the man who marries early to be pointed at with the finger of scorn? is he to be whipped at the cart's tail? is he to be confined for years in a prison? is he to have his children exposed? Are not all direct punishments for an offence of this kind shocking and unnatural to the last degree? And yet, if it be absolutely necessary, in order to prevent the most overwhelming wretchedness, that there should be some restraint on the tendency to early marriages, when the resources of the country are only sufficient to support a slow rate of increase, can the most fertile imagination conceive one at once so natural, so just, so consonant to the laws of God and to the best laws framed by the most enlightened men, as that each individual should be responsible for the maintenance of his own children; that is, that he should be subjected to the natural inconveniences and difficulties arising from the indulgence of his inclinations, and to no other whatever?

That this natural check to early marriages arising from a view of the difficulty attending the support of a large family operates very widely throughout all classes of society in every civilized state, and may be expected to be still more effective, as the lower classes of people continue to improve in knowledge and prudence, cannot admit of the slightest doubt. But the operation of this natural check depends exclusively upon the existence of the laws of property, and succession; and in a state of equality and community of property could only be replaced by some artificial regulation of a very different stamp, and a much more unnaturalCharacter. Of this Mr. Owen is fully sensible, and has in consequence taxed his ingenuity to the utmost to invent some mode, by which the difficulties arising from the progress of population could be got rid of, in the state of society to which he looks forward. His absolute inability to suggest any mode of accomplishing this object that is not unnatural, immoral, or cruel in a high degree, together with the same want of success in every other person, ancient[13] or modern, who has made a similar attempt, seem to shew that the argument against systems of equality founded on the principle of population does not admit of a plausible answer, even in theory. The fact of the tendency of population to increase beyond the means of subsistence may be seen in almost every register of a country parish in the kingdom. The unavoidable effect of this tendency to depress the whole body of the people in want and misery, unless the progress of the population be somehow or other retarded, is equally obvious; and the impossibility of checking the rate of increase in a state of equality, without resorting to regulations that are unnatural, immoral or cruel, forms all argument at once conclusive against every such system.

Chap. IV: Of Emigration.
Chap. V: Of Poor Laws.
Chap. VI: Of Poor Laws, continued.
Chap. VII: Of Poor Laws, continued.
Chap. VIII: Of the Agricultural System.
Chap. IX: Of the Commercial System.

A country which excels in commerce and manufactures, may purchase corn from a great variety of others; and it may be supposed, perhaps, that, proceeding upon this system, it may continue to purchase an increasing quantity, and to maintain a rapidly increasing population, till the lands of all the nations with which it trades are fully cultivated. As this is an event necessarily at a great distance, it may appear that the population of such a country will not be checked from the difficulty of procuring subsistence till after the lapse of a great number of ages.

There are, however, causes constantly in operation, which will occasion the pressure of this difficulty, long before the event here contemplated has taken place, and while the means of raising food in the surrounding countries may still be comparatively abundant.

In the first place, advantages which depend exclusively upon capital and skill, and the present possession of particularChannels of commerce, cannot in their nature be permanent. We know how difficult it is to confine improvements in machinery to a single spot; we know that it is the constant object, both of individuals and countries, to increase their capital; and we know, from the past history of commercial states, that theChannels of trade are not unfrequently taking a different direction. It is unreasonable therefore to expect that any one country, merely by the force of skill and capital, should remain in possession of markets uninterrupted by foreign competition. But, when a powerful foreign competition takes place, the exportable commodities of the country in question must soon fall to prices which will essentially reduce profits; and the fall of profits will diminish both the power and the will to save. Under these circumstances the accumulation of capital will be slow, and the demand for labour proportionably slow, till it comes nearly to a stand; while, perhaps, the new competitors either by raising their own raw materials or by some other advantages, may still be increasing their capitals and population with some degree of rapidity.

But, secondly, even if it were possible for a considerable time to exclude any formidable foreign competition, it is found that domestic competition produces almost unavoidably the same effects. If a machine be invented in a particular country, by the aid of which one man can do the work of ten, the possessors of it will of course at first make very unusual profits; but, as soon as the invention is generally known, so much capital and industry will be brought into this new and profitable employment, as to make its products greatly exceed both the foreign and domestic demand at the old prices. These prices, therefore, will continue to fall, till the stock and labour employed in this direction cease to yield unusual profits. In this case it is evident that, though in an early period of such a manufacture, the product of the industry of one man for a day might have been exchanged for such a portion of food as would support forty or fifty persons; yet, at a subsequent period, the product of the same industry might not purchase the support of ten.

In the cotton trade of this country, which has extended itself so wonderfully during the last twenty-five years, very little effect has hitherto been produced by foreign competition.[30] The very great fall which has taken place in the prices of cotton goods has been almost exclusively owing to domestic competition; and this competition has so glutted both the home and foreign markets, that the present capitals employed in the trade, notwithstanding the very peculiar advantages which they possess from the saving of labour, have ceased to possess any advantage whatever in the general rate of their profits. Although, by means of the admirable machinery used in the spinning of cotton, one boy or girl can now do as much as many grown persons could do formerly; yet neither the wages of the labourer, nor the profits of his master, are higher than in those employments where no machinery is used, and no saving of labour accomplished.

The country has, however, in the mean time, been very greatly benefited. Not only have all its inhabitants been enabled to obtain a superior fabric for clothing, at a less expense of labour and property, which must be considered as a great and permanent advantage; but the high temporary profits of the trade have occasioned a great accumulation of capital, and consequently a great demand for labour; while the extending markets abroad and the new values thrown into the market at home, have created such a demand for the products of every species of industry, agricultural and colonial, as well as commercial and manufacturing, as to prevent a fall of profits.

This country, from the extent of its lands, and its rich colonial possessions, has a large *arena* for the employment of an increasing capital; and the general rate of its profits are not, as it appears, very easily and rapidly reduced by accumulation. But a country, such as we are considering, engaged principally in manufactures, and unable to direct its industry to the same variety of pursuits, would sooner find its rate of profits diminished by an increase of capital, and no ingenuity in machinery which was not continually progressive could save it, after a certain period, from low profits and low wages, and their natural consequences, a check to population.

Thirdly. A country which is obliged to purchase both the raw materials of its manufactures and the means of subsistence for its population from foreign countries, is almost entirely dependent for the increase of its wealth and population on the increasing wealth and demands of the countries with which it trades.

It has been sometimes said, that a manufacturing country is no more dependent upon the country which supplies it with food and raw materials, than the agricultural country is on that which manufactures for it; but this is really an abuse of terms. A country with great resources in land may find it decidedly for its advantage to employ the main part of its capital in cultivation, and to import its manufactures. In so doing, it will often employ the whole of its industry most productively, and most rapidly increase its stock. But, if the slackness of its neighbours in manufacturing, or any other cause, should either considerably check or altogether prevent the importation of manufactures, a country with food and raw materials provided at home cannot be, long at a loss. For a time it would not certainly be so well supplied; but manufacturers and artizans would soon be found, and would soon acquire tolerable skill;[31] and though the capital and population of the country might not, under the new circumstances in which it was placed, increase so rapidly as before, it would still have the power of increasing in both to a great extent.

On the other hand, if food and raw materials were denied to a nation merely manufacturing, it is obvious that it could not longer exist. But not only does the absolute existence of such a nation, on an extreme supposition, depend upon its foreign commerce, but its progress in wealth must be almost entirely measured by the progress and demand of the countries which deal with it. However skilful, industrious and saving such a nation might be, if its customers, from indolence and want of accumulation, would not or could not take off a yearly increasing value of its commodities, the effects of its skill and machinery would be but of very short duration.

That the cheapness of manufactured commodities, occasioned by skill and machinery in one country, is calculated to encourage an increase of raw produce in others, no person can doubt; but we know at the same time that high profits may continue for a considerable period in an indolent and ill-governed state, without producing an increase of wealth; yet, unless such an increase of wealth and demand were produced in the surrounding countries, the increasing ingenuity and exertions of the manufacturing and commercial state would be lost in continually falling prices. It would not only be obliged, as its skill and capital increased, to give a larger quantity of manufactured produce for the raw produce which it received in return; but might be unable, even with the temptation of reduced prices, to stimulate its customers to such purchases as would allow of an increasing importation of food and raw materials; and without such an increasing importation, it is quite obvious that the population must become stationary.

It would come to the same thing, whether this inability to obtain an increasing quantity of food were occasioned by the advancing money price of corn, or the falling money price of manufactures. In either case the effect would be the same; and it is certain that this effect might take place in either way, from increasing competition and accumulation in the manufacturing nation, and the want of them in the agricultural, long before any essential increase of difficulty had occurred in the production of corn.

Fourthly. A nation which is obliged to purchase from others nearly the whole of its raw materials, and the means of its subsistence, is not only dependent entirely upon the demands of its customers, as they may be variously affected by indolence, industry or caprice, but it is subjected to a necessary and unavoidable diminution of demand in the natural progress of these countries towards that proportion of skill and capital which they may reasonably be expected after a certain time to possess. It is generally an accidental and temporary, not a natural and permanent division of labour, which constitutes one state the manufacturer and the carrier of others. While, in these landed nations, agricultural profits continue very high, it may fully answer to them to pay others as their manufacturers and carriers; but when the profits on land fall, or the tenures on which it can be held are not such as to encourage the investment of an accumulating capital, the owner of this capital will naturally look towards commerce and manufactures for its employment; and, according to the just reasoning of Adam Smith and the Economists, finding at home both the materials of manufactures, the means of subsistence, and the power of carrying on their own trade with foreign countries, they will probably be able to conduct the business of manufacturing and carrying for themselves at a cheaper rate than if they allowed it to continue in the hands of others. As long as the agricultural nations continued to apply their increasing capital principally to the land, this increase of capital would be of the greatest possible advantage to the manufacturing and commercial nation. It would be indeed the main cause and great regulator of its progress in wealth and population. But after they had turned their attention to manufactures and commerce, their further increase of capital would be the signal of decay and destruction to the manufactures and commerce, which they had before supported. And thus, in the natural progress of national improvement, and without the competition of superior skill and capital, a purely commercial state must be undersold and driven out of the markets by those who possess the advantage of land.

In the distribution of wealth during the progress of improvement, the interests of an independent state in relation to others are essentially different from those of a particular province, in relation to the kingdom to which it belongs, a point which has not been sufficiently attended to. If agricultural capital increases, and agricultural profits diminish in Sussex, the overflowing stock will go to London, Manchester, Liverpool, or some other place where it can probably be engaged in manufactures or commerce more advantageously than at home. But if Sussex were an independent kingdom, this could not take place; and the corn which is now sent to London must be withdrawn to support manufacturers and traders living within its confines. If England, therefore, had continued to be separated into the seven kingdoms of the Heptarchy, London could not possibly have been what it is; and that distribution of wealth and population which takes place at present, and which we may fairly presume is the most beneficial to the whole of the realm, would have been essentiallyChanged, if the object had been to accumulate the greatest quantity of wealth and population in particular districts, instead of the whole island. But at all times the interest of each independent state is to accumulate the greatest quantity of wealth within its limits. Consequently, the interest of an independent state, with regard to the countries with which it trades, can rarely be the same as the interest of a province with regard to the empire to which it belongs; and the accumulation of capital which would occasion the withdrawing of the exports of corn in the one case, would leave them perfectly undisturbed in the other.

If, from the operation of one or more of the causes above enumerated, the importation of corn into a manufacturing and commercial country should be essentially checked, and should either actually decrease, or be prevented from increasing, it is quite evident that its population must be checked nearly in the same proportion.

Venice presents a striking instance of a commercial state, at once stopped in its progress to wealth and population by foreign competition. The discovery made by the Portuguese of a passage to India by the Cape of Good Hope completely turned theChannel of the Indian trade. The high profits of the Venetians, which had been the foundation of their rapidly increasing wealth, and of their extraordinary preponderance as a naval and commercial power, were not only suddenly reduced; but the trade itself, on which these high profits had been made, was almost annihilated, and their power and wealth were shortly contracted to those more confined limits which suited their natural resources.

In the middle of the 15th century, Bruges in Flanders was the great *entrepôt* of the trade between the north and the south of Europe. Early in the 16th century its commerce began to decline under the competition of Antwerp. Many English and foreign merchants in consequence left the declining city, to settle in that which was rapidly increasing in commerce and wealth. About the middle of the 16th century Antwerp was at the zenith of its power. It contained above a hundred thousand inhabitants, and was universally allowed to be the most illustrious mercantile city, and to carry on the most extensive and richest commerce of any in the north of Europe.

The rising greatness of Amsterdam was favoured by the unfortunate siege and capture of Antwerp by the duke of Parma; and the competition of the extraordinary industry and persevering exertions of the Hollanders not only prevented Antwerp from recovering her commerce, but gave a severe blow to the foreign trade of almost all the other Hanse Towns.

The subsequent decline of the trade of Amsterdam itself was caused partly by the low profits arising from home competition and abundance of capital; partly by excessive taxation, which raised the price of the necessaries of life; but more than either, perhaps, by the progress of other nations possessing greater natural advantages, and being able, even with inferior skill, industry and capital, beneficially to carry on much of that trade which had before fallen almost exclusively into the hands of the Dutch.

As early as 1669 and 1670, when Sir William Temple was in Holland, the effects of abundance of capital and domestic competition were such, that most of the foreign trades were losing ones, except the Indian, and that none of them gave a profit of more than two or three per cent.[32] In such a state of things both the power and the will to save must be greatly diminished. The capital must have been either stationary or declining, or at the best very slowly progressive. In fact, Sir William Temple gives it as his opinion that the trade of Holland had for some years passed its meridian, and begun sensibly to decay.[33] Subsequently, when the progress of other nations was still more marked, it appeared from undoubted documents that most of the trades of Holland, as well as its fisheries, had decidedly fallen off, and that no branch of its commerce had retained its former vigour, except the American and African trades, and that of the Rhine and Maese, which are independent of foreign power and competition.

In 1669, the whole population of Holland and West Friezeland was estimated by John de Witt at 2,400,000.[34] In 1778, the population of the seven provinces was estimated only at 2,000,000;[35] and thus, in the course of above a hundred years, the population, instead of increasing, as is usual, had greatly diminished.

In all these cases of commercial states, the progress of wealth and population seems to have been checked by one or more of the causes above mentioned, which must necessarily affect more or less the power of commanding the means of subsistence.

Universally it may be observed, that if, from any cause or causes whatever, the funds for the maintenance of labour in any country cease to be progressive, the effective demand for labour will also cease to be progressive; and wages will be reduced to that sum, which, under the existing prices of provisions, and the existing habits of the people, will just keep up, and no more than keep up, a stationary population. A state so circumstanced is under a moral impossibility of increasing, whatever may be the plenty of corn, or however high may be the profits of stock in other countries.[36] It may indeed at a subsequent period, and under new circumstances, begin to increase again. If by some happy invention in mechanics, the discovery of some newChannel of trade, or an unusual increase of agricultural wealth and population in the surrounding countries, its exports, of whatever kind, were to become unusually in demand, it might again import an increasing quantity of corn, and might again increase its population. But as long as it is unable to make yearly additions to its imports of food, it will evidently be unable to furnish the means of support to an increasing population; and it will necessarily experience this inability, when, from the state of its commercial transactions, the funds for the maintenance of its labour become stationary, or begin to decline.

Chap. X: Of Systems of Agriculture and Commerce combined.
Chap. XI: Of Corn-Laws. Bounties upon Exportation.
Chap. XII: Of Corn-Laws. Restrictions upon Importation.
Chap. XIII: Of increasing Wealth, as it affects the Condition of the Poor.
Chap. XIV: General Observations.
Appendix I,: 1807
Appendix II,: 1817
Appendix III,: 1825

3 An Essay on the Principle of Population, vol. 1 [1826, 6th ed.]

Thomas Robert Malthus

3.1 Preface to the Second Edition

The Essay on the Principle of Population, which I published in 1798, was suggested, as is expressed in the preface, by a paper in Mr. Godwin's Inquirer. It was written on the impulse of the occasion, and from the few materials which were then within my reach in a country situation. The only authors from whose writings I had deduced the principle, which formed the main argument of the Essay, were Hume, Wallace, Adam Smith, and Dr. Price; and my object was to apply it, to try the truth of those speculations on the perfectibility of man and society, which at that time excited a considerable portion of the public attention.

In the course of the discussion I was naturally led into some examination of the effects of this principle on the existing state of society. It appeared to account for much of that poverty and misery observable among the lower classes of people in every nation, and for those reiterated failures in the efforts of the higher classes to relieve them. The more I considered the subject in this point of view, the more importance it seemed to acquire; and this consideration, joined to the degree of public attention which the Essay excited, determined me to turn my leisure reading towards an historical examination of the effects of the principle of population on the past and present state of society; that, by illustrating the subject more generally, and drawing those inferences from it, in application to the actual state of things, which experience seemed to warrant, I might give it a more practical and permanent interest.

In the course of this inquiry I found that much more had been done than I had been aware of, when I first published the Essay. The poverty and misery arising from a too rapid

T. R. Malthus (✉)
University of Oxford, Oxford, UK

D. Coleman (ed.), *Thomas Robert Malthus – An Essay on the Principle of Population*, Classic Texts in the Sciences,
https://doi.org/10.1007/978-3-031-94503-8_3

increase of population had been distinctly seen, and the most violent remedies proposed, so long ago as the times of Plato and Aristotle. And of late years the subject has been treated in such a manner by some of the French Economists; occasionally by Montesquieu, and, among our own writers, by Dr. Franklin, Sir James Stewart, Mr. Arthur Young, and Mr. Townsend, as to create a natural surprise that it had not excited more of the public attention.

Much, however, remained yet to be done. Independently of the comparison between the increase of population and food, which had not perhaps been stated with sufficient force and precision, some of the most curious and interesting parts of the subject had been either wholly omitted or treated very slightly. Though it had been stated distinctly, that population must always be kept down to the level of the means of subsistence; yet few inquiries had been made into the various modes by which this level is effected; and the principle had never been sufficiently pursued to its consequences, nor had those practical inferences drawn from it, which a strict examination of its effects on society appears to suggest.

These therefore are the points which I have treated most in detail in the following Essay. In its present shape it may be considered as a new work, and I should probably have published it as such, omitting the few parts of the former which I have retained, but that I wished it to form a whole of itself, and not to need a continual reference to the other. On this account I trust that no apology is necessary to the purchasers of the first edition.

To those who either understood the subject before, or saw it distinctly on the perusal of the first edition, I am fearful that I shall appear to have treated some parts of it too much in detail, and to have been guilty of unnecessary repetitions. These faults have arisen partly from want of skill, and partly from intention. In drawing similar inferences from the state of society in a number of different countries, I found it very difficult to avoid some repetitions; and in those parts of the inquiry which led to conclusions different from our usual habits of thinking, it appeared to me that, with the slightest hope of producing conviction, it was necessary to present them to the reader's mind at different times, and on different occasions. I was willing to sacrifice all pretensions to merit of composition, to the chance of making an impression on a larger class of readers.

The main principle advanced is so incontrovertible, that, if I had confined myself merely to general views, I could have intrenched myself in an impregnable fortress; and the work, in this form, would probably have had a much more masterly air. But such general views, though they may advance the cause of abstract truth, rarely tend to promote any practical good; and I thought that I should not do justice to the subject, and bring it fairly under discussion, if I refused to consider any of the consequences which appeared necessarily to flow from it, whatever these consequences might be. By pursuing this plan, however, I am aware that I have opened a door to many objections, and, probably, to much severity of criticism: but I console myself with the refection, that even the errors into which I may have fallen, by affording a handle to argument, and an additional excitement to examination, may be subservient to the important end of bringing a subject so nearly connected with the happiness of society into more general notice.

Throughout the whole of the present work I have so far differed in principle from the former, as to suppose the action of another check to population which does not come under the head either of vice or misery; and, in the latter part I have endeavoured to soften some of the harshest conclusions of the first Essay. In doing this, I hope that I have not violated the principles of just reasoning; nor expressed any opinion respecting the probable improvement of society, in which I am not borne out by the experience of the past. To those who still think that any check to population whatever would be worse than the evils which it would relieve, the conclusions of the former Essay will remain in full force; and if we adopt this opinion we shall be compelled to acknowledge, that the poverty and misery which prevail among the lower classes of society are absolutely irremediable.

I have taken as much pains as I could to avoid any errors in the facts and calculations which have been produced in the course of the work. Should any of them nevertheless turn out to be false, the reader will see that they will not materially affect the general scope of the reasoning.

From the crowd of materials which presented themselves, in illustration of the first branch of the subject, I dare not flatter myself that I have selected the best, or arranged them in the most perspicuous method. To those who take an interest in moral and political questions, I hope that the novelty and importance of the subject will compensate the imperfections of its execution.

London,
June 8, 1803.

3.2 Preface to the Fifth Edition

This Essay was first published at a period of extensive warfare, combined, from peculiar circumstances, with a most prosperous foreign commerce.

It came before the public, therefore, at a time when there would be an extraordinary demand for men, and very little disposition to suppose the possibility of any evil arising from the redundancy of population. Its success, under these disadvantages, was greater than could have been reasonably expected; and it may be presumed that it will not lose its interest, after a period of a different description has succeeded, which has in the most marked manner illustrated its principles, and confirmed its conclusions.

On account, therefore, of the nature of the subject, which, it must be allowed is one of permanent interest, as well as of the attention likely to be directed to it in future, I am bound to correct those errors of my work, of which subsequent experience and information may have convinced me, and to make such additions and alterations as appear calculated to improve it, and promote its utility.

It would have been easy to have added many further historical illustrations of the first part of the subject; but as I was unable to supply the want I once alluded to, of accounts of sufficient accuracy to ascertain what part of the natural power of increase each particular check destroys, it appeared to me that the conclusion which I had before drawn from very ample evidence of the only kind that could be obtained, would hardly receive much additional force by the accumulation of more, precisely of the same description.

In the two first books, therefore, the only additions are a new chapter on France, and one on England, chiefly in reference to facts which have occurred since the publication of the last edition.

In the third book I have given an additional chapter on the Poor-Laws; and as it appeared to me that the chapters on the Agricultural and Commercial Systems, and the Effects of increasing Wealth on the Poor, were not either so well arranged, or so immediately applicable to the main subject, as they ought to be; and as I further wished to make some alterations in the chapter on Bounties upon Exportation, and add something on the subject of Restrictions upon Importation, I have recast and rewritten the chapters which stand the 8th, 9th, 10th, 11th, 12th, 13th, in the present edition; and given a new title, and added two or three passages, to the 14th and last chapter of the same book.

In the fourth book I have added a new chapter to the one entitled *Effects of the Knowledge of the principal Cause of Poverty on Civil Liberty;* and another to the chapter on the *Different Plans of improving the Poor;* and I have made a considerable addition to the Appendix, in reply to some writers on the Principles of Population, whose works have appeared since the last edition.

These are the principal additions and alterations made in the present edition. They consist, in a considerable degree, of the application of the general principles of the Essay to the present state of things.

For the accommodation of the purchasers of the former editions, these additions and alterations will be published in a separate volume.

East-India College,
June 7th, 1817.

3.3 Advertisement to the Sixth Edition.

Jan. 2d, 1826.

The additions to the present edition chiefly consist of some further documents and inferences relating to the state of the population in those countries, in which fresh enumerations, and registers of births, deaths and marriages, have appeared since the publication of my last edition in 1817. They refer principally to England, France, Sweden, Russia, Prussia, and America, and will be found in the chapters which treat of the population of these countries. In the chapter on the Fruitfulness of Marriages an additional table has been given, (vol. i. p. 498.) which, from the per centage increase of population in the interval between those decennial enumerations which are now taking place in some countries, shews the period of their doubling, or the rate at which they are increasing. At the end of the Appendix my reasons for not replying to the late publication of Mr. Godwin are shortly stated. In other parts of the work some inconsiderable alterations and corrections have been made which it is unnecessary to specify; and a few notes have been added, the principal of which is one on the variations in the price of corn in Holland under a free trade, and the error of supposing that the scarcity of one country is generally counterbalanced by the plenty of some other.—Vol. ii. p. 207.

3.4 Book I: Of the Checks to Population in the Less Civilized Parts of the World and in Past Times.

3.4.1 Book I, Chapter I: Statement of the Subject. Ratios of the Increase of Population and Food.

In an inquiry concerning the improvement of society, the mode of conducting the subject which naturally presents itself, is,

1. To investigate the causes that have hitherto impeded the progress of mankind towards happiness; and,
2. To examine the probability of the total or partial removal of these causes in future.

To enter fully into this question, and to enumerate all the causes that have hitherto influenced human improvement, would be much beyond the power of an individual. The principal object of the present essay is to examine the effects of one great cause intimately united with the very nature of man; which, though it has been constantly and powerfully operating since the commencement of society, has been little noticed by the writers who have treated this subject. The facts which establish the existence of this cause have, indeed, been repeatedly stated and acknowledged; but its natural and necessary effects have been almost totally overlooked; though probably among these effects may be reckoned a very considerable portion of that vice and misery, and of that unequal distribution of the bounties of nature, which it has been the unceasing object of the enlightened philanthropist in all ages to correct.

The cause to which I allude, is the constant tendency in all animated life to increase beyond the nourishment prepared for it.

It is observed by Dr. Franklin, that there is no bound to the prolific nature of plants or animals, but what is made by their crowding and interfering with each other's means of subsistence. Were the face of the earth, he says, vacant of other plants, it might be gradually sowed and overspread with one kind only, as for instance with fennel: and were it empty of other inhabitants, it might in a few ages be replenished from one nation only, as for instance with Englishmen.[1]

This is incontrovertibly true. Through the animal and vegetable kingdoms Nature has scattered the seeds of life abroad with the most profuse and liberal hand; but has been

[1] Franklin's Miscell. p. 9.

comparatively sparing in the room and the nourishment necessary to rear them. The germs of existence contained in this earth, if they could freely develope themselves, would fill millions of worlds in the course of a few thousand years. Necessity, that imperious, all-pervading law of nature, restrains them within the prescribed bounds. The race of plants and the race of animals shrink under this great restrictive law; and man cannot by any efforts of reason escape from it.

In plants and irrational animals, the view of the subject is simple. They are all impelled by a powerful instinct to the increase of their species; and this instinct is interrupted by no doubts about providing for their offspring. Wherever therefore there is liberty, the power of increase is exerted; and the superabundant effects are repressed afterwards by want of room and nourishment.

The effects of this check on man are more complicated. Impelled to the increase of his species by an equally powerful instinct, reason interrupts his career, and asks him whether he may not bring beings into the world, for whom he cannot provide the means of support. If he attend to this natural suggestion, the restriction too frequently produces vice. If he hear it not, the human race will be constantly endeavouring to increase beyond the means of subsistence. But as, by that law of our nature which makes food necessary to the life of man, population can never actually increase beyond the lowest nourishment capable of supporting it, a strong check on population, from the difficulty of acquiring food, must be constantly in operation. This difficulty must fall somewhere, and must necessarily be severely felt in some or other of the various forms of misery, or the fear of misery, by a large portion of mankind.

That population has this constant tendency to increase beyond the means of subsistence, and that it is kept to its necessary level by these causes, will sufficiently appear from a review of the different states of society in which man has existed. But, before we proceed to this review, the subject will, perhaps, be seen in a clearer light, if we endeavour to ascertain what would be the natural increase of population, if left to exert itself with perfect freedom; and what might be expected to be the rate of increase in the productions of the earth, under the most favourable circumstances of human industry.

It will be allowed that no country has hitherto been known, where the manners were so pure and simple, and the means of subsistence so abundant, that no check whatever has existed to early marriages from the difficulty of providing for a family, and that no waste of the human species has been occasioned by vicious customs, by towns, by unhealthy occupations, or too severe labour. Consequently in no state that we have yet known, has the power of population been left to exert itself with perfect freedom.

Whether the law of marriage be instituted, or not, the dictate of nature and virtue seems to be an early attachment to one woman; and where there were no impediments of any kind in the way of an union to which such an attachment would lead, and no causes of depopulation afterwards, the increase of the human species would be evidently much greater than any increase which has been hitherto known.

In the northern states of America, where the means of subsistence have been more ample, the manners of the people more pure, and the checks to early marriages fewer, than in any of the modern states of Europe, the population has been found to double itself, for above a century and a half successively, in less than twenty-five years.[2] Yet, even during these periods, in some of the towns, the deaths exceeded the births,[3] a circumstance which clearly proves that, in those parts of the country which supplied this deficiency, the increase must have been much more rapid than the general average.

In the back settlements, where the sole employment is agriculture, and vicious customs and unwholesome occupations are little known, the population has been found to double itself in fifteen years.[4] Even this extraordinary rate of increase is probably short of the utmost power of population. Very severe labour is requisite to clear a fresh country; such situations are not in general considered as particularly healthy; and the inhabitants, probably, are occasionally subject to the incursions of the Indians, which may destroy some lives, or at any rate diminish the fruits of industry.

According to a table of Euler, calculated on a mortality of 1 in 36, if the births be to the deaths in the proportion of 3 to 1, the period of doubling will be only 12 years and 4-5ths.[5] And this proportion is not only a possible supposition, but has actually occurred for short periods in more countries than one.

Sir William Petty supposes a doubling possible in so short a time as ten years.[6]

But, to be perfectly sure that we are far within the truth, we will take the slowest of these rates of increase, a rate in which all concurring testimonies agree, and which has been repeatedly ascertained to be from procreation only.

It may safely be pronounced, therefore, that population, when unchecked, goes on doubling itself every twenty-five years, or increases in a geometrical ratio.

The rate according to which the productions of the earth may be supposed to increase, it will not be so easy to determine. Of this, however, we may be perfectly certain, that the ratio of their increase in a limited territory must be of a totally different nature from the ratio of the increase of population. A thousand millions are just as easily doubled every twenty-five years by the power of population as a thousand. But the food to support the increase from the greater number will by no means be obtained with the same facility. Man is necessarily confined in room. When acre has been added to acre till all the fertile land is occupied, the yearly increase of food must depend upon the melioration of the land already in possession. This is a fund; which, from the nature of all soils, instead of increasing, must be gradually diminishing. But population, could it be supplied with food, would go on with unexhausted vigour; and the increase of one period would furnish the power of a greater increase the next, and this without any limit.

---

[2] It appears, from some recent calculations and estimates, that from the first settlement of America, to the year 1800, the periods of doubling have been but very little above twenty years. See a note on the increase of American population in Book ii. chap. xi.

[3] Price's Observ. on Revers. Pay. vol. i. p. 274. 4th edit.

[4] Id. p. 282.

[5] See this table at the end of chap. iv. book ii.

[6] Polit. Arith. p. 14.

Book I, Chapter II.

From the accounts we have of China and Japan, it may be fairly doubted, whether the best-directed efforts of human industry could double the produce of these countries even once in any number of years. There are many parts of the globe; indeed, hitherto uncultivated, and almost unoccupied; but the right of exterminating, or driving into a corner where they must starve, even the inhabitants of these thinly-peopled regions, will be questioned in a moral view. The process of improving their minds and directing their industry would necessarily be slow; and during this time, as population would regularly keep pace with the increasing produce, it would rarely happen that a great degree of knowledge and industry would have to operate at once upon rich unappropriated soil. Even where this might take place, as it does sometimes in new colonies, a geometrical ratio increases with such extraordinary rapidity, that the advantage could not last long. If the United States of America continue increasing, which they certainly will do, though not with the same rapidity as formerly, the Indians will be driven further and further back into the country, till the whole race is ultimately exterminated, and the territory is incapable of further extension.

These observations are, in a degree, applicable to all the parts of the earth, where the soil is imperfectly cultivated. To exterminate the inhabitants of the greatest part of Asia and Africa, is a thought that could not be admitted for a moment. To civilise and direct the industry of the various tribes of Tartars and Negroes, would certainly be a work of considerable time, and of variable and uncertain success.

Europe is by no means so fully peopled as it might be. In Europe there is the fairest chance that human industry may receive its best direction. The science of agriculture has been much studied in England and Scotland; and there is still a great portion of uncultivated land in these countries. Let us consider at what rate the produce of this island might be supposed to increase under circumstances the most favourable to improvement.

If it be allowed that by the best possible policy, and great encouragements to agriculture, the average produce of the island could be doubled in the first twenty-five years, it will be allowing, probably, a greater increase than could with reason be expected.

In the next twenty-five years, it is impossible to suppose that the produce could be quadrupled. It would be contrary to all our knowledge of the properties of land. The improvement of the barren parts would be a work of time and labour; and it must be evident to those who have the slightest acquaintance with agricultural subjects, that in proportion as cultivation extended, the additions that could yearly be made to the former average produce must be gradually and regularly diminishing. That we may be the better able to compare the increase of population and food, let us make a supposition, which, without pretending to accuracy, is clearly more favourable to the power of production in the earth, than any experience we have had of its qualities will warrant.

Let us suppose that the yearly additions which might be made to the former average produce, instead of decreasing, which they certainly would do, were to remain the same; and that the produce of this island might be increased every twenty-five years, by a quantity equal to what it at present produces. The most enthusiastic speculator cannot suppose a greater increase than this. In a few centuries it would make every acre of land in the island like a garden.

If this supposition be applied to the whole earth, and if it be allowed that the subsistence for man which the earth affords might be increased every twenty-five years by a quantity equal to what it at present produces, this will be supposing a rate of increase much greater than we can imagine that any possible exertions of mankind could make it.

It may be fairly pronounced, therefore, that, considering the present average state of the earth, the means of subsistence, under circumstances the most favourable to human industry, could not possibly be made to increase faster than in an arithmetical ratio.

The necessary effects of these two different rates of increase, when brought together, will be very striking. Let us call the population of this island eleven millions; and suppose the present produce equal to the easy support of such a number. In the first twenty-five years the population would be twenty-two millions, and the food being also doubled, the means of subsistence would be equal to this increase. In the next twenty-five years, the population would be forty-four millions, and the means of subsistence only equal to the support of thirty-three millions. In the next period the population would be eighty-eight millions, and the means of subsistence just equal to the support of half that number. And, at the conclusion of the first century, the population would be a hundred and seventy-six millions, and the means of subsistence only equal to the support of fifty-five millions, leaving a population of a hundred and twenty-one millions totally unprovided for.

Taking the whole earth, instead of this island, emigration would of course be excluded; and, supposing the present population equal to a thousand millions, the human species would increase as the numbers, 1, 2, 4, 8, 16, 32, 64, 128, 256, and subsistence as 1, 2, 3, 4, 5, 6, 7, 8, 9. In two centuries the population would be to the means of subsistence as 256 to 9; in three centuries as 4096 to 13, and in two thousand years the difference would be almost incalculable.

In this supposition no limits whatever are placed to the produce of the earth. It may increase for ever and be greater than any assignable quantity; yet still the power of population being in every period so much superior, the increase of the human species can only be kept down to the level of the means of subsistence by the constant operation of the strong law of necessity, acting as a check upon the greater power.

3.4.2 Book I, Chapter II: Of the General Checks to Population, and the Mode of Their Operation.

The ultimate check to population appears then to be a want of food, arising necessarily from the different ratios according to which population and food increase. But this ultimate check is never the immediate check, except in cases of actual famine.

The immediate check may be stated to consist in all those customs, and all those diseases, which seem to be generated by a scarcity of the means of subsistence; and all those causes, independent of this scarcity, whether of a moral or physical nature, which tend prematurely to weaken and destroy the human frame.

These checks to population, which are constantly operating with more or less force in every society, and keep down the number to the level of the means of subsistence, may be classed under two general heads—the preventive, and the positive checks.

The preventive check, as far as it is voluntary, is peculiar to man, and arises from that distinctive superiority in his reasoning faculties, which enables him to calculate distant consequences. The checks to the indefinite increase of plants and irrational animals are all either positive, or, if preventive, involuntary. But man cannot look around him, and see the distress which frequently presses upon those who have large families; he cannot contemplate his present possessions or earnings, which he now nearly consumes himself, and calculate the amount of each share, when with very little addition they must be divided, perhaps, among seven or eight, without feeling a doubt whether, if he follow the bent of his inclinations, he may be able to support the offspring which he will probably bring into the world. In a state of equality, if such can exist, this would be the simple question. In the present state of society other considerations occur. Will he not lower his rank in life, and be obliged to give up in great measure his former habits? Does any mode of employment present itself by which he may reasonably hope to maintain a family? Will he not at any rate subject himself to greater difficulties, and more severe labour, than in his single state? Will he not be unable to transmit to his children the same advantages of education and improvement that he had himself possessed? Does he even feel secure that, should he have a large family, his utmost exertions can save them from rags and squalid poverty, and their consequent degradation in the community? And may he not be reduced to the grating necessity of forfeiting his independence, and of being obliged to the sparing hand of Charity for support?

These considerations are calculated to prevent, and certainly do prevent, a great number of persons in all civilized nations from pursuing the dictate of nature in an early attachment to one woman.

If this restraint do not produce vice, it is undoubtedly the least evil that can arise from the principle of population. Considered as a restraint on a strong natural inclination, it must be allowed to produce a certain degree of temporary unhappiness; but evidently slight, compared with the evils which result from any of the other checks to population; and merely of the same nature as many other sacrifices of temporary to permanent gratification, which it is the business of a moral agent continually to make.

When this restraint produces vice, the evils which follow are but too conspicuous. A promiscuous intercourse to such a degree as to prevent the birth of children, seems to lower, in the most marked manner, the dignity of human nature. It cannot be without its effect on men, and nothing can be more obvious than its tendency to degrade the female character, and to destroy all its most amiable and distinguishing characteristics. Add to which, that among those unfortunate females, with which all great towns abound, more real distress and aggravated misery are, perhaps, to be found, than in any other department of human life.

When a general corruption of morals, with regard to the sex, pervades all the classes of society, its effects must necessarily be, to poison the springs of domestic happiness, to weaken conjugal and parental affection, and to lessen the united exertions and ardour of parents in the care and education of their children;—effects which cannot take place without a decided diminution of the general happiness and virtue of the society; particularly as the necessity of art in the accomplishment and conduct of intrigues, and in the concealment of their consequences necessarily leads to many other vices.

The positive checks to population are extremely various, and include every cause, whether arising from vice or misery, which in any degree contributes to shorten the natural duration of human life. Under this head, therefore, may be enumerated all unwholesome occupations, severe labour and exposure to the seasons, extreme poverty, bad nursing of children, great towns, excesses of all kinds, the whole train of common diseases and epidemics, wars, plague, and famine.

On examining these obstacles to the increase of population which I have classed under the heads of preventive and positive checks, it will appear that they are all resolvable into moral restraint, vice, and misery.

Of the preventive checks, the restraint from marriage which is not followed by irregular gratifications may properly be termed moral restraint.[7]

Promiscuous intercourse, unnatural passions, violations of the marriage bed, and improper arts to conceal the consequences of irregular connexions, are preventive checks that clearly come under the head of vice.

Of the positive checks, those which appear to arise unavoidably from the laws of nature, may be called exclusively misery; and those which we obviously bring upon ourselves, such as wars, excesses, and many others which it would be in our power to avoid, are of a mixed nature. They are brought upon us by vice, and their consequences are misery.[8]

---

[7] It will be observed, that I here use the term *moral* in its most confined sense. By moral restraint I would be understood to mean a restraint from marriage, from prudential motives, with a conduct strictly moral during the period of this restraint; and I have never intentionally deviated from this sense. When I have wished to consider the restraint from marriage unconnected with its consequences, I have either called it prudential restraint, or a part of the preventive check, of which indeed it forms the principal branch.

[8] As the general consequence of vice is misery, and as this consequence is the precise reason why an action is termed vicious, it may appear that the term misery alone would be here sufficient, and that it is superfluous to use both. But the rejection of the term vice would introduce a considerable confusion into our language and ideas. We want it particularly to distinguish those actions, the general tendency of which is to produce misery, and which are therefore prohibited by the commands of the Creator, and the precepts of the moralist, although, in their immediate or individual effects, they may produce perhaps exactly the contrary. The gratification of all our passions in its immediate effect is happiness, not misery; and, in individual instances, even the remote consequences (at least in this life) may possibly come under the same denomination. There may have been some irregular connexions with women, which have added to the happiness of both parties, and have injured no one. These individual actions, therefore, cannot come under the head of misery. But they are still evidently vicious, because an action is so denominated, which violates an express precept, founded upon its general tendency to produce misery, whatever may be its individual effect; and no person can doubt the general tendency of an illicit intercuorse between the sexes, to injure the happiness of society.

The sum of all these preventive and positive checks, taken together, forms the immediate check to population; and it is evident that, in every country where the whole of the procreative power cannot be called into action, the preventive and the positive checks must vary inversely as each other; that is, in countries either naturally unhealthy, or subject to a great mortality, from whatever cause it may arise, the preventive check will prevail very little. In those countries, on the contrary, which are naturally healthy, and where the preventive check is found to prevail with considerable force, the positive check will prevail very little, or the mortality be very small.

In every country some of these checks are, with more or less force, in constant operation; yet, notwithstanding their general prevalence, there are few states in which there is not a constant effort in the population to increase beyond the means of subsistence. This constant effort as constantly tends to subject the lower classes of society to distress, and to prevent any great permanent melioration of their condition.

These effects, in the present state of society, seem to be produced in the following manner. We will suppose the means of subsistence in any country just equal to the easy support of its inhabitants. The constant effort towards population, which is found to act even in the most vicious societies, increases the number of people before the means of subsistence are increased. The food, therefore, which before supported eleven millions, must now be divided among eleven millions and a half. The poor consequently must live much worse, and many of them be reduced to severe distress. The number of labourers also being above the proportion of work in the market, the price of labour must tend to fall, while the price of provisions would at the same time tend to rise. The labourer therefore must do more work, to earn the same as he did before. During this season of distress, the discouragements to marriage and the difficulty of rearing a family are so great, that the progress of population is retarded. In the mean time, the cheapness, of labour, the plenty of labourers, and the necessity of an increased industry among them, encourage cultivators to employ more labour upon their land, to turn up fresh soil, and to manure and improve more completely what is already in tillage, till ultimately the means of subsistence may become in the same proportion to the population, as at the period from which we set out. The situation of the labourer being then again tolerably comfortable, the restraints to population are in some degree loosened; and, after a short period, the same retrograde and progressive movements, with respect to happiness, are repeated.

This sort of oscillation will not probably be obvious to common view; and it may be difficult even for the most attentive observer to calculate its periods. Yet that, in the generality of old states, some alternation of this kind does exist though in a much less marked, and in a much more irregular manner, than I have described it, no reflecting man, who considers the subject deeply, can well doubt.

One principal reason why this oscillation has been less remarked, and less decidedly confirmed by experience than might naturally be expected, is, that the histories of mankind which we posses are, in general, histories only of the higher classes, We have not many accounts that can be depended upon, of the manners and customs of that part of mankind, where these retrograde and progressive movements chiefly take place. A satisfactory history of this kind, of one people and of one period, would require the constant and minute attention of many observing minds in local and general remarks on the state of the lower classes of society, and the causes that influenced it; and, to draw accurate inferences upon this subject, a succession of such historians for some centuries would be necessary. This branch of statistical knowledge has, of late years, been attended to in some countries,[9] and we may promise ourselves a clearer insight into the internal structure of human society from the progress of these inquiries. But the science may be said yet to be in its infancy, and many of the objects, on which it would be desirable to have information, have been either omitted or not stated with sufficient accuracy. Among these, perhaps, may be reckoned the proportion of the number of adults to the number of marriages; the extent to which vicious customs have prevailed in consequence of the restraints upon matrimony; the comparative mortality among the children of the most distressed part of the community, and of those who live rather more at their ease; the variations in the real price of labour; the observable differences in the state of the lower classes of society, with respect to ease and happiness, at different times during a certain period; and very accurate registers of births, deaths, and marriages, which are of the utmost importance in this subject.

A faithful history, including such particulars, would tend greatly to elucidate the manner in which the constant check upon population acts; and would probable prove the existence of the retrograde and progressive movements that have been mentioned; though the times of their vibration must necessarily be rendered irregular from the operation of many interrupting causes; such as, the introduction or failure of certain manufactures; a greater or less prevalent spirit of agricultural enterprise; years of plenty, or years of scarcity; wars, sickly seasons, poor-laws, emigrations and other causes of a similar nature.

A circumstance which has, perhaps, more than any other, contributed to conceal this oscillation from common view, is the difference between the nominal and real price of labour. It very rarely happens that the nominal price of labour universally falls; but we well know that it frequently remains the same, while the nominal price of provisions has been gradually rising. This, indeed, will generally be the case, if the increase of manufactures and commerce be sufficient to employ the new labourers that are thrown into the market,

[9]The judicious questions which Sir John Sinclair circulated in Scotland, and the valuable accounts which he has collected in that part of the island, do him the highest honour; and these accounts will ever remain an extraordinary monument of the learning, good sense, and general information of the clergy of Scotland. It is to be regretted that the adjoining parishes are not put together in the work, which would have assisted the memory both in attaining and recollecting the state of particular districts. The repetitions and contradictory opinions which occur are not in my opinion so objectionable; as, to the result of such testimony, more faith may be given than we could possibly give to the testimony of any individual. Even were this result drawn for us by some master hand, though much valuable time would undoubtedly be saved, the information would not be so satisfactory. If, with a few subordinate improvements, this work had contained accurate and complete registers for the last 150 years, it would have been inestimable, and would have exhibited a better picture of the internal state of a country than has yet been presented to the world. But this last most essential improvement no diligence could have effected.

and to prevent the increased supply from lowering the money-price.[10] But an increased number of labourers receiving the same money-wages will necessarily, by their competition, increase the money-price of corn. This is, in fact, a real fall in the price of labour; and, during this period, the condition of the lower classes of the community must be gradually growing worse. But the farmers and capitalists are growing rich from the real cheapness of labour. Their increasing capitals enable them to employ a greater number of men; and, as the population had probably suffered some check from the greater difficulty of supporting a family, the demand for labour, after a certain period, would be great in proportion to the supply, and its price would of course rise, if left to find its natural level; and thus the wages of labour, and consequently the condition of the lower classes of society, might have progressive and retrograde movements, though the price of labour might never nominally fall.

In savage life, where there is no regular price of labour, it is little to be doubted that similar oscillations took place. When population has increased nearly to the utmost limits of the food, all the preventive and the positive checks will naturally operate with increased force. Vicious habits with respect to the sex will be more general, the exposing of children more frequent, and both the probability and fatality of wars and epidemics will be considerably greater; and these causes will probably continue their operation till the population is sunk below the level of the food; and then the return to comparative plenty will again produce an increase, and, after a certain period, its further progress will again be checked by the same causes.[11]

But without attempting to establish these progressive and retrograde movements in different countries, which would evidently require more minute histories than we possess, and which the progress of civilization naturally tends to counteract, the following propositions are intended to be proved:—

1. Population is necessarily limited by the means of subsistence.
2. Population invariably increases where the means of subsistence increase, unless prevented by some very powerful and obvious checks.[12]

[10]If the new labourers thrown yearly into the market should find no employment but in agriculture, their competition might so lower the money-price of labour, as to prevent the increase of population from occasioning an effective demand for more corn; or, in other words, if the landlords and farmers could get nothing but an additional quantity of agricultural labour in exchange for any additional produce which they could raise, they might not be tempted to raise it.

[11]Sir James Stuart very justly compares the generative faculty to a spring loaded with a variable weight, (Polit. Econ. vol. i. b. i. c. 4. p. 20.) which would of course produce exactly that kind of oscillation which has been mentioned. In the first book of his Political Economy, he has explained many parts of the subject of population very ably.

[12]I have expressed myself in this cautious manner, because I believe there are some instances, where population does not keep up to the level of the means of subsistence. But these are extreme cases; and, generally speaking, it might be said, that,

3. These checks, and the checks which repress the superior, power of population, and keep its effects on a level with the means of subsistence, are all resolvable into moral restraint, vice and misery.

The first of these propositions scarcely needs illustration. The second and third will be sufficiently established by a review of the immediate checks to population in the past and present state of society.

This review will be the subject of the following chapters.

3.4.3 Book I, Chapter III: Of the Checks to Population in the Lowest Stage of Human Society.

The wretched inhabitants of Tierra del Fuego have been placed, by the general consent of voyagers, at the bottom of the scale of human beings.[13] Of their domestic habits and manners, however, we have few accounts. Their barren country, and the miserable state in which they live, have prevented any intercourse with them that might give such information; but we cannot be at a loss to conceive the checks to population among a race of savages, whose very appearance indicates them to be half starved, and who, shivering with cold, and covered with filth and vermin, live in one of the most inhospitable climates in the world, without having sagacity enough to provide themselves with such conveniencies as might mitigate its severities, and render life in some measure more comfortable.[14]

Next to these, and almost as low in genius and resources, have been placed the natives of Van Diemen's land;[15] but some late accounts have represented the islands of Andaman in the East as inhabited by a race of savages still lower in wretchedness even than these. Every thing that voyagers have related of savage life is said to fall short of the barbarism of this people. Their whole time is spent in search of food: and as their woods yield them few or no supplies of animals, and but little vegetable diet, their principal occupation is that of climbing the rocks, or roving along the margin of the sea, in search of a precarious meal of fish, which, during the tempestuous season, they often seek for in vain. Their stature seldom exceeds five feet; their bellies are protuberant, with high shoulders, large heads, and limbs disproportionably slender. Their countenances exhibit the extreme of wretchedness, a horrid mixture of famine and ferocity; and their extenuated and diseased figures plainly

---

2. Population always increases where the means of subsistence increase 3. The checks which repress the superior power of population, and keep its effects on a level with the means of subsistence, are all resolvable into moral restraint, vice and misery.

Book I, Chapter III.

[13] Cook's First Voy. vol. ii. p. 59.

[14] Cook's Second Voy. vol. ii. p. 187.

[15] Vancouver's Voy., vol. ii. b. iii. c. i. p. 13.

indicate the want of wholesome nourishment. Some of these unhappy beings have been found on the shores in the last stage of famine.[16]

In the next scale of human beings we may place the inhabitants of New Holland, of a part of whom we have some accounts that may be depended upon, from a person who resided a considerable time at Port Jackson, and had frequent opportunities of being a witness to their habits and manners. The narrator of Captain Cook's first voyage having mentioned the very small number of inhabitants, that was seen on the eastern coast of New Holland, and the apparent inability of the country, from its desolate state, to support many more, observes, "By what means the inhabitants of this country are reduced to such a number as it can subsist, is not perhaps very easy to guess; whether, like the inhabitants of New Zealand, they are destroyed by the hands of each other in contests for food; whether they are swept off by accidental famine; or whether there is any cause that prevents the increase of the species, must be left for future adventurers to determine."[17]

The account which Mr. Collins has given of these savages will, I hope, afford in some degree a satisfactory answer. They are described as, in general, neither tall nor well made. Their arms, legs, and thighs, are thin, which is ascribed to the poorness of their mode of living. Those who inhabit the sea-coast depend almost entirely on fish for their sustenance, relieved occasionally by a repast on some large grubs which are found in the body of the dwarf gum-tree. The very scanty stock of animals in the woods, and the very great labour necessary to take them, keep the inland natives in as poor a condition as their brethren on the coast. They are compelled to climb the tallest trees after honey and the smaller animals, such as the flying squirrel and the opossum. When the stems are of great height, and without branches, which is generally the case in thick forests, this is a process of great labour, and is effected by cutting a notch with their stone hatchets for each foot successively, while their left arm embraces the tree. Trees were observed notched in this manner to the height of eighty feet before the first branch, where the hungry savage could hope to meet with any reward for so much toil.[18]

The woods, exclusive of the animals occasionally found in them, afford but little sustenance. A few berries, the yam, the fern root, and the flowers of the different banksias, make up the whole of the vegetable catalogue.[19]

A native with his child, surprised on the banks of the Hawksbury river by some of our colonists, launched his canoe in a hurry, and left behind him a specimen of his food, and of the delicacy of his stomach. From a piece of water-soaked wood, full of holes, he had been extracting and eating a large worm. The smell both of the worm and its habitation was in the highest degree offensive. These worms, in the language of the country, are called Cah-bro; and a tribe of natives dwelling inland, from the circumstance of eating these

[16] Symes's Embassy to Ava, ch. i. p. 129, and Asiatic Researches, vol. iv. p. 401.

[17] Cook's First Voy. vol. iii. p. 240.

[18] Collins's Account of New South Wales, Appendix, p. 549. 4to.

[19] Id. Appen. p. 557. 4to.

loathsome worms, is named Cah-brogal. The wood-natives also make a paste formed of the fern root and the large and small ants, bruised together; and, in the season, add the eggs of this insect.[20]

In a country, the inhabitants of which are driven to such resources for subsistence, where the supply of animal and vegetable food is so extremely scanty, and the labour necessary to procure it is so severe, it is evident, that the population must be very thinly scattered in proportion to the territory. Its utmost bounds must be very narrow. But when we advert to the strange and barbarous customs of these people, the cruel treatment of their women, and the difficulty of rearing children; instead of being surprised that it does not more frequently press to pass these bounds, we shall be rather inclined to consider even these scanty resources as more than sufficient to support all the population that could grow up under such circumstances.

The prelude to love in this country is violence, and of the most brutal nature. The savage selects his intended wife from the women of a different tribe, generally one at enmity with his own. He steals upon her in the absence of her protectors, and having first stupified her with blows of a club, or wooden sword, on the head, back, and shoulders, every one of which is followed by a stream of blood, he drags her through the woods by one arm, regardless of the stones and broken pieces of trees that may lie in his route, and anxious only to convey his prize in safety to his own party. The woman thus treated becomes his wife, is incorporated into the tribe to which he belongs, and but seldom quits him for another. The outrage is not resented by the relations of the female, who only retaliate by a similar outrage when it is in their power.[21]

The union of the sexes takes place at an early age; and instances were known to our colonists of very young girls having been much and shamefully abused by the males.[22]

The conduct of the husband to his wife or wives, seems to be nearly in character with this strange and barbarous mode of courtship. The females bear on their heads the traces of the superiority of the males, which is exercised almost as soon as they find strength in their arms to inflict a blow. Some of these unfortunate beings have been observed with more scars on their shorn heads, cut in every direction, than could well be counted. Mr. Collins feelingly says, "The condition of these women is so wretched, that I have often, on seeing a female child borne on its mother's shoulders, anticipated the miseries to which it was born, and thought it would be a mercy to destroy it."[23] In another place, speaking of Bennilong's wife being delivered of a child, he says, "I here find in my papers a note, that for some offence Bennilong had severely beaten this woman in the morning, a short time before she was delivered."[24]

---

[20]Collins's Account of New South Wales, Appendix, p. 558.

[21]Collins's New South Wales, Appen. p. 559.

[22]Id. Appen. p. 563.

[23]Id. Appen. p. 583.

[24]Collins's New South Wales, Appen. note, p. 562.

Women treated in this brutal manner must necessarily be subject to frequent miscarriages, and it is probable that the abuse of very young girls, mentioned above as common, and the too early union of the sexes in general, would tend to prevent the females from being prolific. Instances of a plurality of wives were found more frequent than of a single wife; but what is extraordinary, Mr. Collins did not recollect ever to have noticed children by more than one. He had heard from some of the natives, that the first wife claimed an exclusive right to the conjugal embrace, while the second was merely the slave and drudge of both.[25]

An absolutely exclusive right in the first wife to the conjugal embrace seems to be hardly probable; but it is possible that the second wife may not be allowed to rear her offspring. At any rate, if the observation be generally true, it proves that many of the women are without children, which can only be accounted for from the very severe hardships which they undergo, or from some particular customs which may not have come to the knowledge of Mr. Collins.

If the mother of a sucking child die, the helpless infant is buried alive in the same grave with its mother. The father himself places his living child on the body of his dead wife, and having thrown a large stone upon it, the grave is instantly filled by the other natives. This dreadful act was performed by Co-le-be, a native well known to our colonists, and who, on being talked to on the subject, justified the proceeding, by declaring that no woman could be found who would undertake to nurse the child, and that therefore it must have died a much worse death than that which he had given it. Mr. Collins had reason to believe that this custom was generally prevalent, and observes, that it may in some measure account for the thinness of the population.[26]

Such a custom, though in itself perhaps it might not much affect the population of a country, places in a strong point of view the difficulty of rearing children in savage life. Women obliged by their habits of living to a constant change of place, and compelled to an unremitting drudgery for their husbands, appear to be absolutely incapable of bringing up two or three children nearly of the same age. If another child be born before the one above it can shift for itself, and follow its mother on foot, one of the two must almost necessarily perish for want of care. The task of rearing even one infant, in such a wandering and laborious life, must be so troublesome, and painful, that we are not to be surprised that no woman can be found to undertake it who is not prompted by the powerful feelings of a mother.

To these causes, which forcibly repress the rising generation, must be added those which contribute subsequently to destroy it; such as the frequent wars of these savages with different tribes, and their perpetual contests with each other; their strange spirit of retaliation and revenge, which prompts the midnight murder, and the frequent shedding of innocent blood; the smoke and filth of their miserable habitations, and their poor mode

[25] Id. Appen. p. 560.

[26] Collins's New South Wales, Appen. p. 607.

of living, productive of loathsome cutaneous disorders; and, above all, a dreadful epidemic like the small-pox, which sweeps off great numbers.[27]

In the year 1789 they were visited by this epidemic, which raged among them with all the appearance and virulence of the small-pox. The desolation, which it occasioned, was almost incredible. Not a living person was to be found in the bays and harbours that were before the most frequented. Not a vestige of a human foot was to be traced on the sands. They had left the dead to bury the dead. The excavations in the rocks were filled with putrid bodies, and in many places the paths were covered with skeletons.[28]

Mr. Collins was informed, that the tribe of Co-le-be, the native mentioned before, had been reduced by the effects of this dreadful disorder to three persons, who found themselves obliged to unite with some other tribe, to prevent their utter extinction.[29]

Under such powerful causes of depopulation, we should naturally be inclined to suppose that the animal and vegetable produce of the country would be increasing upon the thinly scattered inhabitants, and, added to the supply of fish from their shores; would be more than sufficient for their consumption; yet it appears, upon the whole, that the population is in general so nearly on a level with the average supply of food, that every little deficiency from unfavourable weather or other causes, occasions distress. Particular times, when the inhabitants seemed to be in great want, are mentioned as not uncommon, and, at these periods, some of the natives were found reduced to skeletons, and almost starved to death.[30]

3.4.4 Book I, Chapter IV: Of the Checks to Population Among the American Indians.

We may next turn our view to the vast continent of America, the greatest part of which was found to be inhabited by small independent tribes of savages, subsisting, nearly like the natives of New Holland, on the productions of unassisted nature. The soil was covered by an almost universal forest, and presented few of those fruits and esculent vegetables which grow in such profusion in the islands of the South Sea. The produce of a most rude and imperfect agriculture, known to some of the tribe of hunters, was so trifling as to be considered only as a feeble aid to the subsistence acquired by the chase. The inhabitants of this new world therefore might be considered as living principally by hunting and fishing;[31] and the narrow limits to this mode of subsistence are obvious. The supplies derived from fishing could reach only those who were within a certain distance of the lakes,

[27] See generally, the Appendix to Collins's Account of the English Colony in New South Wales.

[28] Collins's New South Wales, Appendix, p. 597.

[29] Collins's New South Wales, Appendix, p. 598.

[30] Id. c. iii. p. 34, and Appen. p. 551.

[31] Robertson's History of America, vol. ii. b. iv. p. 127. et seq. octavo edit. 1780.

the rivers, or the sea-shore; and the ignorance and indolence of the improvident savage would frequently prevent him from extending the benefits of these supplies much beyond the time when they were actually obtained. The great extent of territory required for the support of the hunter has been repeatedly stated and acknowledged.[32] The number of wild animals within his reach, combined with the facility, with which they may be either killed or insnared, must necessarily limit the number of his society. The tribes of hunters, like beasts of prey, whom they resemble in their mode of subsistence, will consequently be thinly scattered over the surface of the earth. Like beasts of prey, they must either drive away or fly from every rival, and be engaged in perpetual contests with each other.[33]

Under such circumstances, that America should be very thinly peopled in proportion to its extent of territory, is merely an exemplification of the obvious truth, that population cannot increase without the food to support it. But the interesting part of the inquiry, that part, to which I would wish particularly to draw the attention of the reader, is, the mode by which the population is kept down to the level of this scanty supply. It cannot escape observation, that an insufficient supply of food to any people does not shew itself merely in the shape of famine, but in other more permanent forms of distress, and in generating certain customs, which operate sometimes with greater force in the prevention of a rising population than in its subsequent destruction.

It was generally remarked, that the American women were far from being prolific.[34] This unfruitfulness has been attributed by some to a want of ardour in the men towards their women, a feature of character, which has been considered as peculiar to the American savage. It is not however peculiar to this race, but probably exists in a great degree among all barbarous nations, whose food is poor and insufficient, and who live in a constant apprehension of being pressed by famine or by an enemy. Bruce frequently takes notice of it, particularly in reference to the Galla and Shangalla, savage nations on the borders of Abyssinia,[35] and Vaillant mentions the phlegmatic temperament of the Hottentots as the chief reason of their thin population.[36] It seems to be generated by the hardships and dangers of savage life, which take off the attention from the sexual passion; and that these are the principal causes of it among the Americans, rather than any absolute constitutional defect, appears probable, from its diminishing nearly in proportion to the degree in which these causes are mitigated or removed. In those countries of America, where, from peculiar situation or further advantages in improvement, the hardships of savage life are less severely felt, the passion between the sexes becomes more ardent. Among some of the

---

[32] Franklin's Miscell. p. 2.

[33] Robertson, b. iv. p. 129.

[34] Robertson, b. iv. p. 106. Burke's America, vol. i, p. 187. Charlevoix, Hist. de la Nouvelle France, tom. iii. p. 304. Lafitau, Mœurs des Sauvages, tom. i. p. 590. In the course of this chapter I often give the same references as Robertson; but never without having examined and verified them myself. Where I have not had an opportunity of doing this, I refer to Robertson alone.

[35] Travels to discover the Source of the Nile, vol. ii. pp. 223, 559.

[36] Voyage dans l'Intérieur de l'Afrique, tom. i. p. 12, 13.

tribes seated on the banks of rivers well stored with fish, or others that inhabit a territory greatly abounding in game or much improved in agriculture, the women are more valued and admired; and as hardly any restraint is imposed on the gratification of desire, the dissoluteness of their manners is sometimes excessive.[37]

If we do not then consider this apathy of the Americans as a natural defect in their bodily frame, but merely as a general coldness, and an infrequency of the calls of the sexual appetite, we shall not be inclined to give much weight to it as affecting the number of children to a marriage; but shall be disposed to look for the cause of this unfruitfulness in the condition and customs of the women in a savage state. And here we shall find reasons amply sufficient to account for the fact in question.

It is justly observed by Dr. Robertson, that, "Whether man has been improved by the progress of arts and civilization, is a question which in the wantonness of disputation has been agitated among philosophers. That women are indebted to the refinement of polished manners for a happy change in their state, is a point which can admit of no doubt."[38] In every part of the world, one of the most general characteristics of the savage is to despise and degrade the female sex.[39] Among most of the tribes in America their condition is so peculiarly grievous, that servitude is a name too mild to describe their wretched state. A wife is no better than a beast of burden. While the man passes his days in idleness or amusement, the woman is condemned to incessant toil. Tasks are imposed upon her without mercy, and services are received without complacence or gratitude.[40] There are some districts in America where this state of degradation has been so severely felt, that mothers have destroyed their female infants, to deliver them at once from a life in which they were doomed to such a miserable slavery.[41]

This state of depression and constant labour, added to the unavoidable hardships of savage life, must be very unfavourable to the office of child-bearing;[42] and the libertinage which generally prevails among the women before marriage, with the habit of procuring abortions, must necessarily render them more unfit for bearing children afterwards.[43] One of the missionaries, speaking of the common practice among the Natchez of changing their wives, adds, unless they have children by them; a proof that many of these marriages were unfruitful, which may be accounted for from the libertine lives of the women before wedlock, which he had previously noticed.[44]

---

[37] Robertson, b. iv. p. 71. Lettres Edif. et Curieuses, tom. vi. p. 48, 322, 330. tom. vii. p. 20. 12mo. edit. 1780. Charlevoix, tom. iii. p. 303, 423. Hennepin, Mœurs des Sauvages, p. 37.

[38] Robertson, b. iv. p. 103.

[39] Robertson, b. iv. p. 103. Lettres Edif, passim. Charlevoix, Hist. Nouv. Fr. tom. iii. p. 287. Voy. de Pérouse, c. ix. p. 402, 4to. London.

[40] Robertson, b. iv. p. 105. Lettres Edif. tom. vi. p. 329. Major Roger's North America, p. 211. Creuxii Hist. Canad. p. 57.

[41] Robertson, b. iv. p. 106. Raynal, Hist. des Indes, tom. iv. c. vii. p. 110, 8vo. 10 vol. 1795.

[42] Robertson, b. iv. p. 106. Creuxii Hist. Canad. p. 57. Lafitau, tom. i. p. 590.

[43] Robertson, b. iv. p. 72. Ellis's Voyage, p. 198. Burke's America, vol. i. p. 187.

[44] Lettres Edif. tom. vii. p. 20, 22.

The causes that Charlevoix assigns of the sterility of the American women, are, the suckling their children for several years, during which time they do not cohabit with their husbands; the excessive labour to which they are always condemned, in whatever situation they may be; and the custom established in many places, of permitting the young women to prostitute themselves before marriage. Added to this, he says, the extreme misery to which these people are sometimes reduced, takes from them all desire of having children.[45] Among some of the ruder tribes it is a maxim not to burthen themselves with rearing more than two of their offspring.[46] When twins are born, one of them is commonly abandoned, as the mother cannot rear them both; and when a mother dies during the period of suckling her child, no chance of preserving its life remains, and, as in New Holland, it is buried in the same grave with the breast that nourished it.[47]

As the parents are frequently exposed to want themselves, the diffculty of supporting their children becomes at times so great, that they are reduced to the necessity of abandoning or destroying them.[48] Deformed children are very generally exposed; and, among some of the tribes in South America, the children of mothers who do not bear their labours well, experience a similar fate, from a fear that the offspring may inherit the weakness of its parent.[49]

To causes of this nature we must ascribe the remarkable exemption of the Americans from deformities of make. Even when a mother endeavours to rear all her children without distinction, such a proportion of the whole number perishes under the rigorous treatment which must be their lot in the savage state, that probably none of those who labour under any original weakness or infirmity can attain the age of manhood. If they be not cut off as soon as they are born, they cannot long protract their lives under the severe discipline that awaits them.[50] In the Spanish provinces, where the Indians do not lead so laborious a life, and are prevented from destroying their children, great numbers of them are deformed, dwarfish, mutilated, blind and deaf.[51]

Polygamy seems to have been generally allowed among the Americans, but the privilege was seldom used, except by the Caciques and chiefs, and now and then by others in some of the fertile provinces of the South, where subsistence was more easily procured. The difficulty of supporting a family confined the mass of the people to one wife;[52] and this difficulty was so generally known and acknowledged, that fathers, before they consented to give their daughters in marriage, required unequivocal proofs in the suitor of his skill in

---

[45] Charlevoix, N. Fr. tom. iii. p. 304.

[46] Robertson, b. iv. p. 107. Lettres Edif. tom. ix. p. 140.

[47] Robertson, b. iv. p. 107. Lettres Edif. tom. viii. p. 86.

[48] Robertson, b. iv. p. 108.

[49] Lafitau, Mœurs des Sauv. tom. i. p. 592.

[50] Charlevoix, tom: iii. p. 303. Raynal, Hist. des Indes, tom. viii. l. xv. p. 22.

[51] Robertson, b. iv. p. 73. Voyage d'Ulloa, tom. i. p. 232.

[52] Robertson, b. iv. p. 102. Lettres Edif. tom. viii. p. 87.

hunting, and his consequent ability to support a wife and children.[53] The women, it is said, do not marry early;[54] and this seems to be confirmed by the libertinage among them before marriage, so frequently taken notice of by the missionaries and other writers.[55]

The customs above enumerated, which appear to have been generated principally by the experience of the difficulties attending the rearing of a family, combined with the number of children that must necessarily perish under the hardships of savage life, in spite of the best efforts of their parents to save them,[56] must, without doubt, most powerfully repress the rising generation.

When the young savage has passed safely through the perils of his childhood, other dangers scarcely less formidable await him on his approach to manhood. The diseases to which man is subject in the savage state, though fewer in number, are more violent and fatal than those which prevail in civilized society. As savages are wonderfully improvident, and their means of subsistence always precarious, they often pass from the extreme of want to exuberant plenty, according to the vicissitudes of fortune in the chase, or to the variety in the produce of the seasons.[57] Their inconsiderate gluttony in the one case, and their severe abstinence in the other, are equally prejudicial to the human constitution; and their vigour is accordingly at some seasons impaired by want, and at others by a superfluity of gross aliment, and the disorders arising from indigestions.[58] These, which may be considered as the unavoidable consequences of their mode of living, cut off considerable numbers in the prime of life. They are likewise extremely subject to consumptions, to pleuritic, asthmatic, and paralytic disorders, brought on by the immoderate hardships and fatigues which they endure in hunting and war, and by the inclemency of the seasons, to which they are continually exposed.[59]

The missionaries speak of the Indians in South America as subject to perpetual diseases for which they know no remedy.[60] Ignorant of the use of the most simple herbs, or of any change in their gross diet, they die of these diseases in great numbers. The Jesuit Fauque says, that, in all the different excursions which he had made, he scarcely found a single individual of an advanced age.[61] Robertson determines the period of human life to be

---

[53] Lettres Edif. tom. ix. p. 364. Robertson, b. iv. p. 115.

[54] Robertson, b. iv. p. 107.

[55] Lettres Edif. passim. Voyage d'Ulloa, tom. i. p. 343. Burke's America, vol. i. p. 187. Charlevoix, tom. iii. p. 303, 304.

[56] Creuxius says, that scarcely one in thirty reaches manhood (Hist. Canad. p. 57); but this must be a very great exaggeration.

[57] Robertson, b. iv. p. 85.

[58] Charlevoix, tom., iii, p. 302, 303.

[59] Robertson, b. iv. p. 86. Charlevoix, tom. iii. p. 364. Lafitau, tom. ii. p. 360, 361.

[60] Lettres Edif. tom. viii. p. 83.

[61] Lettres Edif. tom. vii. p. 317. et seq.

shorter among savages than in well-regulated and industrious communities.[62] Raynal, notwithstanding his frequent declamations in favour of savage life, says of the Indians of Canada, that few are so long lived as our people, whose manner of living is more uniform and tranquil.[63] And Cook and Pérouse confirm these opinions in the remarks which they make on some of the inhabitants of the north-west coast of America.[64]

In the vast plains of South America, a burning sun, operating on the extensive swamps and the inundations that succeed the rainy seasons, sometimes produces dreadful epidemics. The missionaries speak of contagious distempers as frequent among the Indians, and occasioning at times a great mortality in their villages.[65] The smallpox every where makes great ravages, as, from want of care and from confined habitations, very few that are attacked recover from it.[66] The Indians of Paraguay are said to be extremely subject to contagious distempers, notwithstanding the care and attentions of the Jesuits. The small-pox and malignant fevers, which, from the ravages they make, are called plagues, frequently desolate these flourishing missions; and, according to Ulloa, were the cause that they had not increased in proportion to the time of their establishment, and the profound peace which they had enjoyed.[67]

These epidemics are not confined to the south. They are mentioned as if they were not uncommon among the more northern nations;[68] and, in a late voyage to the north-west coast of America, Captain Vancouver gives an account of a very extraordinary desolation apparently produced by some distemper of this kind. From New Dungeness he traversed a hundred and fifty miles of the coast without seeing the same number of inhabitants. Deserted villages were frequent, each of which was large enough to contain all the scattered savages that had been observed in that extent of country. In the different excursions which he made, particularly about Port Discovery, the skulls, limbs, ribs and back-bones, or some other vestiges of the human body, were scattered promiscuously in great numbers; and, as no warlike scars were observed on the bodies of the remaining Indians, and no particular signs of fear and suspicion were noticed, the most probable conjecture seems to be, that this depopulation must have been occasioned by pestilential disease.[69] The small-pox appears to be common and fatal among the Indians on this coast. Its indelible marks were observed on many, and several had lost the sight of one eye from it.[70]

---

[62] B. iv. p. 86.

[63] Raynal, b. xv. p. 23.

[64] Cook, third Voy. vol. iii. ch. ii. p. 520. Voy. de Pérouse, ch. ix.

[65] Lettres Edif. tom. viii. p. 79, 339. tom. ix. p. 125.

[66] Voyage d'Ulloa, tom. i. p. 349.

[67] Id. tom. i. p. 549.

[68] Lettres Edif. tom. vi. p. 335.

[69] Vancouver's Voy. vol. i. b. ii. c. v. p. 256.

[70] Id. c. iv. p. 242.

In general, it may be remarked of savages, that, from their extreme ignorance, the dirt of their persons, and the closeness and filth of their cabins,[71] they lose the advantage which usually attends a thinly peopled country, that of being more exempt from pestilential diseases than those which are fully inhabited. In some parts of America the houses are built for the reception of many different families; and fourscore or a hundred people are crowded together under the same roof. When the families live separately, the huts are extremely small, close and wretched, without windows; and with the doors so low, that it is necessary to creep on the hands and knees to enter them.[72] On the north-west coast of America, the houses are, in general, of the large kind; and Meares describes one of most extraordinary dimensions, belonging to a chief near Nootka Sound in which eight hundred persons ate, sat, and slept.[73] All voyagers agree with respect to the filth of the habitations and the personal nastiness of the people on this coast.[74] Captain Cook describes them as swarming with vermin, which they pick off and eat;[75] and speaks of the state of their habitations in terms of the greatest disgust.[76] Pérouse declares that their cabins have a nastiness and stench to which the den of no known animal in the world can be compared.[77]

Under such circumstances, it may be easily imagined what a dreadful havoc an epidemic must make, when once it appears among them; and it does not seem improbable, that the degree of filth described should generate distempers of this nature, as the air of their houses cannot be much purer than the atmosphere of the most crowded cities.

Those who escape the dangers of infancy and of disease, are constantly exposed to the chances of war; and notwithstanding the extreme caution of the Americans in conducting their military operations, yet, as they seldom enjoy any interval of peace, the waste of their numbers in war is considerable.[78] The rudest of the American nations are well acquainted with the rights of each community to its own domains.[79] And as it is of the utmost consequence to prevent others from destroying the game in their hunting grounds, they guard this national property with a jealous attention. Innumerable subjects of dispute necessarily arise. The neighbouring nations live in a perpetual state of hostility with each other.[80] The very act of increasing in one tribe must be an act of aggression on its

[71] Charlevoix speaks in the strongest terms of the extreme filth and stench of the American cabins, "On ne peut entrer dans leurs cabanes qu'on ne soit impesté:" and the dirt of their meals, he says, "vous feroit horreur." Vol. iii. p. 338.

[72] Robertson, b. iv. p. 182. Voyage d'Ulloa, tom. i. p. 340.

[73] Meares's Voyage, ch. xii. p. 138.

[74] Meares's Voyage, ch. xxiii. p. 252. Vancouver's Voyage, vol. iii. b. vi. c. i. p. 313.

[75] Cook's 3d Voyage, vol. ii. p. 305.

[76] Id. c. iii. p. 316.

[77] Voyage de Pérouse, c. ix. p. 403.

[78] Charlevoix; Hist. de la Nouv. France, tom. iii. 202, 203, 429.

[79] Robertson, b. iv. p. 147.

[80] Robertson, b. iv. p. 147. Lettres Edif. tom. viii. p. 40, 86, 8 passim. Cook's 3d Voy. vol. ii. p. 324. Meares's Voy. ch. xxiv. p. 267.

neighbours; as a larger range of territory will be necessary to support its increased numbers. The contest will in this case naturally continue, either till the equilibrium is restored by mutual losses, or till the weaker party is exterminated, or driven from its country. When the irruption of an enemy desolates their cultivated lands, or drives them from their hunting-grounds; as they have seldom any portable stores, they are generally reduced to extreme want. All the people of the district invaded, are frequently forced to take refuge in woods or mountains, which can afford them no subsistence, and where many of them perish.[81] In such a flight each consults alone his individual safety. Children desert their parents, and parents consider their children as strangers. The ties of nature are no longer binding. A father will sell his son for a knife or a hatchet.[82] Famine and distresses of every kind complete the destruction of those whom the sword had spared; and in this manner whole tribes are frequently extinguished.[83]

Such a state of things has powerfully contributed to generate that ferocious spirit of warfare observable among savages in general, and most particularly among the Americans. Their object in battle is not conquest, but destruction.[84] The life of the victor depends on the death of his enemy; and, in the rancour and fell spirit of revenge with which he pursues him, he seems constantly to bear in mind the distresses that would be consequent on defeat. Among the Iroquois, the phrase by which they express their resolution of making war against an enemy, is, "Let us go and eat that nation." If they solicit the aid of a neighbouring tribe, they invite them to eat broth made of the flesh of their enemies.[85] Among the Abnakis, when a body of their warriors enters an enemy's territory, it is generally divided into different parties, of thirty or forty; and the chief says to each, " To you is given such a hamlet to eat, to you such a village,"[86] &c. These expressions remain in the language of some of the tribes, in which the custom of eating their prisoners taken in war no longer exists. Cannibalism, however, undoubtedly prevailed in many parts of the new world;[87] and, contrary to the opinion of Dr. Robertson, I cannot but think that it must have had its origin in extreme want, though the custom might afterwards be continued from other motives. It seems to be a worse compliment to human nature and to the savage state, to attribute this horrid repast to malignant passions, without the goad of necessity, rather than to the great law of self-preservation, which has at times overcome every other feeling, even among the most humane and civilized people. When once it had prevailed, though only occasionally, from this cause; the fear that a savage might feel of becoming a repast to his

[81] Robertson, b. iv. p. 172. Charlevoix, Nouv. France, tom. iii. p. 203.

[82] Lettres Edif. tom. viii. p. 346.

[83] Robertson, b. iv, p. 172. Account of North America, by Major Rogers, p. 250.

[84] Robertson, b. iv. p. 150.

[85] Robertson b. iv. p. 164.

[86] Lettres Edif. tom. vi. p. 205.

[87] Robertson, b. iv. p. 164.

enemies, might easily raise the passion of rancour and revenge to so high a pitch, as to urge him to treat his prisoners in this way, though not prompted at the time by hunger.

The missionaries speak of several nations, which appeared to use human flesh whenever they could obtain it, as they would the flesh of any of the rarer animals.[88] These accounts may perhaps be exaggerated, though they seem to be confirmed in a great degree by the late voyages to the north-west coast of America, and by Capt. Cook's description of the state of society in the southern island of New Zealand.[89] The people of Nootka Sound appear to be cannibals;[90] and the chief of the district, Maquinna, is said to be so addicted to this horrid banquet, that, in cold blood, he kills a slave every moon to gratify his unnatural appetite.[91]

The predominant principle of self-preservation, connected most intimately in the breast of the savage, with the safety and power of the community to which he belongs, prevents the admission of any of those ideas of honour and gallantry in war, which prevail among more civilized nations. To fly from an adversary that is on his guard, and to avoid a contest where he cannot contend without risk to his own person, and consequently to his community, is the point of honour with the American. The odds of ten to one are necessary to warrant an attack on a person who is armed and prepared to resist; and even then each is afraid of being the first to advance[92] The great object of the most renowned warrior is by every art of cunning and deceit, by every mode of stratagem and surprise that his invention can suggest, to weaken and destroy the tribes of his enemies with the least possible loss to his own. To meet an enemy on equal terms is regarded as extreme folly. To fall in battle, instead of being reckoned an honourable death,[93] is a misfortune, which subjects the memory of a warrior to the imputation of rashness and imprudence. But to lie in wait day after day, till he can rush upon his prey when most secure, and least able to resist him; to steal in the dead of night upon his enemies, set fire to their huts, and massacre the inhabitants, as they fly naked and defenceless from the flames,[94] are deeds of glory, which will be of deathless memory in the breasts of his grateful countrymen.

This mode of warfare is evidently produced by a consciousness of the difficulties attending the rearing of new citizens under, the hardships and dangers of savage life. And these powerful causes of destruction may in some instances be so great as to keep down the population even considerably below the means of subsistence; but the fear that the Americans betray of any diminution of their society, and their apparent wish to increase

[88] Lettres Edif. tom. viii. p. 105, 271. tom. vi. p. 266.

[89] Cautious as Captain Cook always is, he says of the New Zealanders, "it was but too evident that they have a great liking for this kind of food." Second Voyage, vol. i. p. 246. And in the last Voyage, speaking of their perpetual hostilities, he says, "and perhaps the desire of a good meal may be no small incitement." Vol. i. p. 137.

[90] Cook's Third Voyage, vol. ii, p. 271.

[91] Meares's Voyage, ch. xxiv. p. 255.

[92] Lettres Edif. tom. vi. p. 360.

[93] Charlevoix, No. Fr. tom. iii. p. 376.

[94] Robertson, b. iv. p. 155. Lettres Edif. tom. vi. p. 182, 360.

it, are no proofs that this is generally the case. The country could not probably support the addition that is coveted in each society; but an accession of strength to one tribe opens to it new sources of subsistence in the comparative weakness of its adversaries; and, on the contrary, a diminution of its numbers, so far from giving greater plenty to the remaining members, subjects them to extirpation or famine from the irruptions of their stronger neighbours.

The Chiriguanes, originally only a small part of the tribe of Guaranis, left their native country in Paraguay, and settled in the mountains towards Peru. They found sufficient subsistence in their new country, increased rapidly, attacked their neighbours, and by superior valour or superior fortune gradually exterminated them, and took possession of their lands; occupying a great extent of country, and having increased, in the course of some years, from three or four thousand to thirty thousand,[95] while the tribes of their weaker neighbours were daily thinned by famine and the sword.

Such instances prove the rapid increase even of the Americans under favourable circumstances, and sufficiently account for the fear which prevails in every tribe of diminishing its numbers, and the frequent wish to increase, them,[96] without supposing a superabundance of food in the territory actually possessed.

That the causes,[97] which have been mentioned as affecting the population of the Americans, are principally regulated by the plenty or scarcity of subsistence, is sufficiently evinced from the greater frequency of the tribes, and the greater numbers in each, throughout all those parts of the country, where, from the vicinity of lakes or rivers, the superior fertility of the soil, or further advances in improvement, food becomes more abundant. In the interior of the provinces bordering on the Oronoco, several hundred miles may be traversed in different directions without finding a single hut, or observing the footsteps of a single creature. In some parts of North America, where the climate is more rigorous, and the soil less fertile, the desolation is still greater. Vast tracts of some hundred leagues have been crossed through uninhabited plains and forests.[98] The missionaries speak of journeys of twelve days without meeting a single soul,[99] and of immense tracts of country, in which scarcely three or four scattered villages were to be found.[100] Some of

[95] Lettres Edif. tom. viii. p. 243. Les Chiriguanes multiplièrent prodigieusement, et en assez peu d'années leur nombre monta à trente mille ames.

[96] Lafitau, tom. ii. p. 163.

[97] These causes may perhaps appear more than sufficient to keep the population down to the level of the means of subsistence; and they certainly would be so, if the representations given of the unfruitfulness of the Indian women were universally, or even generally true. It is probable that some of the accounts are exaggerated, but it is difficult to say which; and it must be acknowledged, that, even allowing for all such exaggerations, they are amply sufficient to establish the point proposed.

[98] Robertson, b. iv. p. 129, 130.

[99] Lettres Edif. tom. vi. p. 357.

[100] Id. p. 321.

these deserts furnished no game,[101] and were therefore entirely desolate; others, which were to a certain degree stocked with it, were traversed in the hunting seasons by parties, who encamped and remained in different spots, according to the success they met with, and were therefore really inhabited in proportion to the quantity of subsistence which they yielded.[102]

Other districts of America are described as comparatively fully peopled; such as the borders of the great northern lakes, the shores of Mississippi, Louisiana, and many provinces in South America. The villages here were large, and near each other, in proportion to the superior fruitfulness of the territory in game and fish, and the advances made by the inhabitants in agriculture.[103] The Indians of the great and populous empires of Mexico and Peru sprung undoubtedly from the same stock, and originally possessed the same customs as their ruder brethren; but from the moment when, by a fortunate train of circumstances, they were led to improve and extend their agriculture, a considerable population rapidly followed, in spite of the apathy of the men, or the destructive habits of the women. These habits would indeed in a great measure yield to the change of circumstances; and the substitution of a more quiet and sedentary life for a life of perpetual wandering and hardship, would immediately render the women more fruiful, and enable them at the same time to attend to the wants of a larger family.

In a general view of the American continent, as described by historians, the population seems to have been spread over the surface very nearly in proportion to the quantity of food which the inhabitants of the different parts, in the actual state of their industry and improvement, could obtain; and that, with few exceptions, it pressed hard against this limit, rather than fell short of it, appears from the frequent recurrence of distress for want of food in all parts of America.

Remarkable instances occur, according to Dr. Robertson, of the calamities which rude nations suffer by famine. As one of them, he mentions an account given by Alvar Nugnez Cabeça de Vaca, one of the Spanish adventurers, who resided almost nine years among the savages of Florida. He describes them as unacquainted with every species of agriculture, and living chiefly upon the roots of different plants, which they procure with great difficulty, wandering from place to place in search of them. Sometimes they kill game, sometimes they catch fish, but in such small quantities, that their hunger is so extreme as to compel them to eat spiders, the eggs of ants, worms, lizards, serpents, and a kind of unctuous earth; and, I am persuaded, he says, that if in this country there were any stones, they would swallow them. They preserve the bones of fishes and serpents, which they grind into powder and eat. The only season when they do not suffer much from famine, is when a certain fruit like the opuntia, or prickly-pear, is ripe; but they are sometimes obliged to

[101] Id. tom. ix. p. 145.

[102] Id. tom. vi. p. 66, 81, 345. tom. ix. p. 145.

[103] Id. tom. ix. p. 90, 142. Robertson, b. iv. p. 141.

travel far from their usual place of residence, in order to find it. In another place, he observes that they are frequently reduced to pass two or three days without food.[104]

Ellis, in his Voyage to Hudson's Bay, feelingly describes the sufferings of the Indians in that neighbourhood from extreme want. Having mentioned the severity of the climate, he says, "Great as these hardships are which result from the rigour of the cold, yet it may justly be, affirmed that they are much inferior to those which they feel from the scarcity of provisions, and the difficulty they are under of procuring them. A story which is related at the factories, and known to be true, will sufficiently prove this, and give the compassionate reader a just idea of the miseries to which these unhappy people are exposed." He then gives an account of a poor Indian and his wife, who, on the failure of game, having eaten up all the skins which they wore as clothing, were reduced to the dreadful extremity of supporting themselves on the flesh of two of their children.[105] In another place, he says, "It has sometimes happened that the Indians who come in summer to trade at the factories, missing the succours they expected; have been obliged to singe off the hair from thousands of beaver-skins, in order to feed upon the leather."[106]

The Abbé Raynal, who is continually reasoning most inconsistently in his comparisons of savage and civilized life, though in one place he speaks of the savage as morally sure of a competent subsistence, yet, in his account of the nations of Canada, says, that though they lived in a country abounding in game and fish, yet in some seasons and sometimes for whole years, this resource failed them; and famine then occasioned a great destruction among a people who were at too great a distance to assist each other.[107]

Charlevoix, speaking of the inconveniences and distresses to which the missionaries were subject, observes that not unfrequently the evils which he had been describing are effaced by a greater, in comparison of which all the others are nothing. This is famine. It is true, says he, that the savages can bear hunger with as much patience as they shew carelessness in providing against it; but they are sometimes reduced to extremities beyond their power to support.[108]

It is the general custom among most of the American nations, even those which have made some progress in agriculture, to disperse themselves in the woods at certain seasons of the year, and to subsist for some months on the produce of the chase, as a principal part of their annual supplies.[109] To remain in their villages exposes them to certain famine;[110] and in the woods they are not always sure to escape it. The most able hunters sometimes fail

[104] Robertson, note 28 to p. 117, b. iv.

[105] Robertson, p. 196.

[106] P. 194.

[107] Raynal, Histoire des Indes, tom. viii. l. xv. p. 22.

[108] Hist. N. Fr. tom. iii. p. 338.

[109] Lettres Edif. tom. vi. p. 66, 81, 345. ix. 145.

[110] Id. tom. vi. p. 82, 196, 197, 215. ix. 151.

of success, even where there is no deficiency of game;[111] and in their forests, on the failure of this resource, the hunter or the traveller is exposed to the most cruel want.[112] The Indians, in their hunting excursions, are sometimes reduced to pass three or four days without food;[113] and a missionary relates an account of some Iroquois, who, on one of these occasions, having supported themselves as long as they could, by eating the skins which they had with them, their shoes, and the bark of trees, at length, in despair, sacrificed some of the party to support the rest. Out of eleven, five only returned alive.[114]

The Indians, in many parts of South America, live in extreme want,[115] and are sometimes destroyed by absolute famines.[116] The islands, rich as they appeared to be, were peopled fully up to the level of their produce. If a few Spaniards settled in any district, such a small addition of supernumerary mouths soon occasioned a severe dearth of provisions.[117] The flourishing Mexican empire was in the same state in this respect; and Cortez often found the greatest difficulty in procuring subsistence for his small body of soldiers.[118] Even the missions of Paraguay, with all the care and foresight of the Jesuits, and notwithstanding that their population was kept down by frequent epidemics, were by no means totally exempt from the pressure of want. The Indians of the Mission of St. Michael are mentioned as having at one time increased so much, that the lands capable of cultivation in their neighbourhood, produced only half of the grain necessary for their support.[119] Long droughts often destroyed their cattle,[120] and occasioned a failure of their crops; and on these occasions some of the Missions were reduced to the most extreme indigence, and would have perished from famine, but for the assistance of their neighbours.[121]

The late voyages to the north-west coast of America confirm these accounts of the frequent pressure of want in savage life, and shew the uncertainty of the resource of fishing, which seems to afford, in general, the most plentiful harvest of food that is furnished by unassisted nature. The sea on the coast near Nootka Sound is seldom or never so much frozen as to prevent the inhabitants from having access to it. Yet from the very great precautions they use in laying up stores for the winter, and their attention to prepare and preserve whatever food is capable of it for the colder seasons, it is evident that the sea at

[111] Charlevoix, N. Fr. tom. iii. p. 201. Hennepin, Mœurs des Sauv. p. 78.

[112] Lettres Edif. tom. vi. p. 167, 220.

[113] Id. tom. vi. p. 33.

[114] Id. tom. vi. p. 71.

[115] Lettres Edif. tom. vii. p. 383. ix. 140.

[116] Id. tom. viii. p. 79.

[117] Robertson, b. iv. p. 121. Burke's America, vol. i. p. 30.

[118] Robertson, b. viii. p. 212.

[119] Lettres Edif. tom. ix. p. 381.

[120] Id. tom. ix. p. 191.

[121] Id. tom. ix. p. 206, 380.

these times yields no fish; and it appears that they often undergo very great hardships from want of provisions in the cold months.[122] During a Mr. Mackay's stay at Nootka Sound, from 1786 to 1787, the length and severity of the winter occasioned a famine. The stock of dried fish was expended, and no fresh supplies of any kind were to be caught; so that the natives were obliged to submit to a fixed allowance, and the chiefs brought every day to our countrymen the stated meal of seven dried, herrings' heads. Mr. Meares says that the perusal of this gentleman's journal would shock any mind tinctured with humanity.[123]

Captain Vancouver mentions some of the people to the north of Nootka Sound as living very miserably on a paste made of the inner bark of the pine-tree and cockles.[124] In one of the boat excursions, a party of Indians was met with who had some halibut, but, though very high prices were offered, they could not be induced to part with any. This, as Captain Vancouver observes, was singular, and indicated a very scanty supply.[125] At Nootka Sound, in the year 1794, fish had become very scarce and bore an exorbitant price; as, either from the badness of the season or from neglect, the inhabitants had experienced the greatest distress for want of provisions during winter.[126]

Pérouse describes the Indians in the neighbourhood of Port François as living during the summer in the greatest abundance by fishing, but exposed in the winter to perish from want.[127]

It is not therefore, as Lord Kaimes imagines, that the American tribes have never increased sufficiently to tender the pastoral or agricultural state necessary to them;[128] but, from some cause or other, they have not adopted in any great degree these more plentiful modes of procuring subsistence, and therefore have not increased so as to become populous. If hunger alone could have prompted the savage tribes of America to such a change in their habits, I do not conceive that there would have been a single nation of hunters and fishers remaining; but it is evident that some fortunate train of circumstances, in addition to this stimulus, is necessary for the purpose; and it is undoubtedly probable, that these arts of obtaining food will be first invented and improved in those spots which are best suited to them, and where the natural fertility of the situation, by allowing a greater number of people to subsist together, would give the fairest chance to the inventive powers of the human mind.

Among most of the American tribes that we have been considering, so great a degree of equality prevailed that all the members of each community would be nearly equal sharers in the general hardships of savage life and in the pressure of occasional famines. But in many

[122] Meares's Voyage, ch. xxiv. p. 266.

[123] Id. ch. xi. p. 132.

[124] Vancouver's Voyage, vol. ii. b. ii. c. ii. p. 273.

[125] Id. ib. p. 282.

[126] Id. vol. iii. b. vi. c. i. p. 304.

[127] Voyage de Pérouse, ch. ix. p. 400.

[128] Sketches of the History of Man, vol. i. p. 99, 105, 8vo. 2d edit.

of the more southern nations, as in Bogota,[129] and among the Natchez,[130] and particularly in Mexico and Peru, where a great distinction of ranks prevailed, and the lower classes were in a state of absolute servitude,[131] it is probable that, on occasion of any failure of subsistence, these would be the principal sufferers, and that the positive checks to population would act almost exclusively on this part of the community.

The very extraordinary depopulation that has taken place among the American Indians, may appear to some to contradict the theory which is intended to be established; but it will be found that the causes of this rapid diminution may all be resolved into the three great checks to population which have been stated; and it is not asserted, that these checks, operating from particular circumstances with unusual force, may not, in some instances, be more powerful even than the principle of increase.

The insatiable fondness of the Indians for spirituous liquors,[132] which, according to Charlevoix, is a rage that passes all expression,[133] by producing among them perpetual quarrels and contests which often terminate fatally, by exposing them to a new train of disorders which their mode of life unfits them to contend with, and by deadening and destroying the generative faculty in its very source, may alone be considered as a vice adequate to produce the present depopulation. In addition to this, it should be observed that almost every where the connexion of the Indians with Europeans has tended to break their spirit, to weaken or give a wrong direction to their industry, and in consequence to diminish the sources of subsistence. In St. Domingo, the Indians neglected purposely to cultivate their lands in order to starve out their cruel oppressors.[134] In Peru and Chili, the forced industry of the natives was fatally directed to the digging in the bowels of the earth, instead of cultivating its surface; and, among the northern tribes, the extreme desire to purchase European spirits directed the industry of the greatest part of them, almost exclusively, to the procuring of plenty for the purpose of this exchange,[135] which would prevent their attention to the more fruitful sources of subsistence, and at the same time tend rapidly to destroy the produce of the chace. The number of wild animals, in all the known parts of America, is even more diminished than the number of people.[136] The attention to agriculture has every where slackened, rather than increased, as might at first have been expected; from European connexion. In no part of America, either North or South, do we hear of any of the Indian nations living in great plenty, in consequence of their diminished numbers. It

[129] Robertson, b. iv. p. 141.

[130] Lettres Edif. tom. vii. p. 21. Robertson, b. iv, p. 139.

[131] Robertson, b. vii. p. 190, 242.

[132] Major Rogers's Account of North America, p. 210.

[133] Charlevoix, tom. iii. p. 302.

[134] Robertson, b. ii. p. 185. Burke's America, vol. i. b. 300.

[135] Charlevoix, N. Fr. tom. iii. p. 260.

[136] The general introduction of fire-arms among the Indians has probably greatly contributed to the diminution of the wild animals.

may not therefore be very far from the truth, to say that even now, in spite of all the powerful causes of destruction that have been mentioned, the average population of the American nations is, with few exceptions, on a level with the average quantity of food, which in the present state of their industry they can obtain.

3.4.5 Book I, Chapter V: Of the Checks to Population in the Islands of the South Sea.

The Abbé Raynal, speaking of the ancient state of the British isles, and of islanders in general; says of them: "It is among these people that we trace the origin of that multitude of singular institutions which retard the progress of population. Anthropophagy, the castration of males, the infibulation of females, late marriages; the consecration of virginity, the approbation of celibacy, the punishments exercised against girls who become mothers at too early an age,"[137] &c. These customs, caused by a superabundance of population in islands, have been carried, he says, to the continents, where philosophers of our days are still employed to investigate the reason of them. The Abbé does not seem to be aware that a savage tribe in America surrounded by enemies, or a civilized and populous nation hemmed in by others in the same state, is, in many respects, circumstanced like the islander. Though the barriers to a further increase of population be not so well defined, and so open to common observation, on continents as on islands, yet they still present obstacles that are nearly as insurmountable; and the emigrant, impatient of the distresses which he feels in his own country, is by no means secure of finding relief in another. There is probably no island yet known, the produce of which could not be further increased. This is all that can be said of the whole earth. Both are peopled up to their actual produce. And the whole earth is in this respect like an island. But, as the bounds to the number of people on islands, particularly when they are of small extent, are so narrow, and so distinctly marked, that every person must see and acknowledge them, an inquiry into the checks to population on those, of which we have the most authentic accounts, may tend considerably to illustrate the present subject. The question that is asked in Captain Cook's first Voyage, with respect to the thinly scattered savages of New Holland, "By what means the inhabitants of this country are reduced to such a number as it can subsist?"[138] may be asked with equal propriety respecting the most populous islands in the South Sea, or the best peopled countries in Europe and Asia. The question, applied generally, appears to me to be highly curious, and to lead to the elucidation of some of the most obscure, yet important points, in the history of human society. I cannot so clearly and concisely describe the precise aim of the first part of the present work, as by saying that it is an endeavour to answer this question so applied.

[137] Raynal, Histoire des Indes, vol. ii. liv. iii. p. 3. 10 vols. 8vo. 1795.

[138] Cook's First Voyage, vol. iii. p. 240, 4to.

Of the large islands of New Guinea, New Britain, New Caledonia, and the New Hebrides, little is known with certainty. The state of society in them is probably very similar to that which prevails among many of the savage nations of America. They appear to be inhabited by a number of different tribes, who are engaged in frequent hostilities with each other. The chiefs have little authority; and private property being in consequence insecure, provisions have been rarely found on them in abundance.[139] With the large island of New Zealand we are better acquainted; but not in a manner to give us a favourable impression of the state of society among its inhabitants. The picture of it, drawn by Captain Cook in his three different Voyages, contains some of the darkest shades that are any where to be met with in the history of human nature. The state of perpetual hostility, in which the different tribes of these people live with each other, seems to be even more striking than among the savages of any part of America;[140] and their custom of eating human flesh, and even their relish for that kind of food, are established beyond a possibility of doubt.[141] Captain Cook, who is by no means inclined to exaggerate the vices of savage life, says, of the natives in the neighbourhood of Queen Charlotte's Sound, "If I had followed the advice of all our pretended friends, I might have extirpated the whole race; for the people of each hamlet or village, by turns, applied to me to destroy the other. One would have thought it almost impossible that so striking a proof of the divided state in which these miserable people live, could have been assigned."[142] And, in the same chapter, further on, he says, "From my own observations, and the information of Taweiharooa, it appears to me, that the New Zealanders must live under perpetual apprehensions of being destroyed by each other; there being few of their tribe that have not, as they think, sustained wrongs from some other tribes, which they are continually upon the watch to revenge. And, perhaps, the desire of a good meal may be no small incitement.**** Their method of executing their horrible designs is by stealing upon the adverse party in the night; and if they find them unguarded (which, however, I believe, is very seldom the case) they kill every one indiscriminately, not even sparing the women and children. When the massacre is completed, they either feast and gorge themselves on the spot, or carry off as many of the dead bodies as they can, and devour them at home with acts of brutality too shocking to be described.**** To give quarter, or take prisoners, makes no part of the military law, so that the vanquished can only save their lives by flight. This perpetual state of war and destructive method of conducting it, operates so strongly in producing habitual circumspection, that one hardly ever finds a New Zealander off his guard, either by night or by day."[143]

---

[139] See the different accounts of New Guinea and New Britain, in the *Histoire des Navigations aux terres Australes;* and of New Caledonia and the New Hebrides in Cook's Second Voyage, vol. ii. b. iii.

[140] Cook's First Voyage, vol. ii. p. 345. Second Voyage, vol. i. p. 101. Third Voyage, vol. i. p. 161, &c.

[141] Cook's Second Voyage, vol. i. p. 246.

[142] Id. Third Voyage, vol. i. p. 124.

[143] Cook's Third Voyage, vol. i. p. 137.

As these observations occur in the last Voyage; in which the errors of former accounts would have been corrected, and as a constant state of warfare is here represented as prevailing to such a degree that it may be considered as the principal check to the population of New Zealand, little need be added on this subject. We are not informed whether any customs are practised by the women unfavourable to population. If such be known; they are probably never resorted to, except in times of great distress; as each tribe will naturally wish to increase the number of its members in order to give itself greater power of attack and defence. But the vagabond life which the women of the southern island lead, and the constant state of alarm in which they live, being obliged to travel and work with arms in their hands,[144] must undoubtedly be very unfavourable to gestation, and tend greatly to prevent large families.

Yet powerful as these checks to population are, it appears, from the recurrence of seasons of scarcity, that they seldom repress the number of people below the average means of subsistence. "That such seasons there are," (Captain Cook says,) "our observations leave us no room to doubt."[145] Fish is a principal part of their food, which, being only to be procured on the sea-coast, and at certain times,[146] must always be considered as a precarious resource. It must be extremely difficult to dry and preserve any considerable stores in a state of society subject to such constant alarms; particularly, as we may suppose, that the bays and creeks most abounding in fish would most frequently be the subject of obstinate contest to people who were wandering in search of food.[147] The vegetable productions are, the fern root, yams, clams and potatoes.[148] The three last are raised by cultivation, and are seldom found on the southern island, where agriculture is but little known.[149] On the occasional failure of these scanty resources from unfavourable seasons, it may be imagined that the distress must be dreadful. At such periods it does not seem improbable that the desire of a good meal should give additional force to the desire of revenge, and that they should be "perpetually destroying each other by violence, as the only alternative of perishing by hunger."[150]

If we turn our eyes from the thinly scattered inhabitants of New Zealand to the crowded shores of Otaheite and the Society Islands, a different scene opens to our view. All apprehension of dearth seems at first sight to be banished from a country that is described to be fruitful as the garden of the Hesperides.[151] But this first impression would be immediately corrected by a moment's reflection. Happiness and plenty have always been

---

[144] Id. Second Voyage, vol. i. p. 127.

[145] Cook's First Voyage, vol. iii. p. 66.

[146] Id. p. 45.

[147] Id. Third Voyage, vol. i. p. 157.

[148] Id. First Voyage, vol. iii. p. 43.

[149] Id. vol. ii, p. 405.

[150] Cook's First Voyage, vol. iii. p. 45.

[151] Missionary Voyage, Appendix, p. 347.

considered as the most powerful causes of increase. In a delightful climate, where few diseases are known, and the women are condemned to no severe fatigues, why should not these causes operate with a force unparalleled in less favourable regions? Yet if they did, where could the population find room and food in such circumscribed limits? If the numbers in Otaheite, not 40 leagues in circuit, surprised Captain Cook, when he calculated them at two hundred and four thousand,[152] where could they be disposed of in a single century, when they would amount to above three millions, supposing them to double their numbers every twenty-five years.[153] Each island of the group would be in a similar situation. The removal from one to another would be a change of place, but not a change of the species of distress. Effectual emigration, or effectual importation, would be utterly excluded, from the situation of the islands and the state of navigation among their inhabitants.

The difficulty here is reduced to so narrow a compass, is so clear, precise and forcible that we cannot escape from it. It cannot be answered in the usual vague and inconsiderate manner, by talking of emigration, and further cultivation. In the present instance, we cannot but acknowledge, that the one is impossible, and the other glaringly inadequate. The fullest conviction must stare us in the face, that the people on this group of islands could not continue to double their numbers every twenty-five years; and before we proceed to inquire into the state of society on them, we must be perfectly certain that, unless a perpetual miracle render the women barren, we shall be able to trace some very powerful checks to population in the habits of the people.

The successive accounts that we have received of Otaheite and the neighbouring islands, leave us no room to doubt the existence of the Eareeoie societies,[154] which have justly occasioned so much surprise among civilized nations. They have been so often described, that little more need be said of them here, than that promiscuous intercourse and infanticide appear to be their fundamental laws. They consist exclusively of the higher classes; "and" (according to Mr. Anderson)[155] "so agreeable is this licentious plan of life to their disposition, that the most beautiful of both sexes thus commonly spend their youthful days, habituated to the practice of enormities that would disgrace the most savage tribes. **** When an Eareeoie woman is delivered of a child, a piece of cloth dipped in water is

[152]Cook's Second Voyage, vol. i. p. 349.

[153]I feel very little doubt that this rate of increase is much slower than would really take place, supposing every check to be removed. If Otaheite, with its present produce, were peopled only with a hundred persons, the two sexes in equal numbers, and each man constant to one woman; I cannot but think that, for five or six successive periods, the increase would be more rapid than in any instance hitherto known, and that they would probably double their numbers in less than fifteen years.

[154]Cook's First Voyage, vol. ii. p. 207, et seq. Second Voyage, vol. i. p. 352. Third Voyage, vol. ii. p. 157, et seq. Missionary Voyage, Appendix, p. 347, 4to.

[155]Mr. Anderson acted in the capacity of naturalist and surgeon in Cook's last voyage. Captain Cook, and all the officers of the expedition, seem to have had a very high opinion of his talents and accuracy of observation. His accounts, therefore, may be looked upon as of the first authority.

applied to the mouth and nose, which suffocates it."[156] Captain Cook observes, "It is certain that these societies greatly prevent the increase of the superior classes of people, of which they are composed."[157] Of the truth of this observation there can be no doubt.

Though no particular institutions of the same nature have been found among the lower classes; yet the vices which form their most prominent features are but too generally spread. Infanticide is not confined to the Eareeoies. It is permitted to all; and as its prevalence among the higher classes of the people has removed from it all odium, or imputation of poverty, it is probably often adopted rather as a fashion, than a resort of necessity, and appears to be practised familiarly and without reserve.

It is a very just observation of Hume, that the permission of infanticide generally contributes to increase the population of a country.[158] By removing the fears of too numerous a family, it encourages marriage; and the powerful yearnings of nature prevent parents from resorting to so cruel an expedient, except in extreme cases. The fashion of the Eareeoie societies, in Otaheite and its neighbouring islands, may have made them an exception to this observation; and the custom has probably here a contrary tendency.

The debauchery and promiscuous intercourse which prevail among the lower classes of people, though in some instances they may have been exaggerated, are established to a great extent on unquestionable authority. Captain Cook, in a professed endeavour to rescue the women of Otaheite from a too general imputation of licentiousness, acknowledges that there are more of this character here than in any other countries; making at the same time a remark of the most decisive nature, by observing that the women who thus conduct themselves do not in any respect lower their rank in society, but mix indiscriminately with those of the most virtuous character.[159]

The common marriages in Otaheite are without any other ceremony than a present from the man to the parents of the girl. And this seems to be rather a bargain with them for permission to try their daughter, than an absolute contract for a wife. If the father should think that he has not been sufficiently paid for his daughter, he makes no scruple of forcing her to leave her friend, and to cohabit with another person who may be more liberal. The man is always at liberty to make a new choice. Should his consort become pregnant, he may kill the child, and, after that, continue his connexion with the mother, or leave her, according to his pleasure. It is only when he has adopted a child and suffered it to live, that the parties are considered as in the marriage state. A younger wife however may afterwards be joined to the first; but the changing of connexions is much more general than this plan, and is a thing so common that they speak of it with great indifference.[160] Libertinism before marriage seems to be no objection to an union of this kind ultimately.

[156] Cook's Third Voyage, vol, ii. p. 158, 159.

[157] Id. Second Voyage, vol. i. p. 352.

[158] Hume's Essays, vol. i. essay xi. p. 431. 8vo. 1764.

[159] Cook's Second Voyage, vol. i. p. 187.

[160] Id. Third Voyage, vol. ii. p. 157.

The checks to population from such a state of society would alone appear sufficient to counteract the effects of the most delightful climate, and the most exuberant plenty. Yet these are not all. The wars between the inhabitants of the different islands, and their civil contentions among themselves, are frequent, and sometimes carried on in a very destructive manner.[161] Besides the waste of human life in the field of battle, the conquerors generally ravage the enemy's territory, kill or carry off the hogs and poultry, and reduce as much as possible the means of future subsistence. The island of Otaheite, which, in the years 1767 and 1768, swarmed with hogs and fowls, was, in 1773, so ill supplied with these animals, that hardly any thing could induce the owners to part with them. This was attributed by Captain Cook principally to the wars which had taken place during that interval.[162] On Captain Vancouver's visit to Otaheite in 1791, he found that most of his friends, whom he had left in 1777, were dead; that there had been many wars since that time, in some of which the chiefs of the western districts of Otaheite had joined the enemy; and that the king had been for a considerable time completely worsted, and his own districts entirely laid waste. Most of the animals, plants and herbs, which Captain Cook had left, had been destroyed by the ravages of war.[163]

The human sacrifices which are frequent in Otaheite, though alone sufficiently strong to fix the stain of barbarism on the character of the natives, do not probably occur in such considerable numbers as materially to affect the population of the country; and the diseases, though they have been dreadfully increased by European contact, were before peculiarly lenient; and, even for some time afterwards, were not marked by any extraordinary fatality.[164]

The great checks to increase appear to be the vices of promiscuous intercourse, infanticide, and war, each of these operating with very considerable force. Yet, powerful in the prevention and destruction of life as these causes must be, they have not always kept down the population to the level of the means of subsistence. According to Mr. Anderson, "Notwithstanding the extreme fertility of the island, a famine frequently happens, in which it is said many perish. Whether this be owing to the failure of some seasons, to over-population, (which must sometimes almost necessarily happen,) or wars, I have not been able to determine; though the truth of the fact may fairly be inferred from the great economy that they observe with respect to their food, even when there is plenty."[165] After a dinner with a chief at Ulietea, Captain Cook observed, that when the company rose, many of the common people rushed in, to pick up the crumbs which had fallen, and for which they searched the leaves very narrowly. Several of them daily attended the ships, and

[161] Bougainville, Voy. autour du Monde, ch. iii. p. 217. Cook's First Voyage, vol. ii. p. 244. Missionary Voyage, p. 224.

[162] Cook's Second Voyage, vol. i. p. 182, 183.

[163] Vancouver's Voy. vol. i. b. i. c. 6. p. 98. 4to.

[164] Cook's Third Voy. vol. ii. p. 148.

[165] Id. p. 153, 154.

assisted the butchers for the sake of the entrails of the hogs, which were killed. In general, little seemed to fall to their share, except offals. "It must be owned," Captain Cook says, "that they are exceedingly careful of every kind of provision, and waste nothing that can be eaten by man, flesh and fish especially."[166]

From Mr. Anderson's account, it appears that a very small portion of animal food falls to the lot of the lower class of people, and then it is either fish, sea-eggs, or other marine, productions; for, they seldom or never eat pork. The king or principal chief is alone able to furnish this luxury every day; and the inferior chiefs, according to their riches, once a week, fortnight, or month.[167] When the hogs and fowls have been diminished by wars or too great consumption, a prohibition is laid upon these articles of food, which continues in force sometimes for several months, or even for a year or two, during which time of course they multiply very fast, and become again plentiful.[168] The common diet even of the Eareeoies, who are among the principal people of the islands, is, according to Mr. Anderson, made up of at least nine-tenths of vegetable food.[169] And as a distinction of ranks is so strongly marked, and the lives and property of the lower classes of people appear to depend absolutely on the will of their chiefs; we may well imagine that these chiefs will often live in plenty, while their vassals and servants are pinched with want.

From the late accounts of Otaheite in the Missionary Voyage, it would appear, that the depopulating causes above enumerated have operated with most extraordinary force since Captain Cook's last visit. A rapid succession of destructive wars, during a part of that interval, is taken notice of in the intermediate visit of Captain Vancouver;[170] and from the small proportion of women remarked by the Missionaries,[171] we may infer that a greater number of female infants had been destroyed than formerly. This scarcity of women would naturally increase the vice of promiscuous intercourse, and, aided by the ravages of European diseases, strike most effectually at the root of population.[172]

It is probable that Captain Cook, from the data on which he founded his calculation, may have overrated the population of Otaheite; and perhaps the Missionaries have rated it too low;[173] but I have no doubt that the population has very considerably decreased since Captain Cook's visit, from the different accounts that are given of the habits of the people with regard to economy at the different periods. Captain Cook and Mr. Anderson agree in describing their extreme carefulness of every kind of food; and Mr. Anderson, apparently after a very attentive investigation of the subject, mentions the frequent recurrence of

---

[166] Cook's Second Voy. vol. i, p. 176.

[167] Id. Third Voy. vol. ii. p. 154.

[168] Id. p. 155.

[169] Id. p. 148.

[170] Vancouver's Voy. vol. i. b. i. c. 7. p. 137.

[171] Missionary Voyage, p. 192 8 385.

[172] Id. Appen. p. 347.

[173] Id. ch. xiii. p. 212.

famines. The Missionaries, on the contrary, though they strongly notice the distress from this cause in the Friendly Islands and the Marquesas, speak of the productions of Otaheite as being in the greatest profusion; and observe that notwithstanding the horrible waste committed at feastings, and by the Eareeoie society, want is seldom known.[174]

It would appear, from these accounts, that the population of Otaheite is at present repressed considerably below the average means of subsistence, but it would be premature to conclude that it will continue long so. The variations in the state of the island which were observed by Captain Cook in his different visits appear to prove that there are marked oscillations in its prosperity and population.[175] And this is exactly what we should suppose from theory. We cannot imagine that the population of any of these islands has for ages past remained stationary at a fixed number, or that it can have been regularly increasing, according to any rate, however slow. Great fluctuations must necessarily have taken place. Overpopulousness would at all times increase the natural propensity of savages to war; and the enmities occasioned by aggressions of this kind, would continue to spread devastation, long after the original inconvenience, which might have prompted them, had ceased to be felt.[176] The distresses experienced from one or two unfavourable seasons, operating on a crowded population, which was before living with the greatest economy, and pressing hard against the limits of its food, would, in such a state of society, occasion the more general prevalence of infanticide and promiscuous intercourse;[177] and these depopulating causes would in the same manner continue to act with increased force, for some time after the occasion which had aggravated them was at an end. A change of habits to a certain degree, gradually produced by a change of circumstances, would soon restore the population, which could not long be kept below its natural level without the most extreme violence. How far European contact may operate in Otaheite with this extreme violence, and prevent it from recovering its former population, is a point which experience only can determine. But, should this be the case, I have no doubt that, on tracing the causes of it, we shall find them to be aggravated vice and misery.

Of the other islands in the Pacific Ocean we have a less intimate knowledge than of Otaheite; but our information is sufficient to assure us that the state of society in all the principal groups of them is in most respects extremely similar. Among the Friendly and Sandwich islanders, the same feudal system and feudal turbulence, the same extraordinary power of the chiefs and degraded state of the lower orders of society, and nearly the same promiscuous intercourse among a great part of the people, have been found to prevail, as in Otaheite.

---

[174] Missionary Voy. p. 195. Appen. p. 385.

[175] Cook's Second Voy. vol, i. p. 182, 8 seq. and 346.

[176] Missionary Voy. p. 225.

[177] I hope I may never be misunderstood with regard to some of these preventive causes of over-population, and be supposed to imply the slightest approbation of them, merely because I relate their effects. A cause, which may prevent any particular evil, may be beyond all comparison worse than the evil itself.

In the Friendly Islands, though the power of the king was said to be unlimited, and the life and property of the subject at his disposal; yet it appeared that some of the other chiefs acted like petty sovereigns, and frequently thwarted his measures, of which he often complained. "But however independent" (Captain Cook says) "on the despotic power of the king the great men may be, we saw instances enough to prove that the lower orders of people have no property nor safety for their persons, but at the will of the chiefs to whom they respectively belong."[178] The chiefs often beat the inferior people most unmercifully;[179] and, when any of them were caught in a theft on board the ships, their masters, far from interceding for them, would often advise the killing of them,[180] which, as the chiefs themselves appeared to have no great horror of the crime of theft, could only arise from their considering the lives of these poor people as of little or no value.

Captain Cook, in his first visit to the Sandwich Islands, had reason to think that external wars and internal commotions were extremely frequent among the natives.[181] And Captain Vancouver, in his later account, strongly notices the dreadful devastations in many of the islands from these causes. Incessant contentions had occasioned alterations in the different governments since Captain Cook's visit: Only one chief of all that were known at that time was living; and, on inquiry, it appeared that few had died a natural death, most of them having been killed in these unhappy contests.[182] The power of the chiefs over the inferior classes of the people in the Sandwich Islands appears to be absolute. The people, on the other hand, pay them the most implicit obedience; and this state of servility has manifestly a great effect in debasing both their minds and bodies.[183] The gradations of rank seem to be even more strongly marked here than in the other islands, as the chiefs of higher rank behave to those who are lower in this scale in the most haughty and oppressive manner.[184]

It is not known that either in the Friendly or Sandwich Islands infanticide is practised, or that institutions are established similar to the Eareeoie societies in Otaheite. But it seems to be stated on unquestionable authority that prostitution is extensively diffused, and prevails to a great degree among the lower classes of women;[185] which must always operate as a most powerful check to population. It seems highly probable that the *toutous,* or servants, who spend the greatest part of their time in attendance upon the chiefs,[186] do not often marry; and it is evident that the polygamy allowed to the superior people must tend greatly to encourage and aggravate the vice of promiscuous intercourse among the inferior classes.

[178] Cook's Third Voy. vol. i. p. 406.

[179] p. 232.

[180] p. 233.

[181] Cook's Third Voy. vol. ii. p. 247.

[182] Vancouver, vol. i. b. ii. c. ii. p. 187, 188.

[183] Cook's Third Voy. vol. iii. p. 157.

[184] Id.

[185] Cook's Third Voy. vol. i. p. 401. Vol. ii. p. 543. Vol. iii. p. 130. Missionary Voy. p. 270.

[186] Cook's Third Voyage, vol. i. p. 394.

Were it an established fact that in the more fertile islands of the Pacific Ocean very little or nothing was suffered from poverty and want of food, as we could not expect to find among savages in such climates any great degree of moral restraint, the theory on the subject would naturally lead us to conclude, that vice, including war, was the principal check to their population. The accounts which we have of these islands strongly confirm this conclusion. In the three great groups of islands which have been noticed, vice appears to be a most prominent feature. In Easter Island, from the great disproportion of the males to the females,[187] it can scarcely be doubted that infanticide prevails, though the fact may not have come to the knowledge of any of our navigators. Pérouse seemed to think that the women in each district were common property to the men of that district,[188] though the numbers of children which he saw[189] would rather tend to contradict this opinion. The fluctuations in the population of Easter Island appear to have been very considerable since its first discovery by Roggewein in 1722, though it cannot have been much affected by European intercourse. From the description of Pérouse it appeared, at the time of his visit, to be recovering its population, which had been in a very low state, probably either from drought, civil dissensions, or the prevalence in an extreme degree of infanticide and promiscuous intercourse. When Captain Cook visited it in his second voyage, he calculated the population at six or seven hundred,[190] Pérouse at two thousand;[191] and, from the number of children which he observed, and the number of new houses that were building, he conceived that the population was on the increase.[192]

In the Marianne Islands, according to Pere Gobien, a very great number[193] of the young men remained unmarried, living like the members of the Eareeoie society in Otaheite, and distinguished by a similar name.[194] In the island of Formosa, it is said that the women were not allowed to bring children into the world before the age of thirty-five. If they were with-child prior to that period, an abortion was effected by the priestess, and till the husband was forty years of age the wife continued to live in her father's house, and was only seen by stealth.[195]

---

[187] Id. Second Voy. vol. i, p. 289. Voyage de Pérouse, c. iv, p. 323. c. v. p. 336. 4to. 1794.

[188] Pérouse, c. iv. p. 326. c. v. p. 336.

[189] Id. c. v. p. 336.

[190] Cook's Second Voy. vol. i. p. 289.

[191] Pérouse, c. v. p. 336.

[192] Ibid.

[193] Une infinité de jeunes gens.—Hist. des Navigation aux Terres Australes, vol. ii. p. 507.

[194] Cook's Third Voyage, vol. ii. p. 158, note of the Editor.

[195] Harris's Collection of Voyages, 2 vols. folio edit. 1744, vol. i. p. 794. This relation is given by John Albert de Mandesloe, a German traveller of some reputation for fidelity, though I believe, in this instance, he takes his accounts from the Dutch writers quoted by Montesquieu (Esprit des Loix, liv. 23. ch. 17). The authority is not perhaps sufficient to establish the existence of so strange a custom; though I confess it does not appear to me wholly improbable. In the same account it is mentioned, that there is no difference of condition among these people, and that their wars are so bloodless that the

The transient visits which have been made to some other islands, and the imperfect accounts we have of them, do not enable us to enter into any particular detail of their customs; but, from the general similarity of these customs, as far as has been observed, we have reason to think that, though they may not be marked by some of the more atrocious peculiarities which have been mentioned, vicious habits with respect to women, and wars, are the principal checks to their population.

These however are not all. On the subject of the happy state of plenty, in which the natives of the South-Sea Islands have been said to live, I am inclined to think that our imaginations have been carried beyond the truth by the exuberant descriptions which have sometimes been given of these delightful spats. The not unfrequent pressure of want, even in Otaheite, mentioned in Captain Cook's last voyage, has undeceived us with regard to the most fertile of all these islands; and from the Missionary voyage it appears, that, at certain times of the year, when the bread-fruit is out of season, all temporary scarcity. At Oheitahoo, one of the Marquesas, it amounted to hunger, and the very animals were pinched for want of food. At Tongataboo, the principal of the Friendly Islands, the chiefs to secure plenty changed their abodes to other islands,[196] and, at times, many of the natives suffered much from want.[197] In the Sandwich Islands long droughts sometimes occur,[198] hogs and yams are often very scarce,[199] and visitors are received with an unwelcome austerity, very different from the profuse benevolence of Otaheite. In New Caledonia the inhabitants feed upon spiders,[200] and are sometimes reduced to eat great pieces of steatite to appease the cravings of their hunger.[201]

These facts strongly prove that, in whatever abundance the productions of these islands may be found at pertain periods, or however they may be checked by ignorance, wars and other causes, the average population, generally speaking, presses hard against the limits of the average food. In a state of society, where the lives of the inferior orders of the people seem to be considered by their superiors as of little or no value, it is evident that we are very liable to be deceived with regard to the appearances of abundance; and we may easily conceive that hogs and vegetables might be exchanged in great profusion for European

---

death of a single person generally decides them. In a very healthy climate, where the habits of the people were favourable to population and a community of goods was established, as no individual would have reason to fear *particular poverty* from a large family, the government would be in a manner compelled to take upon itself the suppression of the population by law; and, as this would be the greatest violation of every natural feeling, there cannot be a more forcible argument against a community of goods.

[196] Missionary Voy. Appen. p. 385.

[197] Id. p. 270.

[198] Vancouver's Voy. vol. ii. b. iii. c. viii. p. 230.

[199] Id. c. vii. and viii.

[200] Voyage in Search of Pérouse, ch. xiii. p. 420. Eng. transl. 4to.

[201] Id. ch. xiii. p. 400.

commodities by the principal proprietors, while their vassals and slaves were suffering severely from want.

I cannot conclude this general review of that department of human society which has been classed under the name of savage life, without observing that the only advantage in it above civilized life that I can discover, is the possession of a greater degree of leisure by the mass of the people. There is less work to be done, and consequently there is less labour. When we consider the incessant toil to which the lower classes of society in civilized life are condemned, this cannot but appear to us a striking advantage; but it is probably overbalanced by much greater disadvantages. In all those countries where provisions are procured with facility, a most tyrannical, distinction of rank prevails. Blows and violations of property seem to be matters of course; and the lower classes of the people are in a state of comparative degradation, much below what is known in civilized nations. In that part of savage life where a great degree of equality obtains, the difficulty of procuring food and the hardships of incessant war create a degree of labour not inferior to that which is exerted by the lower classes of the people in civilized society, though much more unequally divided.

But though we may compare the labour of these two classes of human society, their privations and sufferings will admit of no comparison. Nothing appears to me to place this in so striking a point of view, as the whole tenor of education among the ruder tribes of savages in America. Every thing that can contribute to teach the most unmoved patience under the severest pains and misfortunes, every thing that tends to harden the heart, and narrow all the sources of sympathy, is most sedulously inculcated on the savage. The civilized man, on the contrary, though he may be advised to bear evil with patience when it comes, is not instructed to be always expecting it. Other virtues are to be called into action besides fortitude. He is taught to feel for his neighbour, or even his enemy, in distress; to encourage and expand his social affections; and, in general, to enlarge the sphere of pleasurable emotions. The obvious inference from these two different modes of education is, that the civilized man hopes to enjoy, the savage expects only to suffer.

The preposterous system of Spartan discipline, and that unnatural absorption of every private feeling in concern for the public, which has sometimes been so absurdly admired, could never have existed but among a people exposed to perpetual hardships and privations from incessant war, and in a state under the constant fear of dreadful reverses of fortune. Instead of considering these phenomena as indicating any peculiar tendency to fortitude and patriotism in the disposition of the Spartans, I should merely consider them as a strong indication of the miserable and almost savage state of Sparta, and of Greece in general at that time. Like the commodities in a market, those virtues will be produced in the greatest quantity, for which there is the greatest demand; and where patience under pain and privations, and extravagant patriotic sacrifices, are the most called for, it is a melancholy indication of the misery of the people, and the insecurity of the state.

3.4.6 Book I, Chapter VI: Of the Checks to Population Among the Ancient Inhabitants of the North of Europe.

A history of the early migrations and settlements of mankind, with the motives which prompted them, would illustrate in a striking manner the constant tendency in the human race to increase beyond the means of subsistence. Without some general law of this nature, it would seem as if the world could never have been peopled. A state of sloth, and not of restlessness and activity, seems evidently to be the natural state of man; and this latter disposition could not have been generated but by the strong goad of necessity, though it might afterwards be continued by habit, and the new associations that were formed from it, the spirit of enterprise, and the thirst of martial glory.

We are told that Abraham and Lot had so great substance in cattle, that the land would not bear them both, that they might dwell together. There was strife between their herdsmen. And Abraham proposed to Lot to separate, and said, "Is not the whole land before thee? If thou wilt take the left hand, then I will go to the right; if thou depart to the right hand, then I will go to the left."[202]

This simple observation and proposal is a striking illustration of that great spring of action which overspread the whole earth with people; and, in the progress of time, drove some of the less fortunate inhabitants of the globe, yielding to irresistible pressure, to seek a scanty subsistence in the burning deserts of Asia and Africa, and the frozen regions of Siberia and North America. The first migrations would naturally find no other obstacles than the nature of the country; but when a considerable part of the earth had been peopled, though but thinly, the possessors of these districts would not yield them to others without a struggle; and the redundant inhabitants of any of the more central spots could not find room for themselves without expelling their nearest neighbours, or at least passing through their territories, which would necessarily give occasion to frequent contests.

The middle latitudes of Europe and Asia seem to have been occupied at an early period of history by nations of shepherds. Thucydides gave it as his opinion, that the civilized states of Europe and Asia, in his time, could not resist the Scythians united. Yet a country in pasture cannot possibly support so many inhabitants as a country in tillage. But what renders nations of shepherds so formidable, is the power which they possess of moving altogether, and the necessity they frequently feel of exerting this power in search of fresh pasture for their herds. A tribe that is rich in cattle has an immediate plenty of food. Even the parent stock may be devoured in case of absolute necessity. The women live in greater ease than among nations of hunters, and are consequently more prolific. The men, bold in their united strength, and confiding in their power of procuring pasture for their cattle by change of place, feel probably but few fears about providing for a family. These combined causes soon produce their natural and invariable effect, an extended population. A more frequent and rapid change of place then becomes necessary. A wider and more extensive

[202] Genesis, ch. xiii.

territory is successively occupied. A broader desolation extends all around them. Want pinches the less fortunate members of the society; and at length the impossibility of supporting such a number together becomes too evident to be resisted. Young scions are then pushed out from the parent stock, and instructed to explore fresh regions, and to gain happier seats for themselves by their swords. "The world is all before them where to choose."

Restless from present distress, flushed with the hope of fairer prospects, and animated with the spirit of hardy enterprise, these daring adventurers are likely to become formidable adversaries to all who oppose them. The inhabitants of countries long settled, engaged in the peaceful occupations of trade and agriculture, would not often be able to resist the energy of men acting under such powerful motives of exertion. And the frequent contests with tribes in the same circumstances with themselves, would be so many struggles for existence, and would be fought with a desperate courage, inspired by the reflection, that death would be the punishment of defeat, and life the prize of victory.

In these savage contests, many tribes must have been utterly exterminated. Many probably perished by hardships and famine. Others; whose leading star had given them a happier direction, became great and powerful tribes, and in their turn sent off fresh adventurers in search of other seats. These would at first owe allegiance to their parent tribe; but in a short time the ties which bound them would be little felt, and they would remain friends, or become enemies, according as their power, their ambition or their convenience, might dictate.

The prodigious waste of human life, occasioned by this perpetual struggle for room and food, would be more than supplied by the mighty power of population, acting in some degree unshackled from the constant habit of migration. A prevailing hope of bettering their condition by change of place, a constant expectation of plunder, a power even, if distressed, of selling their children as slaves, added to the natural carelessness of the barbaric character, would all conspire to raise a population, which would remain to be repressed afterwards by famine and war.

The tribes that possessed themselves of the more fruitful regions, though they might win them and maintain them by continual battles, rapidly increased in number and power, from the increased means of subsistence; till at length the whole territory, from the confines of China to the shores of the Baltic, was peopled by a various race of barbarians, brave, robust, and enterprising, inured to hardships, and delighting in war.[203] While the different fixed governments of Europe and Asia, by superior population and superior skill, were able to oppose an impenetrable barrier to their destroying hordes, they wasted their superfluous

[203] The various branchings, divisions; and contests of the great Tartar nation are curiously described in the Genealogical History of the Tartars by the Khan Abul Ghazi: (translated into English from the French, with additions, in 2 vols. 8vo.) but the misfortune of all history is, that while the particular motives of a few princes and leaders, in their various projects of ambition, are sometimes detailed with accuracy, the general causes which crowd their standards with willing followers are often entirely overlooked.

numbers in contests with each other; but the moment that the weakness of the settled governments, or the casual union of many of these wandering tribes, gave them the ascendant in power, the storm discharged itself on the fairest provinces of the earth; and China, Persia, Egypt and Italy were overwhelmed at different periods in this flood of barbarism.

These remarks are strongly exemplified in the fall of the Roman empire. The shepherds of the north of Europe were long held in check by the vigour of the Roman arms, and the terror of the Roman name. The formidable irruption of the Cimbri in search of new settlements, though signalized by the destruction of five consular armies, was at length arrested in its victorious career by Marius; and the barbarians were taught to repent their rashness by the almost complete extermination of this powerful colony.[204] The names of Julius Cæsar, of Drusus, Tiberius, and Germanicus, impressed on their minds by the slaughter of their countrymen, continued to inspire them with a fear of encroaching on the Roman territory. But they were rather triumphed over than vanquished;[205] and though the armies or colonies which they sent forth were either cut off or forced back into their original seats, the vigour of the great German nation remained unimpaired, and ready to pour forth her hardy sons in constant succession, wherever they could force an opening for themselves by their swords. The feeble reigns of Decius, Gallus, Æmilianus, Valerian, and Gallienus, afforded such an opening, and were in consequence marked by a general irruption of barbarians. The Goths, who were supposed to have migrated in the course of some years from Scandinavia to the Euxine, were bribed to withdraw their victorious troops by an annual tribute. But no sooner was the dangerous secret of the wealth and weakness of the Roman empire thus revealed to the world, than new swarms of barbarians spread devastation through the frontier provinces, and terror as far as the gates of Rome.[206] The Franks, the Allemanni, the Goths, and adventurers of less considerable tribes, comprehended under these general appellations, poured like a torrent on different parts of the empire. Rapine and oppression destroyed the produce of the present and the hope of future harvests. A long and general famine was followed by a wasting plague, which for fifteen years ravaged every city and province of the Roman empire; and, judging from the mortality in some spots, it was conjectured that in a few years war, pestilence, and famine, had consumed the moiety of the human species.[207] Yet the tide of emigration still continued at intervals to roll impetuously from the north; and the succession of martial princes, who repaired the misfortunes of their predecessors, and propped the falling fate of the empire, had to accomplish the labours of Hercules in freeing the Roman territory from these barbarous invaders. The Goths, who, in the year 250 and the following years, ravaged the empire both by sea and land with various success, but in the end with the almost total

[204]Tacitus de Moribus Germanorum, s. 37.

[205]Id. ib.

[206]Gibbon's Decline and Fall of the Roman Empire, vol. i. c. x. p. 407, et seq. 8vo. Edit. 1783.

[207]Id. vol. i. c. x. p. 455, 456.

loss of their adventurous bands,[208] in the year 269 sent out an emigration of immense numbers, with their wives and families, for the purpose of settlement.[209] This formidable body, which was said to consist at first of 320,000 barbarians,[210] was ultimately destroyed and dispersed by the vigour and wisdom of the emperor Claudius. His successor, Aurelian, encountered and vanquished new hosts of the same name that had quitted their settlements in the Ukraine; but one of the implied conditions of the peace was, that he should withdraw the Roman forces from Dacia, and relinquish this great province to the Goths and Vandals.[211] A new and most formidable invasion of the Allemanni threatened soon after to sack the mistress of the world, and three great and bloody battles were fought by Aurelian before this destroying host could be exterminated, and Italy be delivered from its ravages.[212]

The strength of Aurelian had crushed on every side the enemies of Rome. After his death they seemed to revive with an increase of fury and numbers. They were again vanquished on all sides by the active vigour of Probus. The deliverance of Gaul alone from the German invaders is reported to have cost the lives of four hundred thousand barbarians.[213] The victorious emperor pursued his successes into Germany itself; and the princes of the country, astonished at his presence, and dismayed and exhausted by the ill success of their last emigration, submitted to any terms that the conquerors might impose.[214] Probus, and afterwards Diocletian[215] adopted the plan of recruiting the exhausted provinces of the empire by granting lands to the fugitive or captive barbarians, and disposing of their superfluous numbers where they might be the least likely to be dangerous to the state; but such colonizations were an insufficient vent for the population of the north, and the ardent temper of the barbarians would not always bend to the slow labours of agriculture.[216] During the vigorous reign of Diocletian, unable to make an effectual impression on the Roman frontiers, the Goths, the Vandals, the Gepidæ, the Burgundians, and the Allemanni, wasted each other's strength by mutual hostilities, while the subjects of the empire enjoyed the bloody spectacle, conscious that, whoever vanquished, they vanquished the enemies of Rome.[217]

Under the reign of Constantine the Goths were again formidable. Their strength had been restored by a long peace, and a new generation had arisen, which no longer

[208] Id. vol. i. c. x. p. 431.

[209] Gibbon, vol. ii. c. xi. p. 13.

[210] Id. p. 11.

[211] Id. p. 19, A. D. 270.

[212] Id. p. 26.

[213] Id. vol. ii. c. xii. p. 75.

[214] Gibbon, vol. ii. c. xii. p. 79, A. D. 277.

[215] Id. c. xiii. p. 132, A. D. 296.

[216] Id. c. xii. p. 84.

[217] Id. c. xiii. p. 130.

remembered the misfortunes of ancient days.[218] In two successive wars great numbers of them were slain. Vanquished on every side, they were driven into the mountains; and, in the course of a severe campaign, above a hundred thousand were computed to have perished by cold and hunger.[219] Constantine adopted the plan of Probus and his successors in granting lands to those suppliant barbarians who were expelled from their own country. Towards the end of his reign, a competent portion, in the provinces of Pannonia, Thrace, Macedonia, and Italy, was assigned for the habitation and subsistence of three hundred thousand Sarmatians.[220]

The warlike Julian had to encounter and vanquish new swarms of Franks and Allemanni, who, emigrating from their German forests during the cival wars of Constantine, settled in different parts of Gaul, and made the scene of their devastations three times more extensive than that of their conquests.[221] Destroyed and repulsed on every side, they were pursued in five expeditions into their own country;[222] but Julian had conquered, as soon as he had penetrated into Germany; and in the midst of that mighty hive, which had sent out such swarms of people as to keep the Roman world in perpetual dread, the principal obstacles to his progress were almost impassable roads and vast unpeopled forests.[223]

Though thus subdued and prostrated by the victorious arms of Julian, this hydra-headed monster rose again after a few years; and the firmness, vigilance and powerful genius of Valentinian were fully called into action, in protecting his dominions from the different irruptions of the Allemanni, the Burgundians, the Saxons, the Goths, the Quadi, and the Sarmatians.[224]

The fate of Rome was at length determined by an irresistible emigration of the Huns from the east and north, which precipitated on the empire the whole body of the Goths;[225] and the continuance of this powerful pressure on the nations of Germany seemed to prompt them to the resolution of abandoning to the fugitives of Sarmatia their woods and morasses, or at least of discharging their superfluous numbers on the provinces of the Roman empire.[226] An emigration of four hundred thousand persons issued from the same coast of the Baltic, which had poured forth the myriads of Cimbri and Teutones during the vigour of the Republic.[227] When this host was destroyed by war and famine, other adventurers

---

218 Id. c. xiv. p. 254, A. D. 322.

219 Gibbon, vol. iii. c. xviii. p. 125, A. D. 332.

220 Id. p. 127.

221 Id. c. xix. p. 215, A. D. 356.

222 Id. p. 228, and vol. iv. c. xxii. p. 17, from A. D. 357 to 359.

223 Id. vol. iv. c. xxii. p. 17, and vol. iii. c. xix. p. 229.

224 Gibbon vol. iv. c. xxv. from A. D. 364 to 375.

225 Id. vol. iv. c. xxvi. p. 382, et seq. A. D. 376.

226 Id. vol. v. c. xxx. p. 213.

227 Id. p. 214. A. D. 406.

succeeded. The Suevi, the Vandals, the Alani, the Burgundians, passed the Rhine, never more to retreat.[228] The conquerors, who first settled, were expelled or exterminated by new invaders. Clouds of barbarians seemed to collect from all parts of the northern hemisphere. Gathering fresh darkness and terror as they rolled on, the congregated bodies at length obscured the sun of Italy, and sunk the western world in night.

In two centuries from the flight of the Goths across the Danube, barbarians of various names and lineage had plundered and taken possession of Thrace, Pannonia, Gaul, Britain, Spain, Africa, and Italy.[229] The most horrible devastations and an incredible destruction of the human species accompanied these rapid conquests; and famine and pestilence, which always march in the train of war when it ravages with such inconsiderate cruelty, raged in every part of Europe. The historians of the times, who beheld these scenes of desolation, labour and are at a loss for expressions to describe them; but, beyond the power of language, the numbers and the destructive violence of these barbarous invaders were evinced by the total change which took place in the state of Europe.[230] These tremendous effects, so long and so deeply felt throughout the fairest portions of the earth, may be traced in a great degree to the simple cause of the superiority of the power of population to the means of subsistence.

Machiavel, in the beginning of his History of Florence, says, "The people who inhabit the northern parts that lie between the Rhine and the Danube, living in a healthful and prolific climate, often increase to such a degree, that vast numbers of them are forced to leave their native country and go in search of new habitations. When any of those provinces begins to grow too populous and wants to disburden itself, the following method is observed. In the first place, it is divided into three parts, in each of which there is an equal portion of the nobility and commonalty, the rich and the poor. After this they cast lots; and that division on which the lot falls, quits the country and goes to seek its fortune, leaving the other two more room and liberty to enjoy their possessions at home. These emigrations proved the destruction of the Roman Empire."[231] Gibbon is of opinion that Machiavel has represented these emigrations too much as regular and concerted measures;[232] but I think it highly probable that he had not erred much in this respect, and that it was a foresight of the frequent necessity of thus discharging their redundant

---

[228] Id. p. 224.

[229] Robertson's Charles V. vol. i. sect. i. p. 7. 8vo. 1782.

[230] Id. p. 10, 11, 12.

[231] Istorie Fiorentine Machiavelli, l. i. p. 1, 2.

[232] Gibbon, vol. i. c. ix. p 360, note. Paul Diaconus, from whom it is supposed that Machiavel has taken this description, writes thus:—Septentrionalis plaga quantò magis ab æstu solis remota est et nivali frigore gelida, tantò salubrior corporibus hominum et propagandis gentibus magis coaptata. Sicut è contrario, omnis meridiana regio, quò solis est fervori vicinior, eò morbis est abundantior, et educandis minus apta mortalibus. **** Multæque quoque ex eâ, eò quod tantas mortalium turmas germinat, quantas alere vix sufficit, sæpe gentes egressæ sunt, quæ non solum partes Asiæ, sed etiam maxime sibi contiguam Enropam afflixere. (De Gestis Longobardorum, l. i. c. i.)

population, which gave occasion to that law among the Germans, taken notice of by Cæsar and Tacitus, of not permitting their cultivated lands to remain longer than a year under the same possessors.[233] The reasons, which, Cæsar mentions as being assigned for this custom, seem to be hardly adequate; but if we add to them the prospect of emigration in the manner described by Machiavel, the custom will appear to be highly useful, and a double weight will be given to one of the reasons that Cæsar mentions; namely, lest they should be led, by being accustomed to one spot, to exchange the toils of war for the business of agriculture.[234]

Gibbon very justly rejects, with Hume and Robertson, the improbable supposition that the inhabitants of the north were far more numerous formerly than at present;[235] but he thinks himself obliged at the same time to deny the strong tendency to increase in the northern nations,[236] as if the two facts were necessarily connected. For a careful distinction should always be made, between a redundant population and a population actually great. The Highlands of Scotland are probably more redundant in population than any other part of Great Britain; and though it would be admitting a palpable absurdity to allow that the north of Europe, covered in early ages with immense forests, and inhabited by a race of people who supported themselves principally by their herds and flocks,[237] was more populous in those times than in its present state; yet the facts detailed in the Decline and Fall of the Roman Empire, or even the very slight sketch of them that I have given, cannot rationally be accounted for, without the supposition of a most powerful tendency in these people to increase, and to repair their repeated losses by the prolific power of nature.

From the first irruption of the Cimbri, to the final extinction of the western empire, the efforts of the German nations to colonize or plunder were unceasing.[238] The numbers that were cut off during this period by war and famine were almost incalculable, and such as could not possibly have been supported with undiminished vigour by a country thinly peopled, unless the stream had been supplied by a spring of very extraordinary power.

Gibbon describes the labours of Valentinian in securing the Gallic frontier against the Germans; an enemy, he says, whose strength was renewed by a stream of daring volunteers which incessantly flowed from the most distant tribes of the north.[239] An easy adoption of strangers was probably a mode, by which some of the German nations renewed their strength so suddenly,[240] after the most destructive defeats; but this explanation only

[233] De Bello Gallico, vi. 22. De Moribus German. s. xxvi.

[234] De Bello Gallico, vi. 22.

[235] Gibbon, vol. i. c. ix. p. 361.

[236] Id. p. 348.

[237] Tacitus de Moribus German. sect. v.; Cæsar de Bell. Gall. vi. 22.

[238] Cæsar found in Gaul a most formidable colony under Ariovistus, and a general dread prevailing that in a few years all the Germans would pass the Rhine. De Bell. Gall. i. 31.

[239] Gibbon, vol. iv. c. xxv. p. 283.

[240] Id. ib. note.

removes the difficulty a little further off. It makes the earth rest upon the tortoise; but does not tell us on what the tortoise rests. We may still ask what northern reservoir supplied this incessant stream of daring adventurers? Montesquieu's solution of the problem will, I think, hardly be admitted. The swarms of barbarians which issued formerly from the north, appear no more, he says, at present; and the reason he gives is, that the violence of the Romans had driven the people of the south into the north, who, as long as this force continued, remained there; but as soon as it was weakened, spread themselves again over every country.

The same phenomenon appeared after the conquests and tyrannies of Charlemagne and the subsequent dissolution of his empire; and if a prince, he says, in the present days were to make similar ravages in Europe, the nations driven into the north, and resting on the limits of the universe,[241] would there make a stand till the moment when they would inundate or conquer Europe a third time. In a note he observes, "we see to what the famous question is reduced—why the north is no longer so fully peopled as in former times?"

If the famous question, or rather the answer to it, be reduced to this, it is reduced to a miracle; for without some supernatural mode of obtaining food, how these collected nations could support themselves in such barren regions for so long a period as during the vigour of the Roman empire, it is a little difficult to conceive; and one can hardly help smiling at the bold figure of these prodigious crowds making their last determined stand on the limits of the universe, and living, as we must suppose, with the most patient fortitude on air and ice for some hundreds of years, till they could return to their own homes and resume their usual more substantial mode of subsistence.

The whole difficulty, however, is at once removed, if we apply to the German nations at that time a fact which is so generally known to have occurred in America, and suppose that, when not checked by wars and famine, they increased at a rate that would double their numbers in twenty-five or thirty years. The propriety, and even the necessity, of applying this rate of increase to the inhabitants of ancient Germany will strikingly appear from that most valuable picture of their manners which has been left us by Tacitus. He describes them as not inhabiting cities, or even of admitting of contiguous settlements. Every person surrounds his house with a vacant space;[242] a circumstance, which besides its beneficial effect as a security from fire, is strongly calculated to prevent the generation, and check the ravages, of epidemics. "They content themselves almost universally with one wife. Their matrimonial bond is strict and severe, and their manners in this respect deserving of the highest praise.[243] They live in a state of well-guarded chastity, corrupted by no seducing spectacles or convivial incitements. Adultery is extremely rare, and no indulgence is shewn to a prostitute. Neither beauty, youth, nor riches, can procure her a husband: for none there looks on vice with a smile, or calls mutual seduction the way of the world. To limit the

[241] Les nations adossées aux limites de l'univers y tiendroient ferme. Grandeur et Décad. des Rom. c. xvi. p. 187.

[242] Tacitus de Moribus Germ. s. xvi.

[243] Id. s. xviii.

increase of children, or put to death any of the husband's blood, is accounted infamous; and virtuous manners have there more efficacy than good laws elsewhere.[244] Every mother suckles her own children, and does not deliver them into the hands of servants and nurses. The youths partake late of the sexual intercourse, and hence pass the age of puberty unexhausted. Nor are the virgins brought forward. The same maturity, the same full growth, is required; the sexes unite equally matched and robust, and the children inherit the vigour of their parents. The more numerous are a man's kinsmen and relations, the more comfortable is his old age; nor is it any advantage to be childless."[245]

With these manners, and a habit of enterprise and emigration, which would naturally remove all fears about providing for a family, it is difficult to conceive a society with a stronger principle of increase; and we see at once that prolific source of successive armies and colonies, against which the force of the Roman empire so long struggled with difficulty, and under which it ultimately sunk. It is not probable that, for two periods together, or even for one, the population within the confines of Germany ever doubled itself in twenty-five years. Their perpetual wars, the rude state of agriculture, and particularly the very strange custom adopted by most of the tribes of marking their barriers by extensive deserts,[246] would prevent any very great actual increase of numbers. At no one period could the country be called well-peopled, though it was often redundant in population. They abandoned their immense forests to the exercise of hunting, employed in pasturage the most considerable part of their lands, bestowed on the small remainder a rude and careless cultivation, and when the return of famine severely admonished them of the insufficiency of their scanty resources, they accused the sterility of a country which refused to supply the multitude of its inhabitants;[247] but instead of clearing their forests, draining their swamps, and rendering their soil fit to support an extended population, they found it more congenial to their martial habits and impatient dispositions, "to go in quest of food, of plunder, or of glory,"[248] into other countries. These adventurers either gained lands for themselves by their swords or were cut off by the various accidents of war; were received into the Roman armies or dispersed over the Roman territory; or, perhaps, having relieved their country by their absence, returned home laden with spoils, and ready, after having recruited their diminished numbers, for fresh expeditions. The succession of human beings appears to have been most rapid; and as fast as some were disposed of in colonies, or mowed down by the scythe of war and famine, others rose in increased numbers to supply their place.

According to this view of the subject, the North could never have been exhausted; and when Dr. Robertson, describing the calamities of these invasions, says, that they did not cease till the North, by pouring forth successive swarms, was drained of people, and could

[244] Id. s. xix.

[245] Tacitus de Moribus Germ. s. xx.

[246] Cæsar de Bell. Gall, vi. 23.

[247] Gibbon, vol. i. c. ix. p. 360.

[248] Id. vol. i. c. x. p. 417.

no longer furnish instruments of destruction,[249] he will appear to have fallen into the very error which he had before laboured to refute, and to speak as if the northern nations were actually very populous. For they must have been so, if the number of their inhabitants at any one period had been sufficient, notwithstanding the slaughter of war, to people in such a manner Thrace, Pannonia, Gaul, Spain, Africa, Italy and England, as in some parts not to leave many traces of their former inhabitants. The period of the peopling of these countries, however, he himself mentions as two hundred years;[250] and in such a time new generations would arise that would more than supply every vacancy.

The true cause which put a stop to the continuance of northern emigration, was the impossibility any longer of making an impression on the most desirable countries of Europe. They were then inhabited by the descendants of the bravest and most enterprising of the German tribes; and it was not probable that they should so soon degenerate from the valour of their ancestors, as to suffer their lands to be wrested from them by inferior numbers and inferior skill, though perhaps superior hardihood.

Checked for a time by the bravery and poverty of their neighbours by land, the enterprising spirit and overflowing numbers of the Scandinavian nations soon found vent by sea. Feared before the reign of Charlemagne, they were repelled with difficulty by the care and vigour of that great prince; but during the distractions of the empire under his feeble successors, they spread like a devouring flame over Lower Saxony, Friezeland, Holland, Flanders, and the banks of the Rhine as far as Mentz.

After having long ravaged the coasts, they penetrated into the heart of France, pillaged and burnt her fairest towns, levied immense tributes on her monarchs, and at length obtained by grant one of the finest provinces in the kingdom. They made themselves even dreaded in Spain, Italy and Greece, spreading every where desolation and terror. Sometimes they turned their arms against each other, as if bent on their own mutual destruction; at other times they transported colonies to unknown or uninhabited countries, as if they were willing to repair in one place the horrid destruction of the human race occasioned by their furious ravages in another.[251]

The mal-administration and civil wars of the Saxon kings of England produced the same effect as the weakness which followed the reign of Charlemagne in France;[252] and for two hundred years the British isles were incessantly ravaged, and often in part subdued, by these northern invaders. During the eighth, ninth and tenth centuries, the sea was covered with their vessels from one end of Europe to the other;[253] and the countries now the most powerful in arts and arms were the prey of their constant depredations. The growing and consolidating strength of these countries at length removed all further prospect of success

[249] Robertson's Charles V. vol. i. s. i. p. 11.

[250] Id. p. 7.

[251] Mallet, Introd. à l'Histoire de Dannemarc, tom, i. c. x. p. 221, 223, 224. 12mo. 1766.

[252] Id. p. 226.

[253] Mallet, Introd. à l'Histoire de Dannemarc, tom. i. c. x. p. 221.

from such invasions.[254] The nations of the north were slowly and reluctantly compelled to confine themselves within their natural limits and to exchange their pastoral manners, and with them the peculiar facilities of plunder and emigration which they afforded, for the patient labours and slow returns of trade and agriculture. But the slowness of these returns necessarily effected an important change in the manners of the people.

In ancient Scandinavia, during the time of its constant wars and emigrations, few, or none probably, were ever deterred from marrying by the fear of not being able to provide for a family. In modern Scandinavia, on the contrary, the frequency of the marriage union is continually checked by the most imperious and justly-founded apprehensions of this kind. This is most particularly the case in Norway, as I shall have occasion to remark in another place; but the same fears operate in a greater or less degree, though every where with considerable force, in all parts of Europe. Happily the more tranquil state of the modern world does not demand such rapid supplies of human beings; and the prolific powers of nature cannot therefore be so generally called into action.

Mallet, in the excellent account of the northern nations which he has prefixed to his History of Denmark, observes that he had not been able to discover any proofs that their emigrations proceeded from want of room at home;[255] and one of the reasons which he gives, is, that after a great emigration the countries often remained quite deserted and unoccupied for a long time.[256] But instances of this kind, I am inclined to think, were rare, though they might occasionally happen. With the habits of enterprise and emigration which prevailed in those days, a whole people would sometimes move in search of a more fertile territory. The lands, which they before occupied, must of necessity be left desert for a time; and if there were any thing particularly ineligible in the soil or situation, which the total emigration of the people would seem to imply, it might be more congenial to the temper of the surrounding barbarians, to provide for themselves better by their swords than to occupy immediately these rejected lands. Such total emigrations proved the unwillingness to divide; but by no means that they were not straitened for room and food at home.

The other reason, which Mallet gives, is that in Saxony, as well as Scandinavia, vast tracts of land lay in their original uncultivated state, having never been grabbed up or cleared; and that, from the descriptions of Denmark in those times, it appeared that the coasts alone were peopled; but the interior parts formed one vast forest.[257] It is evident that he here falls into the common error of confounding a superfluity of inhabitants with great actual population. The pastoral manners of the people and their habits of war and enterprise

[254] Perhaps the civilized world could not be considered as perfectly secure from another northern or eastern inundation, till the total change in the art of war, by the introduction of gunpowder, gave to improved skill and knowledge the decided advantage over physical force.

[255] Hist. Dan. tom. i. c. ix. p. 206.

[256] Id. p. 205, 206.

[257] Hist. Dan. tom. i. c. ix. p. 207.

prevented them from clearing and cultivating their lands;[258] and then these very forests, by restraining the sources of subsistence within very narrow bounds, contributed to superfluity of numbers; that is, to a population beyond what the scanty supplies of the country could support.

There is another cause not often attended to, why poor, cold and thinly-peopled countries, tend generally to a superfluity of inhabitants, and are strongly prompted to emigration. In warmer and more populous countries, particularly those abounding in great towns and manufactures, an insufficient supply of food can seldom continue long without producing epidemics either in the shape of great and ravaging plagues, or of less violent, though more constant, sicknesses. In poor, cold and thinly-peopled countries, on the contrary, from the antiseptic quality of the air, the misery arising from insufficient or bad food may continue for a considerable time without producing these effects; and consequently this powerful stimulus to emigration continues to operate for a much longer period.[259]

I would by no means, however, be understood to say, that the northern nations never undertook any expeditions, unless prompted by straitened food or circumstances at home. Mallet relates, what was probably true, that it was their common custom to hold an assembly every spring, for the purpose of considering in what quarter they should make war;[260] and among a people who nourished so strong a passion for war, and who considered the right of the strongest as a right divine, occasions for it would never be wanting. Besides this pure and disinterested love of war and enterprise, civil dissensions, the pressure of a victorious enemy, a wish for a milder climate, or other causes, might sometimes prompt to emigration; but, in a general view of the subject, I cannot help considering this period of history as affording a very striking illustration of the principle of population; a principle, which appears to me to have given the original impulse and spring of action, to have furnished the inexhaustible resources, and often prepared the immediate causes of that rapid succession of adventurous irruptions and emigrations, which occasioned the fall of the Roman empire; and afterwards, pouring from the thinly-peopled countries of Denmark and Norway for above two hundred years, ravaged and overran a great part of Europe. Without the supposition of a tendency to increase almost as

---

[258] Nec arare terram aut expectare annum tarsi facile persuaseris, quam vocare hostes et vulnera mereri; pigrum quinimò et iners videtur sudore acquirere quod possis sanguine parare. Tacitus de Mor. Germ. Nothing, indeed, in the history of mankind, is more evident than the extreme difficulty with which habits are changed; and no argument therefore can be more fallacious than to infer that those people are not pinched with want, who do not make a proper use of their lands.

[259] Epidemics return more or less frequently, according to their various soils, situations, air, &c. Hence some return yearly, as in Egypt and Constantinople; others once in four or five years, as about Tripoli and Aleppo; others, scarce once in ten, twelve or thirteen years, as in England; others not in less than twenty years, as in *Norway and the Northern Islands.* Short, History of Air, Seasons; &c. vol. ii. p. 344.

[260] Hist. Dan. c. ix. p. 209.

great as in the United States of America, the facts appear to me not to be accounted for;[261] and with such a supposition, we cannot be at a loss to name the checks to the actual population, when we read the disgusting details of those unceasing wars, and of that prodigal waste of human life, which marked these barbarous periods.

Inferior checks would undoubtedly concur; but we may safely pronounce, that among the shepherds of the North of Europe war and famine were the principal checks that kept the population down to the level of their scanty means of subsistence.

3.4.7 Book I, Chapter VII: Of the Checks to Population Among Modern Pastoral Nations.

The pastoral tribes of Asia, by living in tents and moveable huts, instead of fixed habitations, are still less connected with their territory than the shepherds of the North of Europe. The camp, and not the soil, is the native country of the genuine Tartar. When the forage of a certain district is consumed the tribe makes a regular march to fresh pastures. In the summer it advances towards the north, in the winter returns again to the south; and thus in a time of most profound peace acquires the practical and familiar knowledge of one of the most difficult operations of war. Such habits would strongly tend to diffuse among these wandering tribes the spirit of emigration and conquest. The thirst of rapine, the fear of a too-powerful neighbour, or the inconvenience of scanty pastures, have in all ages been sufficient causes to urge the hordes of Scythia boldly to advance into unknown countries, where they might hope to find a more plentiful subsistence or a less formidable enemy.[262]

In all their invasions, but more particularly when directed against the civilized empires of the south, the Scythian shepherds have been uniformly actuated by a most savage and destructive spirit. When the Moguls had subdued the northern provinces of China, it was proposed, in calm and deliberate council, to exterminate all the inhabitants of that populous country, that the vacant land might be converted to the pasture of cattle. The execution of this horrid design was prevented by the wisdom and firmness of a Chinese mandarin;[263] but the bare proposal of it exhibits a striking picture, not only of the inhuman manner in which the rights of conquest were abused, but of the powerful force of habit among nations of shepherds, and the consequent difficulty of the transition from the pastoral to the agricultural state.

To pursue, even in the most cursory manner, the tide of emigration and conquest in Asia, the rapid increase of some tribes, and the total extinction of others, would lead much too

[261] Gibbon, Robertson and Mallet seem all rather to speak of Jornandes's expression *vagina nationum* as incorrect and exaggerated; but to me it appears exactly applicable, though the other expression, officina gentium, at least their translation of it, *storehouse of nations,* is not accurate.

Book I, Chapter VII.

[262] Gibbon, vol. iv. c. xxvi. p. 348.

[263] Gibbon, vol. vi. c. xxxiv. p. 54.

far. During the periods of the formidable irruptions of the Huns, the wide-extended invasions of the Moguls and Tartars, the sanguinary conquests of Attila, Zingis Khan and Tamerlane, and the dreadful convulsions which attended the dissolution as well as the formation of their empires, the checks to population are but too obvious. In reading of the devastations of the human race in those times, when the slightest motive of caprice or convenience often involved a whole people in indiscriminate massacre,[264] instead of looking for the causes which prevented a further progress in population, we can only be astonished at the force of that principle of increase, which could furnish fresh harvests of human beings for the scythe of each successive conqueror. Our inquiries will be more usefully directed to the present state of the Tartar nations, and the ordinary checks to their increase, when not under the influence of these violent convulsions.

The immense country, inhabited at present by those descendants of the Moguls and Tartars, who retain nearly the same manners as their ancestors, comprises in it almost all the middle regions of Asia, and possesses the advantage of a very fine and temperate climate. The soil is in general of great natural fertility. There are comparatively but few genuine deserts. The wide extended plains without a shrub, which have sometimes received that appellation, and which the Russians call steppes, are covered with a luxuriant grass, admirably fitted for the pasture of numerous herds and flocks. The principal defect of this extensive country is a want of water; but it is said that the parts which are supplied with this necessary article, would be sufficient for the support of four times the number of its present inhabitants, if it were properly cultivated.[265] Every Orda, or tribe, has a particular canton belonging to it, containing both its summer and winter pastures; and the population of this vast territory, whatever it may be, is probably distributed over its surface nearly in proportion to the degree of actual fertility in the different districts.

Volney justly describes this necessary distribution in speaking of the Bedoweens of Syria. "In the barren cantons, that is, those which are ill furnished with plants, the tribes are feeble and very distant from each other, as in the desert of Suez, that of the Red Sea, and the interior part of the Great Desert. When the soil is better covered, as between Damascus and the Euphrates, the tribes are stronger and less distant. And in the cultivable cantons, as the Pachalic of Aleppo, the Hauran, and the country of Gaza, the encampments are numerous and near each other."[266] Such a distribution of inhabitants, according to the quantity of food which they can obtain in the actual state of their industry and habits, may be applied to Grand Tartary, as well as to Syria and Arabia, and is, in fact, equally applicable to the whole earth, though the commerce of civilized nations prevents it from being so obvious as in the more simple stages of society.

The Mahometan Tartars, who inhabit the western parts of Grand Tartary, cultivate some of their lands, but in so slovenly and insufficient a manner as not to afford a principal source

---

[264] Id. p. 55.

[265] Geneal. Hist. of Tartars, vol. ii. sec. i. 8vo. 1730.

[266] Voy. de Volney, tom. i. ch. xxii. p: 351. 8vo. 1787.

of subsistence.[267] The slothful and warlike genius of the barbarian every where prevails, and he does not easily reconcile himself to obtaining by labour what he can hope to acquire by rapine. When the annals of Tartary are not marked by any signal wars and revolutions, its domestic peace and industry are constantly interrupted by petty contests and mutual invasions for the sake of plunder. The Mahometan Tartars are said to live almost entirely by robbing and preying upon their neighbours, as well in peace as in war.[268]

The Usbecks, who possess as masters the kingdom of Chowarasm, leave to their tributary subjects, the Sarts and Turkmans, the finest pastures of their country, merely because their neighbours on that side are too poor or too vigilant to give them hopes of successful plunder. Rapine is their principal resource. They are perpetually making incursions into the territories of the Persians, and of the Usbecks of Great Bucharia; and neither peace nor truce can restrain them, as the slaves and other valuable effects which they carry off form the whole of their riches. The Usbecks and their subjects the Turkmans are perpetually at variance; and their jealousies, fomented often by the princes of the reigning house, keep the country in a constant state of intestine commotion.[269] The Turkmans are always at war with the Curds and the Arabs, who often come and break the horns of their herds, and carry away their wives and daughters.[270]

The Usbecks of Great Bucharia are reckoned the most civilized of all the Mahometan Tartars, yet are not much inferior to the rest in their spirit of rapine.[271] They are always at war with the Persians, and laying waste the fine plains of the province of Chorasan. Though the country which they possess is of the greatest natural fertility, and some of the remains of the ancient inhabitants practise the peaceful arts of trade and agriculture; yet neither the aptitude of the soil, nor the example which they have before them, can induce them to change their ancient habits; and they would rather pillage, rob, and kill their neighbours, than apply themselves to improve the benefits which nature so liberally offers them.[272]

The Tartars of the Casatshia Orda in Turkestan live in a state of continual warfare with their neighbours to the north and east. In the winter they make their incursions towards the Kalmucks, who, about that time, go to scour the frontiers of Great Bucharia and the parts to the south of their country. On the other side they perpetually incommode the Cosacks of the Yaik and the Nogai Tartars. In the summer they cross the mountains of Eagles, and make inroads into Siberia. And though they are often very ill treated in these incursions, and the whole of their plunder is not equivalent to what they might obtain with very little labour from their lands, yet they choose rather to expose themselves to the thousand fatigues and

[267] Geneal. Hist. Tart. vol. ii. p. 382.

[268] Geneal. Hist. Tart. vol. ii. p. 390.

[269] Id. p. 430, 431.

[270] Id. p. 426.

[271] Geneal. Hist. Tart. vol. ii. p. 459.

[272] Id. p. 455.

dangers necessarily attendant on such a life, than apply themselves seriously to agriculture.[273]

The mode of life among the other tribes of Mahometan Tartars presents the same uniform picture, which it would be tiresome to repeat, and for which therefore I refer the reader to the Genealogical History of the Tartars and its valuable notes. The conduct of the author of this history himself, a Chan of Chowarasm, affords a curious example of the savage manner in which the wars of policy, of revenge, or plunder, are carried on in these countries. His invasions of Great Bucharia were frequent; and each expedition was signalized by the ravage of provinces and the utter ruin and destruction of towns and villages. When at any time the number of his prisoners impeded his motions, he made no scruple to kill them on the spot. Wishing to reduce the power of the Turkmans who were tributary to him, he invited all the principal people to a solemn feast, and had them massacred to the number of two thousand. He burnt and destroyed their villages with the most unsparing cruelty, and committed such devastations that the effect of them returned on their authors, and the army of the victors suffered severely from dearth.[274]

The Mahometan Tartars in general hate trade, and make it their business to spoil all the merchants who fall into their hands.[275] The only commerce which is countenanced, is the commerce in slaves. These form a principal part of the booty which they carry off in their predatory incursions, and are considered as a chief source of their riches. Those which they have occasion for themselves, either for the attendance on their herds, or as wives and concubines, they keep, and the rest they sell.[276] The Circassian and Daghestan Tartars, and the other tribes in the neighbourhood of Caucasus, living in a poor and mountainous country, and on that account less subject to invasion, generally overflow with inhabitants; and when they cannot obtain slaves in the common way, steal from one another, and even sell their own wives and children.[277] This trade in slaves, so general among the Mahometan Tartars, may be one of the causes of their constant wars; as, when a prospect of a plentiful supply for this kind of traffic offers itself, neither peace nor alliance can restrain them.[278]

The heathen Tartars, the Kalmucks and Moguls, do not make use of slaves, and are said in general to lead a much more peaceable and harmless life, contenting themselves with the produce of their herds and flocks, which form their sole riches. They rarely make war for the sake of plunder; and seldom invade the territory of their neighbours, unlest to revenge a

---

[273] Geneal. Hist. Tart. vol. ii. p. 573, et seq.

[274] Id. vol. i. ch. xii.

[275] Geneal. Hist. Tart. vol. ii. p. 412.

[276] Id. vol. ii. p. 413.

[277] Id. p. 413, 414, and ch. xii.

[278] "They justify it as lawful to have many wives, because they say they bring us many children, which we can sell for ready-money, or exchange for necessary conveniencies; yet when they have not wherewithal to maintain them, they hold it a piece of charity to murder infants new-born, as also they do such as are sick and past recovery, because they say they free them from a great deal of misery." Sir John Chardin's Travels, Harris's Col. b. iii. c. ii. p. 865.

prior attack. They are not however without destructive wars. The inroads of the Mahometan Tartars oblige them to constant defence and retaliation; and feuds subsist between the kindred tribes of the Kalmucks and Moguls, which, fomented by the artful policy of the emperor of China, are carried on with such animosity as to threaten the entire destruction of one or other of these nations.[279]

The Bedoweens of Arabia and Syria do not live in greater tranquillity than the inhabitants of Grand Tartary. The very nature of the pastoral state seems to furnish perpetual occasions for war. The pastures, which a tribe uses at one period, form but a small part of its possessions. A large range of territory is successively occupied in the course of the year; and, as the whole of this is absolutely necessary for the annual subsistence of the tribe, and is considered as appropriated, every violation of it, though the tribe may be at a great distance, is held to be a just cause of war.[280] Alliances and kindred make these wars more general. When blood is shed, more must expiate it; and as such accidents have multiplied in the lapse of years, the greatest part of the tribes have quarrels between them and live in a state of perpetual hostility.[281] In the times which preceded Mahomet, seventeen hundred battles are recorded by tradition; and a partial truce of two months, which was religiously kept, might be considered, according to a just remark of Gibbon, as still more strongly expressive of their general habits of anarchy and warfare.[282]

The waste of life from such habits might alone appear sufficient to repress their population; but probably their effect is still greater in the fatal check which they give to every species of industry, and particularly to that, the object of which is to enlarge the means of subsistence. Even the construction of a well or a reservoir of water requires some funds and labour in advance; and war may destroy in one day the work of many months and the resources of a whole year.[283] The evils seem mutually to produce each other. A scarcity of subsistence might at first perhaps give occasion to the habits of war; and the habits of war in return powerfully contribute to narrow the means of subsistence.

Some tribes, from the nature of the deserts in which they live, seem to be necessarily condemned to a pastoral life;[284] but even those which inhabit soils proper for agriculture, have but little temptation to practise this art; while surrounded by marauding neighbours. The peasants of the frontier provinces of Syria, Persia and Siberia, exposed, as they are, to the constant incursions of a devastating enemy, do not lead a life that is to be envied by the

---

[279] Geneal. Hist. Tart. vol. ii. p. 545.

[280] Ils se disputeront la terre inculte, comme parmi nous les citoyens se disputent les héritages. Ainsi ils trouveront de fréquentes occasions de guerre pour la nourriture de leurs bestiaux, &c.**** ils auront autant de choses à régler par le droit des gens qu'ils en auront peu à décider par le droit civil. Montes. Esprit des Loix, l. xviii. c. xii.

[281] Voy. de Volney, tom. i. c. xxii. p. 361, 362, 363.

[282] Gibbon, vol. ix. c.1. p. 238, 239.

[283] Voy. de Volney, tom. i. c. xxiii. p. 353.

[284] Voy. de Volney, tom. i. c. xxxiii. p. 350.

wandering Tartar or Arab. A certain degree of security is perhaps still more necessary than richness of soil, to encourage the change from the pastoral to the agricultural state; and where this cannot be attained, the sedentary labourer is more exposed to the vicissitudes of fortune than he who leads a wandering life, and carries all his property with him.[285] Under the feeble, yet oppressive government of the Turks, it is not uncommon for peasants to desert their villages and betake themselves to a pastoral state, in which they expect to be better able to escape from the plunder of their Turkish masters and Arab neighbours.[286]

It may be said, however, of the shepherd, as of the hunter, that if want alone could effect a change of habits, there would be few pastoral tribes remaining. Notwithstanding the constant wars of the Bedoween Arabs, and the other checks to their increase from the hardships of their mode of life, their population presses so hard against the limits of their food, that they are compelled from necessity to a degree of abstinence, which nothing but early and constant habit could enable the human constitution to support. According to Volney, the lower classes of the Arabs live in a state of habitual misery and famine.[287] The tribes of the desert deny that the religion of Mahomet was made for them. "For how," they say, "can we perform ablutions when we have no water; how can we give alms when we have no riches; or what occasion can there be to fast during the month of Ramadan, when we fast all the year?"[288]

The power and riches of a Chaik consist in the number of his tribe. He considers it therefore as his interest to encourage population, without reflecting how it may be supported. His own consequence greatly depends on a numerous progeny and kindred;[289] and in a state of society where power generally procures subsistence, each individual family derives strength and importance from its numbers. These ideas act strongly as a bounty upon population; and, co-operating with a spirit of generosity which almost produces a community of goods,[290] contribute to push it to its utmost verge, and to depress the body of the people in the most rigid poverty.

The habits of polygamy, where there have been losses of men in war, tend perhaps also to produce the same effect. Niebuhr observes that polygamy multiplies families till many of their branches sink into the most wretched misery.[291] The descendants of Mahomet are found in great numbers all over the east, and many of them in extreme poverty. A Mahometan is in some degree obliged to polygamy from a principle of obedience to his prophet, who makes one of the greatest duties of man to consist in procreating children to glorify the Creator. Fortunately, individual interest corrects in some degree, as in many

---

[285] Id. p. 354.

[286] Id. p. 350.

[287] Voy. de Volney, tom. i. c. xxiii. p. 359.

[288] Id. p. 380.

[289] Id. p. 366.

[290] Id. p. 378.

[291] Niebuhr's Travels, vol. ii. c. v. p. 207.

other instances, the absurdity of the legislator; and the poor Arab is obliged to proportion his religious obedience to the scantiness of his resources. Yet still the direct encouragements to population are extraordinarily great; and nothing can place in a more striking point of view the futility and absurdity of such encouragements than the present state of those countries. It is universally agreed that, if their population be not less than formerly, it is indubitably not greater; and it follows as a direct consequence, that the great increase of some families has absolutely pushed others out of existence. Gibbon, speaking of Arabia, observes, that "The measure of population is regulated by the means of subsistence; and the inhabitants of this vast peninsula might be out-numbered by the subjects of a fertile and industrious province."[292] Whatever maybe the encouragements to marriage, this measure cannot be passed. While the Arabs retain their present manners, and the country remains in its present state of cultivation, the promise of Paradise to every man who had ten children would but little increase their numbers, though it might greatly increase their misery. Direct encouragements to population have no tendency whatever to change these manners and promote cultivation. Perhaps indeed they have a contrary tendency; as the constant uneasiness from poverty and want which they occasion must encourage the marauding spirit,[293] and multiply the occasions of war.

Among the Tartars, who from living in a more fertile soil are comparatively richer in cattle, the plunder to be obtained in predatory incursions is greater than among the Arabs. And as the contests are more bloody from the superior strength of the tribes, and the custom of making slaves is general, the loss of numbers in war will be more considerable. These two circumstances united enable some hordes of fortunate robbers to live in a state of plenty, in comparison of their less enterprising neighbours. Professor Pallas gives a particular account of two wandering tribes subject to Russia, one of which supports itself almost entirely by plunder, and the other lives as peaceably as the restlessness of its neighbours will admit. It may be curious to trace the different checks to population that result from these different habits.

The Kirgisiens, according to Pallas,[294] live at their ease in comparison of the other wandering tribes that are subject to Russia. The spirit of liberty and independence which reigns amongst them, joined to the facility with which they can procure a flock sufficient for their maintenance, prevents any of them from entering into the service of others. They all expect to be treated as brothers; and the rich therefore are obliged to use slaves. It may be

---

[292] It is rather a curious circumstance, that a truth so important, which has been stated and acknowledged by so many authors, should so rarely have been pursued to its consequences. People are not every day dying of famine. How then is the population regulated to the measure of the means of subsistence?

[293] Aussi arrive-t'il chaque jour des accidens, des enlèvemens de bestiaux; et cette guerre de maraude est une de celles qui occupent davantage les Arabes. Voy. de Volney, tom. i, c. xxiii. p. 364.

[294] Not having been able to procure the work of Pallas on the history of the Mongol nations, I have here made use of a general abridgment of the works of the Russian travellers, in 4 vols. oct. published at Berne and Lausanne in 1781 and 1784, entitled Découvertes Russes, tom, iii. p. 399.

asked what are the causes which prevent the lower classes of people from increasing till they become poor?

Pallas has not informed us how far vicious customs with respect to women, or the restraints on marriage from the fear of a family may have contributed to this effect; but perhaps the description which he gives of their civil constitution and licentious spirit of rapine, may alone be almost sufficient to account for it. The Chan cannot exercise his authority but through the medium of a council of principal persons, chosen by the people; and even the decrees thus confirmed are continually violated with impunity.[295] Though the plunder and capture of persons, of cattle and of merchandise, which the Kirgisiens exercise on their neighbours the Kazalpacs, the Bucharians, the Persians, the Truchemens, the Kalmucks and the Russians, are prohibited by their laws, yet no person is afraid to avow them. On the contrary, they boast of their successes in this way as of the most honourable enterprises. Sometimes they pass their frontiers alone to seek their fortune, sometimes collect in troops under the command of an able chief, and pillage entire caravans. A great number of Kirgisiens, in exercising this rapine, are either killed or taken into slavery; but about this the nation troubles itself very little. When these ravages are committed by private adventurers, each retains what he has taken, whether cattle or women. The male slaves and the merchandise are sold to the rich, or to foreign traders.[296]

With these habits, in addition to their national wars, which from the fickle and turbulent disposition of the tribe are extremely frequent,[297] we may easily conceive that the checks to population from violent causes may be so powerful as nearly to preclude all others. Occasional famines may sometimes attack them in their wars of devastation,[298] their fatiguing predatory incursions, or from long droughts and mortality of cattle; but in the common course of things the approach of poverty would be the signal for a new marauding expedition; and the poor Kirgisien would either return with sufficient to support him, or lose his life or liberty in the attempt. He who determines to be rich or die, and does not scruple the means, cannot long live poor.

The Kalmucks, who before their emigration in 1771 inhabited the fertile steppes of the Wolga under the protection of Russia, lived in general in a different manner. They were not often engaged in any very bloody wars;[299] and the power of the Chan being absolute,[300] and the civil administration better regulated than among the Kirgisiens, the marauding expeditions of private adventurers were checked. The Kalmuck women are extremely

[295] Découv. Russ. tom. iii. p. 389.

[296] Id. p. 396, 397, 398.

[297] Id. p. 378.

[298] Cette multitude dévaste tout ce qui se trouve sur son passage; ils emmènent avec eux tout le bétail qu'ils ne consomment pas, et réduisent â l'esclavage les femmes, les enfans, et les hommes, qu'ils n'ont pas massacrés. Découv. Russ. tom. iii. p. 390.

[299] Découv. Russ. tom. iii. p. 221. The tribe is described here under the name of Torgots, which was their appropriate appellation. The Russians called them by the more general name of Kalmucks.

[300] Id. p. 327.

prolific. Barren marriages are rare, and three or four children are generally seen playing round every hut. From which (observes Pallas) it may naturally be concluded that they ought to have multiplied greatly during the hundred and fifty years that they inhabited tranquilly the steppes of the Wolga. The reasons which he gives for their not having increased so much as might be expected, are the many accidents occasioned by falls from horses, the frequent petty wars between their different princes and with their different neighbours; and particularly the numbers among the poorer classes who die of hunger, of misery, and every species of calamity, of which the children are most frequently the victims.[301]

It appears that when this tribe put itself under the protection of Russia, it had separated from the Soongares, and was by no means numerous. The possession of the fertile steppes of the Wolga and a more tranquil life soon increased it, and in 1662 it amounted to fifty thousand families.[302] From this period to 1771, the time of its migration, it seems to have increased very slowly. The extent of pastures possessed would not probably admit of a much greater population; as at the time of its flight from these quarters, the irritation of the Chan at the conduct of Russia was seconded by the complaints of the people of the want of pasture for their numerous herds. At this time the tribe amounted to between 55 and 60,000 families. Its fate in this curious migration was what has probably been the fate of many other wandering hordes, who, from scanty pastures or other causes of discontent, have, attempted to seek for fresh seats. The march took place in the winter, and numbers perished on this painful journey from cold, famine, and misery. A great part were either killed or taken by the Kirghises; and those who reached their place of destination, though received at first kindly by the Chinese, were afterwards treated with extreme severity.[303]

Before this migration, the lower classes of the Kalmucks had lived in great poverty and wretchedness, and had been reduced habitually to make use of every animal, plant, or root, from which it was possible to extract nourishment.[304] They very seldom killed any of their cattle that were in health, except indeed such as were stolen.; and these were devoured immediately, for fear of a discovery. Wounded or worn-out horses, and beasts that had died of any disease except a contagious epidemic, were considered as most desirable food. Some of the poorest Kalmucks would eat the most putrid carrion, and even the dung of their cattle.[305] A great number of children perished of course from bad nourishment.[306] In the

[301] Découv. Russ. tom. iii. p. 319, 320, 321.

[302] Id. p. 221. Tooke's View of the Russian Empire, vol. ii. b. ii. p. 30. Another instance of rapid increase presents itself in a colony of baptized Kalmucks, who received from Russia a fertile district to settle in. From 8695, which was its number in 1754, it had increased in 1771 to 14,000. Tooke's View of the Russ. Emp. vol. ii. b. ii. p. 32, 33.

[303] Tooke's View of the Russ. Emp. vol. ii. b. ii. p. 29, 30, 31. Découv. Russ. tom. iii. p. 221.

[304] Découv. Russ. tom. iii. p. 275, 276.

[305] Id. p. 272, 273, 274.

[306] Id. p. 324.

winter all the lower classes suffered severely from cold and hunger.[307] In general, one third of their sheep, and often much more, died in the winter in spite of all their care; and if a frost came late in the season after rain and snow, so that the cattle could not get at the grass, the mortality among their herds became general, and the poorer classes were exposed to inevitable famine.[308]

Malignant fevers, generated principally by their putrid food and the putrid exhalations with which they were surrounded, and the small-pox, which was dreaded like the plague, sometimes thinned their numbers;[309] but in general it appears that their population pressed so hard against the limits of their means of subsistence, that want, with the diseases arising from it, might be considered as the principal check to their increase.

A person travelling in Tartary during the summer months would probably see extensive steppes unoccupied, and grass in profusion spoiling for want of cattle to consume it. He would infer perhaps that the country could support a much greater number of inhabitants, even, supposing them to remain in their shepherd state. But this might be a hasty and unwarranted conclusion. A horse or any other working animal is said to be strong only in proportion to the strength of his weakest part. If his legs be slender and feeble, the strength of his body will be but of little consequence; or if he wants power in his back and haunches, the strength which he may possess in his limbs can never be called fully into action. The same reasoning must be applied to the power of the earth to support living creatures. The profusion of nourishment which is poured forth in the seasons of plenty cannot all be consumed by the scanty numbers that were able to subsist through the season of scarcity. When human industry and foresight are directed in the best manner, the population which the soil can support is regulated by the average produce throughout the year; but among animals, and in the uncivilized states of man, it will be much below this average. The Tartar would find it extremely difficult to collect and carry with him such a quantity of hay as would feed all his cattle well during the winter. It would impede his motions, expose him to the attacks of his enemies, and an unfortunate day might deprive him of the labours of a whole summer; as in the mutual invasions which occur, it seems to be the universal practice, to burn and destroy all the forage and provisions which cannot be carried away.[310] The Tartar therefore provides only for the most valuable of his cattle during the winter, and leaves the rest to support themselves by the scanty herbage which they can pick up. This poor living, combined with the severe cold, naturally destroys a considerable part of them.[311] The population of the tribe is measured by the population of its herds; and the

---

[307] Découv. Russ. tom. iii. p. 310.

[308] Id. p. 270.

[309] Id. p. 311, 312, 313.

[310] On mit le feu à toutes les meules de blé et de fourrage.**** Cent cinquante villages également incendiés. Mémoires du Baron de Tott, tom. i. p. 272. He gives a curious description of the devastation of a Tartar army, and of its sufferings in a winter campaign. Cette journée coûta à l'armée plus de 3,000 hommes, et 30,000 chevaux, qui périrent de froid, p. 267.

[311] Découvertes Russes, vol. iii. p. 261.

average numbers of the Tartars, as of the horses that run wild in the desert, are kept down so low by the annual returns of the cold and scarcity of winter, that they cannot consume all the plentiful offerings of summer.

Droughts and unfavourable seasons have, in proportion to their frequency, the same effects as the winter. In Arabia[312] and a great part of Tartary[313] droughts are not uncommon; and if the periods of their return be not above six or eight years, the average population can never much exceed what the soil can support during these unfavourable times. This is true in every situation; but perhaps, in the shepherd state, man is peculiarly exposed to be affected by the seasons; and a great mortality of parent stock is an evil more fatal and longer felt than the failure of a crop of grain. Pallas and the other Russian travellers speak of epizooties as very common in these parts of the world.[314]

As among the Tartars a family is always honourable, and women are reckoned very serviceable in the management of the cattle and the household concerns, it, is not probable that many are deterred from marriage from the fear of not being able to support a family.[315] At the same time, as all wives are bought of their parents, it must sometimes be out of the power of the poorer classes to make the purchase. The Monk Rubruquis, speaking of this custom; says that, as parents keep all their daughters till they can sell them; their maids are sometimes very stale before they are married.[316] Among the Mahometan Tartars, female captives would supply the place of wives;[317] but among the Pagan Tartars, who make but little use of slaves, the inability to buy wives must frequently operate on the poorer classes as a check to marriage, particularly as their price would be kept up by the practice of polygamy among the rich.[318]

The Kalmucks are said not to be jealous,[319] and from the frequency of the venereal disease among them,[320] we may infer that a certain degree of promiscuous intercourse prevails.

On the whole, therefore, it would appear that in that department of the shepherd life which has been considered in this chapter, the principal checks which keep the population down to the level of the means of subsistence are, restraint from inability to obtain a wife, vicious customs with respect to women, epidemics, wars, famine, and the diseases arising

---

[312] Voy. de Volney, vol. i. c. 23. p. 353.

[313] Découv. Russ. tom. i. p. 467; ii. p. 10, 11, 12, &c.

[314] Id. tom. i. p. 290, &c.; ii. p. 11; iv. p. 304.

[315] Geneal. Hist. of the Tartars, vol. ii. p. 407.

[316] Travels of Wim. Rubruquis, in 1253. Harris's Collection of Voy. b. i. c. ii. p. 561.

[317] Découv. Russ. tom. iii. p. 413.

[318] Pallas takes notice of the scarcity of women or superabundance of males among the Kalmucks, notwithstanding the more constant exposure of the male sex to every kind of accident. Découv. Russ. tom. iii. p. 320.

[319] Id. p: 239.

[320] Id. p. 324.

from extreme poverty. The three first checks and the last appear to have operated with much less force among the shepherds of the north of Europe.

3.4.8 Book I, Chapter VIII: Of the Checks to Population in Different Parts of Africa.

The parts of Africa visited by Park are described by him as neither well cultivated nor well peopled. He found many extensive and beautiful districts entirely destitute of inhabitants; and in general the borders of the different kingdoms were either very thinly peopled or perfectly deserted. The swampy banks of the Gambia, the Senegal, and other rivers towards the coast, appeared to be unfavourable to population, from being unhealthy;[321] but other parts were not of this description; and it was not possible, he says, to behold the wonderful fertility of the soil, the vast herds of cattle proper both for labour and food, and reflect on the means which presented themselves of vast inland navigation, without lamenting that a country so abundantly gifted by nature should remain in its present savage and neglected state.[322]

The causes of this neglected state clearly appear, however, in the description which Park gives of the general habits of the negro nations. In a country divided into a thousand petty states, mostly independent and jealous of each other, it is natural, he says, to imagine that wars frequently originate from very frivolous provocations. The wars of Africa are of two kinds, one called Killi, that which is openly avowed; and the other, Tegria, plundering or stealing. These latter are very common, particularly about the beginning of the dry season, when the labours of harvest are over, and provisions are plentiful.—These plundering excursions always produce speedy retaliation.[323]

The insecurity of property arising from this constant exposure to plunder, must necessarily have a most baneful effect on industry. The deserted state of all the frontier provinces sufficiently proves to what degree it operates. The nature of the climate is unfavourable to the exertion of the negro nations; and, as there are not many opportunities of turning to advantage the surplus produce of their labour, we cannot be surprised that they should in general content themselves with cultivating only so much ground as is necessary for their own support.[324] These causes appear adequately to account for the uncultivated state of the country.

The waste of life in these constant wars and predatory incursions must be considerable; and Park agrees with Buffon in stating, that independently of violent causes, longevity is rare among the negroes. At forty, he says, most of them become grey-haired and covered

---

[321] Park's Interior of Africa, c. xx. p. 261. 4to.

[322] Id. c. xxiii. p. 312.

[323] Park's Interior of Africa, c. xxii. p. 291 8 seq.

[324] Id. c. xxi. p. 280.

with wrinkles, and few of them survive the age of fifty-five or sixty.[325] Buffon attributes this shortness of life to the premature intercourse of the sexes, and very early and excessive debauchery.[326] On this subject perhaps he has been led into exaggerations; but without attributing too much to this cause, it seems agreeable to the analogy of nature to suppose that, as the natives of hot climates arrive much earlier at maturity than the inhabitants of colder countries, they should also perish earlier.

According to Buffon, the negro-women are extremely prolific; but it appears from Park that they are in the habit of suckling their children two or three years, and as the husband during this time devotes the whole of his attention to his other wives, the family of each wife is seldom numerous.[327] Polygamy is universally allowed among the negro nations;[328] and consequently without a greater superabundance of women than we have reason to suppose, many will be obliged to live unmarried. This hardship will principally fall on the slaves, who, according to Park, are in the proportion of three to one to the free men.[329] A master is not permitted to sell his domestic slaves or those born in his own house, except in case of famine, to support himself and family. We may imagine therefore that he will not suffer them to increase beyond the employment which he has for them. The slaves which are purchased, or the prisoners taken in war, are entirely at the disposal of their masters.[330] They are often treated with extreme severity, and in any scarcity of women arising from the polygamy of the free men, would of course be deprived of them without scruple. Few or no women, probably, remain in a state of strict celibacy; but in proportion to the number married, the state of society does not seem to be favourable to increase.

Africa has been at all times the principal mart of slaves. The drains of its population in this way have been great and constant, particularly since their introduction into the European colonies; but perhaps, as Dr. Franklin observes, it would be difficult to find the gap that has been made by a hundred years' exportation of negroes which has blackened half America.[331] For notwithstanding this constant emigration, the loss of numbers from incessant war, and the checks to increase from vice and other causes, it appears that the

---

[325] Park's Africa, c. xxi. p. 284.

[326] L'usage prématuré des femmes est peut-être la cause de la briéveté de leur vie; les enfans sont si débauchés, et si peu contraints par les pères et mères que dès leur plus tendre jeunesse ils se livrent à tout ce que la nature leur suggère; rien n'est si rare que de trouver dans ce peuple quelque fille qui puisse se souvenir du tems auquel elle a cessée d'être vierge. Histoire Naturelle de l'Homme, vol. vi. p. 235. 5th edit. 12mo. 31 vols.

[327] Park's Africa, c. xx. p. 265. As the accounts of Park, and those on which Buffon has founded his observations, are probably accounts of different nations, and certainly at different periods, we cannot infer that either is incorrect because they differ from each other; but as far as Park's observations extend, they are certainly entitled to more credit than any of the travellers which preceded him.

[328] Park's Africa, c. xx. p. 267.

[329] Park's Africa, c. xxii. p. 287.

[330] Id. p. 288.

[331] Franklin's Miscell. p. 9.

population is continually pressing against the limits of the means of subsistence. According to Park, scarce years and famines are frequent. Among the four principal causes of slavery in Africa, he mentions famine next to war;[332] and the express permission given to masters to sell their domestic slaves for the support of their family, which they are not allowed to do on any less urgent occasion,[333] seems to imply the not unfrequent recurrence of severe want. During a great scarcity which lasted for three years in the countries of the Gambia, great numbers of people became slaves. Park was assured by Dr. Laidley that at that time many free men came, and begged with great earnestness to be put upon his slave chain to save them from perishing with hunger.[334] While Park was in Manding, a scarcity of provisions was severely felt by the poor, as the following circumstance painfully convinced him. Every evening during his stay, he observed five or six women come to the Mansa's house and receive each of them a certain quantity of corn. "Observe that boy," said Mansa to him, pointing to a fine child about five years of age—"his mother has sold him to me for forty days' provision for herself and the rest of her family. I have bought another boy in the same manner."[335] In Sooseeta, a small Jallonka village, Mr. Park was informed by the master that he could furnish no provisions, as there had lately been a great scarcity in that part of the country. He assured him that before they had gathered in their present crops all the inhabitants of Kullo had been for twenty-nine days without tasting corn; during which time they had supported themselves entirely on the yellow powder which is found in the pods of the nitta, (so called by the natives,) a species of mimosa, and upon the seeds of the bamboo cane, which when properly pounded and dressed taste very much like rice.[336]

It may be said perhaps that as, according to Park's account, much good land remains uncultivated in Africa, the dearths may be attributed to a want of people; but if this were the case, we can hardly suppose that such numbers would yearly be sent out of the country. What the negro nations really want is security of property, and its general concomitant, industry; and without these, an increase of people would only aggravate their distresses. If, in order to fill up those parts which appeared to be deficient in inhabitants, we were to suppose a high bounty given on children, the effects would probably be, the increase of wars, the increase of the exportation of slaves, and a great increase of misery, but little or no real increase of population.[337]

---

[332] Park's Africa, c. xxii. p. 295.

[333] Id. p. 288, note.

[334] Id. p. 295.

[335] Park's Africa, c. xix. p. 248.

[336] Id. c. xxv. p. 336.

[337] The two great requisites just mentioned for a real increase of population, namely, security of property, and its natural concomitant, industry, cannot be expected to exist among the negro nations, while the traffic in slaves on the coast gives such constant encouragement to the plundering excursions which Park describes. Were this traffic at an end, we might rationally hope that, before the lapse of any long period, future travellers would be able to give us a more favourable picture of the state of society among the African nations, than that drawn by Park.

The customs of some nations, and the prejudices of all, operate in some degree like a bounty of this kind. The Shangalla negroes, according to Bruce, hemmed in on every side by active and powerful enemies, and leading a life of severe labour and constant apprehension, feel but little desire for women. It is the wife, and not the man, that is the cause of their polygamy. Though they live in separate tribes or nations, yet these nations are again subdivided into families. In fighting, each family attacks and defends by itself, and theirs is the spoil and plunder who take it. The mothers therefore, sensible of the disadvantages of a small family, seek to multiply it by all the means in their power; and it is by their importunity, that the husband suffers himself to be overcome.[338] The motives to polygamy among the Galla are described to be the same, and in both nations the first wife courts the alliance of a second for her husband; and the principal argument she makes use of is, that their families may be joined together and be strong, and that her children, by being few in number, may not fall a prey to their enemies in the day of battle.[339] It is highly probable that this extreme desire of having large families defeats its own purpose; and that the poverty and misery, which it occasions, cause fewer children to grow up to maturity, than if the parents confined their attention to the rearing of a smaller number.

Bruce is a great friend to polygamy, and defends it, in the only way in which it is capable of being defended, by asserting, that in the countries in which it principally prevails the proportion of girls to boys born is two or three to one. A fact so extraordinary however cannot be admitted upon the authority of those vague inquiries on which he founds his opinion. That there are considerably more women living than men in these climates, is in the highest degree probable. Even in Europe, where it is known with certainty that more boys are born than girls, the women in general exceed the men in number; and we may imagine that in hot and unhealthy climates, and in a barbarous state of society, the accidents to which the men are exposed must be very greatly increased. The women, by leading a more sedentary life, would suffer less from the effects of a scorching sun and swampy exhalations; they would in general be more exempt from the disorders arising from debauchery; but, above all, they would escape in great measure the ravages of war. In a state of society in which hostilities never cease, the drains of men, from this cause alone, must occasion a great disproportion of the sexes, particularly where it is the custom, as related of the Galla in Abyssinia,[340] to massacre indiscriminately all the males, and save only the marriageable women from the general destruction. The actual disproportion of the sexes arising from these causes probably first gave rise to the permission of polygamy, and has perhaps contributed to make us more easily believe, that the proportion of male and female children in hot climates is very different from what we have experienced it to be in the temperate zone.

---

[338] Bruce's Travels to discover the Source of the Nile, vol. ii. p. 556. 4to.

[339] Bruce's Travels, vol. ii. p. 223.

[340] Bruce's Travels to discover the Source of the Nile; vol. iv. p. 411.

Bruce, with his usual prejudices on this subject, seems to think that the celibacy of a part of the women is fatal to the population of a country. He observes of Jidda that, on account of the great scarcity of provisions, which is the result of an extraordinary concourse of people to a place almost destitute of the necessaries of life, few of the inhabitants can avail themselves of the privilege granted by Mahomet. They cannot therefore marry more than one wife; and from this cause arises, he says, the want of people and the large number of unmarried women.[341] But it is evident that the want of people in this barren spot arises solely from the want of provisions, and that, if each man had four wives, the number of people could not be permanently increased by it.

In Arabia Felix, according to Bruce, where every sort of provision is exceedingly cheap, where the fruits of the ground, the general food of man, are produced spontaneously, the support of a number of wives costs no more than that of so many slaves or servants. Their food is the same, and a blue cotton shirt, a habit common to them all, is not more chargeable for the one than for the other. The consequence is, he says, that celibacy in women is prevented, and the number of people increased in a fourfold ratio by polygamy, to what it is in those countries that are monogamous.[342] And yet, notwithstanding this fourfold increase, it does not appear that any part of Arabia is really very populous.

The effect of polygamy in increasing the number of married women and preventing celibacy is beyond dispute; but how far this may tend to increase the actual population is a very different consideration. It may perhaps continue to press the population harder against the limits of the food; but the squalid and hopeless poverty which this occasions is by no means favourable to industry; and in a climate in which there appears to be many predisposing causes of sickness, it is difficult to conceive that this state of wretchedness does not powerfully contribute to the extraordinary mortality which has been observed in some of these countries.

According to Bruce, the whole coast of the Red Sea, from Suez to Babelmandel, is extremely unwholesome, but more especially between the tropics. Violent fevers, called there Nedad, make the principal figure in this fatal list, and generally terminate the third day in death.[343] Fear frequently seizes strangers upon the first sight of the great mortality which they observe on their first arrival.

Jidda, and all the parts of Arabia adjacent to the eastern coast of the Red Sea, are in the same manner very unwholesome.[344]

In Gondar, fevers perpetually reign, and the inhabitants are all of the colour of a corpse.[345]

---

[341] Id. vol. i. c. xi. p. 280.

[342] Bruce, vol. i. c. xi. p, 281.

[343] Bruce, vol. iii. p. 33.

[344] Id. vol. i. p. 279.

[345] Id. vol. iii. p 178.

In Sirè, one of the finest countries in the world, putrid fevers of the very worst kind are almost constant.[346] In the low grounds of Abyssinia, in general, malignant tertians occasion a great mortality.[347] And every where the small-pox makes great ravages, particularly among the nations bordering on Abyssinia, where it sometimes extinguishes whole tribes.[348]

The effect of poverty, with bad diet, and, its almost constant concomitant, want of cleanliness, in aggravating malignant distempers, is well known; and this kind of wretchedness seems generally to prevail. Of Tchagassa, near Gondar, Bruce observes that the inhabitants, notwithstanding their threefold harvests, are miserably poor.[349] At Adowa, the capital of Tigré, he makes the same remark, and applies it to all the Abyssinian farmers. The land is let yearly to the highest bidder, and in general the landlord furnishes the seed and receives half of the produce; but it is said that he is a very indulgent master who does not take another quarter for the risk he has run; so that the quantity which comes to the share of the husbandman is not more than sufficient to afford a bare sustenance to his wretched family.[350]

The Agows, one of the most considerable nations of Abyssinia in point of number, are described by Bruce as living in a state of misery and penury scarcely to be conceived. We saw a number of women, he says, wrinkled and sun-burnt so as scarcely to appear human, wandering about under a burning sun with one and sometimes two children upon their backs, gathering the seeds of bent grass to make a kind of bread.[351] The Agow women begin to bear children at eleven years old. They marry generally about that age, and there is no such thing as barrenness known among them.[352] In Dixan, one of the frontier towns of Abyssinia, the only trade is that of selling children. Five hundred are exported annually to Arabia; and in times of scarcity, Bruce observes, four times that number.[353]

In Abyssinia polygamy does not regularly prevail. Bruce, indeed, makes rather a strange assertion on this subject; and says that, though we read from the Jesuits a great deal about marriage and polygamy, yet that there is nothing which may be averred more truly than that there is no such thing as marriage in Abyssinia.[354] But, however this may be, it appears clear that few or no women lead a life of celibacy in that country; and that the prolific powers of nature are nearly all called into action, except so far as they are checked by

[346] Id. p. 153.

[347] Id. vol. iv. p. 22.

[348] Id. vol. iii. c. iii. p. 68. c. vii. p. 178; vol. i. c. xiii, p. 353.

[349] Bruce, vol. iii. c. vii. p. 195.

[350] Id. c. v. p. 124.

[351] Id. c. xix. p. 738.

[352] Bruce, vol. iii. c. xix. p. 739.

[353] Id. c. iii. p. 88.

[354] Id. c. xi. p. 306.

promiscuous intercourse. This, however, from the state of manners described by Bruce, must operate very powerfully.[355]

The check to population from war appears to be excessive. For the last four hundred years, according to Bruce, it has never ceased to lay desolate this unhappy country;[356] and the savage manner in which it is carried on surrounds it with tenfold destruction. When Bruce first entered Abyssinia, he saw on every side ruined villages destroyed to their lowest foundations by Ras Michael in his march to Gondar.[357] In the course of the civil wars, while Bruce was in the country, he says, "The rebels had begun to lay waste Dembea, and burnt all the villages in the plain from south to west, making it like a desert between Michael and Fasil.**** The king often ascended to the top of the tower of his palace, and contemplated with the greatest displeasure the burning of his rich villages in Dembea."[358] In another place he says, "The whole country of Degwessa was totally destroyed; men, women and children were entirely extirpated without distinction of age or sex; the houses razed to the ground, and the country about it left as desolate as after the deluge. The villages belonging to the king were as severely treated; an universal cry was heard from all parts, but no one dared to suggest any means of help."[359] In Maitsha, one of the provinces of Abyssinia, he was told that, if ever he met an old man, he might be sure that he was a stranger, as all that were natives died by the lance young.[360]

If the picture of the state of Abyssinia drawn by Bruce, be in any degree near the truth, it places in a strong point of view the force of that principle of increase, which preserves a population fully up to the level of the means of subsistence under the checks of war, pestilential diseases and promiscuous intercourse, all operating in an excessive degree.

The nations which border on Abyssinia are universally short-lived. A Shangalla woman at twenty-two is, according to Bruce, more wrinkled and deformed by age than an European woman at sixty.[361] It would appear, therefore, that in all these countries, as among the northern shepherds in the times of their constant emigrations, there is a very rapid succession of human beings; and the difference in the two instances is, that our northern ancestors died out of their own country, whereas these die at home. If accurate registers of mortality were kept among these nations; I have little doubt that it would appear, that, including the mortality from wars, 1 in 17 or 18 at the least dies annually, instead of 1 in 34, 36, or 40, as in the generality of European states.

The description, which Bruce gives of some parts of the country which he passed through on his return home, presents a picture more dreadful even than the state of

[355] Id. p. 292.

[356] Id. vol. iv. p. 119.

[357] Bruce, vol. iii. c. vii. p. 192.

[358] Id. vol. iv. c. v. p. 112.

[359] Id. vol. iv. p. 258.

[360] Id. c. i. p. 14.

[361] Bruce, vol. ii. p. 559.

Abyssinia, and shews how little population depends on the birth of children, in comparison of the production of food and those circumstances of natural and political situation which influence this produce.

"At half past six," Bruce says, "we arrived at Garigana, a village whose inhabitants had all perished with hunger the year before; their wretched bones being all unburied and scattered upon the surface of the ground where the village formerly stood. We encamped among the bones of the dead; no space could be found free from them."[362]

Of another town or village in his route he observes, "The strength of Teawa was 25 horse. The rest of the inhabitants might be 1200 naked miserable and despicable Arabs, like the rest of those which live in villages.**** Such was the state of Teawa. Its consequence was only to remain till the Daveina Arabs should resolve to attack it, when its corn-fields being burnt and destroyed in a night by a multitude of horsemen, the bones of its inhabitants scattered upon the earth would be all its remains, like those of the miserable village of Garigana."[363]

"There is no water between Tedwa and Beyla. Once Indedidema and a number of villages were supplied with water from wells, and had large crops of Indian corn sown about their possessions. The curse of that country, the the Daveina Arabs, have destroyed Indedidema and all the villages about it; filled up their wells, burnt their crops, and exposed all the in habitants to die by famine."[364]

Soon after leaving Sennaar, he says, "We began to see the effects of the quantity of rain having failed. There was little corn sown, and that so late as to be scarcely above ground. It seems the rains begin later as they pass northward. Many people were here employed in gathering grass-seeds to make a very bad kind of bread. These people appear perfect skeletons, and no wonder, as they live upon such fare. Nothing increases the danger of travelling and prejudice against strangers more, than the scarcity of provisions in the country through which you are to pass."[365]

"Came to Eltic, a straggling village about half a mile from the Nile, in the north of a large bare plain; all pasture, except the banks of the river which are covered with wood. We now no longer saw any corn sown. The people here were at the same miserable employment as those we had seen before, that of gathering grass-seeds."[366]

Under such circumstances of climate and political situation, though a greater degree of foresight, industry and security, might considerably better their condition and increase their population, the birth of a greater number of children without these concomitants would only aggravate their misery, and leave their population where it was.

---

[362] Bruce, vol. iv. p. 349.

[363] Id. p. 353.

[364] Id. p. 411.

[365] Bruce, vol. iv. p. .511.

[366] Id. p. 511.

The same may be said of the once flourishing and populous country of Egypt. Its present depressed state has not been caused by the weakening of the principle of increase, but by the weakening of the principle of industry and foresight, from the insecurity of property consequent on a most tyrannical and oppressive government. The principle of increase in Egypt at present does all that is possible for it to do. It keeps the population fully up to the level of the means of subsistence; and, were its power ten times greater than it really is, it could do no more.

The remains of ancient works, the vast lakes, canals and large conduits for water destined to keep the Nile under control, serving as reservoirs to supply a dry year, and as drains and outlets to prevent the superabundance of water in wet years, sufficiently indicate to us that the former inhabitants of Egypt by art and industry contrived to fertilize a much greater quantity of land from the overflowings of their river, than is done at present; and to prevent, in some measure, the distresses which are now so frequently experienced from a redundant or insufficient inundation.[367] It is said of the governor Petronius, that, effecting by art what was denied by nature, he caused abundance to prevail in Egypt under the disadvantages of such a deficient inundation, as had always before been accompanied by dearth.[368] A flood too great is as fatal to the husbandman as one that is deficient; and the ancients had, in consequence drains and outlets to spread the superfluous waters over the thirsty sands of Lybia, and render even the desert habitable. These works are now all out of repair, and by ill management often produce mischief instead of good. The causes of this neglect, and consequently of the diminished means of subsistence, are obviously to be traced to the extreme ignorance and brutality of the government, and the wretched state of the people. The Mamelukes, in whom the principal power resides, think only of enriching themselves, and employ for this purpose what appears to them to be the simplest method, that of seizing wealth wherever it may be found, of wresting it by violence from the possessor, and of continually imposing new and arbitrary contributions.[369] Their ignorance and brutality, and the constant state of alarm in which they live; prevent them from having any views of enriching the country, the better to prepare it for their plunder. No public works therefore are to be expected from the government, and no individual proprietor dares to undertake any improvement which might imply the possession of capital, as it would probably be the immediate signal of his destruction. Under such circumstances we cannot be surprised that the ancient works are neglected, that the soil is ill cultivated; and that the means of subsistence, and consequently the population, are greatly reduced. But such is the natural fertility of the Delta from the inundations of the Nile, that even without any capital employed upon the land, without a right of succession, and consequently almost without a right of property, it still maintains a considerable population in proportion to its extent, sufficient, if property were secure, and industry well directed, gradually to improve and

[367] Bruce, vol. iii. c. xvii. p. 710.

[368] Voyage de Volney, tom. i. c. iii. p. 33. 8vo.

[369] Voyage de Volney, tom. i. c. xii. p. 170.

extend the cultivation of the country and restore it to its former state of prosperity. It may be safely pronounced of Egypt that it is not the want of population that has checked its industry, but the want of industry that has checked its population.

The immediate causes which keep down the population to the level of the present contracted means of subsistence, are but too obvious. The peasants are allowed for their maintenance only sufficient to keep them alive.[370] A miserable sort of bread made of doura without leaven or flavour, cold water, and raw onions make up the whole of their diet. Meat and fat, of which they are passionately fond, never appear but on great occasions, and among those who are more at their ease. Their habitations are huts made of earth, where a stranger would be suffocated with the heat and smoke; and where the diseases generated by want of cleanliness, by moisture and by bad nourishment, often visit them and commit great ravages. To these physical evils are added a constant state of alarm, the fear of the plunder of the Arabs, and the visits of the Mamelukes, the spirit of revenge transmitted in families, and all the evils of a continual civil war.[371]

In the year 1783 the plague was very fatal; and in 1784 and 1785 a dreadful famine reigned in Egypt, owing to a deficiency in the inundation of the Nile. Volney draws a frightful picture of the misery that was suffered on this occasion. The streets of Cairo, which at first were full of beggars, were soon cleared of all these objects, who either perished or fled. A vast number of unfortunate wretches, in order to escape death, spread themselves over all the neighbouring countries, and the towns of Syria were inundated with Egyptians. The streets and public places were crowded by famished and dying skeletons. All the most revolting modes of satisfying the cravings of hunger were resorted to; the most disgusting food was devoured with eagerness; and Volney mentions the having seen under the walls of ancient Alexandria two miserable wretches seated on the carcase of a camel, and disputing with the dogs its putrid flesh. The depopulation of the two years was estimated at one-sixth of all the inhabitants.[372]

3.4.9 Book I, Chapter IX: Of the Checks to Population in Siberia, Northern and Southern.

The inhabitants of the most northern parts of Asia subsist chiefly by hunting and fishing; and we may suppose therefore that the checks to their increase are of the same nature as those which prevail among the American Indians; except that the check from war is considerably less, and the check from famine perhaps greater, than in the temperate regions

[370] Voyage de Volney, tom. i. c. xii. p. 172.

[371] Volney, tom. i. c. xii. p. 173. This sketch of the state of the peasantry in Egypt given by Volney seems to be nearly confirmed by all other writers on this subject; and particularly in a valuable paper entitled *Considerations générales sur l'Agriculture de l'Egypte, par L. Reynier.* (Mémoires sur l' Egypte, tom. iv. p. 1.)

[372] Voy. de Volney, tom. i. c. xii. s. ii.

of America. M. de Lesseps, who travelled from Kamtschatka to Petersburgh with the papers of the unfortunate Pérouse, draws a melancholy picture of the misery sometimes suffered in this part of the world from a scarcity of food. He observes, while at Bolcheretsk, a village of Kamtschatka; "Very heavy rains are injurious in this country, because they occasion floods which drive the fish from the rivers. A famine, the most distressing to the poor Kamtschadales, is the result; as happened last year in all the villages along the western coast of the peninsula. This dreadful calamity occurs so frequently in this quarter, that the inhabitants are obliged to abandon their dwellings, and repair with their families to the border of the Kamtschatka river where they hope to find better resources, fish being more plentiful in this river. Mr. Kasloff (the Russian officer who conducted M. de Lesseps) had intended to proceed along the western coast; but the news of this famine determined him, contrary to his wishes, to return rather than be driven to the necessity of stopping half way or perishing with hunger."[373] Though a different route was pursued, yet in the course of the journey almost all the dogs, which drew the sledges, died for want of food; and every dog, as soon as he failed, was immediately devoured by the others.[374]

Even at Okotsk, a town of considerable trade, the inhabitants wait with hungry impatience for the breaking up of the river Okhota in the spring. When M. de Lesseps was there, the stock of dried fish was nearly exhausted. Meal was so dear that the common people were unable to purchase it. On drawing the river prodigious numbers of small fish were caught, and the joy and clamour redoubled at the sight. The most famished were first served. M. de Lesseps feelingly says, "I could not refrain from tears on perceiving the ravenousness, of these poor creatures;****whole families contended for the fish, which were devoured raw before my eyes."[375]

Throughout all the northern parts of Siberia, the small-pox is very fatal. In Kamtschatka, according to M. de Lesseps, it has carried off three fourths[376] of the native inhabitants.

Pallas confirms this account; and, in describing the Ostiacks on the Obi, who live nearly in the same manner, observes that this disorder makes dreadful ravages among them, and may be considered as the principal check to their increase.[377] The extraordinary mortality of the small-pox among these people is very naturally accounted for by the extreme heat, filth and putrid air of their underground habitations. Three or four Ostiack families are crowded together in one hut; and nothing can be so disgusting as their mode of living. They never wash their hands, and the putrid remains of the fish, and the excrements of the children, are never cleared away. From this description, says Pallas, one may easily form an idea of the stench, the fœtid vapours and humidity of their Yourts.[378] They have seldom

[373] Travels in Kamtschatka, vol. i. p. 147. 8vo. Eng. trans. 1790.

[374] Id. p. 264.

[375] Id. vol. ii. p. 252, 253.

[376] Travels in Kamtschatka, vol. i. p. 128.

[377] Voy. de Pallas, tom. iv. p. 68. 4to. 5 vols. 1788, Paris.

[378] Id. p. 60.

many children. It is a rare thing to see three or four in one family; and the reason given by Pallas is that so many die young on account of their bad nourishment.[379] To this, perhaps, should be added the state of miserable and laborious servitude to which the women are condemned,[380] which certainly prevents them from being prolific.

The Samoyedes, Pallas thinks, are not quite so dirty as the Ostiacks, because they are more in motion during the winter in hunting; but he describes the state of the women amongst them as a still more wretched and laborious servitude;[381] and consequently the check to population from this cause must be greater.

Most of the natives of these inhospitable regions live nearly in the same miserable manner, which it would be therefore mere repetition to describe. From what has been said, we may form a sufficient idea of the principal checks that keep the actual population down to the level of the scanty means of subsistence which these dreary countries afford.

In some of the southern parts of Siberia, and in the districts adjoining the Wolga, the Russian travellers describe the soil to be of extraordinary fertility. It consists in general of a fine black mould of so rich a nature as not to require or even to bear dressing. Manure only makes the corn grow too luxuriantly, and subjects it to fall to the ground and be spoiled. The only mode of recruiting this kind of land which is practised is, by leaving it one year out of three in fallow; and proceeding in this way, there are some grounds, the vigour of which is said to be inexhaustible.[382] Yet, notwithstanding the facility with which, as it would appear, the most plentiful subsistence might be procured, many of these districts are thinly peopled, and in none of them, perhaps, does population increase in the proportion that might be expected from the nature of the soil.

Such countries seem to be under that moral impossibility of increasing, which is well described by Sir James Steuart.[383] If either from the nature of the government, or the habits of the people, obstacles exist to the settlement of fresh farms or the subdivision of the old ones, a part of the society may suffer want, even in the midst of apparent plenty. It is not enough that a country should have the power of producing food in abundance, but the state of society must be such as to afford the means of its proper distribution; and the reason why population goes on slowly in these countries is, that the small demand for labour prevents that distribution of the produce of the soil, which, while the divisions of land remain the same, can alone make the lower classes of society partakers of the plenty which it affords. The mode of agriculture is described to be extremely simple, and to require very few labourers. In some places the seed is merely thrown on the fallow.[384] The buck-wheat is a common culture; and though it is sown very thin, yet one sowing will last five or six years,

[379] Id. p. 72.

[380] Id. p. 60.

[381] Voy. de Pallas, tom. iv. p. 92.

[382] Id. p. 5.

[383] Polit. Econ. b. i. c. v. p. 30. 4to.

[384] Voy. de Pallas, tom. i. p. 250.

and produce every year twelve or fifteen times the original quantity. The seed which falls during the time of the harvest is sufficient for the next year, and it is only necessary to pass a harrow once over it in the spring. And this is continued till the fertility of the soil begins to diminish. It is observed, very justly, that the cultivation of no kind of grain can so exactly suit the indolent inhabitants of the plains of Siberia.[385]

With such a system of agriculture, and with few or no manufactures, the demand for labour must very easily be satisfied. Corn will undoubtedly be very cheap; but labour will in proportion be still cheaper. Though the farmer may be able to provide an ample quantity of food for his own children, yet the wages of his labourer may not be sufficient to enable him to rear up a family with ease.

If, from observing the deficiency of population compared with the fertility of the soil, we were to endeavour to remedy it by giving a bounty upon children, and thus enabling the labourer to rear up a greater number, what would be the consequence? Nobody would want the work of the supernumerary labourers that were thus brought into the market. Though the ample subsistence of a man for a day might be purchased for a penny, yet nobody would give these people a farthing for their labour. The farmer is able to do all that he wishes, all that he thinks necessary in the cultivation of the soil, by means of his own family and the one or two labourers that he might have before. As these people therefore can give him nothing that he wants, it is not to be expected that he should overcome his natural indolence, and undertake a larger and more troublesome concern, merely to provide them gratuitously with food. In such a state of things, when the very small demand for manufacturing labour is satisfied, what are the rest to do? They are, in fact, as completely without the means of subsistence as if they were living upon a barren sand. They must either emigrate to some place where their work is wanted, or perish miserably of poverty. Should they be prevented from suffering this last extremity by a scanty subsistence given to them, in consequence of a scanty and only occasional use of their labour, it is evident that, though they might exist themselves, they would not be in a capacity to marry and continue to increase the population.

If in the best cultivated and most populous countries of Europe the present divisions of land and farms had taken place, and had not been followed by the introduction of commerce and manufactures, population would long since have come to a stand from the total want of motive to further cultivation, and the consequent want of demand for labour; and it is obvious that the excessive fertility of the country now under consideration would rather aggravate than diminish the difficulty.

It will probably be said that, if there were much good land unused; new settlements and divisions would of course take place, and that the redundant population would raise its own food, and generate the demand for it, as in America.

This would, no doubt, be the case under favourable circumstances; if, for instance, in the first place, the land were of such a nature as to afford all the other materials of capital as

---

[385] Découv. Muss. vol. iv. p. 329. 8vo. 4 vols. Berne.

well as corn; secondly, if such land were to be purchased in small lots, and the property well secured under a free government; and, thirdly, if habits of industry and accumulation generally prevailed among the mass of the people. But the failure of any of these conditions would essentially check; or might altogether stop, the progress of population. Land that would bear the most abundant crops of corn might be totally unfit for extensive and general settlements from a want either of wood or of water. The accumulations of individuals would go most reluctantly and slowly to the land, if the tenures on which farms were held were either insecure or degrading; and no facility of production could effect a permanent increase and proper distribution of the necessaries of life under inveterate habits of indolence and want of foresight.

It is obvious that the favourable circumstances here alluded to have not been combined in Siberia; and even on the supposition of there being no physical defects in the nature of the soil to be overcome, the political and moral difficulties in the way of a rapid increase of population could yield but slowly to the best-directed efforts. In America the rapid increase of agricultural capital is occasioned in a great degree by the savings from the high wages of common labour. The command of thirty or forty pounds at the least, is considered as necessary to enable an active young man to begin a plantation of his own in the back settlements. Such a sum may be saved in a few years without much difficulty in America, where labour is in great demand and paid at a high rate; but the redundant labourer of Siberia would find it extremely difficult to collect such funds as would enable him to build a house, to purchase stock and utensils, and to subsist till he could bring his new land into proper order and obtain an adequate return. Even the children of the farmer, when grown up, would not easily provide these necessary funds. In a state of society where the market for corn is extremely narrow, and the price very low, the cultivators are always poor; and though they may be able amply to provide for their family in the simple article of food, yet they cannot realize a capital to divide among their children, and enable them to undertake the cultivation of fresh land. Though this necessary capital might be very small, yet even this small sum the farmer perhaps cannot acquire; for when he grows a greater quantity of corn than usual, he finds no purchaser for it,[386] and cannot convert it into any permanent article which will enable any of his children to command an equivalent portion of subsistence or labour in future.[387] He often, therefore, contents himself with growing only what is sufficient for the immediate demands of his family, and the narrow market to which he is accustomed. And if he has a large family, many of his children probably fall into the rank of labourers, and their further increase is checked, as in the case of the labourer before described, by a want of the means of subsistence.

[386] Il y a fort peu de débit dans le pays, parceque la plupart des habitans sont cultivateurs, et élèvent eux-mêmes des bestiaux.—Voy. de Pallas, tom. iv. p. 4.

[387] In addition to the causes here mentioned, I have lately been informed that one of the principal reasons why large tracts of rich land lie uncultivated in this part of the world is the swarm of locusts, which at certain seasons covers these districts, and from the ravages of which it is impossible to protect the rising crop.

It is not therefore a direct encouragement to the procreation and rearing of children that is wanted in these countries, in order to increase their population; but the creation of an effectual demand for the produce of the soil, by promoting the means of its distribution. This can only be effected by the introduction of manufactures, and by inspiring the cultivator with a taste for them, and thus enlarging the internal market.

The late empress of Russia encouraged both manufacturers and cultivators; and furnished to foreigners of either description capital free of all interest for a certain term of years.[388] These well directed efforts, added to what had been done by Peter I., had, as might be expected, a considerable effect; and the Russian territories, particularly the Asiatic part of them, which had slumbered for centuries with a population nearly stationary, or at most increasing very languidly, seem to have made a sudden start of late years. Though the population of the more fertile provinces of Siberia be still very inadequate to the richness of the soil; yet in some of them agriculture flourishes in no inconsiderable degree, and great quantities of corn are grown. In a general dearth which happened in 1769, the province of Isetsk was able, notwithstanding a scanty harvest, to supply in the usual manner the founderies and forges of the Ural, besides preserving from the horrors of famine all the neighbouring provinces.[389] And in the territory of Krasnoyarsk, on the shores of the Yenissey, in spite of the indolence and drunkenness of the inhabitants, the abundance of corn is so great that no instance has ever been known of a general failure.[390] Pallas justly observes that, if we consider that Siberia not two hundred years ago was a wilderness utterly unknown, and in point of population even far behind the almost desert tracts of North America, we may reasonably be astonished at the present state of this part of the world, and at the multitude of its Russian inhabitants, who in numbers greatly exceed the natives.[391]

When Pallas was in Siberia, provisions in these fertile districts, particularly in the environs of Krasnoyarsk, were most extraordinarily cheap. A pood, or forty pounds, of wheaten flour, was sold for about two-pence halfpenny, an ox for five or six shillings, and a cow for three or four.[392] This unnatural cheapness, owing to a want of vent for the products of the soil, was perhaps the principal check to industry. In the period which has since elapsed, the prices have risen considerably;[393] and we may conclude therefore that the

[388] Tooke's View of the Russian Empire, vol. ii. p. 242. The principal effect, perhaps, of these importations of foreigners, was the introduction of free men instead of slaves, and of German industry instead of Russian indolence; but the introduction of that part of capital which consists in machinery would be a very great point, and the cheapness of manufactures would soon give the cultivators a taste for them.

[389] Voy. de Pallas, tom. iii. p. 10.

[390] Id. tom. iv. p. 3.

[391] Voy. de Pallas, tom. iv. p. 6.

[392] Id. p. 3.

[393] Tooke's View of the Russian Empire, vol. iii. p 239.

object wanted has been in a great measure attained, and that the population proceeds with rapid strides.

Pallas, however, complains that the intentions of the empress respecting the peopling of Siberia were not always well fulfilled by her subordinate agents, and that the proprietors to whose care this was left, often sent off colonists, in every respect unfit for the purpose in regard to age, disease and want of industrious habits.[394] Even the German settlers in the districts near the Wolga are, according to Pallas, deficient in this last point,[395] and this is certainly a most essential one. It may indeed be safely asserted that the importation of industry is of infinitely more consequence to the population of a country, than the importation of men and women considered only with regard to numbers. Were it possible at once to change the habits of a whole people, and to direct its industry at pleasure, no government would ever be reduced to the necessity of encouraging foreign settlers. But to change long existing habits is of all enterprises the most difficult. Many years must elapse under the most favourable circumstances, before the Siberian boor will possess the industry and activity of an English labourer. And though the Russian government has been incessant in its endeavours to convert the pastoral tribes of Siberia to agriculture, yet many obstinately persist in bidding defiance to any attempts that can be made to wean them from their injurious sloth.[396]

Many other obstacles concur to prevent that rapid growth of the Russian colonies, which the procreative power would permit. Some of the low countries of Siberia are unhealthy from the number of marshes which they contain;[397] and great and wasting epizooties are frequent among the cattle.[398] In the districts near the Wolga, though the soil is naturally rich, yet droughts are so frequent, that there is seldom more than one good harvest out of three.[399] The colonists of Saratof, after they had been settled for some years, were obliged to remove on this account to other districts; and the whole expense of building their houses,

---

[394] Voy. de Pallas, tom. v. p. 5.

[395] Id. p. 253.

[396] Tooke's Russian Empire, vol. iii. p. 313.

[397] Voy. de Pallas, tom. iii. p. 16. Though in countries where the procreative power is never fully called into action, unhealthy seasons and epidemics have but little effect on the average population; yet in new colonies, which are differently circumstanced in this respect, they materially impede its progress. This point is not sufficiently understood. If in countries which were either stationary or increasing very slowly, all the immediate checks to population, which had been observed, were to continue in force, no abundance of food could materially increase the number of people. But the precise way in which such an abundance operates is by diminishing the immediate checks which before prevailed. Those, however, which may remain, either from the difficulty of changing habits, or from any unfavourable circumstances in the soil or climate, will still continue to operate in preventing the procreative power from producing its full effect.

[398] Voy. de Pallas, tom. iii. p. 17. tom. v. p. 411

[399] Id. tom. v. p. 252, et seq.

amounting to above a million of rubles, was remitted to them by the empress.[400] For purposes either of safety or convenience, the houses of each colony are all built contiguous or nearly so, and not scattered about upon the different farms. A want of room is in consequence soon felt in the immediate neighbourhood of the village; while the distant grounds remain in a state of very imperfect cultivation. On observing this in the colony of Kotschesnaia, Pallas proposed that a certain part should be removed by the empress to other districts, that the remainder might be left more at their ease.[401] This proposal seems to prove that spontaneous divisions of this kind did not often take place, and that the children of the colonists might not always find an easy mode of settling themselves, and rearing up fresh families. In the flourishing colony of the Moravian brethren in Sarepta, it is said that the young people cannot marry without the consent of their priests; and that their consent is not in general granted till late.[402] It would appear, therefore, that among the obstacles to the increase of population, even in these new colonies, the preventive check has its share. Population can never increase with great rapidity but when the real price of common labour is very high, as in America; and from the state of society in this part of the Russian territories, and the consequent want of a proper vent for the produce of industry, this effect, which usually accompanies new colonies and is essential to their rapid growth, does not take place in any considerable degree.[403]

3.4.10 Book I, Chapter X: Of the Checks to Population in the Turkish Dominions and Persia.

In the Asiatic parts of the Turkish dominions it will not be difficult, from the accounts of travellers, to trace the checks to population and the causes of its present decay; and as there is little difference in the manners of the Turks, whether they inhabit Europe or Asia, it will not be worth while to make them the subject of distinct consideration.

The fundamental cause of the low state of population in Turkey, compared with its extent of territory, is undoubtedly the nature of the government. Its tyranny, its feebleness,

---

[400] Tooke's Russian Empire, vol. ii. p. 245.

[401] Voy. de Pallas, tom. v. p. 253.

[402] Voy. de Pallas, tom, v. p. 175.

[403] Other causes may concur in restraining the population of Siberia, which have not been noticed by Pallas. In general, it should be observed, with regard to all the immediate checks to population, which I either have had or shall have occasion to mention, that, as it is evidently impossible to ascertain the extent to which each acts, and the proportion of the whole procreative power which it impedes, no accurate inferences respecting the actual state of population can be drawn from them *à priori*. The prevailing checks in two different nations may appear to be exactly the same as to kind, yet if they are different in degree, the rate of increase in each will, of course, be as different as possible. All that can be done, therefore, is to proceed as in physical inquiries; that is, first to observe the facts, and then account for them from the best lights that can be collected.

its bad laws and worse administration of them, together with the consequent insecurity of property, throw such obstacles in the way of agriculture that the means of subsistence are necessarily decreasing yearly, and with them, of course, the number of people. The miri, or general land-tax paid to the sultan; is in itself moderate;[404] but by abuses inherent in the Turkish government, the pachas and their agents have found out the means of rendering it ruinous. Though they cannot absolutely alter the impost which has been established by the sultan, they have introduced a multitude of changes which, without the name, produce all the effects of an augmentation.[405] In Syria, according to Volney, having the greatest part of the land at their disposal, they clog their concessions with burdensome conditions, and exact the half, and sometimes even two-thirds, of the crop. When the harvest is over, they cavil about losses, and as they have the power in their hands, they carry off what they think proper. If the season fail, they still exact the same sum, and expose every thing that the poor peasant possesses to sale. To these constant oppressions are added a thousand accidental extortions. Sometimes a whole village is laid under contribution for some real or imaginary offence. Arbitrary presents are exacted on the accession of each governor; grass, barley and straw are demanded for his horses; and commissions are multiplied, that the soldiers who carry the orders may live upon the starving peasants, whom they treat with the most brutal insolence and injustice.[406]

The consequence of these depredations is that the poorer class of inhabitants, ruined, and unable any longer to pay the miri, become a burden to the village, or fly into the cities; but the miri is unalterable, and the sum to be levied must be found somewhere. The portion of those who are thus driven from their homes falls on the remaining inhabitants, whose burden, though at first light, now becomes insupportable. If they should be visited by two years of drought and famine the whole village is ruined and abandoned; and the tax which it should have paid, is levied on the neighbouring lands.[407]

The same mode of proceeding takes place with regard to the tax on the Christians, which has been raised by these means from three, five, and eleven piastres, at which it was at first fixed, to thirty-five and forty, which absolutely impoverishes those on whom it is levied, and obliges them to leave the country. It has been remarked that these exactions have made a rapid progress during the last forty years; from which time are dated the decline of agriculture, the depopulation of the country and the diminution in the quantity of specie carried into Constantinople.[408]

The food of the peasants is almost every where reduced to a little flat cake of barley or doura, onions, lentils and water. Not to lose any part of their corn, they leave in it all sorts of wild grain, which often produce bad consequences. In the mountains of Lebanon and

[404] Voy. de Volney, tom. ii. c. xxxvii. p. 373. 8vo. 1787.

[405] Voy. de Volney, tom. ii. c. xxxvii. p. 373.

[406] Id. p. 374.

[407] Voy. de Volney, tom. ii. c. xxxvii. p. 375.

[408] Id. p. 376.

Nablous, in time of dearth, they gather the acorns from the oaks, which they eat after boiling or roasting them in ashes.[409]

By a natural consequence of this misery, the art of cultivation is in the most deplorable state. The husbandman is almost without instruments, and those he has are very bad. His plough is frequently no more than the branch of a tree cut below a fork, and used without wheels. The ground is tilled by asses and cows, rarely by oxen, which would bespeak too much riches. In the districts exposed to the Arabs, as in Palestine, the countryman must sow with his musket in his hand; and scarcely does the corn turn yellow before it is reaped, and concealed in subterraneous caverns. As little as possible is employed for seed-corn, because the peasants sow no more than is barely necessary for their subsistence. Their whole industry is limited to a supply of their immediate wants; and to procure a little bread, a few onions, a blue shirt, and a bit of woollen, much labour is not necessary. "The peasant lives therefore in distress; but at least he does not enrich his tyrants, and the avarice of despotism is its own punishment."[410]

This picture, which is drawn by Volney, in describing the state of the peasants in Syria, seems to be confirmed by all other travellers in these countries; and, according to Eton, it represents very nearly the condition of the peasants in the greatest part of the Turkish dominions.[411] Universally, the offices of every denomination are set up to public sale; and in the intrigues of the seraglio, by which the disposal of all places is regulated, every thing is done by means of bribes. The pachas, in consequence, who are sent into the provinces, exert to the utmost their power of extortion; but are always outdone by the officers immediately below them, who, in their turn, leave room for their subordinate agents.[412]

The pacha must raise money to pay the tribute, and also to indemnify himself for the purchase of his office, support his dignity, and make a provision in case of accidents; and as all power, both military and civil, centres in his person from his representing the sultan, the means are at his discretion, and the quickest are invariably considered as the best.[413] Uncertain of tomorrow, he treats his province as a mere transient possession, and endeavours to reap, if possible, in one day the fruit of many years, without the smallest regard to his successor, or the injury that he may do to the permanent revenue.[414]

The cultivator is necessarily more exposed to these extortions than the inhabitant of the towns. From the nature of his employment, he is fixed to one spot, and the productions of agriculture do not admit of being easily concealed. The tenure of the land and the rights of succession are besides uncertain. When a father dies, the inheritance reverts to the sultan, and the children can only redeem the succession by a considerable sum of money. These

---

[409] Id. p. 377.

[410] Voy. de Volney, tom. ii. c. xxxvii. p. 379.

[411] Eton's Turkish Emp. c. viii. 2d edit. 1799.

[412] Eton's Turkish Emp. c. ii. p. 55.

[413] Voy. de Volney, tom. ii. c. xxxiii. p. 347.

[414] Id. p. 350.

considerations naturally occasion an indifference to landed estates. The country is deserted; and each person is desirous of flying to the towns, where he will not only in general meet with better treatment, but may hope to acquire a species of wealth which he can more easily conceal from the eyes of his rapacious masters.[415]

To complete the ruin of agriculture, a maximum is in many cases established, and the peasants are obliged to furnish the towns with corn at a fixed price. It is a maxim of Turkish policy, originating in the feebleness of the government and the fear of popular tumults, to keep the price of corn low in all the considerable towns. In the case of a failure in the harvest, every person who possesses any corn is obliged to sell it at the price fixed, under pain of death; and if there be none in the neighbourhood, other districts are ransacked for it.[416] When Constantinople is in want of provisions, ten provinces are perhaps famished for a supply.[417] At Damascus, during the scarcity in 1784, the people paid only one penny farthing a pound for their bread, while the peasants in the villages were absolutely dying with hunger.[418]

The effect of such a system of government on agriculture need not be insisted upon. The causes of the decreasing means of subsistence are but too obvious; and the checks, which keep the population down to the level of these decreasing resources, may be traced with nearly equal certainty, and will appear to include almost every species of vice and misery that is known.

It is observed in general that the Christian families consist of a greater number of children than the Mahometan families in which polygamy prevails.[419] This is an extraordinary fact; because though polygamy, from the unequal distribution of women which it occasions, be naturally unfavourable to the population of a whole country; yet the individuals who are able to support a plurality of wives, ought certainly, in the natural course of things, to have a greater number of children than those who are confined to one. The way in which Volney principally accounts for this fact is that, from the practice of polygamy, and very early marriages, the Turks are enervated while young, and impotence at thirty is very common.[420] Eton notices an unnatural vice as prevailing in no inconsiderable degree among the common people, and considers it as one of the checks to the population;[421] but the five principal causes of depopulation which he enumerates, are,

1. The plague, from which the empire is never entirely free.
2. Those terrible disorders which almost always follow it, at least in Asia.

---

[415] Voy. de Volney, tom. ii. c. xxxvi. p. 369.

[416] Id. c. xxxviii. p. 38.

[417] Id. c. xxxiii. p. 345.

[418] Id. c. xxxviii. p. 381.

[419] Eton's Turkish Emp. c. vii. p. 275.

[420] Voy. de Volney, tom. ii. c. xl. p. 445.

[421] Eton's Turkish Emp. c. vii, p. 275.

3. Epidemic and endemic maladies in Asia, which make as dreadful ravages as the plague itself, and which frequently visit that part of the empire.
4. Famine.
5. And lastly, the sicknesses which always follow a famine, and which occasion a much greater mortality.[422]

He afterwards gives a more particular account of the devastations of the plague in different parts of the empire, and concludes by observing, that if the number of the Mahometans have decreased, this cause alone is adequate to the effect;[423] and that, things going on in their present train, the Turkish population will be extinct in another century.[424] But this inference, and the calculations which relate to it, are without doubt erroneous. The increase of population in the intervals of these periods of mortality is probably greater than he is aware of. At the same time it must be remarked that in a country where the industry of the husbandman is confined to the supply of his necessary wants, where he sows only to prevent himself from starving, and is unable to accumulate any surplus produce, a great loss of people is not easily recovered; as the natural effects arising from the diminished numbers cannot be felt in the same degree as in countries where industry prevails, and property is secure.

According to the Persian legislator Zoroaster, to plant a tree, to cultivate a field, to beget children, are meritorious acts; but it appears from the accounts of travellers, that many among the lower classes of people cannot easily attain the latter species of merit; and in this instance, as in numberless others, the private interest of the individual corrects the errors of the legislator. Sir John Chardin says, that matrimony in Persia is very expensive, and that only men of estates will venture upon it, lest it prove their ruin.[425] The Russian travellers seem to confirm this account, and observe that the lower classes of people are obliged to defer marriage till late; and that it is only among the rich that this union takes place early.[426]

The dreadful convulsions to which Persia has been continually subject for many hundred years must have been fatal to her agriculture. The periods of repose from external wars and internal commotions have been short and few; and even during the times of profound peace, the frontier provinces have been constantly subject to the ravages of the Tartars.

The effect of this state of things is such as might be expected. The proportion of uncultivated to cultivated land in Persia, Sir John Chardin states to be ten to one;[427] I and the mode in which the officers of the Shah and private owners let out their lands to

[422] Eton's Turkish Emp. c. vii. p. 264.

[423] Id. p. 291.

[424] Id. p. 280.

[425] Sir John Chardin's Travels, Harris's Collect. b. iii. c. ii. p. 870.

[426] Découv. Russ. tom. ii. p. 293.

[427] Chardin's Travels, Harris's Collect. b. iii. c. ii. p. 902.

husbandmen is not that which is best calculated to reanimate industry. The grain in Persia is also very subject to be destroyed by hail, drought, locusts, and other insects,[428] which probably tends rather to discourage the employment of capital in the cultivation of the soil.

The plague does not extend to Persia; but the small-pox is mentioned by the Russian travellers as making very fatal ravages.[429]

It will not be worth while to enter more minutely on the checks to population in Persia, as they seem to be nearly similar to those which have been just described in the Turkish dominions. The superior destruction of the plague, in Turkey, is perhaps nearly balanced by the greater frequency of internal commotions in Persia.

3.4.11 Book I, Chapter XI: Of the Checks to Population in Indostan and Tibet.

In the ordinances of Menu, the Indian legislator, which Sir Wm. Jones has translated, and called the *Institutes of Hindu Law,* marriage is very greatly encouraged, and a male heir is considered as an object of the first importance.

"By a son a man obtains victory over all people; by a son's son he enjoys immortality; and afterwards by the son of that grandson he reaches the solar abode." "Since the son delivers his father from the hell, named Put, he was therefore called puttra, by Brahma himself."[430]

Among the different nuptial rites, Menu has ascribed particular qualities to each.

"A son of a *Bráhmì,* or wife by the first ceremony, redeems from sin, if he perform virtuous acts, ten ancestors, ten descendants and himself, the twenty-first person." "A son born of a wife by the *Daiva* nuptials redeems seven anal seven, in higher and lower degrees; of a wife by the *Arsha,* three and three; of a wife by the *Prájápatya* six and six."[431]

A housekeeper is considered as of the most eminent order. "The divine sages, the manes, the gods, the spirits and guests pray for benefits to masters of families."[432] An elder brother not married before the younger, is mentioned among the persons who are particularly to be shunned.[433]

---

[428] Id.

[429] Découv. Russ, tom. ii. p. 377.

[430] Sir William Jones's Works, vol. iii. c. ix. p. 354. Speaking of the Indian laws, the Abbé Raynal says, "La population est un devoir primitif, un ordre de la nature si sacré, que la loi permet de tromper, de mentir, de se parjurer pour favoriser un mariage." Hist. des Indes, tom. i. l. i. p. 81. 8vo. 10 vols. Paris, 1795.

[431] Sir Wm. Jones's Works, vol. iii. c. iii. p. 124.

[432] Id. p. 130.

[433] Id. p. 141.

Such ordinances would naturally cause marriage to be considered a religious duty; yet it seems to be rather a succession of male heirs, than a very numerous progeny, that is the object so much desired.

"The father having begotten a son, discharges his debt to his own progenitors." "That son alone, by whose birth he discharges the debt, and through whom he attains immortality, was begotten from a sense of duty; all the rest are considered by the wise as begotten from love of pleasure."[434]

A widow is on some occasions allowed to have one son by the brother, or some appointed kinsman of the deceased husband, but on no account a second. "The first object of the appointment being obtained according to law, both brother and sister must live together like a father and daughter by affinity."[435]

In almost every part of the ordinances of Menu, sensuality of all kinds is strongly reprobated, and chastity inculcated as a religious duty.

"A man by the attachment of his organs to sensual pleasures incurs certain guilt; but having wholly subdued them, he hence attains heavenly bliss." "Whatever man may obtain all those gratifications, or whatever man may resign them completely, the resignation of all pleasures is far better than the attainment of them."[436]

It is reasonable to suppose that such passages might, in some degree, tend to counteract those encouragements to increase, which have been before mentioned; and might prompt some religious persons to desist from further indulgences, when they had obtained one son, or to remain more contented than they otherwise would have been in an unmarried state. Strict and absolute chastity seems indeed to supersede the obligation of having descendants.

"Many thousands of Brahmens having avoided sensuality from their early youth, and having left no issue in their families, have ascended nevertheless to Heaven." "And like those abstemious men, a virtuous wife ascends to Heaven though she have no child, if after the decease of her lord she devote herself to pious austerity.[437]

The permission to a brother or other kinsman to raise up an heir for the deceased husband, which has been noticed, extends only to women of the servile class.[438] Those of the higher classes are not even to pronounce the name of another man, but "to continue till death forgiving all injuries, performing harsh duties, avoiding every sensual pleasure, and cheerfully practising the incomparable rules of virtue."[439]

---

[434] Id. vol. iii. c. ix. p. 340.

[435] Sir William Jones's Works, vol. iii. c. ix. p. 343.

[436] Id. vol. iii. c. ii. p. 96.

[437] Sir William Jones's Works, vol. iii. c. v. p. 221.

[438] Id. c. ix. p. 343.

[439] Id. c. v. p. 221.

Besides these strict precepts relating to the government of the passions, other circumstances would perhaps concur to prevent the full effect of the ordinances which encourage marriage.

The division of the people into classes, and the continuance of the same profession in the same family, would be the means of pointing out to each individual, in a clear and distinct manner, his future prospects respecting a livelihood; and from the gains of his father he would be easily enabled to judge whether he could support a family by the same employment. And though, when a man cannot gain a subsistence in the employments appropriate to his class, it is allowable for him, under certain restrictions, to seek it in another; yet some kind of disgrace seems to attach to this expedient; and it is not probable that many persons would marry with the certain prospect of being obliged thus to fall from their class, and to lower in so marked a manner their condition in life.

In addition to this, the choice of a wife seems to be a point of considerable difficulty. A man might remain unmarried for some time, before he could find exactly such a companion as the legislator prescribes. Ten families of a certain description, be they ever so great or ever so rich in kine, goats, sheep, gold and grain, are studiously to be avoided. Girls with too little or too much hair, who are too talkative, who have bad eyes, a disagreeable name or any kind of sickness, who have no brother, or whose father is not well known, are all; with many others, excluded; and the choice will appear to be in some degree confined, when it must necessarily rest upon a girl "whose form has no defect; who has an agreeable name; who walks gracefully, like a phenicopteros or a young elephant; whose hair and teeth are moderate respectively in quantity and size; whose body has exquisite softness."[440]

It is observed, that a woman of the servile class is not mentioned, even in the recital of any ancient story, as the wife of a Brahmen or of a Cshatriya, though in the greatest difficulty to find a suitable match; which seems to imply, that such a difficulty might sometimes occur.[441]

Another obstacle to marriage arising from Hindoo customs is, that an elder brother who does not marry seems in a manner to confine all his other brothers to the same state; for a younger brother, who marries before the elder, incurs disgrace, and is mentioned among the persons who ought to be shunned.[442]

The character, which the legislator draws of the manners and dispositions of the women in India, is extremely unfavourable. Among many other passages expressed with equal severity, he observes, that, "through their passion for men, their mutable temper, their want of settled affection and their perverse nature, let them be guarded in this world ever so well, they soon become alienated from their husbands."[443]

---

[440] Sir William Jones's Works, vol. iii. c. iii. p. 120.

[441] Id. p. 121.

[442] Sir William Jones's Works, vol. iii. c. iii. p. 141.

[443] Id. c. ix. p. 337.

This character, if true, probably proceeded from their never being allowed the smallest degree of liberty,[444] and from the state of degradation to which they were reduced by the practice of polygamy; but however this may be, such passages tend strongly to shew that illicit intercourse between the sexes was frequent, notwithstanding the laws against adultery. These laws are noticed as not relating to the wives of public dancers or singers, or of such base men as lived by the intrigues of their wives;[445] a proof that these characters were not uncommon, and were to a certain degree permitted. Add to this, that the practice of polygamy[446] among the rich would sometimes render it difficult for the lower classes of people to obtain wives; and this difficulty would probably fall particularly hard on those, who were reduced to the condition of slaves.

From all these circumstances combined, it seems probable that among the checks to population in India the preventive check would have its share; but from the prevailing habits and opinions of the people, there is reason to believe that the tendency to early marriages was still always predominant; and in general prompted every person to enter into this state, who could look forward to the slightest chance of being able to maintain a family. The natural consequence of this was, that the lower classes of people were reduced to extreme poverty, and were compelled to adopt the most frugal and scanty mode of subsistence. This frugality was still further increased, and extended in some degree to the higher classes of society, by its being considered as an eminent virtue.[447] The population would thus be pressed hard against the limits of the means of subsistence, and the food of the country would be meted out to the major part of the people in the smallest shares that could support life. In such a state of things every failure in the crops from unfavourable seasons would be felt most severely; and India, as might be expected, has in all ages been subject to the most dreadful famines.

A part of the ordinances of Menu is expressly dedicated to the consideration of times of distress, and instructions are given to the different classes respecting their conduct during these periods. Brahmens pining with hunger and want are frequently mentioned[448] and certain ancient and virtuous characters are described, who had done impure and unlawful acts, but who were considered by the legislator as justified on account of the extremities to which they were reduced.

"Ajígarta, dying with hunger, was going to destroy his own son by selling him for some cattle; yet he was guilty of no crime, for he only sought a remedy against famishing." "Vámadéva, who well knew right and wrong, was by no means rendered impure, though desirous, when oppressed by hunger, of eating the flesh of dogs." "Viswámitra too, than whom none knew better the distinctions between virtue and vice, resolved, when he was

---

[444] Id. c. v. p. 219.

[445] Id. c. viii. p. 325.

[446] Sir William Jones's Works, vol, iii. c: ix. p. 346, 347.

[447] Id. c. iii. p. 133.

[448] Sir William Jones's Works, vol. iii. c. iv. p. 165. c. x. p. 397.

perishing with hunger, to eat the haunch of a dog, which he had received from a *Chaudála.*"[449]

If these great and virtuous men of the highest class, whom all persons were under the obligation of assisting, could be reduced to such extremities, we may easily conjecture what must have been the sufferings of the lowest class.

Such passages clearly prove the existence of seasons of the most severe distress, at the early period when these ordinances were composed; and we have reason to think, that they have occurred at irregular intervals ever since. One of the Jesuits says that it is impossible for him to describe the misery to which he was witness during the two-years' famine in 1737 and 1738;[450] but the description which he gives of it, and of the mortality which it occasioned, is sufficiently dreadful without further detail. Another Jesuit, speaking more generally, says, "Every year we baptize a thousand children, whom their parents can no longer feed, or who, being likely to die, are sold to us by their mothers, in order to get rid of them."[451]

The positive checks to population would of course fall principally upon the Sudrá class, and those still more miserable beings, who are the outcasts of all the classes and are not even suffered to live within the towns.[452]

On this part of the population the epidemics, which are the consequences of indigence and bad nourishment, and the mortality among young children, would necessarily make great ravages and thousands of these unhappy wretches would probably be swept off in a period of scarcity, before any considerable degree of want had reached the middle classes of the society. The Abbé Raynal says (on what authority I know not), that, when the crops of rice fail, the huts of these poor outcasts are set on fire, and the flying inhabitants shot by the proprietors of the grounds, that they may not consume any part of the produce.[453]

The difficulty of rearing a family even among the middle and higher classes of society, or the fear of sinking from their cast, has driven the people in some parts of India to adopt the most cruel expedients to prevent a numerous offspring. In a tribe on the frontiers of Junapore, a district of the province of Benares, the practice of destroying female infants has been fully substantiated. The mothers were compelled to starve them. The reason that the people gave for this cruel practice was the great expense of procuring suitable matches for their daughters. One village only furnished an exception to this rule, and in that village several old maids were living.

It will naturally occur, that the race could not be continued upon this principle: but it appeared that the particular exceptions to the general rule and the intermarriages with other

---

[449] Id. c. x. p. 397, 398.

[450] Lettres Edif. tom. xiv. p. 178.

[451] Id. p. 284.

[452] ir William Jones's Works, vol. iii. c. x. p. 390.

[453] Hist. des Indes, tom. i. liv. i. p. 97. 8vo. 10 vols. Paris, 1795.

tribes were sufficient for this purpose. The East-India Company obliged these people to enter into an engagement not to continue this inhuman practice.[454]

On the coast of Malabar the Nayrs do not enter into regular marriages, and the right of inheritance and succession rests in the mother of the brother, or otherwise goes to the sister's son, the father of the child being always considered as uncertain.

Among the Brahmens, when there are more brothers than one, only the elder or eldest of them marries. The brothers, who thus maintain celibacy, cohabit with Nayr women without marriage in the way of the Nayrs. If the eldest brother has not a son, then the next brother marries.

Among the Nayrs, it is the custom for one Nayr woman to have attached to her two males, or four, or perhaps more.

The lower casts, such as carpenters, ironsmiths, and others, have fallen into the imitation of their superiors, with this difference, that the joint concern in one woman is confined to brothers and male relations by blood, to the end that no alienation may take place in the course of the succession.[455]

Montesquieu takes notice of this custom of the Nayrs on the coast of Malabar, and accounts for it on the supposition that it was adopted in order to weaken the family ties of this cast, that as soldiers they might be more at liberty to follow the calls of their profession: but I should think that it originated more probably in a fear of the poverty arising from a large family, particularly as the custom seems to have been adopted by the other classes.[456]

In Tibet, according to Turner's account of that country, a custom of this kind prevails generally. Without pretending absolutely to determine the question of its origin, Mr. Turner leans to the supposition that it arose from the fear of a population too great for an unfertile country. From travelling much in the east he had probably been led to observe the effects necessarily resulting from an overflowing population, and is in consequence one among the very few writers, who see these effects in their true light. He expresses himself very strongly on this subject, and, in reference to the above custom, says, "It certainly appears, that a superabundant population in an unfertile country must be the greatest of all calamities, and produce eternal warfare or eternal want. Either the most active and the most able part of the community must be compelled to emigrate, and to become soldiers of fortune or merchants of chance; or else, if they remain at home, be liable to fall a prey to famine in consequence of some accidental failure in their scanty crops. By thus linking whole families together in the matrimonial yoke, the too rapid increase of population was perhaps checked, and an alarm prevented, capable of pervading the most fertile region upon the earth, and of giving birth to the most inhuman and unnatural practice, in the richest, the most productive and the most populous country in the world. I allude to the Empire of China, where a mother, not foreseeing the means of raising or providing for a numerous

[454] Asiatic Researches, vol. iv. p. 354.

[455] Asiatic Researches, vol. v. p. 14.

[456] Esprit des Loix, liv. xvi. c. 5.

family, exposes her newborn infant to perish in the fields; a crime, however odious, by no means I am assured unfrequent."[457]

In almost every country of the globe individuals are compelled by considerations of private interest to habits, which tend to repress the natural increase of population; but Tibet is perhaps the only country, where these habits are universally encouraged by the government, and where to repress rather than to encourage population seems to be a public object.

In the first career of life the Bootea is recommended to distinction by a continuance in a state of celibacy; as any matrimonial contract proves almost a certain hinderance to his rise in rank, or his advancement to offices of political importance. Population is thus opposed by the two powerful bars of ambition and religion; and the higher orders of men, entirely engrossed by political or ecclesiastical duties, leave to the husbandman and labourer, to those who till the fields and live by their industry, the exclusive charge of propagating the species.[458]

Hence religious retirement is frequent,[459] and the number of monasteries and nunneries is considerable. The strictest laws exist to prevent a woman from accidentally passing a night within the limits of the one, or a man within those of the other; and a regulation is framed completely to obviate abuse; and establish respect towards the sacred orders of both sexes.

The nation is divided into two distinct and separate classes, those who carry on the business of the world, and those who hold intercourse with heaven. No interference of the laity ever interrupts the regulated duties of the clergy. The latter, by mutual compact, take charge of all spiritual concerns; and the former by their labours enrich and populate the state.[460]

But even among the laity the business of population goes on very coldly. All the brothers of a family, without any restriction of age or of numbers, associate their fortunes with one female, who is chosen by the eldest, and considered as the mistress of the house; and whatever may be the profits of their several pursuits, the result flows into the common store.[461]

The number of husbands is not apparently defined, or restricted within any limits. It sometimes happens that in a small family there is but one male; and the number, Mr. Turner says, may seldom exceed that which a native of rank at Teshoo Loomboo pointed out to him in a family resident in the neighbourhood, in which five brothers were then living together very happily with one female under the same connubial compact. Nor is this sort of league confined to the lower ranks of people alone; it is found also frequently in the most opulent families.[462]

---

[457] Turner's Embassy to Tibet, part ii. c. x. p. 351.

[458] Id. c. i. p. 172.

[459] Ib.

[460] Turner's Embassy, part ii. c. viii. p. 312.

[461] Id. c. x. p. 348, 350.

[462] Turner's Embassy, part ii. c. x. p. 349.

It is evident that this custom, combined with the celibacy of such a numerous body of ecclesiastics, must operate in the most powerful manner as a preventive check to population. Yet, notwithstanding this excessive check, it would appear, from Mr. Turner's account of the natural sterility of the soil, that the population is kept up to the level of the means of subsistence; and this seems to be confirmed by the number of beggars in Teshoo Loomboo. On these beggars, and the charity which feeds them, Mr. Turner's remark, though common, is yet so just and important that it cannot be too often repeated.

"Thus I unexpectedly discovered," he says, "where I had constantly seen the round of life moving in a tranquil regular routine, a mass of indigence and idleness, of which I had no idea. But yet it by no means surprised me, when I considered that, wherever indiscriminate charity exists, it will never want objects on which to exercise its bounty, but will always attract expectants more numerous than it has the means to gratify. No human being can suffer want at Teshoo Loomboo. It is on this humane disposition, that a multitude even of Musselmen, of a frame probably the largest and most robust in the world, place their reliance for the mere maintenance of a feeble life; and besides these, I am informed, that no less than three hundred Hindoos, Goseins, and Sunniasses, are daily fed at this place by the Lama's bounty."[463]

3.4.12 Book I, Chapter XII: Of the Checks to Population in China and Japan.

The account, which has lately been given of the population of China, is so extraordinary as to startle the faith of many readers, and tempt them to suppose, either that some accidental error must have crept into the calculations from an ignorance of the language; or that the mandarin, who gave Sir George Staunton the information, must have been prompted by a national pride, (which is common every where, but particularly remarkable in China,) to exaggerate the power and resources of his country. It must be allowed that neither of these circumstances is very improbable; at the same time it will be found that the statement of Sir George Staunton does not very essentially differ from other accounts of good authority: and, so far from involving any contradiction, is rendered probable by a reference to those descriptions of the fertility of China, in which all the writers who have visited the country agree.

According to Duhalde, in the poll made at the beginning of the reign of Kang-hi, there were found 11,052,872 families, and 59,788,364 men able to bear arms; and yet neither the princes, nor the officers of the court, nor the mandarins, nor the soldiers who had served and been discharged; nor the literati, the licentiates, the doctors, the bonzas, nor young persons under twenty years of age, nor the great multitudes living either on the sea or on rivers in barks, are comprehended in this number.[464]

---

[463] Turner's Embassy, part ii. c. ix. p. 330.

[464] Duhalde's Hist. of China, 2 vols. folio, 1738. vol. i. p. 244.

The proportion, which the number of men of a military age bears to the whole population of any country, is generally estimated as 1 to 4. If we multiply 59,788,364 by 4, the result will be 239,153,456; but in the general calculations on this subject, a youth is considered as capable of bearing arms before he is twenty. We ought therefore to have multiplied by a higher number. The exceptions to the poll seem to include almost all the superior classes of society, and a very great number among the lower. When all these circumstances are taken into consideration, the whole population, according to Duhalde, will not appear to fall very short of the 333,000,000 mentioned by Sir George Staunton.[465]

The small number of families in proportion to the number of persons able to bear arms, which is a striking part of this statement of Duhalde, is accounted for by a custom noticed by Sir George Staunton as general in China. In the enclosure belonging to one dwelling, he observes that a whole family of three generations, with all their respective wives and children, will frequently be found. One small room is made to serve for the individuals of each family, sleeping in different beds, divided only by mats hanging from the ceiling. One common room is used for eating.[466] In China there is besides a prodigious number of slaves,[467] who will of course be reckoned as part of the families to which they belong. These two circumstances may perhaps be sufficient to account for what at first appears to be a contradiction in the statement.

To account for this population, it will not be necessary to recur to the supposition of Montesquieu, that the climate of China is in any peculiar manner favourable to the production of children, and that the women are more prolific than in any other part of the world.[468] The causes which have principally contributed to produce this effect appear to be the following:

First, the excellence of the natural soil, and its advantageous position in the warmest parts of the temperate zone, a situation the most favourable to the productions of the earth. Duhalde has a long chapter on the plenty which reigns in China, in which he observes, that almost all that other kingdoms afford may be found in China; and that China produces an infinite number of things, which are to be found nowhere else. This plenty, he says, may be attributed as well to the depth of the soil, as to the painful industry of its inhabitants, and the great number of lakes, rivers, brooks and canals, wherewith the country is watered.[469]

Secondly, the very great encouragement that from the beginning of the monarchy has been given to agriculture, which has directed the labours of the people to the production of the greatest possible quantity of human subsistence. Duhalde says, that what makes these people undergo such incredible fatigues in cultivating the earth is not barely their private interest, but rather the veneration paid to agriculture, and the esteem which the emperors

[465] Embassy to China, vol. ii. Appen. p. 615. 4to.

[466] Embassy to China, vol. ii. Appen. p. 155. 4to.

[467] Duhalde's China, vol. i. p. 278.

[468] Esprit des Loix, liv. viii. c. xxi.

[469] Duhalde's China, vol. i. p. 314.

themselves have always had for it, from the commencement of the monarchy. One emperor of the highest reputation was taken from the plough to sit on the throne. Another found out the art of draining water from several low countries, which were till then covered with it, of conveying it in canals to the sea, and of using these canals to render the soil fruitful.[470] He besides wrote several books on the manner of cultivating land, by dunging, tilling and, watering it. Many other emperors expressed their zeal for this art and made laws to promote it; but none raised its esteem to a higher pitch than Ven-ti, who reigned 179 years before Christ. This prince, perceiving that his country was ruined by wars, resolved to engage his subjects to cultivate their lands, by the example of ploughing with his own hands the land belonging to his palace, which obliged all the ministers and great men of his court to do the same.[471]

A great festival, of which this is thought to be the origin, is solemnized every year in all the cities of China on the day that the sun enters the fifteenth degree of Aquarius, which the Chinese consider as the beginning of their spring. The emperor goes himself in a solemn manner to plough a few ridges of land, in order to animate the husbandman by his own example; and the mandarins of every city perform the same ceremony.[472] Princes of the blood and other illustrious persons hold the plough after the emperor, and the ceremony is preceded by the spring sacrifice, which the emperor as chief pontiff offers to Shangti, to procure plenty in favour of his people.

The reigning emperor in the time of Duhalde celebrated this festival with extraordinary solemnity, and in other respects shewed an uncommon regard for husbandmen. To encourage them in their labours, he ordered the governors of all the cities to send him notice every year of the person in this profession, in their respective districts, who was most remarkable for his application to agriculture, for unblemished reputation, for preserving union in his own family, and peace with his neighbours, and for his frugality, and aversion to all extravagance.[473] The mandarins in their different provinces encourage with honours the vigilant cultivator, and stigmatize with disgrace the man whose lands are neglected.[474]

In a country, in which the whole of the government is of the patriarchal kind, and the emperor is venerated as the father of his people and the fountain of instruction, it is natural to suppose, that these high honours paid to agriculture would have a powerful effect. In the gradations of rank, they have raised the husbandman above the merchant or mechanic;[475] and the great object of ambition among the lower classes is to become possessed of a small portion of land. The number of manufacturers bears but a very inconsiderable proportion to

[470] Id. p. 274.

[471] Duhalde's China, vol. i. p. 275.

[472] Id. ib.

[473] Id. p. 276.

[474] Lettres Edif. tom. xix. p. 132.

[475] Duhalde's s China, vol. i. p. 272.

that of husbandmen in China;[476] and the whole surface of the empire is, with trifling exceptions, dedicated to the production of food for man alone. There is no meadow, and very little pasture; neither are the fields cultivated in oats, beans or turnips, for the support of cattle of any kind. Little land is taken up for roads, which are few and narrow, the chief communication, being by water. There are no commons or lands suffered to lie waste by the neglect or the caprice or for the sport of great proprietors. No arable land lies fallow. The soil, under a hot and fertilizing sun, yields annually in most instances double crops; in consequence of adapting the culture to the soil, and of supplying its defects by mixture with other earths, by manure, by irrigation and by careful and judicious industry of every kind. The labour of man is little diverted from that industry, to minister to the luxuries of the opulent and powerful, or in employments of no real use. Even the soldiers of the Chinese army, except during the short intervals of the guards which they are called upon to mount, or the exercises or other occasional services which they perform, are mostly employed in agriculture. The quantity of subsistence is increased also by converting more species of animals and vegetables to that purpose, than is usual in other countries.[477]

This account, which is given by Sir George Staunton, is confirmed by Duhalde and the other Jesuits; who agree in describing the persevering industry of the Chinese, in manuring, cultivating and watering their lands, and their success in producing a prodigious quantity of human subsistence.[478] The effect of such a system of agriculture on population must be obvious.

Lastly, the extraordinary encouragements that have been given to marriage, which have caused the immense produce of the country to be divided into very small shares, and have consequently rendered China more populous, in proportion to its means of subsistence, than perhaps any other country in the world.

The Chinese acknowledge two ends in marriage;[479] the first is, that of perpetuating the sacrifices in the temple of their fathers; and the second, the multiplication of the species. Duhalde says, that the veneration and submission of children to parents, which is the grand principle of their political government, continues even after death, and that the same duties are paid to them as if they were living. In consequence of these maxims, a father feels some sort of dishonour, and is not easy in his mind, if he do not marry off all his children; and an elder brother, though he inherit nothing from his father, must bring up the younger children and marry them, lest the family should become extinct, and the ancestors be deprived of the honours and duties they are entitled to from their descendants.[480]

Sir George Staunton observes that whatever is strongly recommended, and generally practised, is at length considered as a kind of religious duty; and that the marriage union as

---

[476] Embassy to China, Staunton, vol. ii. p. 544.

[477] Embassy to China, Staunton, vol. ii. p. 545.

[478] Duhalde, chapter on Agriculture, vol. i. p. 272; chapter on Plenty, p. 314.

[479] Lettres Edif. et Curieuses, tom. xxiii. p. 448.

[480] Duhalde's China, vol. i. p. 303.

such takes place in China, wherever there is the least prospect of subsistence for a future family. This prospect however is not always realized, and the children are then abandoned by the wretched authors of their being;[481] but even this permission given to parents thus to expose their offspring tends undoubtedly to facilitate marriage, and encourage population. Contemplating this extreme resource beforehand, less fear is entertained of entering into the married state; and the parental feelings will always step forwards, to prevent a recurrence to it, except under the most dire necessity. Marriage with the poor is besides a measure of prudence, because the children, particularly the sons, are bound to maintain their parents.[482]

The effect of these encouragements to marriage among the rich, is to subdivide property, which has in itself a strong tendency to promote population. In China there is less inequality in the fortunes than in the conditions of men. Property in land has been divided into very moderate parcels, by the successive distribution of the possessions of every father equally among his sons. It rarely happens that there is but one son to enjoy the whole property of his deceased parents; and from the general prevalence of early marriages, this property is not often increased by collateral succession.[483] These causes constantly tend to level wealth; and few succeed to such an accumulation of it, as to render them independent of any efforts of their own for its increase. It is a common remark among the Chinese, that fortunes seldom continue considerable in the same family beyond the third generation.[484]

The effect of encouragements to marriage on the poor is to keep the reward of labour as low as possible, and consequently to press them down to the most abject state of poverty. Sir George Staunton observes, that the price of labour is generally found to bear as small a proportion every where to the rate demanded for provisions, as the common people can suffer; and that, notwithstanding the advantage of living together in large families, like soldiers in a mess, and the exercise of the greatest economy in the management of these messes, they are reduced to the use of vegetable food, with a very rare and scanty relish of any animal substance.[485]

Duhalde, after describing the painful industry of the Chinese, and the shifts and contrivances unknown in other countries, to which they have recourse in order to gain a subsistence, says, "Yet it must be owned, that, notwithstanding the great sobriety and industry of the inhabitants of China, the prodigious number of them occasions a great deal of misery. There are some so poor that, being unable to supply their children with common necessaries, they expose them in the streets." **** "In the great cities, such as Pekin and Canton, this shocking sight is very common."[486]

[481] Embassy to China, vol. ii. p. 157.

[482] Embassy to China, vol. ii. p. 157.

[483] Id. p. 151.

[484] Id. p. 152.

[485] Embassy to China, Staunton, vol. ii. p. 156.

[486] Duhalde's China, vol. i. p. 277.

The Jesuit Premare, writing to a friend of the same society, says, "I will tell you a fact, which may appear to be a paradox,[487] but is nevertheless strictly true. It is, that the richest and most flourishing empire of the world is notwithstanding, in one sense, the poorest and the most miserable of all. The country, however extensive and fertile it may be, is not sufficient to support its inhabitants. Four times as much territory would be necessary to place them at their ease. In Canton alone, there is, without exaggeration, more than a million of souls, and in a town three or four leagues distant a still greater number. Who then can count the inhabitants of this province? But what is this to the whole empire, which contains fifteen great provinces, all equally peopled? To how many millions would such a calculation amount? A third part of this infinite population would hardly find sufficient rice to support itself properly.

"It is well known, that extreme misery impels people to the most dreadful excesses. A spectator in China, who examines things closely, will not be surprised that mothers destroy or expose many of their children; that parents sell their daughters for a trifle; that the people should be interested; and that there should be such a number of robbers. The surprise is, that nothing still more dreadful should happen; and that in the times of famine which are here but too frequent, millions of people should perish with hunger, without having recourse to those dreadful extremities, of which we read examples in the histories of Europe.

"It cannot be said in China, as in Europe, that the poor are idle, and might gain a subsistence, if they would work. The labours and efforts of these poor people are beyond conception. A Chinese will pass whole days in digging the earth, sometimes up to his knees in water, and in the evening is happy to eat a little spoonful of rice, and to drink the insipid water in which it was boiled. This is all that they have in general."[488]

A great part of this account is repeated in Duhalde; and, even allowing for some exaggeration, it shews in a strong point of view to what degree population has been forced in China, and the wretchedness which has been the consequence of it. The population which has arisen naturally from the fertility of the soil, and the encouragements to agriculture, may be considered as genuine and desirable; but all that has been added by the encouragements to marriage has not only been an addition of so much pure misery in itself, but has completely interrupted the happiness which the rest might have enjoyed.

The territory of China is estimated at about eight times the territory of France.[489] Taking the population of France only at 26 millions, eight times that number will give 208,000,000; and when the three powerful causes of population, which have been stated, are considered, it will not appear incredible, that the population of China should be to the population of France, according to their respective superficies, as 333 to 208, or a little more than 3 to 2.

[487] Lettres Edif. et Curieuses, tom. xvi. p. 394.

[488] Lettres Edif. et Curieuses, tom. xvi. p. 394. et seq.

[489] Embassy to China, Staunton, vol. ii. p. 546.

The natural tendency to increase is every where so great that it will generally be easy to account for the height, at which the population is found in any country. The more difficult as well as the more interesting part of the inquiry is, to trace the immediate causes, which stop its further progress. The procreative power would, with as much facility, double in twenty-five years the population of China, as that of any of the states of America; but we know that it cannot do this, from the palpable inability of the soil to support such an additional number. What then becomes of this mighty power in China? And what are the kinds of restraint, and the forms of premature death, which keep the population down to the level of the means of subsistence?

Notwithstanding the extraordinary encouragements to marriage in China, we should perhaps be led into an error, if we were to suppose that the preventive check to population does not operate. Duhalde says, that the number of bonzas is considerably above a million, of which there are two thousand unmarried at Pekin, besides three hundred and fifty thousand more in their temples established in different places by the emperor's patents, and that the literary bachelors alone are about ninety thousand.[490]

The poor, though they would probably always marry when the slightest prospect opened to them of being able to support a family, and, from the permission of infanticide, would run great risks in this respect; yet they would undoubtedly be deterred from entering into this state, under the certainty of being obliged to expose all their children, or to sell themselves and families as slaves; and from the extreme poverty of the lower classes of people, such a certainty would often present itself. But it is among the slaves themselves, of which, according to Duhalde, the misery in China produces a prodigious multitude, that the preventive check to population principally operates. A man sometimes sells his son, and even himself and wife, at a very moderate price. The common mode is, to mortgage themselves with a condition of redemption, and a great number of men and maid servants are thus bound in a family.[491] Hume, in speaking of the practice of slavery among the ancients, remarks very justly, that it will generally be cheaper to buy a full-grown slave, than to rear up one from a child. This observation appears to be particularly applicable to the Chinese. All writers agree in mentioning the frequency of the dearths in China; and, during these periods, it is probable that slaves would be sold in great numbers for little more than a bare maintenance. It could very rarely therefore answer to the master of a family, to encourage his slaves to breed; and we may suppose, in consequence, that a great part of the servants in China, as in Europe, remain unmarried.

The check to population, arising from a vicious intercourse with the sex, does not appear to be very considerable in China. The women are said to be modest and reserved, and adultery is rare. Concubinage is however generally practised, and in the large towns public

---

[490] Duhalde's China, vol. i. p. 244.

[491] Duhalde's China, vol. i. p. 278. La misère et le grand nombre d'habitants de l'empire y causent cette multitude prodigieuse d'esclaves: presque tous les valets, et généralement toutes les filles de service d'une maison sont esclaves. Lettres Edif. tom. xix, p. 145.

women are registered; but their number is not great, being proportioned, according to Sir George Staunton, to the small number of unmarried persons, and of husbands absent from their families.[492]

The positive checks to population from disease, though considerable, do not appear to be so great as might be expected. The climate is in general extremely healthy. One of the missionaries goes so far as to say, that plagues or epidemic disorders are not seen once in a century;[493] but this is undoubtedly an error, as they are mentioned by others as if they were by no means so unfrequent. In some instructions to mandarins, relating to the burying of the poor, who have in general no regular places of sepulture, it is observed that when epidemic diseases prevail, the roads are found covered with bodies sufficient to infect the air to a great distance;[494] and the expression of years of contagion[495] occurs soon after, in a manner which seems to imply that they are not uncommon. On the first and fifteenth day of every month the mandarins assemble, and give their people a long discourse, wherein every governor acts the part of a father who instructs his family.[496] In one of these discourses, which Duhalde produces, the following passage occurs: "Beware of those years which happen from time to time, when epidemic distempers, joined to a scarcity of corn, make all places desolate. Your duty is then to have compassion on your fellow-citizens, and assist them with whatever you can spare."[497]

It is probable that the epidemics, as is usually the case, fall severely on the children. One of the Jesuits, speaking of the number of infants whom the poverty of their parents condemns to death the moment that they are born, writes thus: "There is seldom a year, in which the churches at Pekin do not reckon five or six thousand of these children purified by the waters of baptism. This harvest is more or less abundant according to the number of catechists which we can maintain. If we had a sufficient number, their cares need not be confined alone to the dying infants that are exposed. There would be other occasions for them to exercise their zeal, particularly at certain times of the year, when the small-pox or epidemic disorders carry off an incredible number of children."[498] It is indeed almost impossible to suppose that the extreme indigence of the lower classes of people should not produce diseases, likely to be fatal to a considerable part of those children whom their parents might attempt to rear in spite of every difficulty.

Respecting the number of infants which are actually exposed, it is difficult to form the slightest guess; but, if we believe the Chinese writers themselves, the practice must be very common. Attempts have been made at different times by the government to put a stop to it,

---

[492] Embassy to China, vol. ii. p. 157.

[493] Lettres Edif. tom. xxii. p. 187.

[494] Lettres Edif. tom. xix. p. 126.

[495] Id. p. 127.

[496] Duhalde's China, vol. i. p. 259.

[497] Id. 256.

[498] Lettres Edif. tom. xix. p. 100.

but always without success. In a book of instructions before alluded to, written by a mandarin celebrated for his humanity and wisdom, a proposal is made for the establishment of a foundling hospital in his district, and an account is given of some ancient establishments of the same kind,[499] which appear to have fallen into disuse. In this book the frequency of the exposure of children and the dreadful poverty which prompts it, are particularly described. "We see," he says, "people so poor, that they cannot furnish the nourishment necessary for their own children. It is on this account that they expose so great a number. In the metropolis; in the capitals of the provinces and in the places of the greatest commerce, their number is the most considerable; but many are found in parts that are less frequented, and even in the country. As the houses in towns are more crowded together, the practice is more obvious; but every where these poor unfortunate infants have need of assistance."[500]

In the same work, part of an edict to prevent the drowning of children runs thus: "When the tender offspring just produced is thrown without pity into the waves, can it be said that the mother has given or that the child has received life, when it is lost as soon as it is begun to be enjoyed? The poverty of the parents is the cause of this crime. They have hardly enough to support themselves, much less are they able to pay a nurse and provide for the expenses necessary for the support of their children. This drives them to despair; and not being able to bring themselves to suffer two people to die that one may live, the mother, to preserve the life of her husband, consents to sacrifice her child. It costs much, however, to the parental feelings; but the resolution is ultimately taken, and they think that they are justified in disposing of the life of their child to prolong their own. If they exposed their children in a secret place, the babe might work upon their compassion with its cries. What do they do then? They throw it into the current of the river, that they may lose sight of it immediately, and take from it at once all chance of life."[501]

Such writings appear to be most authentic documents respecting the general prevalence of infanticide.

Sir George Staunton has stated, from the best information which he could collect, that the number of children annually exposed at Pekin is about two thousand;[502] but it is highly probable that the number varies extremely from year to year, and depends very much upon seasons of plenty or seasons of scarcity. After any great epidemic or destructive famine, the number is probably very small; it is natural that it should increase gradually on the return to a crowded population; and it is without doubt the greatest, when an unfavourable season takes place, at a period in which the average produce is already insufficient to support the overflowing multitude.

---

[499] Ibid. p. 110.

[500] Lettres Edif. tom. xix. p. 111.

[501] Lettres Edif. tom. xix. p. 124.

[502] Embassy to China, vol. ii. p. 159.

These unfavourable seasons do not appear to be unfrequent, and the famines which follow them are perhaps the most powerful of all the positive checks to the Chinese population; though at some periods the checks from wars and internal commotions have not been inconsiderable.[503] In the annals of the Chinese monarchs, famines are often mentioned;[504] and it is not probable that they would find a place among the most important events and revolutions of the empire, if they were not desolating and destructive to a great degree.

One of the Jesuits remarks that the occasions when the mandarins pretend to shew the greatest compassion for the people are, when they are apprehensive of a failure in the crops, either from drought, from excessive rains, or from some other accident, such as a multitude of locusts, which sometimes overwhelms certain provinces.[505] The causes here enumerated are probably those, which principally contribute to the failure of the harvests in China; and the manner in which they are mentioned seems to shew that they are not uncommon.

Meares speaks of violent hurricanes, by which whole harvests are dissipated, and a famine follows. From a similar cause, he says, accompanied by excessive drought, a most dreadful dearth prevailed in 1787, throughout all the southern provinces of China, by which an incredible number of people perished. It was no uncommon thing at Canton to see the famished wretch breathing his last, while mothers thought it a duty to destroy their infant children, and the young to give the stroke of fate to the aged, to save them from the agonies of such a dilatory death.[506]

The Jesuit Parennin, writing to a member of the Royal Academy of Sciences, says, "Another thing that you can scarcely believe is, that dearths should be so frequent in China;"[507] and in the conclusion of his letter he remarks that, if famine did not, from time to time, thin the immense number of inhabitants which China contains, it would be impossible for her to live in peace.[508] The causes of these frequent famines he endeavours to investigate; and begins by observing, very justly, that in a time of dearth China can obtain no assistance from her neighbours, and must necessarily draw the whole of her resources from her own provinces.[509] He then describes the delays and artifices, which often defeat the emperor's intentions to assist, from the public granaries, those parts of the country which are the most distressed. When a harvest fails in any province, either from excessive drought or a sudden inundation, the great mandarins have recourse to the public granaries; but often find them empty, owing to the dishonesty of the inferior mandarins, who have the charge of them. Examinations and researches are then made, and an unwillingness prevails

---

[503] Annals of the Chinese Monarchs. Duhalde's China, vol. i. p. 136.

[504] Ibid.

[505] Lettres Edif. tom. xix. p. 154.

[506] Meares's Voyage, ch. vii. p. 92.

[507] Lettres Edif. et Curieuses, tom. xxii. p. 174.

[508] Id. p. 186.

[509] Id. p. 175.

to inform the court of such disagreeable intelligence. Memorials are however at length presented. These memorials pass through many hands, and do not reach the emperor till after many days. The great officers of state are then ordered to assemble, and to deliberate on the means of relieving the misery of the people. Declarations full of expressions of compassion for the people are in the mean time published throughout the empire. The resolution of the tribunal is at length made known; but numberless other ceremonies delay its execution; while those who are suffering have time to die with hunger, before the remedy arrives. Those who do not wait for this last extremity crawl as well as they can into other districts, where they hope to get support, but leave the greatest part of their number dead on the road.[510]

If, when a dearth occurs, the court do not make some attempt to relieve the people, small parties of plunderers soon collect, and their numbers increase by degrees, so as to interrupt the tranquillity of the province. On this account numerous orders are always given, and movements are continually taking place, to amuse the people till the famine is over; and as the motives to relieve the people are generally rather reasons of state than genuine compassion, it is not probable that they should be relieved at the time, and in the manner, that their wants require.[511]

The last cause of famine, which is mentioned in this investigation, and on which the writer lays considerable stress, is the very great consumption of grain in making spirits;[512] but in stating this as a cause of famine, he has evidently fallen into a very gross error; yet in the Abbé Grosier's general description of China this error has been copied, and the cause above mentioned has been considered as one of the grand sources of the evil.[513] But, in reality, the whole tendency of this cause is in a contrary direction. The consumption of corn in any other way but that of necessary food, checks the population before it arrives at the utmost limits of subsistence; and as the grain may be withdrawn from this particular use in the time of a scarcity, a public granary is thus opened, richer probably than could have been formed by any other means. When such a consumption has been once established, and has become permanent, its effect is exactly as if a piece of land, with all the people upon it, were removed from the country. The rest of the people would certainly be precisely in the same state as they were before, neither better nor worse, in years of average plenty; but in the time of dearth the produce of this land would be returned to them, without the mouths to help them to eat it. China, without her distilleries, would certainly be more populous; but on a failure of the seasons, would have still less resource than she has at present; and, as far as the magnitude of the cause would operate, would in consequence be more subject to famines, and those famines would be more severe.

[510] Lettres Edif. tom. xxii. p. 180.

[511] Id. p. 187.

[512] Id. p. 184.

[513] Vol. i. b. iv. c. iii. p. 396. 8vo. Eng. tran.

The state of Japan resembles in so many respects that of China, that a particular consideration of it would lead into too many repetitions. Montesquieu attributes its populousness to the birth of a greater number of females;[514] but the principal cause of this populousness is, without doubt, as in China, the persevering industry of the natives, directed, as it has always been, principally to agriculture.

In reading the preface to Thunberg's account of Japan, it would seem extremely difficult to trace the checks to the population of a country the inhabitants of which are said to live in such happiness and plenty; but the continuation of his own work contradicts the impression of his preface; and in the valuable history of Japan by Kæmpfer these checks are sufficiently obvious. In the extracts from two historical chronicles published in Japan, which he produces,[515] a very curious account is given of the different mortalities, plagues, famines, bloody wars and other causes of destruction, which have occurred since the commencement of these records. The Japanese are distinguished from the Chinese, in being much more warlike, seditious, dissolute and ambitious: and it would appear, from Kæmpfer's account, that the check to population from infanticide, in China, is balanced by the greater dissoluteness of manners with regard to the sex, and the greater frequency of wars and intestine commotions which prevail in Japan. With regard to the positive checks to population from disease and famine, the two countries seem to be nearly on a level.

3.4.13 Book I, Chapter XIII: Of the Checks to Population Among the Greeks.

It has been generally allowed, and will not indeed admit of a doubt, that the more equal division of property among the Greeks and Romans, in the early period of their history, and the direction of their industry principally to agriculture, must have tended greatly to encourage population. Agriculture is not only, as Hume states,[516] that species of industry, which is chiefly requisite to the subsistence of multitudes, but it is in fact the *sole* species by which multitudes can exist; and all the numerous arts and manufactures of the modern world, by which such numbers appear to be supported, have no tendency whatever to increase population, except so far as they tend to increase the quantity and to facilitate the distribution of the products of agriculture.

In countries where, from the operation of particular causes, property in land is divided into very large shares, these arts and manufactures are absolutely necessary to the existence of any considerable population. Without them modern Europe would be unpeopled. But where property is divided into small shares, the same necessity for them does not exist. The

---

[514] Liv. xxiii. c. xii. It is surprising that Montesquieu, who appears sometimes to understand the subject of population, should at other times make such observations as this.

[515] Book ii.

[516] Essay xi. p. 467, 4to. edit.

division itself attains immediately one great object, that of distribution; and if the demand for men be constant, to fight the battles and support the power and dignity of the state, we may easily conceive that this motive, joined to the natural love of a family, might be sufficient to induce each proprietor to cultivate his land to the utmost, in order that it might support the greatest number of descendants.

The division of people into small states, during the early periods of Greek and Roman history, gave additional force to this motive. Where the number of free citizens did not perhaps exceed ten or twenty thousand, each individual would naturally feel the value of his own exertions; and knowing that the state to which he belonged, situated in the midst of envious and watchful rivals, must depend chiefly on its population for its means of defence and safety, would be sensible that, in suffering the lands which were allotted to him to lie idle, he would be deficient in his duty as a citizen. These causes appear to have produced a considerable attention to agriculture, without the intervention of the artificial wants of mankind to encourage it. Population followed the products of the earth with more than equal pace; and when the overflowing numbers were not taken off by the drains of war or disease, they found vent in frequent and repeated colonization. The necessity of these frequent colonizations, joined to the smallness of the states, which brought the subject immediately home to every thinking person, could not fail to point out to the legislators and philosophers of those times the strong tendency of population to increase beyond the means of subsistence; and they did not, like the statesmen and projectors of modern days, overlook the consideration of a question, which so deeply affects the happiness and tranquillity of society. However we may justly execrate the barbarous expedients which they adopted to remove the difficulty, we cannot but give them some credit for their penetration in seeing it; and in being fully aware that, if not considered and obviated, it would be sufficient of itself to destroy their best-planned schemes of republican equality and happiness.

The power of colonization is necessarily limited; and after the lapse of some time it might be extremely difficult; if not impossible, for a country, not particularly well suited for this purpose, to find a vacant spot proper for the settlement of its expatriated citizens. It was necessary therefore to consider of other resources besides colonization.

It is probable that the practice of infanticide had prevailed from the earliest ages in Greece. In the parts of America where it was found to exist it appears to have originated from the extreme difficulty of rearing many children, in a savage and wandering life, exposed to frequent famines and perpetual wars. We may easily conceive that it had a similar origin among the ancestors of the Greeks or the native inhabitants of the country. And, when Solon permitted the exposing of children, it is probable that he only gave the sanction of law to a custom already prevalent.

In this permission he had without doubt two ends in view. First, that which is most obvious, the prevention of such an excessive population as would cause universal poverty and discontent; and, secondly, that of keeping the population up to the level of what the territory could support, by removing the terrors of too numerous a family, and consequently the principal obstacle to marriage. From the effect of this practice in China we have reason to think that it is better calculated to attain the latter than the former purpose. But if the legislator either did not see this, or if the barbarous habits of the times prompted parents invariably to prefer the murder of their children to poverty, the practice would appear to be very particularly calculated to answer both the ends in view; and to preserve, as completely and as constantly as the nature of the thing would permit, the requisite proportion between the food and the numbers which were to consume it.

On the very great importance of attending to this proportion, and the evils that must necessarily result, of weakness on the one hand, or of poverty on the other, from the deficiency or the excess of population, the Greek political writers strongly insist; and propose in consequence various modes of maintaining the relative proportion desired.

Plato, in the republic which he considers in his books of laws, limits the number of free citizens and of habitations to five thousand and forty; and this number he thinks may be preserved, if the father of every family choose one out of his sons for his successor to the lot of land which he has possessed, and, disposing of his daughters in marriage according to law, distribute his other sons, if he have any, to be adopted by those citizens who are without children. But if the number of children upon the whole be either too great or too few, the magistrate is to take the subject particularly into his consideration, and to contrive so, that the same number of five thousand and forty families should still be maintained. There are many modes, he thinks, of effecting this object. Procreation, when it goes on too fast, may be checked, or, when it goes on too slow, may be encouraged, by the proper distribution of honours and marks of ignominy, and by the admonitions of the elders, to prevent or promote it according to circumstances.[517]

In his Philosophical Republic[518] he enters more particularly into this subject, and proposes that the most excellent among the men should be joined in marriage to the most excellent among the women, and the inferior citizens matched with the inferior females; and that the offspring of the first should be brought up, of the others not. On certain festivals appointed by the laws, the young men and women who are betrothed are to be assembled, and joined together with solemn ceremonies. But the number of marriages is to be determined by the magistrates; that, taking into consideration the drains from wars, diseases and other causes, they may preserve, as nearly as possible, such a proportion of citizens, as will be neither too numerous nor too few, according to the resources and demands of the state. The children, who are thus born from the most excellent of the citizens, are to be carried to certain nurses destined to this office; inhabiting a separate part of the city; but those which are born from the inferior citizens, and any from the others which are imperfect in their limbs, are to be buried in some obscure and unknown place.

He next proceeds to consider the proper age for marriage, and determines it to be twenty for the women, and thirty for the men. Beginning at twenty, the woman is to bear children for the state till she is forty, and the man is to fulfil his duty in this respect from thirty to fifty-five. If a man produce a child into public either before or after this period, the action is

[517] Plato de Legibus, lib. v.

[518] Plato de Republicâ, lib. v.

to be considered in the same criminal and profane light as if he had produced one without the nuptial ceremonies, and instigated solely by incontinence. The same rule should hold, if a man who is of the proper age for procreation be connected with a woman who is also of the proper age, but without the ceremony of marriage by the magistrate; he is to be considered as having given to the state a spurious, profane and incestuous offspring. When both sexes have passed the age assigned for presenting children to the state, Plato allows a great latitude of intercourse; but no child is to be brought to light. Should any infant by accident be born alive, it is to be exposed in the same manner as if the parents could not support it.[519]

From these passages it is evident that Plato fully saw the tendency of population to increase beyond the means of subsistence. His expedients for checking it are indeed execrable; but the expedients themselves, and the extent to which they were to be used, shew his conceptions of the magnitude of the difficulty. Contemplating, as he certainly must do in a small republic, a great proportional drain of people by wars, if he could still propose to destroy the children of all the inferior and less perfect citizens, to destroy also all that were born not within the prescribed ages and with the prescribed forms, to fix the age of marriage late, and after all to regulate the number of these marriages, his experience and his reasonings must have strongly pointed out to him the great power of the principle of increase, and the necessity of checking it.

Aristotle appears to have seen this necessity still more clearly. He fixes the proper age of marriage at thirty-seven for the men, and eighteen for the women, which must of course condemn a great number of women to celibacy, as there never can be so many men of thirty-seven as there are women of eighteen. Yet, though he has fixed the age of marriage for the men at so late a period, he still thinks that there may be too many children, and proposes that the number allowed to each marriage should be regulated; and, if any woman be pregnant after she has produced the prescribed number, that an abortion should be procured before the fœtus has life.

The period of procreating children for the state is to cease with the men at fifty-four or fifty-five, because the offspring of old men, as well as of men too young, is imperfect both in body and mind. When both sexes have passed the prescribed age, they are allowed to continue a connexion; but, as in Plato's republic, no child which may be the result is to be brought to light.[520]

In discussing the merits of the republic proposed by Plato in his books of laws, Aristotle is of opinion that he has by no means been sufficiently attentive to the subject of population; and accuses him of inconsistency in equalizing property without limiting the number of children. The laws on this subject, Aristotle very justly observes, require to be much more definite and precise in a state where property is equalized than in others. Under ordinary governments an increase of population would only occasion a greater subdivision

[519] Plato de Repub. lib. v.

[520] Aristotelis Opera, de Repub. lib. vii. c. xvi.

of landed property; whereas in such a republic the supernumeraries would be altogether destitute, because the lands, being reduced to equal and as it were elementary parts, would be incapable of further partition.[521]

He then remarks that it is necessary in all cases to regulate the proportion of children, that they may not exceed the proper number. In doing this, deaths and barrenness are of course to be taken into consideration. But if, as in the generality of states, every person be left free to have as many children as he pleases, the necessary consequence must be poverty; and poverty is the mother of villany and sedition. On this account Pheidon of Corinth, one of the most ancient writers on the subject of politics, introduced a regulation directly the reverse of Plato's, and limited population without equalizing possessions.[522]

Speaking afterwards of Phaleas of Chalcedon, who proposed, as a most salutary institution, to equalize wealth among the citizens, he adverts again to Plato's regulations respecting property; and observes that those who would thus regulate the extent of fortunes, ought not to be ignorant that it is absolutely necessary at the same time to regulate the number of children. For if children multiply beyond the means of supporting them, the law will necessarily be broken, and families will be suddenly reduced from opulence to beggary,—a revolution always dangerous to public tranquillity.[523]

It appears from these passages that Aristotle clearly saw that the strong tendency of the human race to increase, unless checked by strict and positive laws, was absolutely fatal to every system founded on equality of property; and there cannot surely be a stronger argument against any system of this kind than the necessity of such laws as Aristotle himself proposes.

From a remark which he afterwards makes respecting Sparta, it appears still more clearly that he fully understood the principle of population. From the improvidence of the laws relating to succession, the landed property in Sparta had been engrossed by a few; and the effect was greatly to diminish the populousness of the country. To remedy this evil, and to supply men for continual wars, the kings preceding Lycurgus had been in the habit of naturalizing strangers. It would have been much better however, according to Aristotle, to have increased the number of citizens by a nearer equalization of property. But the law relating to children was directly adverse to this improvement. The legislator, wishing to have many citizens, had encouraged as much as possible the procreation of children. A man who had three sons, was exempt from the night-watch; and he who had four, enjoyed a complete immunity from all public burdens. But it is evident, as Aristotle most justly observes, that the birth of a great number of children, the division of the lands remaining the same, would necessarily cause only an accumulation of poverty.[524]

[521] De Repub. lib. ii. c. vi. Gillies's Aristot. vol. ii. b. ii. p. 87. For the convenience of those who may not choose the trouble of consulting the original, I refer at the same time to Gillies's translation; but some passages he has wholly omitted, and of others he has not given the literal sense, his object being a free version.

[522] De Repub. lib. ii. c. vii. Gillies's Aristot. vol. ii. b. ii. p. 87.

[523] De Repub. lib. ii. c. vii. Gillies's Aristot. vol. ii. b. ii. p. 91.

[524] De Repub. lib. ii. c. ix. Gillies's Aristot. vol. ii. b. ii. p. 107.

He here seems to see exactly the error into which many other legislators besides Lycurgus have fallen; and to be fully aware that to encourage the birth of children, without providing properly for their support, is to obtain a very small accession to the population of a country at the expense of a very great accession of misery.

The legislator of Crete,[525] as well as Solon, Pheidon, Plato and Aristotle, saw the necessity of checking population in order to prevent general poverty; and as we must suppose that the opinions of such men, and the laws founded upon them, would have considerable influence, it is probable that the preventive check to increase, from late marriages and other causes, operated in a considerable degree among the free citizens of Greece.

For the positive checks to population we need not look beyond the wars in which these small states were almost continually engaged; though we have an account of one wasting plague, at least, in Athens; and Plato supposes the case of his republic being greatly reduced by disease.[526] Their wars were not only almost constant, but extremely bloody. In a small army, the whole of which would probably be engaged in close fight, a much greater number in proportion would be slain than in the large modern armies, a considerable part of which often remains untouched;[527] and as all the free citizens of these republics were generally employed as soldiers in every war, losses would be felt very severely, and would not appear to be very easily repaired.

3.4.14 Book I, Chapter XIV: Of the Checks to Population Among the Romans.

The havoc made by war in the smaller states of Italy, particularly during the first struggles of the Romans for power, seems to have been still greater than in Greece. Wallace, in his Dissertation on the Numbers of Mankind, after alluding to the multitudes which fell by the sword in these times, observes, "On an accurate review of the history of the Italians during this period, we should wonder how such vast multitudes could be raised as were engaged in those continual wars till Italy was entirely subdued."[528] And Livy expresses his utter astonishment that the Volsci and Æqui, so often as they were conquered, should have been able to bring fresh armies into the field.[529] But these wonders will perhaps be sufficiently accounted for, if we suppose, what seems to be highly probable, that the constant drains from wars had introduced the habit of giving nearly full scope to the power of population; and that a much larger proportion of births, and of healthy children

[525] Aristot. de Repub. lib. ii. c. x. Gillies's Aristot. vol. ii. b. ii. p. 113.

[526] De Legibus, lib. v.

[527] Hume's Essay, c. xi. p. 951.

[528] Dissertation, p. 62, 8vo. 1763, Edinburgh.

[529] Lib. vi. c. xii.

were rising into manhood and becoming fit to bear arms, than is usual in other states not similarly circumstanced. It was, without doubt, the rapid influx of these supplies, which enabled them, like the ancient Germans, to astonish future historians, by renovating in so extraordinary a manner their defeated and half-destroyed armies.

Yet there is reason to believe that the practice of infanticide prevailed in Italy as well as in Greece from the earliest times. A law of Romulus forbad the exposing of children before they were three years old,[530] which implies that the custom of exposing them as soon as they were born had before prevailed. But this practice was of course never resorted to, unless when the drains from wars were insufficient to make room for the rising generation; and consequently, though it may be considered as one of the positive checks to the full power of increase, yet, in the actual state of things, it certainly contributed rather to promote than impede population.

Among the Romans themselves, engaged as they were in incessant wars from the beginning of their republic to the end of it, many of which were dreadfully destructive, the positive check to population from this cause alone must have been enormously great. But this cause alone, great as it was, would never have occasioned that want of Roman citizens under the emperors which prompted Augustus and Trajan to issue laws for the encouragement of marriage and of children, if other causes, still more powerful in depopulation, had not concurred.

When the equality of property, which had formerly prevailed in the Roman territory, had been destroyed by degrees, and the land had fallen into the hands of a few great proprietors, the citizens, who were by this change successively deprived of the means of supporting themselves, would naturally have no resource to prevent them from starving, but that of selling their labour to the rich, as in modern states: but from this resource they were completely cut off by the prodigious number of slaves, which, increasing by constant influx with the increasing luxury of Rome, filled up every employment both in agriculture and manufactures. Under such circumstances, so far from being astonished that the number of free citizens should decrease, the wonder seems to be that any should exist besides the proprietors. And in fact many could not have existed but for a strange and preposterous custom, which, however, the strange and unnatural state of the city might perhaps require, that of distributing vast quantities of corn to the poorer citizens gratuitously. Two hundred thousand received this distribution in Augustus's time; and it is highly probable that a great part of them had little else to depend upon. It is supposed to have been given to every man of full years; but the quantity was not enough for a family, and too much for an individual.[531] It could not therefore enable them to increase; and, from the manner in which Plutarch speaks of the custom of exposing children among the poor,[532] there is great reason to believe that many were destroyed in spite of the *jus trium liberorum.* The passage in

[530] Dionysius Halicarn. lib. ii. 15.

[531] Hume, Essay, xi. p. 488.

[532] De Amore Prolis.

Tacitus, in which, speaking of the Germans, he alludes to this custom in Rome, seems to point to the same conclusion.[533] What effect, indeed, could such a law have among a set of people, who appear to have been so completely excluded from all the means of acquiring a subsistence, except that of charity, that they would be scarcely able to support themselves, much less a wife and two or three children? If half of the slaves had been sent out of the country, and the people had been employed in agriculture and manufactures, the effect would have been to increase the number of Roman citizens with more certainty and rapidity than ten thousand laws for the encouragement of children.

It is possible that the *jus trium liberorum,* and the other laws of the same tendency, might have been of some little use among the higher classes of Roman citizens; and indeed from the nature of these laws, consisting as they did principally of privileges, it would appear that they were directed chiefly to this part of society. But vicious habits of every possible kind preventive of population[534] seem to have been so generally prevalent at this period, that no corrective laws could have any considerable influence. Montesquieu justly observes, that "the corruption of manners had destroyed the office of censor, which had been established itself to destroy the corruption of manners; but when the corruption of manners becomes general, censure has no longer any force."[535] Thirty-four years after the passing of the law of Augustus respecting marriage, the Roman knights demanded its repeal. On separating the married and the unmarried, it appeared that the latter considerably exceeded in number the former; a strong proof of the inefficacy of the law.[536]

In most countries vicious habits preventive of population appear to be a consequence, rather than a cause, of the infrequency of marriage; but in Rome the depravity of morals seems to have been the direct cause which checked the marriage union, at least among the higher classes. It is impossible to read the speech of Metellus Numidicus in his censorship without indignation and disgust. "If it were possible," he says, "entirely to go without wives, we would deliver ourselves at once from this evil; but as the laws of nature have so ordered it that we can neither live happy with them nor continue the species with out them, we ought to have more regard for our lasting security than for our transient pleasures."[537]

---

[533] De Moribus Germanorum, 19. How completely the laws relating to the encouragement of marriage and of children were despised, appears from a speech of Minucius Felix in Octavio, cap. 30. *"Vos enim video procreatos filios nunc feris et avibus exponere, nunc adstrangulatos misero mortis genere elidere; sent quæ in ipsis visceribus medicaminibus epotis originem futuri hominus extinguant, et parricidium faciunt antequam pariant."* even Pliny attempts to excuse it; "Quoniam aliquarum fecunditas plena liberis tali veniâ indiget." Lib. xxix. c. iv.

[534] Sed facet aurato vix ulla puerpera lecto;
Tantum artes hujus, tantum medicamina possunt,
Quæ steriles facit, atque homines in ventre necandos
Conducit.—Juvenal, Sat. vi. 593.

[535] Esprit des Loix, liv. xxiii. c. 21.

[536] Ibid.

[537] Aulus Genius, lib. i. c. 6.

Positive laws to encourage marriage and population, enacted on the urgency of the occasion, and not mixed with religion, as in China and some other countries, are seldom calculated to answer the end which they aim at, and therefore generally indicate ignorance in the legislator who proposes them; but the apparent necessity of such laws almost invariably indicates a very great degree of moral and political depravity in the state; and in the countries in which they are most strongly insisted on, not only vicious manners will generally be found to prevail, but political institutions extremely unfavourable to industry, and consequently to population.

On this account I cannot but agree with Wallace[538] in thinking that Hume was wrong in his supposition, that the Roman world was probably the most populous during the long peace under Trajan and the Antonines.[539] We well know that wars do not depopulate much while industry continues in vigour; and that peace will not increase the number of people when they cannot find the means of subsistence. The renewal of the laws relating to marriage under Trajan, indicates the continued prevalence of vicious habits and of a languishing industry, and seems to be inconsistent with the supposition of a great increase of population.

It might be said perhaps that the vast profusion of slaves would more than make up for the want of Roman citizens; but it appears that the labour of these slaves was not sufficiently directed to agriculture to support a very great population. Whatever might be the case with some of the provinces, the decay of agriculture in Italy seems to be generally acknowledged. The pernicious custom of importing great quantities of corn to distribute gratuitously among the people had given it a blow, which it never afterwards recovered. Hume observes that "when the Roman authors complain that Italy, which formerly exported corn, became dependent on all the provinces for its daily bread, they never ascribed this alteration to the increase of its inhabitants, but to the neglect of tillage and agriculture."[540] And in another place he says, "All ancient authors tell us that there was a perpetual influx of slaves to Italy from the remoter provinces, particularly Syria, Cilicia, Cappadocia, and the lesser Asia, Thrace, and Egypt; yet the number of people did not increase in Italy; and writers complain of the continual decay of industry and agriculture."[541] It seems but little probable that the peace under Trajan and the Antonines should have given so sudden a turn to the habits of the people as essentially to alter this state of things.

On the condition of slavery it may be observed that there cannot be a stronger proof of its unfavourableness to the propagation of the species in the countries where it prevails, than the necessity of this continual influx. This necessity forms at once a complete refutation of the observation of Wallace, that the ancient slaves were more serviceable in

[538] Dissertation, Appendix, p. 247.

[539] Essay xi. p. 505.

[540] Essay xi. p. 504.

[541] Essay xi. p. 433.

raising up people than the inferior ranks of men in modern times.[542] Though it is undoubtedly true, as he observes, that all our labourers do not marry, and that many of their children die, or become sickly and useless through the poverty and negligence of their parents;[543] yet, notwithstanding these obstacles to increase, there is perhaps scarcely an instance to be produced where the lower classes of society in any country, if free, do not raise up people fully equal to the demand for their labour.

To account for the checks to population which are peculiar to a state of slavery, and which render a constant recruit of numbers necessary, we must adopt the comparison of slaves to cattle which Wallace and Hume have made; Wallace, to shew that it would be the interest of masters to take care of their slaves and rear up their offspring;[544] and Hume, to prove that it would more frequently be the interest of the master to prevent than to encourage their breeding.[545] If Wallace's observation had been just, it is not to be doubted that the slaves would have kept up their own numbers with ease by procreation; and as it is acknowledged that they did not do this, the truth of Hume's observation is clearly evinced. "To rear a child in London till he could be serviceable, would cost much dearer than to buy one of the same age from Scotland or Ireland, where he had been raised in a cottage, covered with rags, and fed on oatmeal and potatoes. Those who had slaves therefore, in all the richer and more populous countries, would discourage the pregnancy of the females, and either prevent or destroy the birth."[546] It is acknowledged by Wallace that the male slaves greatly exceeded in number the females,[547] which must necessarily be an additional obstacle to their increase. It would appear therefore that the preventive check to population must have operated with very great force among the Greek and Roman slaves; and as they were often ill treated, fed perhaps scantily, and sometimes great numbers of them confined together in close and unwholesome *ergastula,* or dungeons,[548] it is probable that the positive checks to population from disease were also severe, and that when epidemics prevailed, they would be most destructive in this part of the society.

The unfavourableness of slavery to the propagation of the species in the country where it prevails, is not however decisive of the question respecting the absolute population of such a country, or the greater question respecting the populousness of ancient and modern nations. We know that some countries could afford a great and constant supply of slaves without being in the smallest degree depopulated themselves; and if these supplies were poured in, as they probably would be, exactly in proportion to the demand for labour in the nation which received them, the question respecting the populousness of this nation would

[542] Dissert. on the Numbers of Mankind, p. 91.

[543] Id. p. 88.

[544] Dissert. on the Numbers of Mankind, p. 89.

[545] Hume, Essay xi. p. 433.

[546] Ibid.

[547] Appendix to Dissertation, p. 182.

[548] Hume, Essay xi. p. 430.

rest precisely on the same grounds as in modern states, and depend upon the number of people which it could employ and support. Whether the practice of domestic slavery therefore prevail or not, it may be laid down as a position not to be controverted, that, taking a sufficient extent of territory to include within it exportation and importation, and allowing some variation for the prevalence of luxury or of frugal habits, the population of these countries will always be in proportion to the food which the earth is made to produce. And no cause, physical or moral, unless, it operate in an excessive and unusual manner,[549] will have any considerable and permanent effect on the population, except in as far as it influences the production and distribution of the means of subsistence.

In the controversy concerning the populousness of ancient and modern nations this point has not been sufficiently attended to; and physical and moral causes have been brought forward on both sides, from which no just inference in favour of either party could be drawn. It seems to have escaped the attention of both writers, that the more productive and populous a country is in its actual state, the less probably will be its power of obtaining a further increase of produce; and consequently the more checks must necessarily be called into action, to keep the population down to the level of this stationary or slowly increasing produce. From finding such checks, therefore, in ancient or modern nations, no inference can be drawn against the absolute populousness of either. On this account, the prevalence of the small-pox, and of other disorders unknown to the ancients, can by no means be considered as an argument against the populousness of modern nations, though to these physical causes both Hume[550] and Wallace[551] allow considerable weight.

In the moral causes which they have brought forward, they have fallen into a similar error. Wallace introduces the positive encouragements to marriage among the ancients as one of the principal causes of the superior populousness of the ancient world;[552] but the necessity of positive laws to encourage marriage certainly rather indicates a want than an abundance of people; and in the instance of Sparta, to which he particularly refers, it appears from the passage in Aristotle, mentioned in the last chapter, that the laws to encourage marriage were instituted for the express purpose of remedying a marked deficiency of people. In a country with a crowded and overflowing population, a legislator would never think of making express laws to encourage marriage and the procreation of children. Other arguments of Wallace will be found upon examination to be almost equally ineffectual to his purpose.

---

[549]The extreme insalubrity of Batavia, and perhaps the plague in some countries, may be considered as physical causes operating in an excessive degree. The extreme and unusual attachment of the Romans to a vicious celibacy, and the promiscuous intercourse in Otaheite, may be considered as moral causes of the same nature. Such instances, and others of the same kind, which might probably be found, make it necessary to qualify the general proposition as in the text.

[550]Essay xi. p. 425.

[551]Dissertation, p. 80.

[552]Id. p. 93.

Some of the causes which Hume produces are in the same manner unsatisfactory, and rather make against the inference which he has in view than for it. The number of footmen, housemaids and other persons remaining unmarried in modern states, he allows to be an argument against their populousness.[553] But the contrary inference of the two appears to be the more probable. When the difficulties attending the rearing a family are very great, and consequently many persons of both sexes remain single, we may naturally enough infer that population is stationary, but by no means that it is not absolutely great; because the difficulty of rearing a family may arise from the very circumstance of a great absolute population, and the consequent fulness of all the channels to a livelihood; though the same difficulty may undoubtedly exist in a thinly-peopled country, which is yet stationary in its population. The number of unmarried persons in proportion to the whole number, may form some criterion by which we can judge whether population be increasing, stationary, or decreasing; but will not enable us to determine any thing respecting absolute populousness. Yet even in this criterion we are liable to be deceived. In some of the southern countries early marriages are general, and very few women remain in a state of celibacy; yet the people not only do not increase, but the actual number is perhaps small. In this case the removal of the preventive check is made up by the excessive force of the positive check. The sum of all the positive and preventive checks taken together, forms undoubtedly the immediate cause which represses population; but we never can expect to obtain and estimate accurately this sum in any country; and we can certainly draw no safe conclusion from the contemplation of two or three of these checks taken by themselves, because it so frequently happens that the excess of one check is balanced by the defect of some other. Causes, which affect the number of births or deaths, may or may not affect the average population, according to circumstances; but causes, which affect the production and distribution of the means of subsistence, must necessarily affect population; and it is therefore upon these latter causes alone (independently of actual enumerations) that we can with certainty rely.

All the checks to population, which have been hitherto considered in the course of this review of human society, are clearly resolvable into moral restraint, vice, and misery.

Of that branch of the preventive check which I have denominated moral restraint, though it has certainly had some share in repressing the natural power of population, yet, taken in its strict sense, it must be allowed to have operated feebly, compared with the others. Of the other branch of the preventive check, which comes under the head of vice, though its effect appears to have been very considerable in the later periods of Roman history, and in some other countries; yet, upon the whole, its operation seems to have been inferior to the positive checks. A large portion of the procreative power appears to have been called into action, the redundancy from which was checked by violent causes. Among these, war is the most prominent and striking feature; and after this may be ranked famines and violent diseases. In most of the countries considered, the population seems to have

[553] Essay xi.

been seldom measured accurately according to the average and permanent means of subsistence, but generally to have vibrated between the two extremes; and consequently the oscillations between want and plenty are strongly marked, as we should naturally expect among less civilized nations.

3.5 Book II: Of the Checks to Population in the Different States of Modern Europe.

3.5.1 Book II, Chapter I: Of the Checks to Population in Norway.

In reviewing the states of modern Europe, we shall be assisted in our inquiries by registers of births, deaths and marriages, which, when they are complete and correct, point out to us with some degree of precision whether the prevailing checks to population are of the positive or preventive kind. The habits of most European nations are of course much alike, owing to the similarity of the circumstances in which they are placed; and it is to be expected therefore that their registers should sometimes give the same results. Relying however too much upon this occasional coincidence, political calculators have been led into the error of supposing that there is, generally speaking, an invariable order of mortality in all countries: but it appears, on the contrary, that this order is extremely variable; that it is very different in different places of the same country, and within certain limits depends upon circumstances, which it is in the power of man to alter.

Norway, during nearly the whole of the last century, was in a peculiar degree exempt from the drains of people by war. The climate is remarkably free from epidemic sicknesses; and, in common years, the mortality is less than in any other country in Europe, the registers of which are known to be correct.[554] The proportion of the annual deaths to the whole population, on an average throughout the whole country, is only as 1 to 48.[555] Yet the population of Norway never seems to have increased with great rapidity. It has made a start within the last ten or fifteen years; but till that period its progress must have been very slow, as we know that the country was peopled in very early ages, and in 1769 its population was only 723,141.[556]

Before we enter upon an examination of its internal economy, we must feel assured that, as the positive checks to its population have been so small, the preventive checks must have been proportionably great; and we accordingly find from the registers that the proportion of

[554]The registers for Russia give a smaller mortality; but it is supposed that they are defective. It appears, however, that in England and Wales during the ten years ending with 1820, the mortality was still less than in Norway.

[555]Thaarup's Statistik der Danischen Monarchie, vol. ii. p. 4.

[556]Id. Table II. p. 5.

yearly marriages to the whole population is as 1 to 130,[557] which is a smaller proportion of marriages than appears in the registers of any other country, except Switzerland.

One cause of this small number of marriages is the mode in which the enrolments for the army have been conducted till within a very few years. Every man in Denmark and Norway born of a farmer or labourer is a soldier.[558] Formerly the commanding officer of the district might take these peasants at any age he pleased; and he in general preferred those that were from twenty-five to thirty, to such as were younger. After being taken into the service, a man could not marry without producing a certificate, signed by the minister of the parish, that he had substance enough to support a wife and family; and even then it was further necessary for him to obtain the permission of the officer. The difficulty, and sometimes the expense, of obtaining this certificate and permission, generally deterred those who were not in very good circumstances, from thinking of marriage till their service of ten years was expired; and as they might be enrolled at any age under thirty-six, and the officers were apt to take the oldest first, it would often be late in life before they could feel themselves at liberty to settle.

Though the minister of the parish had no legal power to prevent a man from marrying who was not enrolled for service, yet it appears that custom had in some degree sanctioned a discretionary power of this kind, and the priest often refused to join a couple together when the parties had no probable means of supporting a family.

Every obstacle, however, of this nature, whether arising from law or custom, has now been entirely removed. A full liberty is given to marry at any age, without leave either of the officer or priest; and in the enrolments for the army all those of the age of twenty are taken first, then all those of twenty-two, and so on till the necessary number is completed.

The officers in general disapprove of this change. They say that a young Norwegian has not arrived at his full strength and does not make a good soldier at twenty. And many are of opinion that the peasants will now marry too young, and that more children will be born than the country can support.

But, independently of any regulations respecting the military enrolments, the peculiar state of Norway throws very strong obstacles in the way of early marriages. There are no large manufacturing towns to take off the overflowing population of the country; and as

[557] Thaarup's Statistik der Danischen Monarchie, vol. ii, p. 4. The proportion of yearly marriages to the whole population is one of the most obvious criterions of the operation of the preventive check, though not quite a correct one. Generally speaking, the preventive check is greater than might be inferred from this criterion; because in the healthy countries of Europe, where a small proportion of marriages takes place, the greater number of old people living at the time of these marriages will be more than counterbalanced by the smaller proportion of persons under the age of puberty. In such a country as Norway, the persons from 20 to 50, that is, of the most likely age to marry, bear a greater proportion to the whole population than in most of the other countries of Europe; and consequently the actual proportion of marriages in Norway, compared with that of others, will not express the full extent in which the preventive check operates.

[558] The few particulars, which I shall mention relating to Norway, were collected during a summer excursion in that country in the year 1799.

each village naturally furnishes from itself a supply of hands more than equal to the demand, a change of place in search of work seldom promises any success. Unless therefore an opportunity of foreign emigration offer, the Norwegian peasant generally remains in the village in which he was born, and as the vacancies in houses and employments must occur very slowly, owing to the small mortality that takes place, he will often see himself compelled to wait a considerable time, before he can attain a situation, which will enable him to rear a family.

The Norway farms have in general a certain number of married labourers employed upon them, in proportion to their size, who are called housemen. They receive from the farmer a house, and a quantity of land nearly sufficient to maintain a family; in return for which they are under the obligation of working for him at a low and fixed price, whenever they are called upon. Except in the immediate neighbourhood of the towns, and on the sea-coast, the vacancy of a place of this kind is the only prospect which presents itself of providing for a family. From the small number of people, and the little variety of employment, the subject is brought distinctly within the view of each individual; and he must feel the absolute necessity of repressing his inclinations to marriage, till some such vacancy offer. If, from the plenty of materials, he should be led to build a house for himself, it could not be expected, that the farmer, if he had a sufficient number of labourers before, should give him an adequate portion of land with it; and though he would in general find employment for three or four months in the summer, yet there would be little chance of his earning enough to support a family during the whole year. It is probable, that it was in cases of this kind, where the impatience of the parties prompted them to build, or propose to build a house themselves, and trust to what they could earn, that the parish priests exercised the discretionary power of refusing to marry.

The young men and women therefore are obliged to remain with the farmers as unmarried servants, till a houseman's place becomes vacant: and of these unmarried servants there is in every farm, and every gentleman's family, a much greater proportion, than the work would seem to require. There is but little division of labour in Norway. Almost all the wants of domestic economy are supplied in each separate household. Not only the common operations of brewing, baking, and washing, are carried on at home, but many families make or import their own cheese and butter, kill their own beef and mutton, import their own grocery stores; and the farmers and country people in general spin their own flax and wool, and weave their own linen and woollen clothes. In the largest towns, such as Christiana and Drontheim, there is nothing that can be called a market. It is extremely difficult to get a joint of fresh meat; and a pound of fresh butter is an article not to be purchased, even in the midst of summer. Fairs are held at certain seasons of the year, and stores of all kinds of provisions that will keep are laid in at these times; and, if this care be neglected, great inconveniencies are suffered, as scarcely any thing is to be bought retail. Persons who make a temporary residence in the country, or small merchants not possessed of farms, complain heavily of this inconvenience; and the wives of merchants, who have large estates, say, that the domestic economy of a Norway family is so extensive and complicated, that the necessary superintendance of it requires their whole attention, and that they can find no time for any thing else.

It is evident, that a system of this kind must require a great number of servants. It is said besides, that they are not remarkable for diligence, and that to do the same quantity of work more are necessary than in other countries. The consequence is, that in every establishment the proportion of servants will be found two or three times as great as in England; and a farmer in the country, who in his appearance is not to be distinguished from any of his labourers, will sometimes have a household of twenty persons, including his own family.

The means of maintenance to a single man are, therefore, much less confined than to a married man; and under such circumstances the lower classes of people cannot increase much, till the increase of mercantile stock, or the division and improvement of farms, furnishes a greater quantity of employment for married labourers. In countries more fully peopled this subject is always involved in great obscurity. Each man naturally thinks, that he has as good a chance of finding employment as his neighbour; and that, if he fail in one place, he shall succeed in some other. He marries, therefore, and trusts to fortune; and the effect too frequently is, that redundant population occasioned in this manner is repressed by the positive checks of poverty and disease. In Norway the subject is not involved in the same obscurity. The number of additional families, which the increasing demand for labour will support, is more distinctly marked. The population is so small, that even in the towns it is difficult to fall into any considerable error on this subject; and in the country the division and improvement of an estate, and the creation of a greater number of housemen's places, must be a matter of complete notoriety. If a man can obtain one of these places, he marries, and is able to support a family; if he cannot obtain one, he remains single. A redundant population is thus prevented from taking place, instead of being destroyed after it has taken place.

It is not to be doubted, that the general prevalence of the preventive check to population, owing to the state of society which has been described, together with the obstacles thrown in the way of early marriages from the enrolments for the army, have powerfully contributed to place the lower classes of people in Norway in a better situation, than could be expected from the nature of the soil and climate. On the sea-coast, where, on account of the hopes of an adequate supply of food from fishing, the preventive check does not prevail in the same degree, the people are very poor and wretched; and, beyond comparison, in a worse state than the peasants in the interior of the country.

The greatest part of the soil in Norway is absolutely incapable of bearing corn, and the climate is subject to the most sudden and fatal changes. There are three nights about the end of August, which are particularly distinguished by the name of iron nights, on account of their sometimes blasting the promise of the fairest crops. On these occasions the lower classes of people necessarily suffer; but as there are scarcely any independent labourers, except the housemen that have been mentioned, who all keep cattle, the hardship of being obliged to mix the inner bark of the pine with their bread is mitigated by the stores of cheese, of salt butter, of salt meat, salt fish, and bacon, which they are generally enabled to lay up for the winter provision. The period in which the want of corn presses the most severely is generally about two months before harvest; and at this time the cows, of which the poorest housemen have generally two or three, and many five or six, begin to give milk, which must be a great assistance to the family, particularly to the younger part of it. In the summer of the year 1799, the Norwegians appeared to wear a face of plenty and content, while their neighbours the Swedes were absolutely starving; and I particularly remarked, that the sons of housemen and the farmers' boys were fatter, larger, and had better calves to their legs, than boys of the same age and in similar situations in England.

It is also without doubt owing to the prevalence of the preventive check to population, as much as to any peculiar healthiness of the air, that the mortality in Norway is so small. There is nothing in the climate or the soil, that would lead to the supposition of its being in any extraordinary manner favourable to the general health of the inhabitants; but as in every country the principal mortality takes place among very young children, the smaller number of these in Norway, in proportion to the whole population, will naturally occasion a smaller mortality than in other countries, supposing the climate to be equally healthy.

It may be said, perhaps, and with truth, that one of the principal reasons of the small mortality in Norway is, that the towns are inconsiderable and few, and that few people are employed in unwholesome manufactories. In many of the agricultural villages of other countries, where the preventive check to population does not prevail in the same degree, the mortality is as small as in Norway. But it should be recollected, that the calculation in this case is for those particular villages alone; whereas in Norway the calculation of one in forty-eight is for the whole country. The redundant population of these villages is disposed of by constant emigrations to the towns, and the deaths of a great part of those that are born in the parish do not appear in the registers. But in Norway all the deaths are within the calculation, and it is clear, that, if more were born than the country could support, a great mortality must take place in some form or other. If the people were not destroyed by disease, they would be destroyed by famine. It is indeed well known, that bad and insufficient food will produce disease and death in the purest air and the finest climate. Supposing therefore no great foreign emigration, and no extraordinary increase in the resources of the country, nothing but the more extensive prevalence of the preventive check to population in Norway can secure to her a smaller mortality than in other countries, however pure her air may be, or however healthy the employments of her people.

Norway seems to have been anciently divided into large estates or farms, called Gores; and as, according to the law of succession, all the brothers divide the property equally, it is a matter of surprise, and a proof how slowly the population has hitherto increased, that these estates have not been more subdivided. Many of them are indeed now divided into half gores and quarter gores, and some still lower; but it has in general been the custom, on the death of the father, for a commission to value the estate at a low rate, and if the eldest son can pay his brothers' and sisters'[559] shares, according to this valuation, by mortgaging his estate or otherwise, the whole is awarded to him: and the force of habit and natural

[559] A daughter's portion is the half of a son's portion.

indolence too frequently prompt him to conduct the farm after the manner of his forefathers, with few or no efforts at improvement.

Another great obstacle to the improvement of farms in Norway is a law, which is called Odel's right, by which any lineal descendant can repurchase an estate, which had been sold out of the family, by paying the original purchase-money. Formerly collateral as well as lineal descendants had this power, and the time was absolutely unlimited, so that the purchaser could never consider himself as secure from claims. Afterwards the time was limited to twenty years, and in 1771, it was still further limited to ten years, and all the collateral branches were excluded. It must however be an uninterrupted possession of ten years; for if, before the expiration of this term, a person who has a right to claim under the law give notice to the possessor, that he does not forego his claim, though he is not then in a condition to make the purchase, the possessor is obliged to wait six years more, before he is perfectly secure. And as in addition to this the eldest in the lineal descent may reclaim an estate, that had been repurchased by a younger brother, the law, even in its present amended state, must be considered as a very great bar to improvement; and in its former state, when the time was unlimited and the sale of estates in this way was more frequent, it seems as if it must have been a most complete obstacle to the melioration of farms, and obviously accounts for the very slow increase of population in Norway for many centuries.

A further difficulty in the way of clearing and cultivating the land arises from the fears of the great timber merchants respecting the woods. When a farm has been divided among children and grandchildren, as each proprietor has a certain right in the woods, each in general endeavours to cut as much as he can; and the timber is thus felled before it is fit, and the woods spoiled. To prevent this, the merchants buy large tracts of woods of the farmers, who enter into a contract, that the farm shall not be any further subdivided or more housemen placed upon it; at least that, if the number of families be increased, they should have no right in the woods. It is said, that the merchants who make these purchases are not very strict, provided the smaller farmers and housemen do not take timber for their houses. The farmers who sell these tracts of wood are obliged by law, to reserve to themselves the right of pasturing their cattle, and of cutting timber sufficient for their houses, repairs, and firing.

A piece of ground round a houseman's dwelling cannot be enclosed for cultivation, without an application, first, to the proprietors of the woods, declaring, that the spot is not fit for timber; and afterwards to a magistrate of the district, whose leave on this occasion is also necessary, probably for the purpose of ascertaining, whether the leave of the proprietor had been duly obtained.

In addition to these obstacles to improved cultivation, which may be considered as artificial, the nature of the country presents an insuperable obstacle to a cultivation and population in any respect proportioned to the surface of the soil. The Norwegians, though not in a nomadic state, are still in a considerable degree in the pastoral state, and depend very much upon their cattle. The high grounds that border on the mountains, are absolutely unfit to bear corn; and the only use, to which they can be put, is to pasture cattle upon them for three or four months during the summer. The farmers accordingly send all their cattle to these grounds at this time of the year, under the care of a part of their families; and it is here, that they make all their butter and cheese for sale, or for their own consumption. The great difficulty is to support their cattle during the long winter, and for this purpose it is necessary, that a considerable proportion of the most fertile land in the vallies should be mowed for hay. If too much of it were taken into tillage, the number of cattle must be proportionably diminished, and the greatest part of the higher grounds would become absolutely useless; and it might be a question in that case, whether the country upon the whole would support a greater population.

Notwithstanding, however, all these obstacles, there is a very considerable capacity of improvement in Norway, and of late years it had been called into action. I heard it remarked by a professor at Copenhagen, that the reason why the agriculture of Norway had advanced so slowly was, that there were no gentlemen farmers to set examples of improved cultivation, and break the routine of ignorance and prejudice in the conduct of farms, that had been handed down from father to son for successive ages. From what I saw of Norway I should say, that this want is now in some degree supplied. Many intelligent merchants, and well informed general officers, are at present engaged in farming. In the country round Christiana, very great improvements have taken place in the system of agriculture; and even in the neighbourhood of Drontheim the culture of artificial grasses has been introduced, which, in a country where so much winter feed is necessary for cattle, is a point of the highest importance. Almost every where the cultivation of potatoes has succeeded, and they are growing more and more into general use, though in the distant parts of the country they are not yet relished by the common people.

It has been more the custom of late years than formerly to divide farms; and as the vent for commodities in Norway is not perhaps sufficient to encourage the complete cultivation of large farms, this division of them has probably contributed to the improvement of the land. It seems indeed to be universally agreed, among those who are in a situation to be competent judges, that the agriculture of Norway in general has advanced considerably of late years; and the registers show, that the population has followed with more than equal pace. On an average of ten years, from 1775 to 1784, the proportion of births to deaths was 141 to 100.[560] But this seems to have been rather too rapid an increase; as the following year, 1785, was a year of scarcity and sickness, in which the deaths considerably exceeded the births; and for four years afterwards, particularly in 1789, the excess of births was great. But in five years from 1789 to 1794, proportion of births and deaths was nearly to 100.[561]

[560] Thaarupt's Statistik der Danischen Monarchie, vol. ii. p. 4.

[561] Id. table I. p. 4. In the Tableau Statistique des Etats Danois, since published, it appears that the whole number of births for the five years subsequent to 1794, was 138,799, of deaths 94,530, of marriages 34,313. These numbers give the proportion of births to deaths as 146 to 100, of births to marriages as 4 to 1, and of deaths to marriages as 275 to 100. The average proportion of yearly births is stated to be 1/35, and of yearly deaths 1/49 of the whole population. vol. ii. ch. viii.

Book II, Chapter II.

Many of the most thinking and best informed persons express their apprehensions on this subject, and on the probable result of the new regulations respecting the enrolments of the army, and the apparent intention of the court of Denmark to encourage at all events the population. No very unfavourable season has occurred in Norway since 1785; but it is feared that, in the event of such a season, the most severe distress might be felt from the rapid increase that has of late taken place.

Norway is, I believe, almost the only country in Europe where a traveller will hear any apprehensions expressed of a redundant population, and where the danger to the happiness of the lower classes of people from this cause is in some degree seen and understood. This obviously arises from the smallness of the population altogether, and the consequent narrowness of the subject. If our attention were confined to one parish, and there were no power of emigrating from it, the most careless observer could not fail to remark that, if all married at twenty, it would be perfectly impossible for the farmers, however carefully they might improve their land, to find employment and food for those that would grow up; but when a great number of these parishes are added together in a populous kingdom, the largeness of the subject, and the power of moving from place to place, obscure and confuse our view. We lose sight of a truth, which before appeared completely obvious; and in a most unaccountable manner, attribute to the aggregate quantity of land a power of supporting people beyond comparison greater than the sum of all its parts.

3.5.2 Book II, Chapter II: Of the Checks to Population in Sweden.

Sweden, in many respects, in a state similar to that of Norway. A very large proportion of its population is in the same manner employed in agriculture; and in most parts of the country the married labourers who work for the farmers, like the housemen of Norway, have a certain portion of land for their principal maintenance; while the young men and women that are unmarried live as servants in the farmers' families. This state of things however is not so complete and general as in Norway; and from this cause, added to the greater extent and population of the country, the superior size of the towns and the greater variety of employment, it has not occasioned in the same degree the prevalence of the preventive check to population; and consequently the positive check has operated with more force, or the mortality has been greater.

According to a paper published by M. Wargentin in the *Mémoires abrégés de l'Académie Royale des Sciences de Stockholm,*[562] the yearly average mortality in all Sweden, for nine years ending in 1663, was to the population as 1 to 34¾.[563] M. Wargentin furnished Dr. Price with a continuation of these tables; and an average of

[562] Vol. i. 4to. Printed at Paris, 1772.

[563] Id. p. 27.

21 years gives a result of 1 to 34 3/5; nearly the same.[564] This is undoubtedly a very great mortality, considering the large proportion of the population in Sweden which is employed in agriculture. It appears, from some calculations in Cantzlaer's account of Sweden, that the inhabitants of the towns are to the inhabitants of the country only as 1 to 13;[565] whereas in well-peopled countries, the proportion is often as 1 to 3, or above.[566] The superior mortality of towns therefore cannot much affect the general proportion of deaths in Sweden.

The average mortality of villages according to Sussmilch is 1 in 40.[567] In Prussia and Pomerania, which include a number of great and unhealthy towns, and where the inhabitants of the towns are to the inhabitants of the country as 1 to 4, the mortality is less than 1 in 37.[568] The mortality in Norway, as has been mentioned before, is 1 in 48, which is in a very extraordinary degree less than in Sweden, though the inhabitants of the towns in Norway bear a greater proportion to the inhabitants of the country than in Sweden.[569] The towns in Sweden are indeed larger and more unhealthy than in Norway; but there is no reason to think that the country is naturally more unfavourable to the duration of human life. The mountains of Norway are in general not habitable. The only peopled parts of the country are the valleys. Many of these valleys are deep and narrow clefts in the mountains; and the cultivated spots in the bottom, surrounded as they are by almost perpendicular cliffs of a prodigious height,[570] which intercept the rays of the sun for many hours, do not seem as if they could be so healthy as the more exposed and drier soil of Sweden.

It is difficult therefore entirely to account for the mortality of Sweden, without supposing that the habits of the people, and the continual cry of the government for an increase of subjects, tend to press the population too hard against the limits of subsistence, and consequently to produce diseases, which are the necessary effect of poverty and bad nourishment; and this, from observation, appears to be really the case.

---

[564] Price's Observ. on Revers. Paym. vol. ii. p. 126, 4th edit.

[565] Mémoires pour servir à la connoissance des affaires politiques et économiques du Royaume de Suède, 4to. 1776, ch. vi. p. 187. This work is considered as very correct in its information, and is in great credit at Stockholm.

[566] Sussmilch's Gottliche Ordnung, vol. i. c: ii. sect. xxxiv. edit. 1798.

[567] Id. sect. xxxv. p. 91.

[568] Id. vol. iii. p. 60.

[569] Thaarup's Statistik der Danischen Monarchie, vol. ii. tab. II. p. 5. 1765.

[570] Some of these valleys are strikingly picturesque. The principal road from Christiana to Drontheim leads for nearly 180 English miles through a continued valley of this kind, by the side of a very fine river, which in one part stretches out into the extensive lake Miosen. I am inclined to believe that there is not any river in all Europe, the course of which, affords such a constant succession of beautiful and romantic scenery. It goes under different names in different parts. The verdure in the Norway valleys is peculiarly soft, the foliage of the trees luxuriant, and in summer no traces appear of a northern climate.

Sweden does not produce food sufficient for its population. Its annual want in the article of grain, according to a calculation made from the years 1768 and 1772, is 440,000 tuns.[571] This quantity or near it, has in general been imported from foreign countries, besides pork, butter and cheese to a considerable amount.[572]

The distillation of spirits in Sweden is supposed to consume above 400,000 tuns of grain; and when this distillation has been prohibited by government, a variation in defect appears in the tables of importations;[573] but no great variations in excess are observable to supply the deficiencies in years of scanty harvests, which it is well known occur frequently. In years the most abundant, when the distillation has been free, it is asserted that 388,000 tuns have in general been imported.[574] It follows therefore that the Swedes consume all the produce of their best years, and nearly 400,000 more; and that in their worst years their consumption must be diminished by nearly the whole deficiency in their crops. The mass of the people appears to be too poor to purchase nearly the same quantity of corn at a very advanced price. There is no adequate encouragement therefore to corn merchants to import in great abundance; and the effect of a deficiency of one-fourth or one-third in the crops is, to oblige the labourer to content himself with nearly three-fourths or two-thirds of the corn which he used before, and to supply the rest by the use of any substitutes, which Necessity, the mother of Invention, may suggest. I have said nearly; because it is difficult to suppose that the importations should not be something greater in years of scarcity than in common years, though no marked difference of this kind appears in the tables published by Cantzlaer. The greatest importation, according to these; tables, was in the year 1768, when it amounted to 590,265 tuns of grain;[575] but even this greatest importation is only 150,000 tuns above the average wants of the country; and what is this, to supply a deficiency of one-fourth or one-third of a crop? The whole importation is indeed in this respect trifling.

The population of Sweden, at the time when Cantzlaer wrote, was about two millions and a half.[576] He allows four tuns of grain to a man.[577] Upon this supposition the annual wants of Sweden would be ten millions of tuns, and four or five hundred thousand would go but a little way in supplying a deficiency of two millions and a half or three millions; and if we take only the difference from the average importation it will appear that the assistance which the Swedes receive from importation in a year of scarcity is perfectly futile.

---

[571] Mémoires du Royaume de Suède, table xvii. p. 174.

[572] Id. c. vi. p. 198.

[573] Id. table xlii. p. 418, c. vi. p. 201. I did not find out exactly the measure of the Swedish tun. It is rather less than our sack, or half-quarter.

[574] Id. c. vi. p. 201.

[575] Mémoires du Royaume de Suède, table xlii. p. 418.

[576] Id. ch. vi. p. 184.

[577] Id, p. 196.

The consequence of this state of things is, that the population of Sweden is in a peculiar manner affected by every variation of the seasons; and we cannot be surprised at a very curious and instructive remark of M. Wargentin, that the registers of Sweden shew that the births, marriages and deaths increase and decrease according to the state of the harvests. From the nine years of which he had given tables, he instances the following:

		Marriages.	Births.	Deaths.
Barren years.	1757	18,799	81,878	68,054
	1758	19,584	83,299	74,370
Abundant years.	1759	23,210	85,579	62,662
	1760	23,383	90,635	60,083.[578]

Here it appears that in the year 1760 the births were to the deaths as 15 to 10; but in the year 1758 only as 11 to 10. By referring to the enumerations of the population in 1757 and 1760,[579] which M. Wargentin has given, it appears that the number of marriages in the year 1760 in proportion to the whole population was as 1 to 101; in the year 1757, only as 1 to about 124. The deaths in 1760 were to the whole population as 1 to 39, in 1757 as 1 to 32, and in 1758 as 1 to 31.

In some observations on the Swedish registers, M. Wargentin says that in the unhealthy years about 1 in 29 have died annually, and in the healthy years one in 39; and that taking a middle term the average mortality might be considered at 1 in 36.[580] But this inference does not appear to be just; as a mean between 29 and 39 would give 34; and indeed the tables, which he has himself brought forward, contradict an average mortality of 1 in 36, and prove that it is about 1 in 34¾.

The proportion of yearly marriages to the whole population appears to be on an average nearly as 1 to 112, and to vary between the extremes of 1 to 101, and 1 to 124, according to the temporary prospect of a support for a family. Probably indeed it varies between much greater extremes, as the period from which these calculations are made is merely for nine years.

In another paper which M. Wargentin published in the same collection, he again remarks that in Sweden the years, which are the most fruitful in produce, are the most fruitful in children.[581]

If accurate observations were made in other countries, it is highly probable that differences of the same kind would appear, though not to the same extent.[582] With regard

578 Mémoires Abrégés de l'Académie de Stockholm, p. 29.

579 Id. p. 21, 22.

580 Mémoires Abrégés de l'Académie de Stockholm, p. 29.

581 Id. p. 31.

582 This has been confirmed with regard to England, by the abstracts of parish registers which have lately been published. The years 1795 and 1800 are marked by a diminution of marriages and births, and an increase of deaths.

to Sweden, they clearly prove that its population has a very strong tendency to increase; and that it is not only always ready to follow with the greatest alertness any average increase in the means of subsistence, but that it makes a start forwards at every temporary and occasional increase of food; by which means it is continually going beyond the average increase, and is repressed by the periodical returns of severe want, and the diseases arising from it.

Yet notwithstanding this constant and striking tendency to overflowing numbers, strange to say! the government and the political economists of Sweden are continually calling, out for population! population! Cantzlaer observes, that the government, not having the power of inducing strangers to settle in the country, or of augmenting at pleasure the number of births, has occupied itself since 1748 in every measure which appeared proper to increase the population of the country.[583] But suppose that the government really possessed the power of inducing strangers to settle, or of increasing the number of births at pleasure, what would be the consequence? If the strangers were not such as to introduce a better system of agriculture they would either be starved themselves, or cause more of the Swedes to be starved; and if the yearly number of births were considerably increased, it appears to me perfectly clear, from the tables of M. Wargentin, that the principal effect would be merely an increase of mortality. The actual population might perhaps even be diminished by it; as, when epidemics have once been generated by bad nourishment and crowded houses, they do not always stop when they have taken off the redundant population, but take off with a part, and sometimes a very considerable part, of that which the country might be able properly to support.

In all very northern climates, in which the principal business of agriculture must necessarily be compressed into the small space of a few summer months, it will almost inevitably happen that during this period a want of hands is felt; but, this temporary want should be carefully distinguished from a real, and effectual demand for labour, which includes the power of giving employment and support through the whole year, and not merely for two or three months. The population of Sweden in the natural course of its increase will always be ready fully to answer this effectual demand; and a supply beyond it, whether from strangers or an additional number of births, can only be productive of misery.

It is asserted by Swedish authors that a given number of men and of days produces in Sweden only a third part of what is produced by the same number of each in some other countries;[584] and heavy accusations are in consequence brought against the national industry. Of the general grounds for such accusations, a stranger cannot be a competent judge; but in the present instance it appears to me that more ought to be attributed to the climate and soil than to an actual want of industry in the natives. For a large portion of the year their exertions are necessarily cramped by the severity of the climate; and during the time when they are able to engage in agricultural operations, the natural indifference of the

[583] Mémoires du Royaume de Suède, c. vi. p. 188.

[584] Mémoires du Royaume de Suède, ch. vi. p. 191.

soil and the extent of surface required for a given produce, inevitably employ a greater proportional quantity of labour. It is well known in England that a farm of large extent, consisting of a poor soil, is worked at a much greater expense for the same produce than a small one of rich land. The natural poverty of the soil in Sweden, generally speaking, cannot be denied.[585]

In a journey up the western side of the country, and afterwards in crossing it from Norway to Stockholm, and thence up the eastern coast to the passage over to Finland, I confess that I saw fewer marks of a want of national industry than I should have expected: As far as I could judge, I very seldom saw any land uncultivated, which would have been cultivated in England; and I certainly saw many spots of land in tillage, which never would have been touched with a plough here. These were lands in which every five or ten yards there were large stones or rocks, round which the plough must necessarily be turned, or be lifted over them; and the one or the other is generally done according to their size. The plough is very light, and drawn by one horse; and in ploughing among the stumps of the trees when they are low, the general practice is to lift it over them. The man who holds the plough does this very nimbly, with little or no stop to the horse:

Of the value of those lands for tillage, which are at present covered with immense forests, I could be no judge; but both the Swedes and the Norwegians are accused of clearing these woods away too precipitately, and without previously considering what is likely to be the real value of the land when cleared. The consequence is, that for the sake of one good crop of rye; which may always be obtained from the manure afforded by the ashes of the burnt trees, much growing timber is sometimes spoiled, and the land perhaps afterwards becomes almost entirely useless. After the crop of rye has been obtained, the common practice is to turn cattle in upon the grass, which may accidentally grow up. If the land be naturally good, the feeding of the cattle prevents fresh firs from rising; but if it be bad, the cattle of course, cannot remain long in it, and the seeds, with which, every wind is surcharged, sow the ground again thickly with firs.

On observing many spots of this kind both in Norway and Sweden, I could not help being struck with the idea, that, though for other reasons it was very little probable, such appearances certainly made it seem possible that these countries might have been better peopled formerly than at present; and that lands, which are now covered with forests, might have produced corn a thousand years ago. Wars, plagues, or that greater depopulator than either, a tyrannical government, might have suddenly destroyed or expelled the greatest part of the inhabitants; and a neglect of the land for twenty or thirty years in Norway or Sweden would produce a very strange difference in the face of the country. But this is merely an idea which I could not help mentioning, but which the reader already knows has not had weight enough with me to make me suppose the fact in any degree probable.

---

[585]Cantzlaer mentions the returns from land *effectivement ensemencé* as only three grains for one. ch. vi. p. 196.

To return to the agriculture of Sweden. Independently of any deficiency in the national industry, there are certainly some circumstances in the political regulations of the country which tend to impede the natural progress of its cultivation. There are still some burdensome Corvées remaining, which the possessors of certain lands are obliged to perform for the domains of the crown.[586] The posting of the country is undoubtedly very cheap and convenient to the traveller; but it is conducted in a manner to occasion a great waste of labour to the farmer, both in men and horses. It is calculated by the Swedish economists that the labour, which would be saved by the abolition of this system alone, would produce annually 300,000 tuns of grain.[587] The very great distance of the markets in Sweden, and the very incomplete division of labour, which is almost a necessary consequence of it, occasion also a great waste of time and exertion. And if there be no marked want of diligence and activity among the Swedish peasants, there is certainly a want of knowledge as to the best modes of regulating the rotation of their crops, and of manuring and improving their lands.[588]

If the government were employed in removing these impediments, and in endeavours to encourage and direct the industry of the farmers, and to circulate the best information on agricultural subjects, it would do much more for the population of the country than by the establishment of five hundred foundling hospitals.

According to Cantzlaer, the principal measures in which the government had been engaged for the encouragement of the population, were the establishment of colleges of medicine, and of lying-in and foundling hospitals.[589] The establishment of colleges of medicine for the cure of the poor gratis, may, in many cases, be extremely beneficial, and was so probably in the particular circumstances of Sweden; but the example of the hospitals of France, which have the same object, may create a doubt whether even such establishments are universally to be recommended. Lying-in hospitals, as far as they have an effect, are probably rather prejudicial than otherwise; as, according to the principle on which they are generally conducted, their tendency is certainly to encourage vice. Foundling hospitals, whether they attain their professed and immediate object or not, are in every view hurtful to the state; but the mode in which they operate I shall have occasion to discuss more particularly in another chapter.

The Swedish government, however, has not been exclusively employed in measures of this nature. By an edict in 1776, the commerce of grain was rendered completely free throughout the whole interior of the country; and with regard to the province of Scania, which grows more than its consumption, exportation free of every duty was allowed.[590] Till this period the agriculture of the southern provinces had been checked by the want of

---

[586] Mémoires du Royaume de Suède, ch. vi. p. 202.

[587] Mémoires du Royaume de Suede, ch. vi. p. 204.

[588] Id. ch. vi.

[589] Id. p. 188.

[590] Mémoires du Royaume de Suède, ch. vi. p. 204.

vent for their grain, on account of the difficulty of transport, and the absolute prohibition of selling it to foreigners at any price. The northern provinces are still under some difficulties in this respect; though, as they never grow a quantity sufficient for their consumption, these difficulties are not so much felt.[591] It may be observed however, in general, that there is no check more fatal to improving cultivation than any difficulty in the vent of its produce, which prevents the farmer from being able to obtain in good years a price for his corn not much below the general average.

But what perhaps has contributed more than any other cause to the increasing population of Sweden is the abolition of a law in 1748, which limited the number of persons to each henman or farm.[592] The object of this law appears to have been, to force the children of the proprietors to undertake the clearing and cultivation of fresh lands, by which it was thought that the whole country would be sooner improved. But it appears from experience that these children, being without sufficient funds for such undertakings, were obliged to seek their fortune in some other way; and great numbers, in consequence, are said to have emigrated. A father may now, however, not only divide his landed property into as many shares as he thinks proper, but these divisions are particularly recommended by the government; and considering the immense size of the Swedish henmans, and the impossibility of their being cultivated completely by one family, such divisions must in every point of view be highly useful.

The population of Sweden in 1751 was 2,229,661.[593] In 1799, according to an account which I received in Stockholm from Professor Nicander, the successor to M. Wargentin, it was 3,043,731. This is a very considerable addition to the permanent population of the country, which has followed a proportional increase in the produce of the soil, as the imports of corn are not greater than they were formerly, and there is no reason to think that the condition of the people is, on an average, worse.

This increase, however, has not gone forwards without periodical checks, which, if they have not for a time entirely stopped its progress, have always retarded the rate of it. How often these checks have recurred during the last fifty years, I am not furnished with sufficient data to be able to say; but I can mention some of them. From the paper of M. Wargentin,[594] already quoted in this chapter, it appears that the years 1757 and 1758 were barren, and comparatively mortal years. If we were to judge from the increased importation of 1768,[595] this would also appear to be an unproductive year. According to the additional tables with which M. Wargentin furnished Dr. Price, the years 1771, 1772 and 1773, were particularly mortal.[596] The year 1759 must have been very highly so, as in the

[591] Id. ibid.

[592] Mémoires du Royaume de Suède, ch. vi. p. 177.

[593] Mémoires du Royaume de Suède, ch. vi. p. 184.

[594] Mémoires de l'Académie de Stockholm, p. 29.

[595] Mémoires du Royaume de Suède, table xlii.

[596] Price's Observ. on Revers. Pay. vol. ii, p. 125.

accounts which received from Professor Nicander, this year alone materially affected the average proportion of births to deaths for the twenty years ending in 1795. This proportion, including the year 1789, was 100 to 77; but abstracting it, was 100 to 75; which is a great difference for one year to make in an average of twenty. To conclude the catalogue, the year 1799, when I was in Sweden must have been a very fatal one. In the provinces bordering on Norway, the peasants called it the worst that they had ever remembered. The cattle had all suffered extremely during the winter, from the drought of the preceding year; and in July, about a month before the harvest, a considerable portion of the people was living upon bread made of the inner bark of the fir, and of dried sorrel, absolutely without any mixture of meal to make it more palatable and nourishing. The sallow looks and melancholy countenances of the peasants betrayed the unwholesomeness of their nourishment. Many had died; but the full effects of such a diet had not then been felt. They would probably appear afterwards in the form of some epidemic sickness.

The patience, with which the lower classes of people in Sweden bear these severe pressures is perfectly astonishing, and can only arise from their being left entirely to their own resources, and from the belief that they are submitting to the great law of necessity, and not to the caprices of their rulers. Most of the married labourers, as has before been observed, cultivate a small portion of land; and when, from an unfavourable season, their crops fail, or their cattle die, they see the cause of their want, and bear it as the visitation of Providence. Every man will submit with becoming patience to evils which he believes to arise from the general laws of nature; but when the vanity and mistaken benevolence of the government and the higher classes of society have, by a perpetual interference with the concerns of the lower classes, endeavoured to persuade them, that all the good which they enjoy is conferred upon them by their rulers and rich benefactors, it is very natural that they should attribute all the evil which they suffer to the same sources; and patience under such circumstances cannot reasonably be expected. Though to avoid still greater evils, we may be allowed to repress this impatience by force, if it shew itself in overt acts; yet the impatience itself appears to be clearly justified in this case: and those are in a great degree answerable for its consequences, whose conduct has tended evidently to encourage it.

Though the Swedes had supported the severe dearth of 1799 with extraordinary resignation; yet afterwards, on an edict of the government to prohibit the distillation of spirits, it is said that there were considerable commotions in the country. The measure itself was certainly calculated to benefit the people; and the manner in which it was received, affords a curious proof of the different temper with which people bear an evil arising from the laws of nature, or a privation caused by the edicts of a government.

The sickly periods in Sweden, which have retarded the rate of its increase in population, appear in general to have arisen from the unwholesome nourishment occasioned by severe want. And this want has been caused by unfavourable seasons, falling upon a country which was without any reserved store, either in its general exports or in the liberal division of food to the labourer in common years; and which was therefore peopled fully up to its produce, before the occurrence of the scanty harvest. Such a state of things is a clear proof that, if, as some of the Swedish economists assert, their country ought to have a population of nine or ten millions,[597] they have nothing further to do than to make it produce food sufficient for such a number; and they may rest perfectly assured that they will not want mouths to eat it, without the assistance of lying-in and foundling hospitals.

Notwithstanding the mortal year of 1789, it appeared from the accounts which I received from professor Nicander, that the general healthiness of the country had increased. The average mortality for the twenty years ending 1795 was 1 in 37, instead of 1 in less than 35, which had been the average of the preceding twenty years. As the rate of increase had not been accelerated in the twenty years ending in 1795, the diminished mortality must have been occasioned by the increased operation of the preventive check. Another calculation which I received from the professor seemed to confirm this supposition. According to M. Wargentin, as quoted by Sussmilch,[598] 5 standing marriages produced yearly 1 child; but in the latter period; the proportion of standing marriages to annual births was as 5 1/10, and subtracting illegitimate children, as 5 3/10 to 1; a proof that in the latter period the marriages had not been quite so early and so prolific.

From subsequent accounts it appears that the healthiness of Sweden has continued to increase, from which we may fairly infer that the condition, of the mass of the people has been improving.

In all Sweden and Finland during the five years ending with 1805, the mean number of the living at all ages was, males 1,564,611; females 1,683,457; both, 2,348,068. Annual average deaths of males 40,147; of females 39,266; that is, the annual mortality of males was 1 of 38.97 38.97; of females 1 of 42.87; mean, 1 of 40.92.[599]

The annual average births of males were 55,119; of females 52,762; both, 107,882; that is, the proportion of male births to the male population was 1 of 28.38; of female births to the female population 1 of 31.92; mean, 1 of 30.15.

From a valuable table formed by Mr. Milne on these and other data, it appears that, according to the law of mortality which prevailed in Sweden during the five years ending with 1805, the expectation of life at birth would be for males 37.820, for females 41.019; both, 39.385: and that half of the males would live to very nearly 43 years of age, half of the females nearly to 48 years of age, and half of all the births taken together to 45 years.

A proportion of births as 1 to 30.15, and of deaths as 1 to 40.92, would give a yearly excess of births to the population, as 1 to 114.5, which, if continued, would (according to Table II. at the end of Ch. xi. Bk. ii.) give a rate of increase such as to double the population in less than 80 years.

---

[597] Mémoires du Royaume de Suède, ch. vi. p. 1 196.

[598] Göttliche Ordnung, vol. i. C. vi. s. 120, p. 231.

[599] Transactions of the Royal Academy of Sciences at Stockbolm for the year 1809, and Supplement to the Encyclopædia Britannica, Article, Mortality, by Mr. Milne, Actuary to the Sun Life Assurance Society. The period of five years here noticed was free from any remarkable epidemics, and vaccination had commenced in 1804.

Book II, Chapter III.

In the Revue Encyclopédique for March, 1825, a short account is given of the result of a commission to inquire into the progress of population in Sweden since 1748, from which it appears, that Sweden properly so called, exclusive of Finland, contained then 1,736,483 inhabitants; in 1773, 1,958,797; in 1798, 2,352,298; and in 1823, 2,687,457. In 1823, there had been 56,054 deaths, and 98,259 births. The excess of the births in that year alone was therefore 42,205, and it is stated that, supposing the same excess in the next year, 1824, the average annual excess of the last fifteen years would be 23,333. This would be in the proportion of 1 to 108 of the average population, an excess which, if continued, would double the population in about 75 years. According to the foregoing numbers, the proportion of the births to the population was in 1823 as 1 to 27.3, of the deaths as 1 to 47.9. The healthiness of the country, therefore, and the rate of its increase in population, has continued to advance since 1805. This increase is attributed to the progress of agriculture and industry, and the practice of vaccination.

The gradual diminution of mortality since the middle of the last century is very striking.

3.5.3 Book II, Chapter III: Of the Checks to Population in Russia.

The lists of births, deaths and marriages in Russia, present such extraordinary results that it is impossible not to receive them with a considerable degree of suspicion; at the same time the regular manner in which they have been collected, and their agreement with each other in different years, entitle them to attention.

In a paper presented in 1768, by B. F. Herman, to the academy of Petersburg, and published in the *Nova Acta Academiæ, tom.* iv., a comparison is made of the births, deaths and marriages in the different provinces and towns of the empire, and the following proportions are given:

In Petersburg the births are to the burials as	13 to 10
In the government of Moscow	21 — 10
District of Moscow excepting the town	21 — 10
Tver	26 — 10
Novogorod	20 — 10
Pskovsk	22 — 10
Resan	20 — 10
Veronesch	29 — 10
Archbishopric of Vologda	23 — 10
Kostroma	20 — 10
Archangel	13 — 10
Tobolsk	21 — 10
Town of Tobolsk	13 — 10
Reval	11 — 10
Vologda	12 — 10

Some of these proportions it will be observed are extraordinarily high. In Veronesch, for instance, the births are to the deaths nearly as 3 to 1, which is as great a proportion, I believe, as ever was known in America. The average result however of these proportions has been, in some degree, confirmed by subsequent observations. Mr. Tooke, in his View of the Russian Empire, makes the general proportion of births to burials throughout the whole country, as 225 to 100,[600] which is 2¼ to 1; and this proportion is taken from the lists of 1793.[601]

From the number of yearly marriages, and yearly births, M. Herman draws the following conclusions:

		Children.
In Petersburg one marriage yields		4
In the government of Moscow about		3
Tver		3
Novogorod		3
Pskovsk		3
Resan		3
Veronesch		4
Vologda		4
Kostroma		3
Archangel		4
Reval		4
Government of Tobolsk		4
Town of Tobolsk,	from 1768 to 1778,	3
	from 1779 to 1783,	5
	in 1783	3

M. Herman observes that the fruitfulness of marriages in Russia does not exceed that of other countries, though the mortality is much less, as appears from the following proportions drawn from a rough calculation of the number of inhabitants in, each government:

		Dies annually.
In Petersburg		1 in 28
In the government of Moscow		1 — 32
District of Moscow		1 — 74
Tver		1 — 75
Novogorod		1 — 68 6/7

(continued)

[600] Vol. ii. b. iii. p. 162.

[601] Id. p. 145.

Pskovsk		1 — 70 4/5
Resan		1 — 50
Veronesch		1 — 79
Archbishopric of Vologda		1 — 65
Kostroma		1 — 59
Archangel		1 — 28 3/5
Reval		1 — 29
Government of Tobolsk		1 — 44
Town of Tobolsk		1 — 32
	in 1783	1 — 22¼

It may be concluded, M. Herman says, that in the greatest number of the Russian provinces the yearly mortality is 1 in 60.[602]

This average number is so high, and some of the proportions in the particular provinces are so extraordinary, that it is impossible to believe them accurate. They have been nearly confirmed, however, by subsequent lists, which, according to Mr. Tooke, make the general mortality in all Russia 1 in 58.[603] But Mr. Tooke himself seems to doubt the accuracy of this particular department of the registers; and I have since heard, from good authority, that there is reason to believe that the omissions in the burials are in all the provinces much greater than the omissions in the births; and consequently that the very great excess of births, and very small mortality, are more apparent than real. It is supposed that many children, particularly in the Ukraine, are privately interred by their fathers without information to the priest. The numerous and repeated levies of recruits take off great numbers, whose deaths are not recorded. From the frequent emigrations of whole families to different parts of the empire and the transportation of malefactors to Siberia, great numbers necessarily die on journeys or in parts where no regular lists are kept; and some omissions are attributed to the neglect of the parish priests, who have an interest in recording the births but not the deaths.

To these reasons I should add, that the population of each province is probably estimated by the number of boors belonging to each estate in it; but it is well known that a great part of them have leave to reside in the towns. Their births therefore appear in the province, but their deaths do not. The apparent mortality of the towns is not proportionably increased by this emigration, because it is estimated according to actual enumeration. The bills of mortality in the towns express correctly the numbers dying out of a certain number known to be actually present in these towns; but the bills of mortality in the provinces, purporting to express the numbers dying out of the estimated population of the province, do really only express the numbers dying out of a much smaller population, because a considerable part of the estimated population is absent.

602 Nova Acta Academiæ, tom. iv.

603 View of the Russian Empire, vol. ii. b. iii. p. 148.

In Petersburg, it appeared by an enumeration in 1784, that the number of males was 126,827, and of females only 65,619.[604] The proportion of males was therefore very nearly double, arising from the numbers who came to the town to earn their capitation tax, leaving their families in the country, and from the custom among the nobles of retaining a prodigious number of their boors as household servants in Petersburg and Moscow.

The number of births in proportion to the whole population in Russia is not different from a common average in other countries, being about 1 in 26.[605]

According to the paper of M. Herman already quoted, the proportion of boys dying within the first year is at Petersburg 1/5, in the government of Tobolsk 1/10, in the town of Tobolsk 1/3, in the Archbishopric of Vologda 1/14, in Novogorod 3/31, in Voronesch 1/24, in Archangel 1/5. The very small mortality of infants in some of these provinces, particularly as the calculation does not seem to be liable to much error, makes the smallness of the general mortality more credible. In Sweden throughout the whole country, the proportion of infants which die within the first year is 1/5 or more.[606]

The proportion of yearly marriages in Russia to the whole population is, according to M. Herman, in the towns, about 1 in 100, and in the provinces about 1 in 70 or 80. According to Mr. Tooke, in the fifteen governments of which he had lists, the proportion was 1 in 92.[607]

This is not very different from other countries. In Petersburg indeed the proportion was 1 in 140;[608] but this is clearly accounted for by what has already been said of the extraordinary number of the males in comparison of the females.

The registers for the city of Petersburg are supposed to be such as can be entirely depended upon; and these tend to prove the general salubrity of the climate. But there is one fact recorded in them, which is directly contrary to what has been observed in all other countries. This is a much greater mortality of female children than of male. In the period from 1781 to 1785, of 1000 boys born 147 only died within the first year, but of the same number of girls 310.[609] The proportion is as 10 to 21, which is inconceivable, and must indeed have been in some measure accidental, as in the preceding periods the proportion was only as 10 to 14; but even this is very extraordinary, as it has been generally remarked, that in every stage of life, except during the period of childbearing, the mortality among females is less than among males. The climate of Sweden does not appear to be very different from that of Russia; and M. Wargentin observes, with respect to the Swedish tables, that it appears from them that the smaller mortality of females is not merely owing to

[604] Mémoires par W. L. Krafft, Nova Acta Academiæ, tom. iv.

[605] Tooke's View of the Russ. Emp. vol. ii. b. iii. p. 147.

[606] Mémoires Abrégés dé l'Académie de Stockholm, p. 28.

[607] View of Russ. Emp. vol. ii. b. iii. p. 146.

[608] Mémoire par W. L. Krafft, Nova Acta Academiæ, tom. iv.

[609] Mémoire par W. L. Krafft, Nova Acta Academiæ, tom. iv.

a more regular and less laborious life, but is a natural law, which operates constantly from infancy to old age.[610]

According to M. Krafft,[611] the half of all that are born at Petersburg live to 25; which shews a degree of healthiness in early life very unusual for so large a town; but after twenty, a mortality much greater than in any other town in Europe takes place, which is justly attributed to the immoderate use of brandy.[612] The mortality between 10 and 15 is so small, that only 1 in 47 males, and 1 in 29 females, die during this period. From 20 to 25 the mortality is so great, that 1 in 9 males and 1 in 13 females die. The tables show that this extraordinary mortality is occasioned principally by pleurisies, high fevers, and consumptions. Pleurisies destroy ¼, high fevers 1/3, and consumptions 1/6, of the whole population. The three together take off 5/7 of all that die.

The general mortality during the period from 1781 to 1785 was, according to M. Krafft, 1 in 37. In a former period it had been 1 in 35, and in a subsequent period, when epidemic diseases prevailed, it was 1 in 29.[613] This average mortality is small for a large town; but there is reason to think, from a passage in M. Krafft's memoir,[614] that the deaths in the hospitals, the prisons, and in the *Maison des Enfans trouvés,* are either entirely omitted, or not given with correctness; and undoubtedly the insertion of these deaths might make a great difference in the apparent healthiness of the town.

In the *Maison des Enfans trouvés* alone the mortality is prodigious. No regular lists are published, and verbal communications are always liable to some uncertainty. I cannot therefore rely upon the information which I collected on the subject; but from the most careful inquiries which I could make of the attendants at the house in Petersburg, I understood that 100 a month was the common average. In the preceding winter, which was the winter of 1788, it had not been uncommon to bury 18 a day. The average number received in the day is about 10; and though they are all sent into the country to be nursed three days after they have been in the house, yet, as many of them are brought in a dying state, the mortality must necessarily be great. The number said to be received appears, indeed, almost incredible; but from what I saw myself, I should be inclined to believe, that both this and the mortality before mentioned might not be far from the truth. I was at the house about noon, and four children had been just received, one of which was evidently dying, and another did not seem as if it would long survive.

A part of the house is destined to the purpose of a lying-in hospital, where every woman that comes is received, and no questions are asked. The children thus born are brought up by nurses in the house, and are not sent into the country like the others. A mother, if she choose it, may perform the office of nurse to her own child in the house, but is not permitted

[610] Mémoires Abrégés de l'Académie de Stockholm; p. 28.

[611] Nova Acta Academiæ, tom. iv.

[612] Tooke's View of the Russian Empire, vol. ii. b. iii. p. 155.

[613] Id. p. 151.

[614] Id. note, p. 150.

to take it away with her. A child brought to the house may at any time be reclaimed by its parents, if they can prove themselves able to support it; and all the children are marked and numbered on being received, that they may be known and produced to, the parents when required, who, if they cannot reclaim them, are permitted to visit them.

The country nurses receive only two roubles a month, which, as the current paper rouble is seldom worth more than half a crown, is only about fifteen pence a week; yet the general expenses are said to be 100,000 roubles a month. The regular revenues belonging to the institution are not nearly equal to this sum; but the government takes on itself the management of the whole affair, and consequently bears all the additional expenses. As children are received without any limit, it is absolutely necessary that the expenses should also be unlimited. It is evident that the most dreadful evils must result from an unlimited reception of children, and only a limited fund to support then. Such institutions, therefore, if managed properly, that is, if the extraordinary mortality do not prevent the rapid accumulation of expense, cannot exist long except under the protection of a very rich government; and even under such protection the period of their failure cannot be very distant.

At six or seven years old the children who have been sent into the country return to the house, where they are taught all sorts of trades and manual operations. The common hours of working are from 6 to 12, and from 2 till 4. The girls leave the house at 18, and the boys at 20 or 21. When the house is too full, some of those which have been sent into the country are not brought back.

The principal mortality, of course, takes place among the infants who are just received, and the children which are brought up in the house; but there is a considerable mortality amongst those who are returned from the country, and are in the firmest stages of life. I was in some degree surprised at hearing this, after having been particularly struck with the extraordinary degree of neatness, cleanliness and sweetness, which appeared to prevail in every department. The house itself had been a palace, and all the rooms were large, airy, and even elegant. I was present while 180 boys were dining. They were all dressed very neatly; the table-cloth was clean, and each had a separate napkin to himself. The provisions appeared to be extremely good, and, there was not the smallest disagreeable smell in the room. In the dormitories there was a separate bed for each child; the bedsteads were of iron without tester or curtains, and the coverlids and sheets particularly clean.

This degree of neatness, almost inconceivable in a large institution, was to be attributed principally to the present Empress Dowager, who interested herself in all the details of the management and, when at Petersburg, seldom passed a week without inspecting them in person. The mortality which takes place in spite of all these attentions, is a clear proof, that the constitution in early youth cannot support confinement and work for eight hours in the day. The children had all rather a pale and sickly countenance, and if a judgment had been formed of the national beauty from the girls and boys in this establishment, it would have been most unfavourable.

It is evident, that, if the deaths belonging to this institution be omitted, the bills of mortality for Petersburg cannot give a representation in any degree near the truth of the real state of the city with respect to healthiness. At the same time it should be recollected, that some of the observations which attest its healthiness, such as the number dying in a thousand, &c., are not influenced by this circumstance; unless indeed we say, what is perhaps true, that nearly all those who would find any difficulty in rearing their children send them to the foundling hospital; and the mortality among the children of those who are in easy circumstances, and live in comfortable houses and airy situations, will of course be much less than a general average taken from all that are born.

The *Maison des Enfans trouvés* at Moscow is conducted exactly upon the same principle as that at Petersburg; and Mr. Tooke gives an account of the surprising loss of children, which it had sustained in twenty years, from the time of its first establishment to the year 1786. On this occasion he observes that if we knew precisely the number of those who died immediately after reception, or who brought in with them the germ of dissolution, a small part only of the mortality would probably appear to be fairly attributable to the foundling hospital; as none would be so unreasonable as to lay the loss of these certain victims to death to the account of a philanthropic institution, which enriches the country from year to year with an ever-increasing number of healthy, active, and industrious burghers.[615]

It appears to me, however, that the greatest part of this premature mortality is clearly to be attributed to these institutions, miscalled philanthropic. If any reliance can be placed on the accounts which are given of the infant mortality in the Russian towns and provinces, it would appear to be unusually small. The greatness of it, therefore, at the foundling hospitals, may justly be laid to the account of institutions which encourage a mother to desert her child, at the very time when of all others it stands most in need of her fostering care. The frail tenure by which an infant holds its life will not allow of a remitted attention, even for a few hours.

The surprising mortality which takes place at these two foundling hospitals of Petersburg and Moscow, which are managed in the best possible manner, (as all who have seen them with one consent assert,) appears to me incontrovertibly to prove, that the nature of these institutions is not calculated to answer the immediate end that they have in view; which I conceive to be the preservation of a certain number of citizens to the state who might otherwise perhaps perish from poverty or false shame. It is not to be doubted that if the children received into these hospitals had been left to the management of their parents, taking the chance of all the difficulties in which they might be involved, a much greater proportion of them would have reached the age of manhood, and have become useful members of the state.

When we look a little deeper into this subject, it will appear that these institutions not only fail in their immediate object, but by encouraging in the most marked manner habits of licentiousness, discourage marriage, and thus weaken the main spring of population. All the well-informed men, with whom I conversed on this subject at Petersburg, agreed

[615]View of the Russian Empire, vol. ii. b. iii. p. 201.

invariably that the institution had produced this effect in a surprising degree. To have a child was considered as one of the most trifling faults which a girl could commit. An English merchant at Petersburg told me, that a Russian girl living in his family, under a mistress who was considered as very strict, had sent six children to the foundling hospital without the loss of her place.

It should be observed, however, that generally speaking six children are not common in this kind of intercourse. Where habits of licentiousness prevail, the births are never in the same proportion to the number of people as in the married state; and therefore the discouragement to marriage, arising from this licentiousness, and the diminished number of births, which is the consequence of it, will much more than counterbalance any encouragement to marriage from the prospect held out to parents of disposing of the children which they cannot support.

Considering the extraordinary mortality which occurs in these institutions, and the habits of licentiousness which they have an evident tendency to create, it may perhaps be truly said, that, if a person wished to check population, and were not solicitous about the means, he could not propose a more effectual measure, than the establishment of a sufficient number of foundling hospitals, unlimited as to their reception of children. And with regard to the moral feelings of a nation, it is difficult to conceive that they must not be sensibly impaired by encouraging mothers to desert their offspring, and endeavouring to teach them that their love for their new-born infants is a prejudice which it is the interest of their country to eradicate. An occasional child-murder from false shame, is saved at a very high price, if it can only be done by the sacrifice of some of the best and most useful feelings of the human heart in a great part of the nation.

On the supposition that foundling hospitals attained their proposed end, the state of slavery in Russia would perhaps render them more justifiable in that country than in any other; because every child brought up at the foundling hospitals becomes a free citizen, and in this capacity is likely to be more useful to the state than if it had merely increased the number of slaves belonging to an individual proprietor. But in countries not similarly circumstanced, the most complete success in institutions of this kind would be a glaring injustice to other parts of the society. The true encouragement to marriage is the high price of labour, and an increase of employments which require to be supplied with proper hands; but if the principal part of these employments, apprenticeships, &c., be filled up by foundlings, the demand for labour among the legitimate part of the society must be proportionally diminished, the difficulty of supporting a family increased, and the best encouragement to marriage removed.

Russia has great natural resources. Its produce is, in its present state, above its consumption; and, it wants nothing but greater freedom of industrious exertion, and an adequate vent for its commodities in the interior parts of the country, to occasion an increase of population astonishingly rapid. The principal obstacle to this, is the vassalage, or rather slavery, of the peasants, and the ignorance and indolence which almost necessarily accompany such a state. The fortune of a Russian nobleman is measured by the number of boors that he possesses, which in general are saleable like cattle, and not *adscripti glebæ*.

His revenue arises from a capitation tax on all the males. When the boors upon an estate are increasing, new divisions of land are made at certain intervals; and either more is taken into cultivation, or the old shares are subdivided. Each family is awarded such a portion of land as it can properly cultivate, and will enable it to pay the tax. It is evidently the interest of the boor not to improve his lands much, and appear to get considerably more than is necessary to support his family and pay the polltax; because the natural consequence will be, that in the next division which takes place, the farm which he before possessed will be considered as capable of supporting two families, and he will be deprived of the half of it. The indolent cultivation that such a state of things must produce is easily conceivable. When a boor is deprived of much of the land which he had before used, he makes complaints of inability to pay his tax, and demands permission for himself or his sons to go and earn it in the towns. This permission is in general eagerly sought after, and is granted without much difficulty by the Seigneurs, in consideration of a small increase of the poll-tax. The consequence is, that the lands in the country are left half cultivated, and the genuine spring of population impaired in its source.

A Russian nobleman at Petersburg, of whom I asked some questions respecting the management of his estate, told me, that he never troubled himself to inquire whether it was properly cultivated or not, which he seemed to consider as a matter in which he was not in the smallest degree concerned. *Cela m'est égal,* says he, *cela me fait ni bien ni mal.* He gave his boors permission to earn their tax how and where they liked, and as long as he received it he was satisfied. But it is evident that by this kind of conduct he sacrificed the future population of his estate, and the consequent future increase of his revenues, to considerations of indolence and present convenience.

It is certain, however, that of late years many noblemen have attended more to the improvement and population of their estates, instigated principally by the precepts and example of the empress Catharine, who made the greatest exertions to advance the cultivation of the country. Her immense importations of German settlers not only contributed to people her state with free citizens instead of slaves, but, what was perhaps of still more importance, to set an example of industry, and of modes of directing that industry, totally unknown to the Russian peasants.

These exertions have been attended, upon the whole, with great success; and it is not to be doubted that, during the reign of the late empress and since, a very considerable increase of cultivation and of population has been going forward in almost every part of the Russian empire.

In the year 1763, an enumeration of the people, estimated by the poll-tax, gave a population of 14,726,696; and the same kind of enumeration in 1783 gave a population of 25,677,000, which, if correct, shews a very extraordinary increase; but it is supposed that the enumeration in 1783 was more correct and complete than the one in 1763. Including the provinces not subject to the poll-tax, the general calculation for 1763 was 20,000,000, and for 1796, 36,000,000.[616]

[616]Tooke's View of the Russian Empire, vol. ii. book iii. sect. i. p. 126, et seq.

In a subsequent edition of Mr. Tooke's View of the Russian Empire, a table of the births, deaths and marriages in the Greek church, is given for the year 1799, taken from a respectable German periodical publication, and faithfully extracted from the general returns received by the synod. It contains all the eparchies except Bruzlaw, which, from the peculiar difficulties attending a correct list of mortality in that eparchy, could not be inserted. The general results are,

	Males.	Females.	Totals.
Births,	531,015.	460,900.	991,915.
Deaths,	275,582.	264,807.	540,389.
Marriages, 257,513.			
Overplus of births	Males,	255,432	451,525.
	Females,	196,093	

To estimate the population Mr. Tooke multiplies the deaths by 58. But as this table has the appearance of being more correct than those which preceded it, and as the proportion of deaths compared with the births is greater in this table than in the others, it is probable that 58 is too great a multiplier. It may be observed, that in this table the births are to the deaths nearly as 183 to 100, the births to marriages as 385 to 100, and the deaths to the marriages as 210 to 100.

These are all more probable proportions than the results of the former tables.

The population of Russia, including the wandering tribes, and the acquired territories, was in 1822 estimated at 54,476,931. But the most interesting part of the population to examine, is that where lists of the births, deaths and marriages can be obtained.

The following table, which is given in the Encyclopædia Britannica, under the head of Russia, is formed from the reports published by the Synod, including only the members of the Orthodox Greek Church, the most numerous body of the people.

	1806	1810	1816	1820
Marriages	299,057	320,389	39,683	317,805
Births	1,361,286	1,374,926	1,457,606	1,570,399
Deaths	818,585	903,380	820,383	917,680

The population belonging to the Greek Church is estimated at 40,351,000.

If the average excess of the births above the deaths be applied to the 14 years ending with 1820, it will appear that, from this excess alone, the population had increased in that period, 8,064,616; and if the population in 1820 were 40,351,000, the population in 1806 was 32,286,384. Comparing the average excess of births with the average population during the 14 years, it will be found that the proportion is as 1 to 63, which (according to Table II. at the end of the 11th Chapter of this Book) would double the population in less than 44 years; a most rapid rate of increase.

The proportion of births to marriages is a little above 4½ to 1; of births to deaths, as 5 to 3; of marriages to the population, as 1 to 114; of births to the population as 1 to 25.2; and of deaths to the population, or the mortality, as 1 to 41.9.

Most of these proportions are essentially different from those mentioned in the earlier part of this chapter; but there is good reason to believe that they are more accurate; and they certainly accord better with the very rapid increase of population which is known to be going on in Russia.

The apparent increase of mortality is to be attributed rather to the former inaccuracy of the registers, than to increased unhealthiness. It is now allowed that the registers before 1796 were very imperfectly kept.

3.5.4 Book II, Chapter IV: Of the Checks to Population in the Middle Parts of Europe.

I have dwelt longer on the northern states of Europe than their relative importance might to some appear to demand, because their internal economy is in many respects essentially different from our own, and a personal though slight acquaintance with these countries has enabled me to mention a few particulars which have not yet been before the public. In the middle parts of Europe the division of labour, the distribution of employments and the proportion of the inhabitants of the country, differ so little from what is observable in England, that it would be in vain to seek for the checks to their population in any peculiarity of habits and manners sufficiently marked to admit of description. I shall therefore endeavour to direct the reader's attention principally to some inferences drawn from the lists of births, marriages and deaths in different countries; and these data will, in many important points, give us more information respecting their internal economy than we could receive from the most observing traveller.

One of the most curious and instructive points of view, in which we can consider lists of this kind, appears to me to be the dependence of the marriages on the deaths. It has been justly observed by Montesquieu, that, wherever there is a place for two persons to live comfortably, a marriage will certainly ensue:[617] but in most of the countries in Europe, in the present state of their population, experience will not allow us to expect any sudden and great increase in the means of supporting a family. The place therefore for the new marriage must, in general, be made by the dissolution of an old one; and we find in consequence, that, except after some great mortality, from whatever cause it may have proceeded, or some sudden change of policy peculiarly favourable to cultivation and trade, the number of annual marriages is regulated principally by the number of annual deaths. They reciprocally influence each other. There are few countries in which the common people have so much foresight, as to defer marriage till they have a fair prospect of being able to support properly

[617] Esprit des Loix, liv. xxii. c. x.

all their children. Some of the mortality therefore, in almost every country, is forced by the too great frequency of marriage; and in every country a great mortality, whether arising principally from this cause or occasioned by the number of great towns and manufactories and the natural unhealthiness of the situation, will necessarily produce a great frequency of marriage.

A most striking exemplification of this observation occurs in the case of some villages in Holland. Sussmilch has calculated the mean proportion of annual marriages compared with the number of inhabitants as between 1 in 107 and 1 in 113, in countries which have not been thinned by plagues or wars, or in which there is no sudden increase in the means of subsistence.[618] And Crome, a later statistical writer, taking a mean between 1 in 92 and 1 in 122, estimates the average proportion of marriages to inhabitants as 1 to 108.[619] But in the registers of 22 Dutch villages, the accuracy of which, according to Sussmilch, there is no reason to doubt, it appears that out of 64 persons there is 1 annual marriage.[620] This is a most extraordinary deviation from the mean proportion. When I first saw this number mentioned, not having then adverted to the mortality in these villages, I was much astonished; and very little satisfied with Sussmilch's attempt to account for it, by talking of the great number of trades, and the various means of getting a livelihood in Holland;[621] as it is evident that, the country having been long in the same state, there would be no reason to expect any great accession of new trades and new means of subsistence, and the old ones would of course all be full. But the difficulty was in a great measure solved, when it appeared that the mortality was between 1 in 22 and 1 in 23,[622] instead of being 1 in 36, as is usual when the marriages are in the proportion of 1 to 108. The births and deaths were nearly equal. The extraordinary number of marriages was not caused by the opening of any new sources of subsistence, and therefore produced no increase of population. It was merely occasioned by the rapid dissolution of the old marriages by death, and the consequent vacancy of some employment by which a family could be supported.

It might be a question, in this case, whether the too great frequency of marriage, that is, the pressure of the population too hard against the limits of subsistence, contributed most to produce the mortality; or the mortality, occasioned naturally by the employments of the people and unhealthiness of the country, the frequency of marriage. In the present instance I should, without doubt, incline to the latter supposition; particularly as it seems to be generally agreed, that the common people in Holland before the Revolution were, upon the whole, in a good state. The great mortality probably arose partly from the natural

[618] Sussmilch, Göttliche Ordnung, vol. i. c. iv. sect. lvi. p. 126.

[619] Crome, ueber die Grösse and Bevölkerung der Europ. Staaten, p. 88, Leips. 1785.

[620] Sussmilch, Göttliche Ordnung, vol. i. c. iv. sect. lviii. p. 127. Such a proportion of marriages could not, however, be supplied in a country like Holland, from the births within the territory, but must be caused principally by the influx of foreigners: and it is known that such an influx, before the Revolution, was constantly taking place. Holland, indeed, has been called the grave of Germany.

[621] Sussmilch, Götittliche Ordnung, vol. i. c. iv, sect, lviii. p. 128.

[622] Sussmilch, Göttliche Ordnung, vol. i, c. ii. sect. xxxvi. p. 92.

marshiness of the soil and the number of canals, and partly from the very great proportion of the people engaged in sedentary occupations, and the very small number in the healthy employments of agriculture.

A very curious and striking contrast to these Dutch villages, tending to illustrate the present subject, will be recollected in what was said respecting the state of Norway. In Norway the mortality is 1 in 48, and the marriages are 1 in 130. In the Dutch villages the mortality 1 in 23, and the marriages 1 in 64. The difference both in the marriages and deaths is above double. They maintain their relative proportions in a very exact manner, and shew how much the deaths and marriages mutually depend upon each other; and that, except where some sudden start in the agriculture of a country enlarges the means of subsistence, an increase of marriages must be accompanied by an increase of mortality, and *vice versâ*.

In Russia this sudden start in agriculture has in a great measure taken place; and consequently, though the mortality is very small, yet the proportion of marriages is not so. But in the progress of the population of Russia, if the proportion of marriages remain the same as at present, the mortality will inevitably increase; or if the mortality remain nearly the same, the proportion of marriages will diminish.

Sussmilch has produced some striking instances of this gradual decrease in the proportional number of marriages, in the progress of a country to a greater degree of cleanliness, healthiness and population, and a more complete occupation of all the means of gaining a livelihood.

In the town of Halle, in the year 1700, the number of annual marriages was to the whole population as 1 to 77. During the course of the 55 following years, this proportion changed gradually, according to Sussmilch's calculation, to 1 in 167.[623] This is a most extraordinary difference, and, if the calculation were quite accurate, would prove to what a degree the check to marriage had operated, and how completely it had measured itself to the means of subsistence. As however the number of people is estimated by calculation and not taken from enumerations, this very great difference in the proportions may not be perfectly correct, or may be occasioned in part by other causes.

In the town of Leipsic, in the year 1620, the annual marriages were to the population as 1 to 82; from the year 1741 to 1756 they were as 1 to 120.[624]

In Augsburg, in 1510, the proportion of marriages to the population was 1 to 86; in 1750 as 1 to 123.[625]

In Dantzic, in the year 1705, the proportion was as 1 to 89; in 1745 as 1 to 118.[626]

In the dukedom of Magdeburgh, in 1700, the proportion was as 1 to 87; from 1752 to 1755 as 1 to 125.

[623] Sussmilch, Göttliche Ordnung, vol. i. c, iv. sect. lxii. p. 132.

[624] Id. sect. lxiii. p. 134.

[625] Id. sect. lxiv. p. 134.

[626] Id. sect. lxv. p. 135.

In the principality of Halberstadt in 1690, the proportion was as 1 to 88; in 1756 as 1 to 112.

In the dukedom of Cleves, in 1705, the proportion was 1 to 83; in 1755, 1 to 100.

In the Churmark of Brandenburgh, in 1700; the proportion was 1 to 76; in 1755, 1 to 108.[627]

More instances of this kind might be produced; but these are sufficient to shew that in countries, where from a sudden increase in the means of subsistence, arising either from a great previous mortality or from improving cultivation and trade, room has been made for a great proportion of marriages, this proportion will annually decrease as the new employments are filled up, and there is no further room for an increasing population.

But in countries which have long been fully peopled, in which the mortality continues the same, and in which no new sources of subsistence are opening, the marriages being regulated principally by the deaths, will generally bear nearly the same proportion to the whole population at one period as at another. And the same constancy will take place even in countries where there is an annual increase in the means of subsistence, provided this increase be uniform and permanent. Supposing it to be such, as for half a century to allow every year of a fixed proportion of marriages beyond those dissolved by death, the population would then be increasing, and perhaps rapidly; but it is evident, that the proportion of marriages to the whole population might remain the same during the whole period.

This proportion Sussmilch has endeavoured to ascertain in different countries and different situations. In the villages of the Churmark of Brandenburgh, one marriage out of 109 persons takes place annually:[628] and the general proportion for agricultural villages he thinks may be taken at between 1 in 108 and 1 in 115.[629] In the small towns of the Churmark, where the mortality is greater, the proportion is 1 to 98;[630] in the Dutch villages mentioned before, 1 to 64; in Berlin 1 to 110;[631] in Paris 1 to 137.[632] According to Crome, in the *unmarrying cities* of Paris and Rome the proportion is only 1 to 60.[633]

All general proportions however of every kind should be applied with considerable caution, as it seldom happens that the increase of food and of population is uniform; and when the circumstances of a country are varying, either from this cause or from any change in the habits of the people with respect to prudence and cleanliness, it is evident that a proportion which is true at one period will not be so at another.

---

[627] Sussmilch, Göttliche Ordnung, vol. i. c. iv. sect. lxxi. p. 140.

[628] Sussmilch, Göttliche Ordnung, vol. i. c. iv. sect. lvi. p. 125.

[629] Id. sect. lxxv. p. 147.

[630] Id. sect. lx. p. 129.

[631] Ibid.

[632] Id. sect. lxix. p. 137.

[633] Crome, ueber die Grösse und Bevölkerung der Europaischen Staaten, p. 89.

Nothing is more difficult than to lay down rules on these subjects that do not admit of exceptions. Generally speaking, it might be taken for granted that an increased facility in the means of gaining a livelihood, either from a great previous mortality or from improving cultivation and trade, would produce a greater proportion of annual marriages; but this effect might not perhaps follow. Supposing the people to have been before in a very depressed state, and much of the mortality to have arisen from the want of foresight which usually accompanies such a state, it is possible that the sudden improvement of their condition might give them more of a decent and proper pride; and the consequence would be, that the proportional number of marriages might remain nearly the same, but they would all rear more of their children, and the additional population that was wanted would be supplied by a diminished mortality, instead of an increased number of births.

In the same manner, if the population of any country had been long stationary, and would not easily admit of an increase, it is possible that a change in the habits of the people, from improved education or any other cause, might diminish the proportional number of marriages; but as fewer children would be lost in infancy from the diseases consequent on poverty, the diminution in the number of marriages would be balanced by the diminished mortality, and the population would be kept up to its proper level by a smaller number of births.

Such changes therefore in the habits of a people should evidently be taken into consideration.

The most general rule that can be laid down on this subject is, perhaps, that any *direct* encouragements to marriage must be accompanied by an increased mortality. The natural tendency to marriage is in every country so great, that without any encouragements whatever a proper place for a marriage will always be filled up. Such encouragements therefore must either be perfectly futile, or produce a marriage where there is not a proper place for one; and the consequence must necessarily be increased poverty and mortality. Montesquieu, in his Lettres Persannes, says, that, in the past wars of France, the fear of being enrolled in the militia tempted a great number of young men to marry without the proper means of supporting a family, and the effect was the birth of a crowd of children, "que l'on cherche encore en France, et que la misère, la famine et les maladies en out fait disparoître."[634]

After so striking an illustration of the necessary effects of direct encouragements to marriage, it is perfectly astonishing that, in his Esprit des Loix he should say that Europe is still in a state to require laws, which favour the propagation of the human species.[635]

Sussmilch adopts the same ideas; and though he contemplates the case of the number of marriages coming necessarily to a stand when the food is not capable of further increase, and examines some countries in which the number of contracted marriages is exactly measured by the number dissolved by death, yet he still thinks that it is one of the principal

[634]Lettre cxxii.

[635]Esprit des Loix, liv. xxiii. c. xxvi.

duties of government to attend to the number of marriages. He cites the examples of Augustus and Trajan, and thinks that a prince or a statesman would really merit the name of father of his people, if, from the proportion of 1 to 120 or 125, he could increase the marriages to the proportion of 1 to 80 or 90:[636] But as it clearly appears, from the instances which he himself produces, that, in countries which have been long tolerably well peopled, death is the most powerful of all the encouragements to marriage; the prince or statesman, who should succeed in thus greatly increasing the number of marriages, might, perhaps, deserve much more justly the title of destroyer, than father, of his people.

The proportion of yearly births to the whole population must evidently depend principally upon the proportion of the people marrying annually; and therefore, in countries which will not admit of a great increase of population, must, like the marriages, depend principally on the deaths. Where an actual decrease of population is not taking place, the births will always supply the vacancies made by death, and exactly so much more as the increasing resources of the country will admit. In almost every part of Europe, during the intervals of the great plagues, epidemics or destructive wars, with which it is occasionally visited, the births exceed the deaths; but as the mortality varies very much in different countries and situations, the births will be found to vary in the same manner, though from the excess of births above deaths which most countries can admit, not in the same degree.

In 39 villages of Holland, where the deaths are about 1 in 23, the births are also about l in 23.[637] In 15 villages round Paris, the births bear the same, or even a greater, proportion to the whole population, on account of a still greater mortality; the births are 1 in 22 7/10, and the deaths the same.[638] In the small towns of Brandenburgh which are in an increasing state, the mortality is 1 in 29, and the births 1 in 24 7/16.[639] In Sweden, where the mortality is about 1 in 35, the births are 1 in 28.[640] In 1056 villages of Brandenburgh in which the mortality is about 1 in 39 or 40, the births are about 1 in 30.[641] In Norway, where the mortality is 1 in 48, the births are 1 in 34.[642] In all these instances, the births are evidently measured by the deaths, after making a proper allowance for the excess of births which the state of each country will admit.

Statistical writers have endeavoured to obtain a general measure of mortality for all countries taken together; but, if such a measure could be obtained, I do not see what good purpose it could answer. It would be but of little use in ascertaining the population of Europe, or of the world; and it is evident, that in applying it to particular countries or particular places, we might be led into the grossest errors. When the mortality of the human

[636] Sussmilch, Göttliche Ordnung, vol. i. c. iv. sect, lxxviii. p. 151.

[637] Sussmilch, Göttliche Ordnung, vol. i. c. vi. s. cxvi. p. 225.

[638] Ibid. and c. ii. s. xxvii. p. 93.

[639] Id. c. ii. s. xxviii. p. 80, and c. vi. s. cxvi. p. 225.

[640] Id. c. vi. s. cxvi. p. 225.

[641] Ibid.

[642] Thaarup's Statistik, vol. ii. p. 4.

race in different countries and different situations, varies so much as from 1 in 20 to 1 in 60, no general average could be used with safety in a particular case, without such a knowledge of the circumstances of the country, with respect to the number of towns, the habits of the people and the healthiness of the situation, as would probably supersede the necessity of resorting to any general proportion, by the knowledge of the particular proportion suited to the country.

There is one leading circumstance, however, affecting the mortality of countries, which may be considered as very general, and which is, at the same time, completely open to observation. This is the number of towns, and the proportion of town to country inhabitants. The unfavourable effects of close habitations and sedentary employments on the health are universal; and therefore on the number of people living in this manner, compared with the number employed in agriculture, will much depend the general mortality of the state.

Upon this principle it has been calculated that when the proportion of the people in the towns to those in the country is as 1 to 3, then the mortality is about 1 in 36: which rises to 1 in 35, or 1 in 33, when the proportion of townsmen to villagers is 2 to 5, or 3 to 7; and falls below 1 in 36, when this proportion is 2 to 7, or 1 to 4. On these grounds the mortality in Prussia is 1 in 38; in Pomerania, 1 in 37½; in the Neumark 1 in 37; in the Churmark 1 in 35; according to the lists for 1756.[643]

The nearest average measure of mortality for all countries, taking towns and villages together, is, according to Sussmilch, 1 in 36.[644] But Crome thinks that this measure, though it might possibly have suited the time at which Sussmilch wrote, is not correct at present, when in most of the states of Europe both the number and size of the towns have increased.[645] He seems to be of opinion indeed, that this mortality was rather below the truth in Sussmilch's time, and that now 1 in 30 would be found to be nearer the average measure. It is not improbable that Sussmilch's proportion is too small, as he had a little tendency, with many other statistical writers, to throw out of his calculations epidemic years; but Crome has not advanced proofs sufficient to establish a general measure of mortality, in opposition to that proposed by Sussmilch. He quotes Busching, who states the mortality of the whole Prussian monarchy to be 1 in 30.[646] But it appears that this inference was drawn from lists for only three years, a period much too short to determine any general average. This proportion, for the Prussian monarchy, is indeed completely contradicted by subsequent observations mentioned by Crome. According to lists for five years, ending in 1784, the mortality was only 1 in 37.[647] During the same periods, the births were to the deaths as 131 to 100. In Silesia the mortality from 1781 to 1784 was 1 in 30; and the births

[643] Sussmilch, Göttliche Ordnung, vol. iii. p. 60.

[644] Vol. i. c. ii. s. xxxv. p. 91.

[645] Crome, über die Gröss und Bevölkerung der Europaischen Staaten, p. 116.

[646] Crome, über die Bevölkerung der Europaisch. Staat. p. 118.

[647] Id. p. 120.

to deaths as 128 to 100. In Gelderland the mortality from 1776 to 1781 was 1 in 27, and the births 1 in 26. These are the two provinces of the monarchy, in which the mortality is the greatest. In some others it is very small. From 1781 to 1784 the average mortality in Neufchatel and Ballengin was only 1 in 44, and the births 1 in 31. In the principality of Halberstadtz, from 1778 to 1784, the mortality was still less, being only 1 in 45 or 46, and the proportion of births to deaths 137 to 100.[648]

The general conclusion which Crome draws is, that the states of Europe may be divided into three classes, to which a different measure of mortality ought to be applied. In the richest and most populous states, where the inhabitants of the towns are to the inhabitants of the country in so high a proportion as 1 to 3, the mortality may be taken as 1 in 30. In those countries which are in a middle state with regard to population and cultivation, the mortality may be considered as 1 in 32. And in the thinly-peopled northern states, Sussmilch's proportion of 1 in 36 may be applied.[649]

These proportions seem to make the general mortality too great, even after allowing epidemic years to have their full effect in the calculations. The improved habits of cleanliness, which appear to have prevailed of late years in most of the towns of Europe, have probably, in point of salubrity, more than counterbalanced their increased size.

In a census which was made in 1817, of the population of Prussia in its present enlarged state, the number of inhabitants was found to be 10,536,571, of which 5,244,308 were males, and 5,320,535 were females. The births were 454,031, the deaths 306,484, and the marriages 112,034. Of the births 53,576, or 1/8.4 were illegitimate. The proportion of males to females born was as 20 to 19. Of the illegitimate children 3 out of every 10 died in the first year after birth; of the legitimate 2 out of 10.[650]

The numbers here stated give a proportion of births to deaths, as 149 to 100; of births to marriages as 4 to 1; of births to the population as 1 to 23.2; of deaths to the population, of males, as 1 to 33; of females, as 1 to 36; of both together, as 1 to 34½; and of marriages to the population as 1 to 94. The proportion of the excess of the births above the deaths to the population is as 1 to 62; an excess which, if continued, would double the population in about 43 years. As it is not however stated how long these proportions have continued, no very certain conclusions can be drawn from them; but there is little, doubt that the population is proceeding with great rapidity.

3.5.5 Book II, Chapter V: Of the Checks to Population in Switzerland.

The situation of Switzerland is in many respects so different from the other states of Europe, and some of the facts that have been collected respecting it are so curious, and

[648] Id. p. 122.

[649] Crome's Europaischen Staaten, p. 127.

[650] Supplement to the Encyclopædia Britannica, Article Prussia.

tend so strongly to illustrate the general principles of this work, that it seems to merit a separate consideration.

About 35 or 40 years ago, a great and sudden alarm appears to have prevailed in Switzerland respecting the depopulation of the country; and the transactions of the Economical Society of Berne, which had been established some years before, were crowded with papers deploring the decay of industry, arts, agriculture and manufactures, and the imminent danger of a total want of people. The greater part of these writers considered the depopulation of the country as a fact so obvious, as not to require proof. They employed themselves, therefore, chiefly in proposing remedies, and, among others, the importation of midwives, the establishment of foundling hospitals, the portioning of young virgins, the prevention of emigration, and the encouragement of foreign settlers.[651]

A paper containing very valuable materials was, however, about this time published by M. Muret, minister of Vevay, who, before he proceeded to point out remedies, thought it necessary to substantiate the existence of the evil. He made a very laborious and careful research into the registers of the different parishes, up to the time of their first establishment, and compared the number of births which had taken place during three different periods of 70 years each, the first ending in 1620, the second in 1690, and the third in 1760.[652] Finding upon this comparison, that the number of births was rather less in the second than in the first period, (and by the help of supposing some omissions in the second period, and, some redundances in the third,) that the number of births in the third was also less than in the second, he considered the evidence for a continued depopulation of the country from the year 1550 as incontrovertible.

Admitting all the premises, the conclusion is not perhaps so certain as he imagined it to be and from other facts which appear in his memoir, I am strongly disposed to believe, that Switzerland, during this period, came under the case supposed in the last chapter; and that the improving habits of the people with respect to prudence, cleanliness &c., had increased gradually the general healthiness of the country, and, by enabling them to rear up to manhood a greater proportion of their children, had furnished the requisite population with a smaller number of births. Of course, the proportion of annual births to the whole population, in the latter period, would be less than in the former.

From accurate calculations of M. Muret, it appears, that during the last period the mortality was extraordinarily small, and the proportion of children reared from infancy to puberty extraordinarily great.[653] In the former periods this could not have been the case in the same degree. M. Muret himself observes, that "the ancient depopulation of the country was to be attributed to the frequent plagues which, in former times, desolated it;" and adds, "if it could support itself, notwithstanding the frequency of so dreadful an evil, it is a proof

[651] See the different Memoirs for the year 1766.

[652] Mémoires, &c. par la Société Economique de Berne. Année 1766, première partie, p. 15 et seq. octavo. Berne.

[653] Mémoires, &c. par la Société Economique de Berne, table xiii. p. 120. Année 1766.

of the goodness of the climate, and of the certain resources which the country could furnish, for a prompt recovery of its population."[654] He neglects to apply this observation as he ought, and forgets that such a prompt repeopling could not take place without an unusual increase of births, and that, to enable a country to support itself against such a source of destruction, a greater proportion of births to the whole population would be necessary than at other times.

In one of his tables he gives a list of all the plagues that have prevailed in Switzerland since the year 1312, from which it appears that this dreadful scourge desolated the country, at short intervals, during the whole of the first period, and extended its occasional ravages to within 22 years of the termination of the second.[655]

It would be contrary to every rule of probability to suppose that, during the frequent prevalence of this disorder, the country could be particularly healthy, and the general mortality extremely small. Let us suppose it to have been such as at present takes place in many other countries, which are exempt from this calamity, about 1 in 32, instead of 1 in 45, as in the last period. The births would of course keep their relative proportion, and instead of 1 in 36,[656] be about 1 in 26. In estimating the population of the country by the births, we should thus have two very different multipliers for the different periods; and though the absolute number of births might be greater in the first period, yet the fact would by no means imply a greater population.

In the present instance, the sum of the births in 17 parishes, during the first 70 years, is given as 49,860, which annually would be about 712. This, multiplied by 26, would indicate a population of 18,512. In the last period the sum of the births is given as 43,910,[657] which will be about 626 annually. This, multiplied by 36, will indicate a population of 22,536; and if the multipliers be just, it will thus appear, that instead of the decrease which was intended to be proved, there had been a considerable increase.

That I have not estimated the mortality too high during the first period, I have many reasons for supposing, particularly a calculation respecting the neighbouring town of Geneva, in which it appears that, in the 16th century, the probability of life, or the age to which half of the born live, was only 4.883, rather less than four years and 9/10ths; and the mean life 18.511, about 18 years and a half. In the 17th century, the probability of life was 11.607, above 11 years and a half; the mean life 23.358. In the 18th century the probability of life had increased to 27.183, 27 years and nearly a fifth, and the mean life to 32 years and a fifth.[658]

It is highly probable that a diminution of mortality, of the same kind, though perhaps not in the same degree, should have taken place in Switzerland; and we know from the registers

[654] Id. p. 22.

[655] Mémoires, &c. par la Société Econ. de Berne. Année 1766, première partie, table IV. p. 22.

[656] Id. table I. p. 21.

[657] Id. table I. p. 16.

[658] See a paper in the Bibliothèque Britannique, published at Geneva, tom. iv. p. 328.

of other countries which have been already noticed, that a greater mortality naturally produces a greater proportion of births.

Of this dependence of the births on the deaths M. Muret himself produces many instances; but not being aware of the true principle of population, they only serve to astonish him, and he does not apply them.

Speaking of the want of fruitfulness in the Swiss women, he says, that Prussia, Brandenburgh, Sweden, France, and indeed every country, the registers of which he had seen, give a greater proportion of baptisms to the number of inhabitants, than the Pays de Vaud, where this proportion is only as 1 to 36.[659] He adds, that from calculations lately made in the Lyonois, it appeared, that in Lyons itself the proportion of baptisms was 1 in 28, in the small towns 1 in 25, and in the parishes 1 in 23 or 24. What a prodigious difference, he exclaims, between the Lyonois and the Pays de Vaud, where the most favourable proportion, and that only in two small parishes of extraordinary fecundity, is not above 1 in 26, and in many parishes it is considerably less than 1 in 40![660] *The same difference,* he remarks, takes place in the *mean life.* In the Lyonois it is a little above 25 years, while in the Pays de Vaud the lowest mean life, and that only in a single marshy and unhealthy parish, is 29½ years, and in many places it is above 45 years.[661]

"But whence comes it," he says, "that the country where children escape the best from the dangers of infancy, and where the mean life, in whatever way the calculation is made, is higher than in any other, should be precisely that in which the fecundity is the smallest? How comes it again that, of all our parishes, the one which gives the mean life the highest, should also be the one where the tendency to increase is the smallest?

"To resolve this question, I will hazard a conjecture, which, however, I give only as such. Is it not, that in order to maintain in all places the proper equilibrium of population, God has wisely ordered things in such a manner, as that the force of life in each country should be in the inverse ratio of its fecundity?[662]

"In fact, experience verifies my conjecture. Leyzin, a village in the Alps, with a population of 400 persons, produces but a little above eight children a year. The Pays de Vaud, in general, in proportion to the same number of inhabitants, produces 11, and the Lyonois 16. But if it happen, that at the age of 20 years, the 8, the 11, and the 16, are reduced to the same number, it will appear that the force of life gives in one place what fecundity does in another. And thus the most healthy countries, having less fecundity, will not overpeople themselves, and the unhealthy countries, by their extraordinary fecundity, will be able to sustain their population."

We may judge of the surprize of M. Muret, at finding from the registers, that the most healthy people were the least prolific, by his betaking himself to a miracle in order to

---

[659] Mémoires, &c. par la Société Econ. de Berne. Année 1766, première partie, p. 47, 48.

[660] Id. p. 48.

[661] Ib.

[662] Mémoires, &c. par la Société Econ. de Berne. Année 1766, première partie, p. 48, et seq.

account for it. But the difficulty does not seem, in the present instance, to be worthy of such an interference. The fact may be accounted for, without resorting to so strange a supposition as that the fruitfulness of women should vary inversely as their health.

There is certainly a considerable difference in the healthiness of different countries, arising partly from the soil and situation, and partly from the habits and employment of the people. When, from these or any other causes whatever, a great mortality takes place, a proportional number of births immediately ensues, owing both to the greater number of yearly marriages from the increased demand for labour, and the greater fecundity of each marriage from being contracted at an earlier, and naturally a more prolific age.

On the contrary, when from opposite causes the healthiness of any country or parish is extraordinarily great; if, from the habits of the people, no vent for an overflowing population be found in emigration, the absolute necessity of the preventive check will be forced so strongly on their attention, that they must adopt it or starve; and consequently the marriages being very late, the number annually contracted will not only be small in proportion to the population, but each individual marriage will naturally be less prolific.

In the parish of Leyzin, noticed by M. Muret, all these circumstances appear to have been combined in an unusual degree. Its situation in the Alps, but yet not too high, gave it probably the most pure and salubrious air; and the employment of the people, being all pastoral, were consequently of the most healthy nature. From the calculations of M. Muret, the accuracy of which there is no reason to doubt, the probability of life in this parish appeared to be so extraordinarily high as 61 years.[663] And the average number of the births being for a period of 30 years almost accurately equal to the number of deaths,[664] clearly proved that the habits of the people had not led them to emigrate, and that the resources of the parish for the support of population had remained nearly stationary. We are warranted therefore in concluding, that the pastures were limited, and could not easily be increased either in quantity or quality. The number of cattle, which could be kept upon them, would of course be limited; and in the same manner the number of persons required for the care of these cattle.

Under such circumstances, how would it be possible for the young men who had reached the age of puberty, to leave their fathers' houses and marry, till an employment of herdsman, dairyman, or something of the kind, became vacant by death? And as, from the extreme healthiness of the people, this must happen very slowly, it is evident that the majority of them must wait during a great part of their youth in their bachelor state, or run the most obvious risk of starving themselves and their families. The case is still stronger than in Norway, and receives a particular precision from the circumstance of the births and deaths being so nearly equal.

If a father had unfortunately a larger family than usual, the tendency of it would be rather to decrease than increase the number of marriages. He might perhaps with economy be just

[663] Mémoires, &c. par la Société Econ. de Berne, Année 1766, table v. p. 64.

[664] Id. table I. p. 15.

able to support them all at home, though he could not probably find adequate employment for them on his small property; but it would evidently be long before they could quit him, and the first marriage among the sons would probably be after the death of the father; whereas, if he had had only two children, one of them might perhaps have married without leaving the parental roof, and the other on the death of the father. It may be said perhaps in general, that the absence or presence of four grown-up unmarried people will make the difference of there being room or not, for the establishment of another marriage and a fresh family.

As the marriages in this parish would, with few exceptions, be very late, and yet from the extreme healthiness of the situation be very slowly dissolved by the death of either of the parties, it is evident that a very large proportion of the subsisting marriages would be among persons so far advanced in life, that most of the women would have ceased to bear children; and in consequence the whole number of subsisting marriages was found to be to the number of annual births in the very unusual proportion of 12 to 1. The births were only about a 49th part of the population; and the number of persons above sixteen was to the number below that age nearly as 3 to 1.[665]

As a contrast to this parish, and a proof how little the number of births can be depended upon for an estimate of population, M. Muret produces the parish of St. Cergue in the Jura, in which the subsisting marriages were to the annual births only in the proportion of 4 to 1, the births were a 26th part of the population, and the number of persons above and below sixteen just equal.[666]

Judging of the population of these parishes from the proportion of their annual births, it would appear, he says, that Leyzin did not exceed St. Cergue by above one-fifth at most; whereas, from actual enumeration, the population of the former turned out to be 405, and of the latter only 171.[667]

I have chosen, he observes, the parishes where the contrast is the most striking; but though the difference be not so remarkable in the rest, yet it will always be found true that from one place to another, even at very small distances, and in situations apparently similar, the proportions will vary considerably.[668]

It is strange that, after making these observations, and others of the same tendency, which I have not produced, he should rest the whole proof of the depopulation of the Pays de Vaud on the proportion of births. There is no good reason for supposing that this proportion should not be different at different periods, as well as in different situations. The extraordinary contrast in the fecundity of the two parishes of Leyzin and St. Cergue depends upon causes within the power of time and circumstances to alter. From the great proportion of infants which was found to grow up to maturity in St. Cergue, it appeared that

[665] Mémoires, &c. par la Société Econ. de Berne. Année 1766, p. 11 and 12.

[666] Ibid.

[667] Id. p. 11.

[668] Id. p. 13.

its natural healthiness was not much inferior to that of Leyzin.[669] The proportion of its births to deaths was 7 to 4;[670] but as the whole number of its inhabitants did not exceed 171, it is evident that this great excess of births could not have been regularly added to the population during the last two centuries. It must have arisen therefore either from a sudden increase of late years in the agriculture or trade of the parish, or from a habit of emigration. The latter supposition I conceive to be the true one; and it seems to be confirmed by the small proportion of adults which has already been noticed. The parish is situated in the Jura, by the side of the high road from Paris to Geneva, a situation which would evidently tend to facilitate emigration; and in fact, it seems to have acted the part of a breeding parish for the towns and flat countries; and the annual drain of a certain portion of the adults made room for all the rest to marry, and to rear a numerous offspring.

A habit of emigration in a particular parish, will not only depend on situation, but probably often on accident. I have little doubt that three or four very successful emigrations have frequently given a spirit of enterprise to a whole village; and three or four unsuccessful ones a contrary spirit. If a habit of emigration were introduced into the village of Leyzin, it is not to be doubted that the proportion of births would be immediately changed; and at the end of twenty years an examination of its registers might give results as different from those at the time of M. Muret's calculations, as they were then from the contrasted parish of St. Cergue. It will hence appear that other causes besides a greater mortality will concur, to make an estimate of population, at different periods, from the proportion of births, liable to great uncertainty.

The facts which M. Muret has collected are all valuable, though his inferences cannot always be considered in the same light. He made some calculations at Vevay, of a nature really to ascertain the question respecting the fecundity of marriages, and to shew the incorrectness of the usual mode of estimating it, though without this particular object in view at the time. He found that 375 mothers had yielded 2,093 children, all born alive; from which it followed, that each mother had produced 5 10/12, or nearly six children.[671] These, however, were all actually mothers, which every wife is not; but allowing for the usual proportion of barren wives at Vevay, which he had found to be 20 out of 478, it will still appear that the married women one with another produced above 5 1/3 children.[672] And yet this was in a town, the inhabitants of which he seems to accuse of not entering into the marriage state at the period when nature calls them, and, when married, of not having all the children which they might have.[673] The general proportion of the annual marriages to the

[669] Mémoires, &c. par la Société Econ. de Berne. Année 1766, table xiii. p. 120.

[670] Id. table I. p. 11.

[671] Mémoires, &c. par la Société Econ. de Berne. Année 1766, p. 29, et seq.

[672] On account of second and third marriages, the fecundity of marriages must always be less than the fecundity of married women. The mothers alone are here considered, without reference to the number of husbands.

[673] Mémoires, &c. par la Société Econ. de Berne. Année 1766, p. 32.

annual births in the Pays de Vaud is as 1 to 3.9,[674] and of course, according to the common mode of calculation, the marriages would appear to yield 3.9 children each.

In a division of the Pays de Vaud into eight different districts, M. Muret found, that in seven towns the mean life was 36 years; and the probability of life, or the age to which half of the born live, 37. In 36 villages, the mean life was 37, and the probability of life 42. In nine parishes of the Alps the mean life was 40, and the probability of life 47. In seven parishes of the Jura these two proportions were 38 and 42: in 12 corn parishes, 37 and 40; in 18 parishes among the great vineyards, 34 and 37; in six parishes of mixed vines and hills, 33 9/10 and 36; and in one marshy parish, 29 and 24.[675]

From another table it appears, that the number of persons dying under the age of 15 was less than 1/5 in the extraordinary parish of Leyzin; and less than ¼ in many other parishes of the Alps and the Jura. For the whole of the Pays de Vaud it was less than 1/3.[676]

In some of the largest towns, such as Lausanne and Vevay, on account of the number of strangers settling in them, the proportion of adults to those under 16 was nearly as great as in the parish of Leyzin, and not far from 3 to 1. In the parishes from which there were not many emigrations, this proportion was about 2 to 1. And in those which furnished inhabitants for other countries, it approached more towards an equality.[677]

The whole population of the Pays de Vaud, M. Muret estimated at 113,000, of which 76,000 were adults. The proportion of adults therefore to those under the age of sixteen, for the whole country, was 2 to 1. Among these 76,000 adults, there were 19,000 subsisting marriages, and consequently 38,000 married persons; and the same number of persons unmarried, though of the latter number 9,000, according to M. Muret, would probably be widows or widowers.[678] With such an average store of unmarried persons, notwithstanding the acknowledged emigrations, there was little ground for the supposition that these emigrations had essentially affected the number of annual marriages, and checked the progress of population.

The proportion of annual marriages to inhabitants in the Pays de Vaud, according to M. Muret's tables, was only 1 to 140,[679] which is even less than in Norway.

All these calculations of M. Muret imply the operation of the preventive check to population in a considerable degree, throughout the whole of the district which he considered; and there is reason to believe, that the same habits prevail in other parts of Switzerland, though varying considerably from place to place, according as the situation or the employments of the people render them more or less healthy, or the resources of the country make room or not for an increase.

---

674 Id. table I. p. 21.

675 Mémoires, &c. par la Société de Berne. Année 1766, table viii. p. 92, et seq.

676 Id. table xiii. p. 120.

677 Id. table xii.

678 Mémoires, &c. par la Société de Berne. Année 1766, première partie, p. 27.

679 Id. première partie, tab. I.

In the town of Berne, from the year 1583 to 1654, the sovereign council had admitted into the Bourgeoisie 487 families, of which 379 became extinct in the space of two centuries, and in 1783 only 108 of them remained. During the hundred years from 1684 to 1784, 207 Bernoise families became extinct. From 1624 to 1712, the Bourgeoisie was given to 80 families. In 1623, the sovereign council united the members of 112 different families, of which 58 only remain.[680]

The proportion of unmarried persons in Berne, including widows and widowers, is considerably above the half of the adults; and the proportion of those below sixteen to those above, is not far from 1 to 3.[681] These are strong, proofs of the powerful operation of the preventive check.

The peasants in the canton of Berne have always had the reputation of being rich, and without doubt it is greatly to be attributed to this cause. A law has for some time prevailed, which makes it necessary for every peasant to prove himself in possession of the arms and accoutrements necessary for the militia, before he can obtain permission to marry. This at once excludes the very poorest from marriage; and a very favourable turn may be given to the habits of many others, from a knowledge that they cannot accomplish the object of their wishes, without a certain portion of industry and economy. A young man who, with this end in view, had engaged in service either at home or in a foreign country, when he had gained the necessary sum, might feel his pride rather raised, and not be contented merely with what would obtain him permission to marry, but go on till he could obtain something like a provision for a family.

I was much disappointed, when in Switzerland, at not being able to procure any details respecting the smaller cantons; but the disturbed state of the country made it impossible. It is to be presumed, however, that as they are almost entirely in pasture, they must resemble in a great measure the alpine parishes of the Pays de Vaud in the extraordinary health of the people, and the absolute necessity of the preventive check; except where these circumstances may have been altered by a more than usual habit of emigration, or by the introduction of manufactures.[682]

The limits to the population of a country strictly pastoral are strikingly obvious. There are no grounds less susceptible of improvement than mountainous pastures. They must necessarily be left chiefly to nature; and when they have been adequately stocked with cattle, little more can be done. The great difficulty in these parts of Switzerland, as in

[680] Statistique de la Suisse, Durand, tom. iv. p. 405, 8vo. 4 vols. Lausanne, 1796.

[681] Beschreibung von Bern, vol. ii. tab. I. p. 35, 2 vols. 8vo. Bern. 1796.

[682] M. Prevost, of Geneva, in his translation of this work, gives some account of the small Canton of Glavis, in which the cotton-manufacture had been introduced. It appears that it had been very prosperous at first, and had occasioned a habit of early marriages, and a considerable increase of population; but subsequently wages became extremely low, and a fourth part of the population was dependent upon charity for their support. The proportions of the births and deaths to the population, instead of being 1 to 36, and 1 to 45, as in the Pays de Vaud, had become as 1 to 26, and 1 to 35. And, according to a later account in the last translation, the proportion of the births to the population, during the 14 years from 1805 to 1819, was as 1 to 24, and of the deaths as 1 to 30.

Norway, is to procure a sufficient quantity of fodder for the winter support of the cattle which have been fed on the mountains in the summer. For this purpose grass is collected with the greatest care. In places inaccessible to cattle, the peasant sometimes makes hay with crampons on his feet; in some places grass not three inches high is cut three times a year; and in the valleys, the fields are seen shaven as close as a bowling-green, and all the inequalities clipped as with a pair of scissors. In Switzerland as in Norway, for the same reasons, the art of mowing seems to be carried to its highest pitch of perfection. As, however, the improvement of the lands in the valleys must depend principally upon the manure arising from the stock, it is evident that the quantity of hay and the number of cattle will be mutually limited by each other; and as the population will of course be limited by the produce of the stock, it does not seem possible to increase it beyond a certain point, and that at no great distance. Though the population, therefore, in the flat parts of Switzerland, has increased during the last century, there is reason to believe that it has been stationary in the mountainous parts. According to M. Muret it has decreased very considerably in the Alps of the Pays de Vaud; but his proofs of this fact have been noticed as extremely uncertain. It is not probable, that the Alps are less stocked with cattle than they were formerly; and if the inhabitants be really rather fewer in number, it is probably owing to the smaller proportion of children, and to the improvement which has taken place in the mode of living.

In some of the smaller cantons, manufactures have been introduced, which, by furnishing a greater quantity of employment, and at the same time a greater quantity of exports for the purchase of corn, have of course considerably increased their population. But the Swiss writers seem generally to agree, that the districts where they have been established, have upon the whole suffered in point of health, morals and happiness.

It is the nature of pasturage to produce food for a much greater number of people than it can employ. In countries strictly pastoral, therefore, many persons will be idle, or at most be very inadequately occupied. This state of things naturally disposes to emigration, and the principal reason why the Swiss have been so much engaged in foreign service. When a father has more than one son, those who are not wanted on the farm are powerfully tempted to enrol themselves as soldiers, or to emigrate in some other way, as the only chance of enabling them to marry.

It is possible, though not probable, that a more than usual spirit of emigration, operating upon a country, in which, as it has appeared, the preventive check prevailed to a very considerable degree, might have produced a temporary check to increase at the period when there was such an universal cry about depopulation. If this were so, it without doubt contributed to improve the condition of the lower classes of people. All the foreign travellers in Switzerland, soon after this time, invariably take notice of the state of the Swiss peasantry as superior to that of other countries. In a late excursion to Switzerland, I was rather disappointed not to find it so superior as I had been taught to expect. The greatest part of the unfavourable change might justly be attributed to the losses and sufferings of the people during the late troubles; but a part perhaps to the ill-directed efforts of the different governments to increase the population, and to the ultimate consequences even of efforts well directed, and for a time calculated to advance the comforts and happiness of the people.

I was very much struck with an effect of this last kind, in an expedition to the *Lac de Joux* in the Jura. The party had scarcely arrived at a little inn at the end of the lake, when the mistress of the house began to complain of the poverty and misery of all the parishes in the neighbourhood. She said that the country produced little, and yet was full of inhabitants; that boys and girls were marrying who ought still to be at school; and that, while this habit of early marriages continued, they should always be wretched and distressed for subsistence.

The peasant, who afterwards conducted us to the source of the Orbe, entered more fully into the subject, and appeared to understand the principle of population almost as well as any man I ever met with. He said, that the women were prolific, and the air of the mountains so pure and healthy, that very few children died, except from the consequences of absolute want; that the soil, being barren, was inadequate to yield employment and food for the numbers that were yearly growing up to manhood; that the wages of labour were consequently very low, and totally insufficient for the decent support of a family; but that the misery and starving condition of the greater part of the society did not operate properly as a warning to others, who still continued to marry; and to produce a numerous offspring which they could not support. This habit of early marriages might really, he said, be called *le vice du pays;* and he was so strongly impressed with the necessary and unavoidable wretchedness that must result from it, that he thought a law ought to be made, restricting men from entering into the marriage state before they were forty years of age, and then allowing it only with *"des vieilles files,"* who might bear them two or three children instead of six or eight.

I could not help being diverted with the earnestness of his oratory on this subject, and particularly with his concluding proposition. He must have seen and felt the misery arising from a redundant population most forcibly, to have proposed so violent a remedy. I found upon inquiry that he had himself married very young.

The only point in which he failed, as to his philosophical knowledge of the subject, was in confining his reasonings too much to barren and mountainous countries, and not extending them to the plains. In fertile situations, he thought, perhaps, that the plenty of corn and employment might remove the difficulty, and allow of early marriages. Not having lived much in the plains, it was natural for him to fall into this error; particularly, as in such situations the difficulty is not only more concealed from the extensiveness of the subject; but is in reality less, from the greater mortality naturally occasioned by low grounds, towns, and manufactories.

On inquiring into the principal cause of what he had named the *predominant vice* of his country, he explained it with great philosophical precision. He said, that a manufacture for the polishing of stones had been established some years ago, which for a time had been in a very thriving state, and had furnished high wages and employment to all the neighbourhood; that the facility of providing for a family, and of finding early employment for children, had greatly encouraged early marriages; and that the same habit had continued, when, from a change of fashion, accident, and other causes, the manufacture was almost at an end. Very great emigrations, he said, had of late years taken place; but the breeding system went on so fast, that they were not sufficient to relieve the country of its superabundant mouths, and the effect was such as he had described to me, and as I had in part seen.

In other conversations which I had with the lower classes of people in different parts of Switzerland and Savoy, I found many, who, though not sufficiently skilled in the principle of population to see its effects on society, like my friend of the *Lac de Joux,* yet saw them clearly enough as affecting their own individual interests; and were perfectly aware of the evils which they should probably bring upon themselves by marrying before they could have a tolerable prospect of being able to maintain a family. From the general ideas which I have found to prevail on these subjects, I should by no means say that it would be a difficult task to make the common people comprehend the principle of population, and its effect in producing low wages and poverty.

Though there is no absolute provision for the poor in Switzerland, yet each parish generally possesses some seignioral rights and property in land for the public use, and is expected to maintain its own poor. These funds, however, being limited, will of course often be totally insufficient; and occasionally voluntary collections are made for this purpose. But, the whole of the supply being comparatively scanty and uncertain, it has not the same bad effects as the parish-rates of England. Of late years much of the common lands belonging to parishes have been parcelled out to individuals, which has of course tended to improve the soil, and increase the number of people; but from the manner in which it has been conducted, it has operated perhaps too much as a systematic encouragement of marriage, and has contributed to increase the number of poor. In the neighbourhood of the richest *communes,* I often observed the greatest number of beggars.

There is reason to believe, however, that the efforts of the Economical Society of Berne to promote agriculture were crowned with some success; and that the increasing resources of the country have made room for an additional population, and furnished an adequate support for the greatest part, if not the whole, of that increase which has of late taken place.

In 1764 the population of the whole canton of Berne, including the Pays de Vaud, was estimated at 336,689. In 1791, it had increased to 414,420. From 1764 to 1777, its increase proceeded at the rate of 2,000 each year; and, from 1778 to 1791, at the rate of 3,109 each year.[683]

3.5.6 Book II, Chapter VI: Of the Checks to Population in France.

As the parochial registers in France, before the revolution, were not kept with particular care nor for any great length of time, and as the few which have been produced exhibit no

[683] Beschreibung von Bern, vol. ii. p. 40.

very extraordinary results, I should not have made this country the subject of a distinct chapter, but for a circumstance attending the revolution, which has excited considerable surprise. This is, the undiminished state of the population in spite of the losses sustained during so long and destructive a contest.[684]

A great national work, founded on the reports of the prefects in the different departments, is at present in some state of forwardness at Paris, and when completed may reasonably be expected to form a very valuable accession to the materials of statistical science in general. The returns of all the prefects are not however yet complete; but I was positively assured by the person who has the principal superintendence of them, that enough is already known to be certain that the population of the old territory of France has rather increased than diminished during the revolution.

Such an event, if true, very strongly confirms the general principles of this work; and assuming it for the present as a fact, it may tend to throw some light on the subject, to trace a little in detail the manner in which such an event might happen.

In every country there is always a considerable body of unmarried persons, formed by the gradual accumulation of the excess of the number rising annually to the age of puberty above the number of persons annually married. The stop to the further accumulation of this body is when its number is such, that the yearly mortality equals the yearly accessions that are made to it. In the Pays de Vaud, as appeared in the last chapter, this body, including widows and widowers, persons who are not actually in the state of marriage, equals the whole number of married persons. But in a country like France, where both the mortality and the tendency to marriage are much greater than in Switzerland, this body does not bear so large a proportion to the population.

According to a calculation in an *Essai d'une Statistique Générale,* published at Paris in 1800, by M. Peuchet, the number of unmarried males in France between 18 and 50 is estimated at 1,451,063; and the number of males, whether married or not, between the same ages, at 5,000,000.[685] It does not appear at what period exactly this calculation was made; but as the author uses the expression *en tems ordinaire,* it is probable that he refers to the period before the revolution. Let us suppose, then, that this number of 1,451,063 expresses the collective body of unmarried males of a military age at the commencement of the revolution.

The population of France before the beginning of the war was estimated by the Constituent Assembly at 26,363,074;[686] and there is no reason to believe that this calculation was too high. Necker, though he mentions the number of 24,800,000, expresses his firm belief that the yearly births at that time amounted to above a million, and consequently, according to his multiplier of 25¾, the whole population was nearly 26 millions;[687] and this calculation was made ten years previous to the estimate of the Constituent Assembly.

[684] This chapter was written in 1802, and refers to the state of France before the peace of Amiens.

[685] P. 32, 8vo. 78 pages.

[686] A. Young's Travels in France, vol. i. c. xvii. p. 466, 4to. 1792.

[687] De l'Administration des Finances, tom. i. c. ix. p. 256, 12mo. 1785.

Taking then the annual births at rather above a million, and estimating that rather above 2/5 would die under 18, which appears to be the case from some calculations of M. Peuchet,[688] it will follow, that above 600,000 persons will annually arrive at the age of 18.

The annual marriages, according to Necker, are 213,774;[689] but as this number is an average of ten years, taken while the population was increasing, it is probably too low. If we take 220,000, then 440,000 persons will be supposed to marry out of the 600,000 rising to a marriageable age; and, consequently; the excess of those rising to the age of 18 above the number wanted to complete the usual proportion of annual marriages, will be 160,000, or 80,000 males. It is evident, therefore, that the accumulated body of 1,451,063 unmarried males, of a military age, and the annual supply of 80,000 youths of 18, might be taken for the service of the state, without affecting in any degree the number of annual marriages. But we cannot suppose that the 1,451,063 should be taken all at once; and many soldiers are married, and in a situation not to be entirely useless to the population. Let us suppose 600,000 of the corps of unmarried males to be embodied at once; and this number to be kept up by the annual supply of 150,000 persons, taken partly from the 80,000, rising annually to the age of 18, and not wanted to complete the number of annual marriages, and partly from the 851,063 remaining of the body of unmarried males, which existed at the beginning of the war: it is evident, that from these two sources 150,000 might be supplied each year, for ten years, and yet allow of an increase in the usual number of annual marriages of above 10,000.

It is true that in the course of the ten years many of the original body of unmarried males will have passed the military age; but this will be balanced, and indeed much more than balanced, by their utility in the married life. From the beginning it should be taken into consideration, that though a man of fifty be generally considered as past the military age, yet, if he marry a fruitful subject, he may by no means be useless to the population; and in fact the supply of 150,000 recruits each year would be taken principally from the 300,000 males rising annually to 18; and the annual marriages would be supplied in a great measure from the remaining part of the original body of unmarried persons. Widowers and bachelors of forty and fifty, who in the common state of things might have found it difficult to obtain an agreeable partner, would probably see these difficulties removed in such a scarcity of husbands; and the absence of 600,000 persons would of course make room for a very considerable addition to the number of annual marriages. This addition in all probability took place. Many among the remaining part of the original body of bachelors, who might otherwise have continued single, would marry under this change of circumstances; and it is known that a very considerable portion of youths under 18, in order to avoid the military conscriptions, entered prematurely into the married state. This was so much the case, and contributed so much to diminish the number of unmarried persons, that in the

[688] Essai, p. 31.

[689] De l'Administration des Finances, tom. i. c. ix. p. 255.

beginning of the year 1798 it was found necessary to repeal the law, which had exempted married persons from conscriptions; and those who married subsequently to this new regulation were taken indiscriminately with the unmarried. And though after this the levies fell in part upon those who were actually engaged in the peopling of the country; yet the number of marriages untouched by these levies might still remain greater than the usual number of marriages before the revolution; and the marriages which were broken by the removal of the husband to the armies would not probably have been entirely barren.

Sir Francis d'Ivernois, who had certainly a tendency to exaggerate, and probably has exaggerated considerably, the losses of the French nation, estimates the total loss of the troops of France, both by land and sea, up to the year 1799, at a million and a half.[690] The round numbers which I have allowed for the sake of illustrating the subject, exceed Sir Francis d'Ivernois's estimate by six hundred thousand. He calculates however a loss of a million of persons more, from the other causes of destruction attendant on the revolution; but as this loss fell indiscriminately on all ages and both sexes, it would not affect the population in the same degree, and will be much more than covered by the 600,000 men in the full vigour of life, which remain above Sir Francis's calculation. It should be observed also, that in the latter part of the revolutionary war the military conscriptions were probably enforced with still more severity in the newly-acquired territories than in the old state; and as the population of these new acquisitions is estimated at five or six millions, it would bear a considerable proportion of the million and a half supposed to be destroyed in the armies.

The law which facilitated divorces to so great a degree in the early part of the revolution was radically bad both in a moral and political view, yet, under the circumstance of a great scarcity of men, it would operate a little like the custom of polygamy, and increase the number of children in proportion to the number of husbands. In addition to this, the women without husbands do not appear all to have been barren; as the proportion of illegitimate births is now raised to 1/11 of the whole number of births, from 1/47,[691] which it was before the revolution; and though this be a melancholy proof of the depravation of morals, yet it would certainly contribute to increase the number of births; and as the female peasants in France were enabled to earn more than usual during the revolution, on account of the scarcity of hands, it is probable that a considerable portion of these children would survive.

Under all these circumstances, it cannot appear impossible, and scarcely even improbable, that the population of France should remain undiminished, in spite of all the causes of

[690]Tableau des Pertes, &c. c. ii. p. 7.—M. Garnier, in the notes to his edition of Adam Smith, calculates that only about a sixtieth part of the French population was destroyed in the armies. He supposes only 500,000 embodied at once, and that this number was supplied by 400,000 more in the course of the war; and allowing for the number which would die naturally, that the additional mortality occasioned by the war was only about 45,000 each year. Tom. v. note xxx. p. 284. If the actual loss were no more than these statements make it, a small increase of births would have easily repaired it; but I should think that these estimates are probably as much below the truth, as Sir Francis d'Ivernois's are above.

[691]Essai de Peuchet, p. 28.

destruction which have operated upon it during the course of the revolution, provided the agriculture of the country has been such as to continue the means of subsistence unimpaired. And it seems now to be generally acknowledged that, however severely the manufactures of France may have suffered, her agriculture has rather increased than diminished. At no period of the war can we suppose that the number of embodied troops exceeded the number of men employed before the revolution in manufactures. Those who were thrown out of work by the destruction of these manufactures, and who did not go to the armies, would of course betake themselves to the labours of agriculture; and it was always the custom in France for the women to work much in the fields, which custom was probably increased during the revolution. At the same time, the absence of a large portion of the best and most vigorous hands would raise the price of labour; and as, from the new land brought into cultivation, and the absence of a considerable part of the greatest consumers[692] in foreign countries, the price of provisions would not rise in proportion, this advance in the real price of labour would not only operate as a powerful encouragement to marriage, but would enable the peasants to live better, and to rear a greater number of their children.

At all times the number of small farmers and proprietors in France was great; and though such a state of things is by no means favourable to the clear surplus produce or disposable wealth of a nation; yet sometimes it is not unfavourable to the absolute produce, and it has always a strong tendency to encourage population. From the sale and division of many of the large domains of the nobles and clergy, the number of landed proprietors has considerably increased during the revolution; and as a part of these domains consisted of parks and chases, new territory has been given to the plough. It is true that the land-tax has been not only too heavy, but injudiciously imposed. It is probable, however, that this disadvantage has been nearly counterbalanced by the removal of the former oppressions, under which the cultivator laboured; and that the sale and division of the great domains may be considered as a clear advantage on the side of agriculture, or at any rate of the gross produce, which is the principal point with regard to mere population.

These considerations make it appear probable that the means of subsistence have at least remained unimpaired, if they have not increased, during the revolution; and a view of the cultivation of France in its present state certainly rather tends to confirm this supposition.

We shall not therefore be inclined to agree with Sir Francis d'Ivernois in his conjecture that the annual births in France have diminished by one-seventh during the revolution.[693] On the contrary, it is more probable that they have increased by this number. The average proportion of births to the population in all France, before the revolution, was, according to Necker, as 1 to 25¾.[694] It has appeared in the reports of some of the prefects which have

[692] Supposing the increased number of children at any period to equal the number of men absent in the armies, yet these children, being all very young, could not be supposed to consume a quantity equal to that which would be consumed by the same number of grown-up persons.

[693] Tableau des Pertes, &c. c. ii. p. 14.

[694] De l'Administration des Finances, tom. i. c. ix. p. 254.

been returned, that the proportion in many country places was raised to 1 to 21, 22, 22½, and 23;[695] and though these proportions might, in some degree, be caused by the absence of a part of the population in the armies, yet I have little doubt that they are principally to be attributed to the birth of a greater number of children than usual. If, when the reports of all the prefects are put together, it should appear, that the number of births has not increased in proportion to the population, and yet that the population is undiminished; it will follow, either that Necker's multiplier for the births was too small, which is extremely probable, as from this cause he appears to have calculated the population too low; or that the mortality, among those not exposed to violent deaths has been less than usual; which, from the high price of labour and the desertion of the towns for the country, is not unlikely.

According to Necker and Moheau; the mortality in France, before, the revolution, was 1 in 30 or 31 1/8.[696] Considering that the proportion of the population which lives in the country is to that in the towns as 3½ to 1,[697] this mortality is extraordinarily great, caused probably by the misery arising from an excess of population; and from the remarks of Arthur Young on the state of the peasantry in France,[698] which are completely sanctioned by Necker,[699] this appears to have been really the case. If we suppose that, from the removal of a part of this redundant population, the mortality has decreased from 1 in 30 to 1 in 35, this favourable change would go a considerable way in repairing the breaches made by war on the frontiers.

The probability is, that both the causes mentioned have operated in part. The births have increased, and the deaths of those remaining in the country have diminished; so that, putting the two circumstances together, it will probably appear, when the results of all the reports of the prefects are known; that, including those who have fallen in the armies and by violent means, the deaths have not exceeded the births in the course of the revolution.

The returns of the prefects are to be given for the year IX. of the republic, and to be compared with the year 1789; but if the proportion of births to the population be given merely for the individual year IX. it will not skew the average proportion of births to the population during the course of the revolution. In the confusion occasioned by this event, it is not probable that any very exact registers should have been kept; but from theory I should be inclined to expect that soon after the beginning of the war, and at other periods during the course of it, the proportion of births to the whole population would be greater than in 1800 and 1801.[700] If it should appear by the returns, that the number of annual marriages

[695] Essai de Peuchet, p. 28.

[696] De l'Administration des Finances, tom. i. c. ix. p. 255. Essai de Peuchet, p. 29.

[697] Young's Travels in France, vol. i. c. xvii. p. 466.

[698] See generally c. xvii. vol. i. and the just observations on these subjects interspersed in many other parts of his very valuable Tour.

[699] De l'Administration des Finances, tom, i. c. ix. p. 262, et seq.

[700] In the *Statistique Générale et Particulière de la France, et de ses Colonies,* lately, published, the returns of the prefects for the year IX. are given, and seem to justify this conjecture. The births are 955,450, the deaths 821,871, and the marriages 202,177. These numbers hardly equal Necker's

has not increased during the revolution, the circumstance will be obviously accounted for by the extraordinary increase in the illegitimate births mentioned before in this chapter, which amount at present to one-eleventh of all the births, instead of one-forty-seventh, according to the calculation of Necker before the revolution.[701]

Sir Francis d'Ivernois observes, "that those have yet to learn the first principles of political arithmetic, who imagine that it is in the field of battle and the hospitals that an account can be taken of the lives which a revolution or a war has cost. The number of men it has killed is of much less importance than the number of children which it has prevented, and will still prevent, from coming into the world. This is the deepest wound which the population of France has received."—"Supposing," he says, "that, of the whole number of men destroyed, only two millions had been united to as many females: according to the calculation of Buffon, these two millions of couples ought to bring into the world twelve millions of children, in order to supply, at the age of thirty-nine, a number equal to that of their parents. This is a point of view, in which the consequences of such a destruction of

---

estimates; and yet all the calculations in this work, both with respect to the whole population and its proportion to a square league, make the old territory of France more populous now than at the beginning of the revolution. The estimate of the population, at the period of the Constituent Assembly, has already been mentioned; and at this time the number of persons to a square league was reckoned 996. In the year VI. of the republic, the result of the Bureau de Cadastre gave a population of 26,048,254, and the number to a square league 1,020. In the year VII. Dépère calculated the whole population of France at 33,501,094, of which 28,810,694 belonged to ancient France; the number to a square league 1,101; but the calculations, it appears, were founded upon the first estimate made by the Constituent Assembly, which was afterwards rejected as too high. In the year IX. and X. the addition of Piedmont and the isle of Elba raised the whole population to 34,376,313; the number to a square league 1,086. The number belonging to Old France is not stated. It seems to have been about 28,000,000.

The part of this work relating to population is not drawn up with much knowledge of the subject. One remark is very curious. It is observed that the proportion of marriages to the population is as 1 to 110, and of births as 1 to 25; from which it is inferred, that one-fourth of the born live to marry. If this inference were just, France would soon he depopulated.

In calculating, the value of lives, the author makes use of Buffou's tables, which are entirely incorrect, being founded principally on registers taken from the villages round Paris. They make the probability of life at birth only a little above eight years; which, taking the towns and the country together, is very short of the just average.

Scarcely any thing worth noticing has been added in this work to the details given in the Essay of Peuchet, which I have already frequently referred to. On the whole I have not seen sufficient grounds to make me alter any of my conjectures in this chapter, though probably they are not well-founded. Indeed, in adopting Sir F. d'Ivernois' calculations respecting the actual loss of men during the revolution, I never thought myself borne out by facts; but, the reader will be aware that I adopted them rather for the sake of illustration than from supposing them strictly true.

[701] Essai de Peuchet, p. 28. It is highly probable that this increase of illegitimate births occasioned a more than usual number of children to be exposed in those dreadful receptacles, *les Hôpitaux des Enfans trouvés,* as noticed by Sir Francis d'Ivernois; but probably this cruel custom was confined to particular districts, and the number exposed, upon the whole, might bear no great proportion to the sum of all the births.

men become almost incalculable; because they have much more effect with regard to the twelve millions of children, which they prevent from coming into existence, than with regard to the actual loss of the two millions and a half of men for whom France mourns. It is not till a future period that she will be able to estimate this dreadful breach."[702]

And yet, if the foregoing reasonings are well-founded, France may not have lost a single birth by the revolution. She has the most just reason to mourn the two millions and a half of individuals which she may have lost, but not their posterity; because; if these individuals had remained in the country, a proportionate number of children, born of other parents, which are now living in France, would not have come into existence. If, in the best governed country in Europe, we were to mourn the posterity which is prevented from coming into being, we should always wear the habit of grief.

It is evident that the constant tendency of the births in every country to supply the vacancies made by death, cannot, in a moral point of view, afford the slightest shadow of excuse for the wanton sacrifice of men. The positive evil that is committed in this case, the pain, misery, and widespreading desolation and sorrow, that are occasioned to the existing inhabitants, can by no means be counterbalanced by the consideration, that the numerical breach in the population will be rapidly repaired. We can have no other right, moral or political, except that of the most urgent necessity, to exchange the lives of beings in the full vigour of their enjoyments, for an equal number of helpless infants.

It should also be remarked that, though the numerical population of France may not have suffered by the revolution, yet, if her losses have been in any degree equal to the conjectures on the subject, her military strength cannot be unimpaired. Her population at present must consist of a much greater proportion than usual of women and children; and the body of unmarried persons, of a military age, must be diminished in a very striking manner. This indeed is known to be the case, from the returns of the prefects which have already been received.

It has appeared that the point at which the drains of men will begin essentially to affect the population of a country is, when the original body of unmarried persons is exhausted, and the annual demands are greater than the excess of the number of males, rising annually to the age of puberty, above the number wanted to complete the usual proportion of annual marriages. France was probably at some distance from this point at the conclusion of the war; but in the present state of her population, with an increased proportion of women and children, and a great diminution of males of a military age, she could not make the same gigantic exertions, which were made at one period, without trenching on the sources of her population.

At all times the number of males of a military age in France was small in proportion to the population, on account of the tendency to marriage,[703] and the great number of

[702] Tableau des Pertes, &c. c. ii. p. 13, 14.

[703] The proportion of marriages to the population in France, according to Necker, is 1 to 113, tom. i. c. ix. p. 255.

children. Necker takes particular notice of this circumstance. He observes, that the effect of the very great misery of the peasantry is to produce a dreadful mortality of infants under three or four years of age; and the consequence is, that the number of young children will always be in too great a proportion to the number of grown-up people. A million of individuals, he justly observes, will in this case neither present the same military force nor the same capacity of labour, as an equal number of individuals in a country where the people are less miserable.[704]

Switzerland, before the revolution, could have brought into the field, or have employed in labour appropriate to grown-up persons, a much greater proportion of her population than France at the same period.[705]

---

[704] De l'Administration des Finances, tom, i. c. ix. p. 263.

[705] Since I wrote this chapter, I have had an opportunity of seeing the *Analyse des Procès Verbaux des Conseils Généraux de Département,* which gives a very particular and highly curious account of the internal state of France for the year VIII. With respect to the population, out of 69 departments, the reports from which are given, in 16 the population is supposed to be increased; in 42 diminished; in 9 stationary; and in 2 the active population is said to be diminished, but the numerical to remain the same. It appears, however, that most of these reports are not founded on actual enumerations; and without such positive data, the prevailing opinions on the subject of population, together with the necessary and universally acknowledged fact of a very considerable diminution in the males of a military age, would naturally dispose people to think that the numbers upon the whole must be diminished. Judging merely from appearances, the substitution of a hundred children for a hundred grown-up persons would certainly not produce the same impression with regard to population. I should not be surprised, therefore, if, when the enumerations for the year IX. are completed, it should appear that the population upon the whole has not diminished. In some of the reports *l'aisance générale répandue sur le peuple,* and *la division des grands propriétés,* are mentioned as the causes of increase; and almost universally, *les mariages prématurés,* and *les mariages multipliés par la crainte des loix militaires,* are particularly noticed.

The land-tax, or *contribution foncière,* is universally complained of; indeed it appears to be extremely heavy, and to fall very unequally. It was intended to be only a fifth of the neat produce; but, from the unimproved state of agriculture in general, the number of small proprietors, and particularly the attempt to cultivate too much surface in proportion to the capital employed, it often amounts to a fourth, a third, or even a half. When property is so much divided that the rent and profit of a farm must be combined, in order to support a family upon it, a land-tax must necessarily greatly impede cultivation; though it has little or no effect of this kind when farms are large, and let out to tenants, as is most frequently the case in England. Among the impediments to agriculture mentioned in the reports, the too great division of lands from the new laws of succession is noticed. The partition of some of the great domains would probably contribute to the improvement of agriculture; but; subdivisions of the nature here alluded to would certainly have a contrary effect, and would tend most particularly to diminish neat produce, and make a land-tax both oppressive and unproductive. If all the land in England were divided into farms of 20*l.* a-year, we should probably be more populous than we are at present; but as a nation we should be extremely poor, and should be under a total inability of maintaining the same number of manufactures or collecting the same taxes as at present. All the departments demand a diminution of the *contribution foncièr* as absolutely necessary to the prosperity of agriculture.

For the state of population in Spain, I refer the reader to the valuable and entertaining travels of Mr. Townsend in that country, in which he will often find the principle of population very happily illustrated. I should have made it the subject of a distinct chapter, but was fearful of extending this part of the work too much, and of falling almost unavoidably into too many repetitions, from the necessity of drawing the same kind of

---

Of the state of the hospitals and charitable establishments, of the prevalence of beggary and the mortality among the exposed children, a most deplorable picture is drawn in almost all the reports; from which we should at first be disposed to infer a greater degree of poverty and misery among all the lower classes of people in general. It appears, however, that the hospitals and charitable establishments lost almost the whole of their revenues during the revolution; and this sudden subtraction of support from a great number of people who had no other reliance, together with the known failure of manufactures in the towns, and the very great increase of illegitimate children, might produce all the distressing appearances described in the reports, without impeaching the great fact of the meliorated condition of agricultural labourers in general, necessarily arising from the acknowledged high price of labour and comparative cheapness of corn; and it is from this part of the society that the effective population of a country is principally supplied. If the poor's rates of England were suddenly abolished, there would undoubtedly be the most complicated distress among those who were before supported by them; but I should not expect that either the condition of the labouring part of the society in general, or the population of the country, would suffer from it. As the proportion of illegitimate children in France has risen so extraordinarily as from 1/47 of all the births to 1/11, it is evident that more might be abandoned in hospitals, and more out of these die than usual, and yet a more than usual number be reared at home, and escape the mortality of those dreadful receptacles. It appears that from the low state of the funds in the hospitals the proper nurses could not be paid, and numbers of children died from absolute famine. Some of the hospitals at last very properly refused to receive any more.

The reports, upon the whole, do not present a favourable picture of the internal state of France; but something is undoubtedly to be attributed to the nature of these reports, which, consisting as they do of observations explaining the state of the different departments, and of particular demands, with a view to obtain assistance or relief from government, it is to be expected that they should lean rather to the unfavourable side. When the question is respecting the imposition of new taxes, or the relief from old ones, people will generally complain of their poverty. On the subject of taxes, indeed, it would appear, as if the French government must be a little puzzled. For though it very properly recommended to the *Conseils généraux* not to indulge in vague complaints, but to mention specific grievances, and propose specific remedies, and particularly not to advise the abolition of one tax without suggesting another; yet all the taxes appear to me to be reprobated, and most frequently in general terms, without the proposal of any substitute.

La contribution foncièr, la taxe mobiliaire, les barrières, les droits de douane, all excite bitter complaints; and the only new substitute that struck me was a tax upon game, which, being at present almost extinct in France, cannot be expected to yield a revenue sufficient to balance all the rest. The work, upon the whole, is extremely curious; and as shewing the wish of the government to know the state of each department, and to listen to every observation and proposal for its improvement, is highly creditable to the ruling power. It was published for a short time; but the circulation of it was soon stopped and confined to the ministers, *les conseils généraux,* &c. Indeed the documents are evidently more of a private than of a public nature, and certainly have not the air of being intended for general circulation.

Book II, Chapter VII.

inference from so many different countries. I could expect, besides, to add very little to what has been so well done by Mr. Townsend.

3.5.7 Book II, Chapter VII: Of the Checks to Population in France (continued).

I have not thought it advisable to alter the conjectural calculations and suppositions of the preceding chapter, on account of the returns of the prefects for the year IX., as well as some returns published since by the government in 1813, having given a smaller proportion of births than I had thought probable; first, because these returns do not contain the early years of the revolution, when the encouragement to marriage and the proportion of births might be expected to be the greatest; and secondly, because they still seem fully to establish the main fact, which it was the object of the chapter to account for, namely, the undiminished population of France, notwithstanding the losses sustained during the revolution; although it may have been effected rather by a decreased proportion of deaths than an increased proportion of births.

According to the returns of the year IX., the proportions of the births, deaths, and marriages, to the whole population, are as follows:—

Births.	Deaths.	Marriages.
1 in 33	1 in 38½	1 in 157.[706]

But these are in fact only the proportions of one year, from which no certain inference can be drawn. They are also applied to a population between three and four millions greater than was contained in ancient France, which population may have always had a smaller proportion of births, deaths, and marriages; and further, it appears highly probable from some of the statements in the *Analyse des Procès Verbaux,* that the registers had not been very carefully kept. Under these circumstances, they cannot be considered as proving what the numbers imply.

In the year XI., according to the Statistique Elémentaire by Peuchet, published subsequently to his *Essai,* an inquiry was instituted under the orders of M. Chaptal for the express purpose of ascertaining the average proportion of births to the population;[707] and such an inquiry, so soon after the returns of the year IX., affords a clear proof that these

[706] See a valuable note of M. Prevost of Geneva to his translation of this work, vol. ii. p. 88. M. Prevost thinks it probable that there are omissions in the returns of the births, deaths, and marriages, for the year IX. He further shews that the proportion of the population to the square league for Old France should be 1014, and not 1086. But if there is reason to believe that there are omissions in the registers, and that the population is made too great, the real proportions will be essentially different from those which are here given.

[707] P. 331. Paris, 1805.

returns were not considered by the minister as correct. In order to accomplish the object in view, choice was made of those communes in 30 departments distributed over the whole surface of prance, which were likely to afford the most accurate returns. And these returns for the year VIII., IX., and, X., gave a proportion of births as 1 in 28.35; of deaths, as 1 in 30.09; and of marriages, as 1 in 132.078.

It is observed by M. Peuchet that the proportion of population to the births is here much greater than had been formerly assumed, but he thinks that, as this calculation had been made from actual enumerations, it should be adopted in preference.

The returns published by the government in 1813 make the population of ancient France 28,786,911, which, compared with 28,000,000, the estimated population of the year IX., shew an increase of about 800,000 in the 11 years, from 1802 to 1813.

No returns of marriages are given, and the returns of births and deaths are given only for fifty departments.

In these fifty departments, during the ten years beginning with 1802 and ending with 1811, the whole number of births amounted to 5,478,669, and of deaths to 9,696,857, which, on a population of 16,710,719, indicates a proportion of births, as 1 in 30½, and of deaths as 1 in 35½.

It is natural to suppose that these fifty departments were chosen on account of their shewing the greatest increase. They contain indeed nearly the whole increase that had taken place in all the departments from the time of the enumeration in the year IX.; and consequently the population of the other departments must have been almost stationary. It may further be reasonably conjectured that the returns of marriages were not published on account of their being considered as unsatisfactory, and shewing a diminution of marriages, and an increased proportion of illegitimate births.

From these returns, and the circumstances accompanying them, it may be concluded, that whatever might have been the real proportion of births before the revolution, and for six or seven subsequent years, when the *mariages prématurés* are alluded to in the Procès Verbaux, and proportions of births as 1 in 21, 22, and 23, are mentioned in the Statistique Générale, the proportions of births, deaths, and marriages, are now all considerably less than they were formerly supposed to be.[708]

It has been asked, whether, if this fact be allowed, it does not clearly follow that the population was incorrectly estimated before the revolution, and that it has been diminished rather than increased since 1792? To this question I should distinctly answer, that it does not follow. It has been seen, in many of the preceding chapters, that the proportions of births, deaths, and marriages, are extremely different in different countries, and there is the

[708] In the year 1792 a law was passed extremely favourable to early marriages. This was repealed in the year XI., and a law substituted which threw great obstacles in the way of marriage, according to Peuchet (p. 234.) These two laws will assist in accounting for a small proportion of births and marriages in the ten years previous to 1813, consistently with the possibility of a large proportion in the first six or seven years after the commencement of the revolution.

strongest reason for believing that they are very different in the same country at different periods, and under different circumstances.

That changes of this kind have taken place in Switzerland has appeared to be almost certain. A similar effect from increased healthiness in our own country may be considered as an established fact. And if we give any credit to the best authorities that can be collected on the subject, it can scarcely be doubted that the rate of mortality has diminished, during the last one or two hundred years, in almost every country in Europe. There is nothing therefore that ought to surprise us in the mere fact of the same population being kept up, or even a decided increase taking place, under a smaller proportion of births, deaths and marriages. And the only question is, whether the actual circumstances of France seem to render such a change probable.

Now it is generally agreed that the condition of the lower classes of people in France before the revolution was very wretched. The wages of labour were about 20 sous, or ten pence a day, at a time when the wages of labour in England were nearly seventeen pence, and the price of wheat of the same quality in the two countries was not very different Accordingly Arthur Young represents the labouring classes of France, just at the commencement of the revolution, as "76 per cent. worse fed, worse clothed, and worse supported, both in sickness and health, than the same classes in England."[709] And though this statement is perhaps rather too strong, and sufficient allowance is not made for the real difference of prices, yet his work every where abounds with observations which shew the depressed condition of the labouring classes in France at that time, and imply the pressure of the population very hard against the limits of subsistence.

On the other hand, it is universally allowed that the condition of the French peasantry has been decidedly improved by the revolution and the division of the national domains. All the writers who advert to the subject notice a considerable rise in the price of labour, partly occasioned by the extension of cultivation, and partly by the demands of the army. In the Statistique Elémentaire of Peuchet, common labour is stated to have risen from 20 to 30 sous,[710] while the price of provisions appears to have remained nearly the same; and Mr. Birbeck, in his late Agricultural Tour in France,[711] says that the price of labour without board is twenty *pence* a day, and that provisions of all kinds are full as cheap again as in England. This would give the French labourer the same command of subsistence as an English labourer would have with three shillings and four pence a day. But at no time were the wages of common day-labour in England so high as three shillings and four pence.

Allowing for some errors in these statements, they are evidently sufficient to establish a very marked improvement in the condition of the lower classes of people in France. But it is next to a physical impossibility that such a relief from the pressure of distress should take place without a diminution in the rate of mortality; and if this diminution in the rate of

[709] Young's Travels in France, vol. i. p. 437.

[710] P. 391.

[711] P. 13.

mortality has not been accompanied by a rapid increase of population, it must necessarily have been accompanied by a smaller proportion of births. In the interval between 1802 and 1813 the population seems to have increased, but to have increased slowly. Consequently a smaller proportion of births, deaths, and marriages, or the more general operation of prudential restraint, is exactly what the circumstances would have led us to expect. There is perhaps no proposition more incontrovertible than this, that, in two countries, in which the rate of increase, the natural healthiness of climate, and the state of towns and manufactures are supposed to be nearly the same, the one in which the pressure of poverty is the greatest will have the greatest proportion of births, deaths, and marriages.

It does not then by any means follow, as has been supposed, that because since 1802 the proportion of births in France has been as 1 in 30, Necker ought to have used 30 as his multiplier instead of 25¾. If the representations given of the state of the labouring classes in France before and since the revolution be in any degree near the truth, as the march of the population in both periods seems to have been nearly the same, the present proportion of births could not have been applicable at the period when Necker wrote. At the same time it is by no means improbable that he took too low a multiplier. It is hardly credible under all circumstances that the population of France should have increased in the interval between 1785 and 1802 so much as from 25½ millions to 28. But if we allow that the multiplier might at that time have been 27 instead of 25¾, it will be allowing as much as is in any degree probable, and yet this will imply an increase of nearly two millions from 1785 to 1813; an increase far short of the rate that has taken place in England, but still sufficient amply to shew the force of the principle of population in overcoming obstacles apparently the most powerful.

With regard to the question of the increase of births in the six or seven first years after the commencement of the revolution, there is no probability of its ever being determined.

In the confusion of the times, it is scarcely possible to suppose that the registers should have been regularly kept; and as they were not collected in the year IX., there is no chance of their being brought forward in a correct state at a subsequent period.

Subsequent to the last edition of this work, further details have appeared respecting the population of France.

Since 1817, regular returns have been made of the annual births, deaths, and marriages, over the whole of the territory comprised in the limits of France, as settled in 1814 and 1815; and an enumeration was made of the population in 1820.

In the *Annuaire* of the *Bureau des Longitudes* for 1825, the numbers of births, deaths, and marriages are given for six years ending with 1822. The sum of these are,

Births.	Deaths.	Marriages.	Excess of births above deaths.
5,747,249	4,589,089	1,313,502	1,158,160

The annual average:

Births.	Deaths.	Marriages.	Average Excess of births:
957,875	764,848	218,917	193,027

The population in 1820, according to an enumeration in each department, was 30,451,187.

From these numbers it appears that the proportion of annual births to the population is as 1 to 31.79; or nearly 1/32; the annual mortality as 1 to 39.81, or nearly 1/40; the proportion of annual marriages to the population is as 1 to 139; the proportion of births to deaths as 125.23 to 100, or very nearly as 5 to 4; and the proportion of marriages to births as 1 to 4.37. The proportion of illegitimate to legitimate births is as 1 to 14.6; the proportion of male to female births as 16 to 15; and the proportion of the annual excess of the births above the deaths to the whole population, which, if the returns are accurate, determines the rate of increase as 1 to 157.

To what degree the returns of the births, deaths, and marriages in the 6 years ending with 1822 are accurate, it is impossible to say. There is a regularity in them which has a favourable appearance. We well know, however, that with the same appearance of regularity there are great omissions in the births and deaths of our own registers. This is at once proved by the circumstance of the excess of the births above the deaths in the interval between two enumerations falling considerably short of the increase of population which appears by such enumerations to have taken place. The enumerations in France during the last twenty-five years have not been so regular, or so much to be depended upon, as those in England. The one in 1813, before noticed, may, however, be compared with that in 1820, and if they are both equally near the truth, it will appear that the population of France during the seven years from 1813 to 1820 must have increased considerably faster than during the six years ending with 1822, as determined by the excess of the births above the deaths. The whole of this excess during these six years, as above stated, was 1,158,160, the annual average of which is 193,027, which, compared with the mean population, or the population of 1820, reduced by the increase of a year, will give a proportion of annual increase to the population, as 1 to about 156; and this proportion of the annual excess of the births above the deaths, to the population, will, according to Table II. at the end of Ch. xi. Book ii, give a rate of increase which would double the population in about 108 years.

On the other hand, as the population of old France in 1813 was 28,786,911, and in 1820 30,451,187, the difference or the increase of population during the seven years being 1,664,276, the annual average increase will be 237,753, instead of 193,026; and this greater annual increase, compared with the mean population of the seven years, will be as 1 to 124, instead of 1 to 156, and the rate of increase will be such as would double the population in about 86 years, instead of 108, showing the probability of considerable omissions in the returns of births and deaths in the 6 years ending with 1822. If, indeed, the two enumerations can be considered as equally near the truth, as there is no reason for supposing that any great difference in the proportion of births could have occurred in the three years preceding 1817, it follows that the French registers require the same kind of correction, though not to the same extent, as our own. In a subsequent chapter I have supposed that the returns of the births for England and Wales are deficient 1/6, and of the burials 1/12. This correction applied to the French returns would exceed what is necessary to account for the increase between 1813 and 1820. But if we suppose the births to be deficient 1/10, and the deaths 1/20, the proportion of the births to the population will then be 1/29.1, and the proportion of the deaths 1/38.1. These proportions will make the annual excess of the births above the deaths, compared with the population, as 1 to a little above 123, which, after a slight allowance for deaths abroad, will give the same period of doubling, or the same rate of increase as that which took place in France between 1813 and 1820, supposing both enumerations to be equally hear the truth.

It is worthy of remark, that, after making the above allowances for omissions in the returns of births and deaths, the proportion of deaths appears to be smaller than in any of the registers before collected; and as the proportion of the births is also smaller than either before the revolution or in the returns from the 30 departments in the years VIII., IX. and X. before noticed; and as there is every reason to believe that there were great omissions in the general returns of the year IX. and that the omissions in the returns from the 50 departments in 1813, were not fewer than in the later registers, it may fairly be presumed that the proportion of births has diminished notwithstanding the increased rate at which the population has been proceeding of late years. This increased rate appears to be owing to a diminished mortality, occasioned by the improved situation of the labouring classes since the revolution, and aided probably by the introduction of vaccination. It shews that an acceleration in the rate of increase is quite consistent with a diminution in the proportion of births, and that such a diminution is likely to take place under a diminished mortality from whatever cause or causes arising.

As a curious and striking proof of the error into which we should fall, in estimating the population of countries at different periods by the increase of births, it may be remarked that, according to Necker, the annual births in France on an average of six years, ending with 1780, were 958,586. The births for the same number of years ending with 1822, were, as above stated, 957,875. Estimating therefore the population by the births, it would appear that in 42 years it had rather diminished than increased, whereas, by enumerations, there is every reason to believe that it has increased in that time nearly four millions.

3.5.8 Book II, Chapter VIII: Of the Checks to Population in England.

The most cursory view of society in this country must convince us, that throughout all ranks the preventive check to population prevails in a considerable degree. Those among the higher classes, who live principally in towns, often want the inclination to marry, from the facility with which they can indulge themselves in an illicit intercourse with the sex. And others are deterred from marrying by the idea of the expenses that they must retrench, and the pleasures of which they must deprive themselves, on the supposition of having a family. When the fortune is large, these considerations are certainly trivial; but a preventive foresight of this kind has objects of much greater weight for its contemplation as we go lower.

A man of liberal education, with an income only just sufficient to enable him to associate in the rank of gentlemen, must feel absolutely certain that, if he marry and have a family, he shall be obliged to give up all his former connexions. The woman, whom a man of education would naturally make the object of his choice, is one brought up in the same habits and sentiments with himself, and used to the familiar intercourse of a society totally different from that to which she must be reduced by marriage. Can a man easily consent to place the object of his affection in a situation so discordant, probably, to her habits and inclinations? Two or three steps of descent in society, particularly at this round of the ladder, where education ends and ignorance begins, will not be considered by the generality of people as a chimerical, but a real evil. If society be desirable, it surely must be free, equal and reciprocal society, where benefits are conferred as well as received, and not such as the dependent finds with his patron, or the poor with the rich.

These considerations certainly prevent many in this rank of life from following the bent of their inclinations in an early attachment. Others, influenced either by a stronger passion or a weaker judgment, disregard these considerations; and it would be hard, indeed, if the gratification of so delightful a passion as virtuous love did not sometimes more than counterbalance all its attendant evils. But I fear it must be acknowledged that the more general consequences of such marriages are rather calculated to justify than disappoint the forebodings of the prudent.

The sons of tradesmen and farmers are exhorted not to marry, and generally find it necessary to comply with this advice, till they are settled in some business or farm, which may enable them to support a family. These events may not perhaps occur till they are far advanced in life. The scarcity of farms is a very general complaint; and the competition in every kind of business is so great, that it is not possible that all should be successful. Among the clerks in counting-houses, and the competitors for all kinds of mercantile and professional employment, it is probable that the preventive check to population prevails more than in any other department of society.

The labourer who earns eighteen pence or two shillings a day, and lives at his ease as a single man, will hesitate a little before he divides that pittance among four or five, which seems to be not more than sufficient for one. Harder fare and harder labour he would perhaps be willing to submit to for the sake of living with the woman he loves; but he must feel conscious, that, should he have a large family and any ill fortune whatever, no degree of frugality, no possible exertion of his manual strength, would preserve him from the heart-rending sensation of seeing his children starve, or of being obliged to the parish for their support. The love of independence is a sentiment that surely none would wish to see eradicated; though the poor-laws of England, it must be confessed, are a system of all others the most calculated gradually to weaken this sentiment, and in the end will probably destroy it completely.

The servants who live in the families of the rich have restraints yet stronger to break through in venturing upon marriage. They possess the necessaries, and even the comforts of life, almost in as great plenty as their masters. Their work is easy and their food luxurious, compared with the work and food of the class of labourers; and their sense of dependence is weakened by the conscious power of changing their masters if they feel themselves offended. Thus comfortably situated at present, what are their prospects if they marry? Without knowledge or capital, either for business or farming, and unused and therefore unable to earn a subsistence by daily labour, their only refuge seems to be a miserable alehouse, which certainly offers no very enchanting prospect of a happy evening to their lives. The greater number of them, therefore, deterred by this uninviting view of their future situation, content themselves with remaining single where they are.

If this sketch of the state of society in England be near the truth, it will be allowed that the preventive check to population operates with considerable force throughout all the classes of the community. And this observation is further confirmed by the abstracts from the registers returned in consequence of the Population Act[712] passed in 1800.

The results of these abstracts shew, that the annual marriages in England and Wales are to the whole population as 1 to 123 1/5,[713] a smaller proportion of marriages than is to be found in any of the countries which have been examined, except Norway and Switzerland.

In the earlier part of the last century, Dr. Short estimated this proportion at about 1 to 115.[714] It is probable that this calculation was then correct; and the present diminution in the proportion of marriages, notwithstanding an increase of population more rapid than formerly, owing to the more rapid progress of commerce and agriculture, is partly a cause, and partly a consequence, of the diminished mortality observed of late years.

The returns of the marriages, pursuant to the late act, are supposed to be less liable to the suspicion of inaccuracy than any other parts of the registers.

Dr. Short, in his *New Observations on Town and Country Bills of Mortality,* says, he will "conclude with the observation of an eminent Judge of this nation, that the growth and increase of mankind is more stinted from the cautious difficulty people make to enter on marriage, from the prospect of the trouble and expenses in providing for a family, than from any thing in the nature of the species." And, in conformity to this idea, Dr. Short proposes to lay heavy taxes and fines on those who live single, for the support of the married poor.[715]

The observation of the eminent Judge is, with regard to the numbers which are prevented from being born, perfectly just; but the inference, that the unmarried ought to be punished, does not appear to be equally so. The prolific power of nature is very far

---

[712]This chapter was written in 1802, just after the first enumeration, the results of which were published in 1801.

[713]Observ. on the Results of the Population Act, p. 11, printed in 1801. The answers to the Population Act have at length happily rescued the question of the population of this country from the obscurity in which it had been so long involved, and have afforded some very valuable data to the political calculator. At the same time it must be confessed that they are not so complete as entirely to exclude reasonings and conjectures respecting the inferences which are to be drawn from them. It is earnestly to be hoped that the subject may not be suffered to drop after the present effort. Now that the first difficulty is removed, an enumeration every ten years might be rendered easy and familiar; and the registers of births, deaths and marriages might be received every year, or at least every five years. I am persuaded, that more inferences are to be drawn respecting the internal state of a country from such registers than we have yet been in the habit of supposing.

[714]New Observ. on Bills of Mortality, p. 265. 8v0. 1750.

[715]New Observ. on Bills of Mortality, p. 247. 8vo. 1750.

indeed from being called fully into action in this country. And yet when we contemplate the insufficiency of the price of labour to maintain a large family, and the amount of mortality which arises directly and indirectly from poverty; and add to this the crowds of children, which are cut off prematurely in our great towns, our manufactories and our workhouses; we shall be compelled to acknowledge, that, if the number born annually were not greatly thinned by this premature mortality, the funds for the maintenance of labour must increase with much greater rapidity than they have ever done hitherto in this country, in order to find work and food for the additional numbers that would then grow up to manhood.

Those, therefore, who live single, or marry late, do not by such conduct contribute in any degree to diminish the actual population; but merely to diminish the proportion of premature mortality, which would otherwise be excessive; and consequently in this point of view do not seem to deserve any very severe reprobation or punishment.

The returns of the births and deaths are supposed, on good grounds, to be deficient; and it will therefore be difficult to estimate, with any degree of accuracy, the proportion which they bear to the whole population.

If we divide the existing population of England and Wales by the average of burials for the five years ending in 1800, it would appear, that the mortality was only 1 in 49;[716] but this is a proportion so extraordinarily small, considering the number of our great towns and manufactories, that it cannot be considered as approaching to the truth.

Whatever may be the exact proportion of the inhabitants of the towns to the inhabitants of the country, the southern part of this island certainly ranks in that class of states, where this proportion is greater than 1 to 3; indeed there is ample reason to believe, that it is greater than 1 to 2. According to the rule laid down by Crome, the mortality ought consequently to be above 1 in 30;[717] according to Sussmilch, above 1 in 33.[718] In the *Observations on the Results of the Population Act,*[719] many probable causes of deficiency in the registry of the burials are pointed out; but no calculation is offered respecting the sum of these deficiencies, and I have no data whatever to supply such a calculation. I will only observe, therefore, that if we suppose them altogether to amount to such a number as will make the present annual mortality about 1 in 40, this must appear to be the lowest proportion of deaths that can well be supposed, considering the circumstances of the country; and, if true, would indicate a most astonishing superiority over the generality of other states, either in the habits of the people with respect to prudence and cleanliness, or in natural healthiness of situation.[720] Indeed, it seems to be nearly ascertained that both these

[716]The population is taken at 9,168,000, and the annual deaths at 186,000. (Obs. on the Results of Pop. Act. p. 6 and 9.)

[717]Ueber die Bevölkerung der Europaischen Staaten, p. 127.

[718]Sussmilch, Göttliche Ordnung, vol. iii. p. 60.

[719]P. 6.

[720]It is by no means surprising, that our population should have been underrated formerly, at least by any person who attempted to estimate it from the proportion of births or deaths. Till the late Population Act no one could have imagined that the actual returns of annual deaths, which might

causes, which tend to diminish mortality, operate in this country to a considerable degree. The small proportion of annual marriages before mentioned indicates that habits of prudence, extremely favourable to happiness, prevail through a large part of the community in spite of the poor-laws; and it appears from the clearest evidence, that the generality of our country parishes are very healthy. Dr. Price quotes an account of Dr. Percival, collected from the ministers of different parishes and taken from positive enumerations, according to which, in some villages, only a 45th, a 50th, a 60th, a 66th, and even a 75th, part dies annually. In many of these parishes the births are to the deaths above 2 to 1, and 4n a single parish above 3 to 1.[721] These however are particular instances, and cannot be applied to the agricultural part of the country in general. In some of the flat situations, and particularly those near marshes, the proportions are found very different, and in a few the deaths exceed the births. In the 54 country parishes, the registers of which Dr. Short collected, choosing them purposely in a great variety of situations, the average mortality was as high as 1 in 37.[722] This is certainly much above the present mortality of our agricultural parishes in general. The period which Dr. Short took, included some considerable epidemics, which may possibly have been above the usual proportion. But sickly seasons should always be included, or we shall fall into great errors. In 1056 villages of Brandenburgh, which Sussmilch examined, the mortality for six good years was 1 in 43; for 10 mixed years about 1 in 38½.[723] In the villages of England which Sir F. M. Eden mentions, the mortality seems to be about 1 in 47 or 48;[724] and in the late returns pursuant to the Population Act, a still greater degree of healthiness appears. Combining these observations together, if we take 1 in 46 or 1 in 48, as the average mortality of the agricultural part of the country, including sickly seasons, this will be the lowest that can be supposed with any degree of probability. But this proportion will certainly be raised to 1 in 40, when we blend it with the mortality of the towns and the manufacturing part of the community, in order to obtain the average for the whole kingdom.

The mortality in London, which includes so considerable a part of the inhabitants of this country, was, according to Dr. Price, at the time he made his calculations, 1 in 20¾; in

---

naturally have been expected to be as accurate in this country as in others, would turn out to be less than a 49th part of the population. If the actual returns for France, even so long ago as the ten years ending with 1780, had been multiplied by 49, she would have appeared at that time to have a population of above 40 millions. The average of annual deaths was 818,491. Necker, de l'Administration des Finances, tom. i. c. ix. p. 255. 12mo. 1785.

[721] Price's Observ. on Revers. Paym. vol. ii. note, p. 10. First additional Essay, 4th edit. In particular parishes, private communications are perhaps more to be depended upon than public returns; because in general those clergymen only are applied to, who are in some degree interested in the subject, and of course take more pains to be accurate.

[722] New Observations on Bills of Mortality, table ix. p. 133.

[723] Göttliche Ordnung, vol. i. c. ii. s. xxi. p. 74.

[724] Estimate of the Number of Inhabitants in G. Britain.

Norwich 1 in 24; in Northampton 1 in 26½; in Newbury 1 in 27½;[725] in Manchester 1 in 28; in Liverpool 1 in 27½,[726] &c. He observes that the number dying annually in towns is seldom so low as 1 in 28, except in consequence of a rapid increase produced by an influx of people at those periods of life when the fewest die, which is the case with Manchester and Liverpool,[727] and other very flourishing manufacturing towns. In general he thinks that the mortality in great towns may be stated at from 1 in 19[728] to 1 in 22 and 23; in moderate towns, from 1 in 24 to 1 in 28; and in the country villages, from 1 in 40 to 1 in 50.[729]

The tendency of Dr. Price to exaggerate the unhealthiness of towns may perhaps be objected to these statements; but the objection seems to be only of weight with regard to London. The accounts from the other towns, which are given, are from documents which his particular opinions could not influence.[730] It should be remarked; however, that there is good reason to believe, that not only London, but the other towns in England, and probably also country villages, were at the time of these calculations less healthy than at present. Dr. William Heberden observes, that the registers of the ten years from 1759 to 1768,[731] from which Dr. Price calculated the probabilities of life in London, indicate a much greater degree of unhealthiness than the registers of late years. And the returns pursuant to the Population Act, even after allowing for great omissions in the burials, exhibit in all our provincial towns, and in the country, a degree of healthiness much greater than had before been calculated. At the same time I cannot but think that 1 in 31, the proportion of mortality for London mentioned in the *Observations on the Results of the Population Act,*[732] is smaller than the truth. Five thousand are not probably enough to allow for the omissions in the burials; and the absentees in the employments of war and commerce are not sufficiently adverted to. In estimating the proportional mortality the resident population alone should be considered.

There certainly seems to be something in great towns, and even in moderate towns, peculiarly unfavourable to the very early stages of life; and the part of the community, on which the mortality principally falls, seems to indicate that it arises more from the closeness and foulness of the air, which may be supposed to be unfavourable to the tender lungs of

---

[725] Price's Observ. on Revers. Paym. vol. i. note, p. 272.

[726] Id. vol. ii. First additional Essay, note, p. 4.

[727] Price's Observ. on Revers. Paym. vol. ii. First additional Essay, note, p. 4.

[728] The mortality at Stockholm was, according to Wargentin, 1 in 19.

[729] Observ. on Revers. Paym. vol. ii. First additional Essay, p. 4.

[730] An estimate of the population or mortality of London, before the late enumeration, always depended much on conjecture and opinion, on account of the great acknowledged deficiencies in the registers; but this was not the case in the same degree with the other towns here named. Dr. Price, in allusion to a diminishing population, on which subject it appears that he has so widely erred, says very candidly, that perhaps he may have been insensibly influenced to maintain an opinion once advanced.

[731] Increase and Decrease of Diseases, p. 32, 4to. 180

[732] P. 13.

children, and the greater confinement which they almost necessarily experience, than from the superior degree of luxury and debauchery usually and justly attributed to towns. A married pair with the best constitutions, who lead the most regular and quiet life, seldom find that their children enjoy the same health in towns as in the country.

In London, according to former calculations, one half of the born died under three years of age; in Vienna and Stockholm under two; in Manchester under five; in Norwich under five: in Northampton under ten.[733] In country villages, on the contrary, half of the born live till thirty, thirty-five, forty, forty-six, and above. In the parish of Ackworth, in Yorkshire, it appears, from a very exact account kept by Dr. Lee of the ages at which all died there for 20 years, that half of the inhabitants live to the age of 46;[734] and there is little doubt, that, if the same kind of account had been kept in some of those parishes before mentioned, in which the mortality is so small as 1 in 60, 1 in 66, and even 1 in 75, half of the born would be found to have lived to 50 or 55.

As the calculations respecting the ages to which half of the born live in towns depend more upon the births and deaths which appear in the registers, than upon any estimates of the number of people, they are on this account less liable to uncertainty, than the calculations respecting the proportion of the inhabitants of any place which dies annually.

To fill up the void occasioned by this mortality in towns, and to answer all further demands for population, it is evident that a constant supply of recruits from the country is necessary; and this supply appears in fact to be always flowing in from the redundant births of the country. Even in those towns where the births exceed the deaths, this effect is produced by the marriages of persons not born in the place. At a time when our provincial towns were increasing much less rapidly than at present, Dr. Short calculated that 9/19 of the married were strangers.[735] Of 1618 married men, and 1618 married women, examined at the Westminster infirmary, only 329 of the men and 495 of the women had been born in London.[736]

Dr. Price supposes that London with its neighbouring parishes, where the deaths exceed the births, requires a supply of 10,000 persons annually. Graunt, in his time, estimated the supply for London alone at 6,000;[737] and he further observes, that, let the mortality of the city be what it will, arising from plague, or any other great cause of destruction, it always fully repairs its loss in two years.[738]

As all these demands, therefore, are supplied from the country, it is evident that we should fall into a very great error, if we were to estimate the proportion of births to deaths

[733] Price's Observ. on Revers. Paym. vol. i. p. 264-266. 4th edit.

[734] Id. vol. i. p. 268.

[735] New Observations on Bills of Mortality, p. 76.

[736] Price's Observ. on Revers. Paym. vol. ii. p. 17.

[737] Short's New Observ. Abstract from Graunt, p. 277.

[738] Id. p. 276.

for the whole kingdom, by the proportion observed in country parishes, from which there must be such numerous emigrations.

We need not, however, accompany Dr. Price in his apprehensions that the country will be depopulated by these emigrations, at least as long as the funds for the maintenance of agricultural labour remain unimpaired. The proportion of births, as well as the proportion of marriages, clearly proves, that, in spite of our increasing towns and manufactories, the demand on the country for people is by no means very pressing.

If we divide the present population of England and Wales by the average number of baptisms for the last five years,[739] it will appear, that the baptisms are to the population as 1 to very nearly 36;[740] but it is supposed, with reason, that there are great omissions in the baptisms.

Dr. Short estimated the proportion of births to the population of England as one to 28.[741] In the agricultural report of Suffolk, the proportion of births to the population was calculated at 1 to 30. For the whole of Suffolk, according to the late returns, this proportion is not much less than 1 to 33.[742] According to a correct account of thirteen villages from actual enumerations, produced by Sir F. M. Eden, the proportion of births to the population was as 1 to 33; and according to another account on the same authority; taken from towns and manufacturing parishes, as 1 to 27¾.[743] If, combining all these circumstances, and adverting at the same time to the acknowledged deficiency in the registry of births, and the known increase of our population of late years, we suppose the true proportion of the births to the population to be as 1 to 30; then assuming the present mortality to be 1 in 40, as before suggested, we shall nearly keep the proportion of baptisms to burials which appears in the late returns. The births will be to the deaths as 4 to 3 or 13 1/3 to 10, a proportion more than sufficient to account for the increase of population which has taken place since the American war, after allowing for those who may be supposed to have died abroad.

In the *Observations on the Results of the Population Act* it is remarked that the average duration of life in England appears to have increased in the proportion of 117 to 100,[744] since the year 1780. So great a change, in so short a time, if true, would be a most striking phenomenon. But I am inclined to suspect that the whole of this proportional diminution of

[739] This was written before the omitted returns were added in 1810. These additions make the births in 1800 amount to 263,000, instead of 255,426, and increase the proportion of registered births to 1 in 35.—See the next chapter.

[740] Average medium of baptisms for the last five years 255,426. Pop. 9,198,000. (Observ. on Results, p. 9.)

[741] New Observ. p. 267.

[742] In private inquiries, dissenters and those who do not christen their children, will not of course be reckoned in the population; consequently such inquiries, as far as they extend, will more accurately express the true proportion of births; and we are fairly justified in making use of them, in order to estimate the acknowledged deficiency of births in the public returns.

[743] Estimate of the Number of Inhabitants in Great Britain, &c. p. 27.

[744] P. 6.

burials does not arise from increased healthiness, but is occasioned, in part, by the greater number of deaths which must necessarily have taken place abroad, owing to the very rapid increase of our foreign commerce since this period; and to the great number of persons absent on naval and military employments, and the constant supply of fresh recruits necessary to maintain undiminished so great a force. A perpetual drain of this kind would certainly have a tendency to produce the effect observed in the returns, and might keep the burials stationary, while the births and marriages were increasing with some rapidity. At the same time, as the increase of population since 1780 is incontrovertible, and the present mortality extraordinarily small, I should still be disposed to believe, that much the greater part of the effect is to be attributed to increased healthiness.

A mortality of 1 in 36 is perhaps too small a proportion of deaths for the average of the whole century; but a proportion of births to deaths as 12 to 10, calculated on a mortality of 1 in 36, would double the population of a country in 125 years, and is therefore as great a proportion of births to deaths, as can be true for the average of the whole century. None of the late calculations imply a more rapid increase than this.

We must not suppose, however, that this proportion of births to deaths, or any assumed proportion of births and deaths to the whole population, has continued nearly uniform throughout the century. It appears from the registers of every country which have been kept for any length of time, that considerable variations occur at different periods. Dr. Short, about the middle of the century, estimated the proportion of births to deaths as 11 to 10;[745] and if the births were at the same time a twenty-eighth part of the population, the mortality was then as high as 1 in 30 4/5. We now suppose that the proportion of births to deaths is above 13 to 10; but if we were to assume this proportion as a criterion by which to estimate the increase of population for the next hundred years, we should probably fall into a very gross error. We cannot reasonably suppose that the resources of this country should increase for any long continuance with such rapidity as to allow of a permanent proportion of births to deaths as 13 to 10, unless indeed this proportion were principally caused by great foreign drains.

From all the data that could be collected, the proportion of births to the whole population of England and Wales, has been assumed to be as 1 to 30; but this is a smaller proportion of births than has appeared in the course of this review to take place in any other country except Norway and Switzerland; and it has been hitherto usual with political calculators, to consider a great proportion of births as the surest sign of a vigorous and flourishing state. It is to be hoped, however, that this prejudice will not last long. In countries circumstanced like America or Russia, or in other countries after any great mortality, a large proportion of births is a favourable symptom; but in the average state of a well-peopled territory there cannot well be a worse sign than a large proportion of births, nor can there well be a better sign than a small proportion.

[745] New Observ. tables II. and iii. p. 22 and 44; Price's Observ. on Revers. Paym. vol. ii. p. 311.

Sir Francis d'Ivernois very justly observes, that, "if the various states of Europe kept and published annually an exact account of their population, noting carefully in a second column the exact age at which the children die, this second column would shew the relative merit of the governments, and the comparative happiness of their subjects. A simple arithmetical statement would then perhaps be more conclusive than all the arguments that could be adduced."[746] In the importance of the inferences to be drawn from such tables, I fully agree with him; and to make these inferences, it is evident, that we should attend less to the column expressing the number of children born, than to the column expressing the number which survived the age of infancy and reached manhood; and this number will almost invariably be the greatest, where the proportion of the births to the whole population is the least. In this point, we rank next after Norway and Switzerland, which, considering the number of our great towns and manufactories, is certainly a very extraordinary fact. As nothing can be more clear, than that all our demands for population are fully supplied, if this be done with a small proportion of births, it is a decided proof of a very small mortality, a distinction on which we may justly pride ourselves. Should it appear from future investigations that I have made too great an allowance for omissions both in the births and in the burials, I shall be extremely happy to find that this distinction, which, other circumstances being the same, I consider as the surest test of happiness and good government, is even greater than I have supposed it to be. In despotic, miserable, or naturally unhealthy countries, the proportion of births to the whole population will generally be found very great.

On an average of the five years ending in 1800, the proportion of births to marriages is 347 to 100. In 1760, it was 362 to 100, from which an inference is drawn, that the registers of births, however deficient, were certainly not more deficient formerly than at present.[747] But a change of this nature, in the appearance of the registers, might arise from causes totally unconnected with deficiencies. If from the acknowledged greater healthiness of the latter part of the century, compared with the middle of it, a greater number of children survived the age of infancy, a greater proportion of the born would of course live to marry, and this circumstance would produce a greater present proportion of marriages compared with the births. On the other hand, if the marriages were rather more prolific formerly than at present, owing to their being contracted at an earlier age, the effect would be a greater proportion of births compared with the marriages. The operation of either or both of these causes would produce exactly the effect observed in the registers: and consequently from the existence of such an effect no inference can justly be drawn against the supposed increasing accuracy of the registers. The influence of the two causes just mentioned on the proportions of annual births to marriages will be explained in a subsequent chapter.

With regard to the general question, whether we have just grounds for supposing that the registry of births and deaths was more deficient in the former part of the century than in the

[746] Tableau des Pertes, &c. c. ii. p. 16.

[747] Observ. on the Results of the Population Act, p. 8.

latter part; I should say, that the late returns tend to confirm the suspicion of former inaccuracy, and to shew that the registers of the earlier part of the century, in every point of view, afford very uncertain data on which to ground any estimates of past population. In the years 1710, 1720, and 1730, it appears from the returns that the deaths exceeded the births; and taking the six periods ending in 1750,[748] including the first half of the century, if we compare the sum of the births with the sum of the deaths, the excess of the births is so small, as to be perfectly inadequate to account for the increase of a million, which, upon a calculation from the births alone, is supposed to have taken place in that time.[749] Consequently, either the registers are very inaccurate, and the deficiencies in the births greater than in the deaths; or these periods, each at the distance of ten years, do not express the just average. These particular years may have been more unfavourable with respect to the proportion of births to deaths than the rest; indeed one of them, 1710, is known to have been a year of great scarcity and distress. But if this suspicion, which is very probable, be admitted, so as to affect the six first periods, we may justly suspect the contrary accident to have happened with regard to the three following periods ending with 1780; in which thirty years it would seem, by the same mode of calculation, that an increase of a million and a half had taken place.[750] At any rate it must be allowed, that the three separate years, taken in this manner, can by no means be considered as sufficient to establish a just average; and what rather encourages the suspicion, that these particular years might be more than usually favourable with regard to births is, that the increase of births from 1780 to 1785 is unusually small,[751] which would naturally be the case without supposing a slower progress than before, if the births in 1780 had been accidentally above the average.

On the whole, therefore, considering the probable inaccuracy of the earlier registers, and the very great danger of fallacy in drawing general inferences from a few detached years, I do not think that we can depend upon any estimates of past population, founded on a calculation from the births, till after the year 1780, when every following year is given; and a just average of the births may be obtained. As a further confirmation of this remark I will just observe, that in the final summary of the abstracts from the registers of England and Wales it appears, that in the year 1790, the total number of births was 248,774, in the year 1795, 247,218, and in 1800, 247,147.[752] Consequently if we had been estimating the population from the births, taken at three separate periods of five years, it would have appeared, that the population during the last ten years had been regularly decreasing, though we have very good reason to believe, that it has increased considerably.

[748] Population Abstracts, Parish Registers. Final summary, p. 455.

[749] Observ. on the Results of the Population Act, p. 9.

[750] Ibid.

[751] Observ. on the Results of the Population Act, p, 9.

[752] Population Abstracts, Parish Registers, p. 455.

In the *Observations on the Results of the Population Act,*[753] a table is given of the population of England and Wales throughout the last century, calculated from the births; but for the reasons given above, little reliance can be placed upon it; and for the population at the revolution, I should be inclined to place more dependence on the old calculations from the number of houses.

It is possible, indeed, though not probable, that these estimates of the population at the different periods of the century may not be very far from the truth, because opposite errors may have corrected each other; but the assumption of the uniform proportion of births on which they are founded is false on the face of the calculations themselves. According to these calculations, the increase of population was more rapid in the period from 1760 to 1780, than from 1780 to 1800; yet it appears, that the proportion of deaths about the year 1780 was greater than in 1800 in the ratio of 117 to 100. Consequently the proportion of births before 1780 must have been much greater than in 1800, or the population in that period could not possibly have increased faster. This overthrows at once the supposition of any thing like uniformity in the proportion of births.

I should indeed have supposed from the analogy of other countries, and the calculations of Mr. King and Dr. Short, that the proportion of births at the beginning and in the middle of the century was greater than at the end. But this supposition would, in a calculation from the births, give a smaller population in the early part of the century than is given in the *Results of the Population Act,* though there are strong reasons for supposing that the population there given is too small. According to Davenant, the number of houses in 1690 was 1,319,215, and there is no reason to think that this calculation erred on the side of excess. Allowing only five to a house instead of 5 3/5, which is supposed to be the proportion at present, this would give a population of above six millions and a half, and it is perfectly incredible, that from this time to the year 1710, the population should have diminished nearly a million and a half. It is far more probable that the omissions in the births should have been much greater than at present, and greater than in the deaths; and this is further confirmed by the observation before alluded to, that in the first half of the century the increase of population, as calculated from the births, is much greater than is warranted by the proportion of births to deaths. In every point of view, therefore, the calculations from the births are little to be depended on.

It must indeed have appeared to the reader, in the course of this work, that registers of births or deaths, excluding any suspicion of deficiencies, must at all times afford very uncertain data for an estimate of population. On account of the varying circumstances of every country, they are both precarious guides. From the greater apparent regularity of the births, political calculators have generally adopted them as the ground of their estimates in preference to the deaths. Necker, in estimating the population of France, observes, that an epidemic disease, or an emigration, may occasion temporary differences in the deaths, and

[753] P. 9.

that therefore the number of births is the most certain criterion.[754] But the very circumstance of the apparent regularity of the births in the registers will now and then lead into great errors. If in any country we can obtain registers of burials for two or three years together, a plague or mortal epidemic will always shew itself, from the very sudden increase of the deaths during its operation, and the still greater diminution of them afterwards. From these appearances, we should of course be directed, not to include the whole of a great mortality in any very short term of years. But there would be nothing of this kind to guide us in the registers of births; and after a country had lost an eighth part of its population by a plague, an average of the five or six subsequent years might shew an increase in the number of births, and our calculations would give the population the highest at the very time that it was the lowest. This appears very strikingly in many of Sussmilch's tables, and most particularly in a table for Prussia and Lithuania, which I shall insert in a subsequent chapter; where, in the year following to the loss of one third of the population, the births were considerably increased, and in an average of five years but very little diminished; and this at a time when, of course, the country could have made but a very small progress towards recovering its former population.

We do not know indeed of any extraordinary mortality which has occurred in England since 1700; and there are reasons for supposing that the proportions of the births and deaths to the population during the last century have not experienced such great variations as in many countries on the continent; at the same time it is certain that the sickly seasons which are known to have occurred, would, in proportion to the degree of their fatality, produce similar effects; and the change which has been observed in the mortality of late years, should dispose us to believe, that similar changes might formerly have taken place respecting the births, and should instruct us to be extremely cautious in applying the proportions, which are observed to be true at present, to past or future periods.

3.5.9 Book II, Chapter IX: Of the Checks to Population in England (continued).

The returns of the Population Act in 1811 undoubtedly presented extraordinary results. They shewed a greatly accelerated rate of progress, and a greatly improved healthiness of the people, notwithstanding the increase of the towns and the increased proportion of the population engaged in manufacturing employments. They thus furnished another striking instance of the readiness with which population starts forwards, under almost any weight, when the resources of a country are rapidly increasing.

The amount of the population in 1800, together with the proportions of births, deaths and marriages, given in the registers, had made it appear that the population had been for

[754] De l'Administration des Finances, tom. i. c. ix. p. 252. 12mo. 1783.

some time increasing at a rate rather exceeding what would result from a proportion of births to deaths as 4 to 3, with a mortality of 1 in 40.

These proportions would add to the population of a country every year 120th part; and if they were to continue, would, according to table II., ch. xi. double the population in every successive period of 83½ years. This is a rate of progress which in a rich and well-peopled country might reasonably be expected to diminish rather than to increase. But instead of any such diminution, it appears that as far as 1810 it had been considerably accelerated.

In 1810, according to the returns from each parish, with the additions of 1/30 for the soldiers, sailors, &c., the population of England and Wales was estimated at 10,488,000,[755] which, compared with 9,168,000, the population of 1800, estimated in a similar manner, shews an increase in the ten years of 1,320,000.

The registered baptisms during ten years were 2,878,906, and the registered burials 1,950,189. The excess of the births is therefore 928,717, which falls very considerably short of the increase shewn by the two enumerations. This deficiency could only be occasioned either by the enumeration in 1800 being below the truth, or by the inaccuracy of the registers of births and burials, or by the operation of these two causes combined; as it is obvious that, if the population in 1800 were estimated correctly, and the registers contained all the births and burials, the difference must exceed rather than fall short of the real addition to the population; that is, it would exceed it exactly by the number of persons dying abroad in the army, navy, &c.

There is reason to believe that both causes had a share in producing the effect observed, though the latter, that is, the inaccuracy of the registers, in much the greatest degree.

In estimating the population throughout the century,[756] the births have been assumed to bear the same proportion at all times to the number of people. It has been seen that such an assumption might often lead to a very incorrect estimate of the population of a country at different and distant periods. As the population however is known to have increased with great rapidity from 1800 to 1810, it is probable that the proportion of births did not essentially diminish during that period. But if, taking the last enumeration as correct, we compare the births of 1810 with the births of 1800, the result will imply a larger population in 1800 than is given in the enumeration for that year.

Thus the average of the last five years' births to 1810 is 297,000, and the average of the five years' births to 1800 is 263,000. But 297,000 is to 263,000 as 10,488,000, the population of 1810, to 9,287,000, which must therefore have been the population in 1800, if the proportion of births be assumed to be the same, instead of 9,198,000, the result of the enumeration. It is further to be observed that the increase of population from 1795 to 1800 is according to the table unusually small, compared with most of the

---

[755] See the Population Abstracts published in 1811, and the valuable Preliminary Observations by Mr. Rickman.

[756] See a table of the population throughout the century, in page xxv. of the Preliminary Observations to the Population Abstracts, printed in 1811.

preceding periods of five years. And a slight inspection of the registers will shew that the proportion of births for five years from 1795, including the diminished numbers of 1796 and 1800, was more likely to be below than above the general average. For these reasons, together with the general impression on the subject, it is probable that the enumeration in 1800 was short of the truth, and perhaps the population at that time may be safely taken at as much as 9,287,000 at the least, or about 119,000 greater than the returns gave it.

But even upon this supposition, neither the excess of births above the deaths in the whole of the ten years, nor the proportion of births to deaths, as given in the registers, will account for an increase from 9,287,000, to 10,488,000. Yet it is not probable that the increase has been much less than is shewn by the proportion of the births at the two periods. Some allowance must therefore necessarily be made for omissions in the registers of births and deaths, which are known to be very far from correct, particularly the registers of births.

There is reason to believe that there are few or no omissions in the register of marriages; and if we suppose the omissions in the births to be one-6th, this will preserve a proportion of the births to the marriages as 4 to 1, a proportion which appears to be satisfactorily established upon other grounds;[757] but if we are warranted in this supposition, it will be fair to take the omissions in the deaths at such a number as will make the excess of the births above the deaths in the ten years accord with the increase of population estimated by the increase of the births.

The registered births in the ten years, as was mentioned before, are 2,878,906, which increased by one-6th will be 3,358,723. The registered burials are 1,950,189, which increased by one-12th will be 2,112,704. The latter subtracted from the former will give 1,246,019 for the excess of births, and the increase of population in the ten years, which number added to 9,287,000, the corrected population of 1800, will give 10,533,019, forty-five thousand above the enumeration of 1810, leaving almost exactly the number which in the course of the ten years appears to have died abroad. This number has been calculated generally at about 4¼, per cent. on the male births; but in the present case there are the means of ascertaining more accurately the number of males dying abroad during the period in question. In the last population returns the male and female births and deaths are separated; and from the excess of the male births above the female births, compared with the male and female deaths, it appears that forty-five thousand males died abroad.[758]

The assumed omissions therefore in the births and burials seem to answer so far very well.

It remains to see whether the same suppositions will give such a proportion of births to deaths, with such a rate of mortality, as will also account for an increase of numbers in ten years from 9,287,000 to 10,488,000.

If we divide the population of 1810 by the average births of the preceding five years, with the addition of one-6th, it will appear that the proportion of births to the population is

---

[757] See the Preliminary Observations on the Population Abstracts, p. xxvi.

[758] See Population Abstracts, 1811, page 196 of the Parish Register Abstract.

as 1 to 30. But it is obvious that if the population be increasing with some rapidity, the average of births for five years, compared with the population at the end of such period, must give the proportion of births too small. And further, there is always a probability that a proportion which is correct for five years may not be correct for ten years. In order to obtain the true proportion applicable to the progress of population during the period in question, we must compare the annual average of the births, for the whole term, with the average or mean population of the whole term.

The whole number of births, with the addition of 1/6, is, as before stated, 3,358,723, and the annual average during the ten years 335,872. The mean population, or the mean between 10,488,000 (the population of 1810) and 9,287,000 (the corrected population of 1800) is 9,887,000; and the latter number divided by the average of the births will give a proportion of births to the population as 1 to rather less than 29½, instead of 30, which will make a considerable difference.

In the same manner, if we divide the population of 1810 by the average of the burials for the preceding five years, with the addition of one-12th, the mortality will appear to be as 1 in nearly 50; but upon the same grounds as with regard to the births, an average of the burials for five years, compared with the population at the end of such term, must give the proportion of burials too small; and further, it is known, in the present case, that the proportion of burials to the population by no means continued the same during the whole time. In fact the registers clearly shew an improvement in the healthiness of the country, and a diminution of mortality progressively through the ten years; and while the average number of annual births increased from 263,000 to 287,000, or more than one-8th, the burials increased only from 192,000 to 196,000 or one-48th. It is obviously necessary then for the purpose in view to compare the average mortality with the average or mean population.

The whole number of burials in the ten years, with the addition of one-12th, is, as was before stated, 2,112,704, and the mean population 9,887,000. The latter, divided by the former, gives the annual average of burials compared with the population as 1 to rather less than 47. But a proportion of births as 1 to 29½, with a proportion of deaths as 1 to 47, will add yearly to the numbers of a country one-79th of the whole, and in ten years will increase the population from 9,287,000 to 10,531,000, leaving 43,000 for the deaths abroad, and agreeing very nearly with the calculation founded on the excess of births.[759]

[759] A general formula for estimating the population of a country at any distance from a certain period, under given circumstances of births and mortality, may be found in Bridge's Elements of Algebra, p. 225.

$\text{Log. } A = \text{log. } P + n \times \text{log. } 1 + \frac{m-b}{mb}$

In the present case, $P = 9{,}287{,}000$; $n = 10$; $m = 47$; $b = 29\frac{1}{2}$.

$\frac{m-b}{mb} = \frac{1}{79}$ and $+ 1 + \frac{m-b}{mb} = \frac{80}{79}$

The log. of $\frac{80}{79} = 00546$; $\therefore n \times \text{log.} 1 + \frac{m-b}{mb} = 05460$. Log. P. = 6.96787, which added to 05460 = 7.02247 the log of A, the number answering to which is 10,531,000.

We may presume therefore that the assumed omissions in the births and deaths from 1800 to 1810 are not far from the truth.

But if these omissions of one-6th for the births, and one-12th for the burials, may be considered as nearly right for the period between 1800 and 1810, it is probable that they may be applied without much danger of error to the period between 1780 and 1800, and may serve to correct some of the conclusions founded on the births alone. Next to an accurate enumeration, a calculation from the excess of births above the deaths is the most to be depended upon. Indeed when the registers contain all the births and deaths, and there are the means of setting out from a known population, it is obviously the same as an actual enumeration; and where a nearly correct allowance can be made for the omissions in the registers, and for the deaths abroad, a much nearer approximation to it may be obtained in this way than from the proportion of births to the whole population, which is known to be liable to such frequent variations.

The whole number of births returned in the twenty years, from 1780 to 1800, is 5,014,899, and of the burials 3,840,455. If we add one-6th to the former, and one-12th to the latter, the two numbers will be 5,850,715, and 4,160,492; and subtracting the latter from the former, the excess of the births above the deaths will be 1,690,223. Adding this excess to the population of 1780, as calculated in Mr. Rickman's tables, from the births, which is 7,953,000, the result will be 9,643,000, a number which, after making a proper allowance for the deaths abroad, is very much above the population of 1800, as before corrected, and still more above the number which is given in the table as of the enumeration.

But if we proceed upon the safer ground just suggested, and, taking the corrected population of 1800 as established, subtract from it the excess of the births during the twenty years, diminished by the probable number of deaths abroad, which in this case will be about 124,000, we shall have the number 7,721,000 for the population of 1780, instead of 7,953,000; and there is good reason to believe that this is nearer the truth;[760] and that not only in 1780, but in many of the intermediate periods, the estimate from the births has represented the population as greater, and increasing more irregularly, than would be found to be true, if recourse could be had to enumerations. This has arisen from the proportion of births to the population being variable, and, on the whole, greater in 1780, and at other periods during the course of the twenty years, than it was in 1800.

In 1795, for instance, the population is represented to be 9,055,000, and in 1800, 9,168,000;[761] but if we suppose the first number to be correct, and add the excess of the births above the deaths in the five intervening years, even without making any allowance for omissions in the registers, we shall find that the population in 1800 ought to have been 9,398,000, instead of 9,168,000; or if we take the number returned for 1800 as correct, it will appear, by subtracting from it the excess of births during the five preceding years, that

[760]The very small difference between the population of 1780 and 1785, as given in the table, seems strongly to imply that one of the two estimates is erroneous.

[761]Population Abstracts, 1811. Preliminary View, p. xxv.

the population in 1795 ought to have been 8,825,000, instead of 9,055,000. Hence it follows, that the estimate from the births in 1795 cannot be correct.

To obtain the population at that period, the safest way is to apply the before-mentioned corrections to the registers, and, having made the allowance of 4¼ per cent. on the male births for the deaths abroad, subtract the remaining excess of the births from the corrected returns of 1800. The result in this case will be 8,831,086 for the population of 1795, implying an increase in the five years of 455,914, instead of only 1,13,000, as shewn by the table calculated from the births.

If we proceed in the same manner with the period from 1790 to 1795, we shall find that the excess of births above the deaths (after the foregoing corrections have been applied, and an allowance has been made of 4¼ per cent. upon the male births for the deaths abroad), will be 415,669, which, subtracted from 8,831,086, the population of 1795, as above estimated, leaves 8,415,417 for the population of 1790.

Upon the same principle, the excess of the births above the deaths in the interval between 1785 and 1790 will turn out to be 416,776. The population in 1785 will therefore be 7,998,641. And in like manner the excess of the births above the deaths in the interval between 1780 and 1785 will be 277,544, and the population in 1780 7,721,097.

The two tables therefore, of the population, from 1780 to 1810, will stand thus:

Table, calculated from the births alone, in the Preliminary Observations to the Population Abstracts, printed in 1811.		Table, calculated from the excess of the births above the deaths, after an allowance made for the omissions in the registers, and the deaths abroad.	
Population in		Population in	
1780	7,953,000	1780	7,721,000
1785	8,016,000	1785	7,998,000
1790	8,675,000	1790	8,415,000
1795	9,055,000	1795	8,831,000
1800	9,168,000	1800	9,287,000
1805	9,828,000	1805	9,837,000
1810	10,488,000	1810	10,488,000

In the first table, or table calculated from the births alone, the additions made to the population in each period of five years are as follow;—

From 1780 to 1785	63,000
From 1785 to 1790	659,000
From 1790 to 1795	380,000
From 1795 to 1800	113,000
From 1800 to 1805	660,000
From 1805 to 1810	660,000

In the second table, or table calculated from the excess of the births above the deaths, after the proposed corrections have been applied, the additions made to the population in each period of five years will stand thus:—

From 1780 to 1785	277,000
From 1785 to 1790	417,000
From 1790 to 1795	416,000
From 1795 to 1800	456,000
From 1800 to 1805	550,000
From 1805 to 1810	651,000

The progress of the population, according to this latter table, appears much more natural and probable than according to the former.

It is in no respect likely that, in the interval between 1780 and 1785, the increase of the population should only have been 63,000, and in the next period 659,000; or that, in the interval between 1795 and 1800, it should have been only 113,000, and in the next period 660,000. But it is not necessary to dwell on probabilities; the most distinct proofs may be brought to shew that, whether the new table be right or not, the old table must be wrong. Without any allowances being made for omissions in the registers, the excess of the births above the deaths, in the period from 1780 to 1785, shews an increase of 193,000, instead of 63,000. And, on the other hand, no allowances for omissions in the registers, that could with the slightest degree of probability be supposed, would make the excess of births above the deaths in the period from 1785 to 1790 equal to 659,000. Making no allowance for omissions, this excess only amounts to 317,306; and if we were to suppose the omissions in the births one-4th, instead of one-6th, and that there were no omissions in the registers of burials, and that no one died abroad, the excess would still fall short of the number stated by many thousands.

The same results would follow, if we were to estimate the progress of population during these periods by the proportion of births to deaths, and the rate of mortality. In the first period the increase would turn out to be very much greater than the increase stated, and in the other very much less.

Similar observations may be made with regard to some of the other periods in the old table, particularly that between 1795 and 1800, which has been already noticed.

It will be found on the other hand, that, if the proportion of births to deaths during each period be estimated with tolerable accuracy and compared with the mean population, the rate of the progress of the population determined by this criterion will, in every period, agree very nearly with the rate of progress determined by the excess of the births above the deaths, after applying the proposed corrections. And it is further worthy of remark that, if the corrections proposed should be in some degree inaccurate, as is probable, the errors arising from any such inaccuracies are likely to be very much less considerable than those which must necessarily arise from the assumption on which the old table is founded; namely, that the births bear at all times the same proportion to the population.

Of course I do not mean to reject any estimates of population formed in this way, when no better materials are to be found; but, in the present case, the registers of the burials as well as baptisms are given every year, as far back as 1780, and these registers, with the firm ground of the last enumeration to stand upon, afford the means of giving a more correct table of the population from 1780 than was before furnished, and of shewing at the same time the uncertainty of estimates from the births alone, particularly with a view to the progress of population during particular periods. In estimating the whole population of a large country, two or three hundred thousand are not of much importance; but, in estimating the rate of increase during a period of five or ten years, an error to this amount is quite fatal. It will be allowed, I conceive, to make an essential difference in our conclusions respecting the rate of increase for any five years which we may fix upon, whether the addition made to the population during the term in question is 63,000 or 277,000, 115,000 or 456,000, 659,000 or 417,000.

With regard to the period of the century previous to 1780, as the registers of the baptisms and burials are not returned for every year, it is not possible to apply the same corrections. And it will be obvious that, in the table calculated from the births previous to this period, when the registers are only given for insulated years at some distance from each other, very considerable errors may arise, not merely from the varying proportion of the births to the population, on averages of five years, but from the individual years produced not representing with tolerable correctness these averages.[762] A very slight glance at the valuable table of baptisms, burials and marriages, given in the Preliminary Observations to the Population Abstracts,[763] will shew how very little dependence ought to be placed upon inferences respecting the population drawn from the number of births, deaths or marriages in individual years. If, for instance, we were estimating the population in the two years 1800 and 1801, compared with the two following years 1802 and 1803, from the proportion of marriages to the population, assuming this proportion to be always the same, it would appear that, if the population in the first two years were nine millions, in the second two years immediately succeeding it would be considerably above twelve millions, and thus it would seem to have increased above three millions, or more than one-third, in this short interval. Nor would the result of an estimate, formed from the births for the two years 1800 and 1801, compared with the two years 1803 and 1804, be materially different; at least such an estimate would indicate an increase of two millions six hundred thousand in three years.

The reader can hardly be surprised at these results, if he recollects that the births, deaths and marriages bear but a small proportion to the whole population; and that consequently variations in either of these, which may take place from temporary causes, cannot possibly

[762] From the one or other of these causes, I have little doubt, that the numbers in the table, for 1760 and 1780, which imply so rapid an increase of population in that interval, do not bear the proper relation to each other. It is probable that the number given for 1770 is too great.

[763] P. 20.

be accompanied by similar variations in, the whole mass of the population. An increase in the births of one-third, which might occur in a single year, instead of increasing the population one-third, would only perhaps increase it one-eightieth or ninetieth.

It follows therefore, as I stated in the last chapter, that the table of the population for the century previous to 1780, calculated from the returns of the births alone, at the distance of ten years each, can only be considered as a very rough approximation towards the truth, in the absence of better materials, and can scarcely in any degree be depended upon for the comparative rate of increase at particular periods.

The population in 1810, compared with that of 1800, corrected as proposed in this chapter, implies a less rapid increase than the difference between the two enumerations; and it has further appeared that the assumed proportion of births to deaths as 47 to 29½ is rather below than above the truth. Yet this proportion is quite extraordinary for a rich and well-peopled territory. It would add to the population of a country one-79th every year, and, were it to continue, would, according to table II. ch. xi. of this book, double the number of inhabitants in less than fifty-five years.

This is a rate of increase, which in the nature of things cannot be permanent. It has been occasioned by the stimulus of a greatly-increased demand for labour, combined with a greatly-increased power of production, both in agriculture and manufactures. These are the two elements which form the most effective encouragement to a rapid increase of population. What has taken place is a striking illustration of the principle of population, and a proof that in spite of great towns, manufacturing occupations, and the gradually-acquired habits of an opulent anti luxuriant people, if the resources of a country will admit of a rapid increase, anal if these resources are so advantageously distributed as to occasion a constantly-increasing demand for labour, the population will not fail to keep pace with them.

Since the publication of the last edition of this work in 1817, a third census of the population has taken place, and the results are highly worthy of our attention.

According to the enumeration in 1821, and the corrected returns of 1811, and 1801, as given in the preliminary observations to the published account by Mr. Rickman, the population of Great Britain was, in 1801, 10,942,646; in 1811, 12,596,803, and in 1821, 14,391,631.

These numbers taken as first stated, and including the very large numbers of males added in 1811 for the army and navy, give an increase of 15 per cent. in the ten years, from 1800 to 1811, and only 14¼ per cent. from 1810 to 1821.[764] But it is calculated that out of the 640,500 males added for the army, navy, and merchant service, above one-third must have been Irish and foreigners. Adding therefore only 1/30 to the resident population in 1801 and 1811, and on account of the peace allowing only 1/50 for the absent males in 1821, the population of England and Wales at the three different periods, without reference to any supposed deficiency in the first enumeration, will stand thus: in 1801, 9,168,000; in

[764] Preliminary Observations, p. viii.

1811, 10,502,500; and in 1821, 12,218,500, giving an increase in the interval between 1800 and 1811 of 14¼ per cent. and in the interval between 1810 and 1821, of 16 1/3 percent. The first of these two rates of increase would double the population in 51 and the other in 46 years. As, however, there must always be some uncertainty respecting the proportion of the persons employed in the army, navy and merchant service, properly belonging to the resident population, and as the male population is on other accounts more frequently on the move than the female, it has been judiciously proposed to estimate the rate of increase by the female population alone. The number of females in Great Britain was in 1801, 5,492,354; in 1811, 6,262,716; and in 1821, 7,253,728, giving an increase in the first period of 14.02 per cent. and in the second of 15.82.[765]

The increase of Scotland taken by itself was in the first period 13 per cent. and in the second 14½. The increase of England and Wales exclusive of Scotland appears to be almost exactly the same; particularly in the second period, whether we estimate it from the females alone, or from the whole population, with the proposed allowances for the army and navy, &c. a proof that these allowances are not far from the truth. At the same time, it should perhaps be remarked, that if, on account of the war, during the greater part of the period from 1800 to 1821, there must have been a greater portion of the male population destroyed than usual, the increase of the whole population ought not to be so great in proportion as the increase of the females; and that if such an increase appears, it is probably owing to too great a number of males having been added to the resident population for the army and navy, or to an influx from Scotland and Ireland.

The numbers above-mentioned, and the rates of increase, have been stated as given by Mr. Rickman in the Preliminary Observations to the Population Abstracts. But in the former part of this chapter, I assumed on what appeared to me to be sufficient grounds that the first enumeration was not so correct as that of 1811, and it is probable that the enumeration of 1811 is not quite so correct as that of 1821. In this case the rates of increase in the two periods will not be so great as above stated, but still they will appear to be very extraordinary.

According to the assumed estimate the population, as given in the enumeration of 1801, was about 119,000 short of the truth; and if on this ground we take the female population of the census in 1801 as deficient 60,000, and suppose that in 1811 it was deficient 30,000, the numbers of females in England and Wales at the different periods will stand thus: In 1801, 4,687,867; in 1811, 5,313,219; and in 1821, 6,144,709; giving an increase of 13.3 per cent. in the period from 1800 to 1811, and of 15.6 per cent. in the period from 1800 to 1821; making the rate of increase in the former period such as, if continued, would double the population in about 55 years, and in the latter, such as would double it in 48 years. Taking the whole 20 years together, the rate of increase would be such as, if continued, would double the population in about 51 years.

[765] Ibid.

This is no doubt a most extraordinary rate of increase, considering the actual population of the country compared with its territory, and the number of its great towns and manufactories. It is less however than that which is stated in the Preliminary Observations to the Population Abstracts. Yet even according to this slower rate of increase it is necessary to suppose that the omissions in the parish registers, particularly in regard to the births, have latterly rather increased than diminished; and this is rendered probable by a statement of Mr. Rickman in the Preliminary Observations. He says, "the question respecting unentered baptisms and burials showed a difference of nearly four to one in the degree of deficiency in the year 1811, the annual average number of unentered baptisms (as stated at the end of the several counties) having been 14,860; of burials, (setting aside London) 3,899; at present the proportion is five to one in the degree of deficiency; the annual average number of unentered baptisms (as stated at the end of the several counties) being 23,066; of burials, (setting aside London) 4,657." And he goes on to say, "Nor does this represent the full amount or proportion of unentered baptisms, the clergy of the most populous places, especially where many of the inhabitants are dissenters, usually declining to hazard an estimate." A burial ground, on the contrary, is a visible object, and among the persons connected with it, the clergyman can usually procure an account (more or less accurate) of the number of interments.

On these grounds it would appear probable that, owing to the increasing number of dissenters, or other causes, the omissions in the registers of births had been lately increasing, rather than diminishing. Yet it has been thought that since the Act of 1812 the registers of births have been more carefully kept; and it is certain that, in the 10 years ending with 1820, the proportion of births to marriages is greater, though the proportions of births and marriages to the whole population are both less than they were either in 1800, or in the ten years ending with 1810. Under these circumstances, it may be advisable to wait for further documents before any fresh conclusion is drawn respecting the probable amount of omissions in the births and burials. What may be considered as certain is, that, whereas the supposed admissions of one sixth in the births and one twelfth in the burials, with a proper allowance for the deaths abroad, are more than sufficient to account for the increase of population during the twenty years from 1781 to 1801, according to the numbers stated by Mr. Rickman, they are not sufficient to account for the increase of population in the 20 years from 1801 to 1821, according to the enumerations.

I have heard it surmised that the enumerations, particularly the two last, may by possibility exceed rather than fall short of the truth, owing to persons being reckoned more than once, from their having different places of residence. It must be allowed, that this supposition would account for the fact of the diminished proportions of births and marriages to the whole population, notwithstanding the apparent increase of that population with extraordinary rapidity. But the same diminished proportions would take place owing to a diminished mortality; and as a diminished mortality has been satisfactorily established on other grounds, it will fairly account for much of what appears. And if anything can justly be attributed to over enumerations, it must be of trifling amount.

That there are great omissions both in the births and burials, and greater in the former than in the latter, it is quite impossible to doubt. The testimony of all the clergy concerned in making the returns was, according to Mr. Rickman, uniform in this respect. And if we suppose only the same proportion of omissions from 1801 to 1821 as we supposed from 1781 to 1801, and commence with the census of 1801, on the presumption that the number of double entries in that enumeration would be balanced probably by the number of deficiencies, it will appear that the excess of the births alone, excluding the deaths abroad, would bring the population to within 184,404 of the enumeration of 1821, and including the allowance for deaths abroad, (which, in this case, from a comparison of the excess of male births with the male and female deaths, appears to be 128,651,) to within 313,055.

On the supposition of such an amount of double entries unbalanced by deficiencies in the two last returns, the enumerations would still shew a very extraordinary increase of population. The rate of increase in the period from 1801 to 1811 would be nearly 13 per cent. (12.88) which would double the population in about 57 years; and in the period from 1811 to 1821, it would be very nearly 15 per cent. (14.95), which would double the population in 50 years.

Under the uncertainty in which we must remain at present as to whether the enumerations partially err in defect or in excess, I have not thought it advisable to alter the amended table of the population from 1781 to 1811, given in the former part of this chapter. It is founded on a principle so very much safer than an estimate for the births alone, that it must at any rate shew the progress of the population more correctly than that given in the Preliminary Observations.

The more indeed the population returns are considered, the more uncertain will appear all estimates of the past population founded on the assumptions that the proportion of the births will always be nearly the same. If the population since the year 1801 were to be estimated in the same way as Mr. Rickman has estimated it before that year, it would appear that the population in 1821, instead of being, according to the enumeration, 12,218,500, would only be 11,625,334, that is, 593,166 or nearly 600,000 short of the enumeration of 1821. And the reason is, that the proportion of births to the population, which, estimated in the way suggested by Mr. Rickman, and without allowing for omissions, was, in 1821, only as 1 to 36.58, was, in 1801, as much as 1 to 34.8.

Supposing the enumerations to be correct, the varying proportions of the births (without allowance for omissions, and comparing the population at the end of each term with the average births for the five preceding years,) would be for 1801 as 1 to 34.8, for 1811 as 1 to 35.3, and for 1821 as 1 to 36.58.

Similar and even greater variations will be found to take place in regard to the proportions of the marriages to the population.

In 1801, the proportion was 1 to 122.2, in 1811, 1 to 126.6, in 1821, 1 to 131.1; and if, assuming that, for the 20 years ending with 1820, the marriages, in which it is supposed that there are very few omissions, would remain in the same proportion to the population as in 1801, we had estimated the population by the marriages, the numbers in 1821, instead of being 12,218,500, would only have been 11,377,548, that is, 840,952 short of the enumeration of 1821.

It appears, then, that if we can put any trust in our enumerations,[766] no reliance can be placed on an estimate of past population founded on the proportions of the births, deaths, or marriages: The same causes which have operated to alter so essentially these proportions during the 20 years for which we have enumerations may have operated in an equal degree before; and it will be generally found true, that the increasing healthiness of a country will not only diminish the proportions of deaths, but the proportions of births and marriages.

3.5.10 Book II, Chapter X: Of the Checks to Population in Scotland and Ireland.

An examination, in detail, of the statistical account of Scotland, would furnish numerous illustrations of the principle of population; but I have already extended this part of the work so much, that I am fearful of tiring the patience of my readers; and shall therefore confine my remarks in the present instance to a few circumstances which have happened to strike me.

On account of the acknowledged omissions in the registers of births, deaths and marriages in most of the parishes of Scotland, few just inferences can be drawn from them. Many give extraordinary results. In the parish of Crossmichael[767] in Kircudbright, the mortality appears to be only 1 in 98, and the yearly marriages 1 in 192. These proportions would imply the most unheard-of healthiness, and the most extraordinary operation of the preventive check; but there can be but little doubt that they are principally occasioned by the omissions in the registry of burials, and the celebration of a part of the marriages in other parishes.

In general, however, it appears, from registers which are supposed to be accurate, that in the country parishes the mortality is small; and that the proportions of 1 in 45, 1 in 50, and 1 in 55, are not uncommon. According to a table of the probabilities of life, calculated from the bills of mortality in the parish of Kettle by Mr. Wilkie, the expectation of an infant's life is 46.6,[768] which is very high, and the proportion which dies in the first year is only one 10th. Mr. Wilkie further adds, that from 36 parish accounts, published in the first volume, the expectation of an infant's life appears to be 40.3. But in a table which he has produced in the last volume, calculated for the whole of Scotland from Dr. Webster's survey, the expectation at birth appears to be only 31 years.[769] This, however, he thinks, must be too low, as it exceeds but little the calculations for the town of Edinburgh.

---

[766]The migrations into England from Ireland and Scotland may account for some portion of the excess of the enumerations above what is warranted by the excess of the births above the deaths.

[767]Statistical Account of Scotland, vol. i. p. 167.

[768]Statistical Account of Scotland, vol. ii. p. 407.

[769]Id. vol. xxi. p. 383.

The Scotch registers appeared to be in general so incomplete, that the returns of 99 parishes only are published in the Population Abstracts of 1801; and, if any judgment can be formed from these, they shew a very extraordinary degree of healthiness, and a very small proportion of births. The sum of the population of these parishes in 1801 was 217,873;[770] the average of burials, for five years ending in 1800, was about 3,815; and of births 4,928:[771] from which it would appear that the mortality in these parishes was only 1 in 56, and the proportion of births 1 in 44. But these proportions are so extraordinary that it is difficult to conceive that they approach near the truth. Combining them with the calculations of Mr. Wilkie, it will not appear probable that the proportion of deaths and births in Scotland should be smaller than what has been allowed for England and Wales; namely, 1 in 40 for the deaths, and 1 in 30 for the births; and it seems to be generally agreed that the proportion of births to deaths is 4 to 3.[772]

With respect to the marriages, it will be still more difficult to form a conjecture. They are registered so irregularly, that no returns of them are given in the Population Abstract. I should naturally have thought, from the Statistical Account, that the tendency to marriage in Scotland was upon the whole greater than in England; but if it be true that the births and deaths bear the same proportion to each other, and to the whole population, in both countries, the proportion of marriages cannot be very different. It should be remarked, however, that supposing the operation of the preventive check to be exactly the same in both countries, and the climates to be equally salubrious, a greater degree of want and poverty would take place in Scotland, before the same mortality was produced as in England, owing to the smaller proportion of towns and manufactories in the former country than in the latter.

From a general view of the statistical accounts the result seems clearly to be, that the condition of the lower classes of people in Scotland has been considerably improved of late years. The price of provisions has risen, but almost invariably the price of labour has risen, in a greater proportion; and it is remarked in most parishes, that more butcher's meat is consumed among the common people than formerly; that they are both better lodged and better clothed; and that their habits with respect to cleanliness are decidedly improved.

A part of this improvement is probably to be attributed to the increase of the preventive check. In some parishes a habit of later marriages is noticed; and in many places, where it is not mentioned, it may be fairly inferred from the proportion of births and marriages and other circumstances. The writer of the account of the parish of Elgin,[773] in enumerating the general causes of depopulation in Scotland, speaks of the discouragement of marriage from

[770] Population Abstracts, Parish Registers, p. 459.

[771] Id. p. 458.

[772] Statistical Account of Scotland, vol. xxi. p. 383. The comparison with England here, refers to the time of the first enumeration. There is little doubt that the mortality of Scotland has diminished, and the proportion of births to deaths increased since 1800.

[773] Vol. v. p. 1.

the union of farms, and the consequent emigration of the flower of their young men, of every class and description, very few of whom ever return. Another cause that he mentions is the discouragement to marriage from luxury; at least, he observes, till people are advanced in years, and then a puny race of children are produced. "Hence how many men of every description remain single? and how many young women of every rank are never married, who in the beginning of this century, or even so late as 1745, would have been the parents of a numerous and healthy progeny?"

In those parts of the country where the population has been rather diminished by the introduction of grazing, or an improved system. of husbandry which requires fewer hands, this effect has chiefly taken place; and I have little doubt that in estimating the decrease of the population since the end of the last, or the beginning of the present century, by the proportion of births at the different periods, they have fallen into the error which has been particularly noticed, with regard to Switzerland and France, and have in consequence made the difference greater than it really is.[774]

The general inference on this subject which I should draw from the different accounts is, that the marriages are rather later than formerly. There are however some decided exceptions. In those parishes where manufactures have been introduced, which afford employment to children as soon as they have reached their 6th or 7th year, a habit of marrying early naturally follows; and while the manufacture continues to flourish and increase, the evil arising from it is not very perceptible; though humanity must confess with a sigh, that one of the reasons why it is not so perceptible is, that room is made for fresh families by the unnatural mortality which takes place among the children so employed.

There are other parts of Scotland however, particularly the Western Isles, and some parts of the Highlands, where population has considerably increased from the subdivision of possessions; and where perhaps the marriages may be earlier than they were formerly, though not caused by the introduction of manufactures. Here the poverty which follows is but too conspicuous: In the account of Delting in Shetland,[775] it is remarked that the people marry very young, and are encouraged to do this by their landlords, who wish to have as many men on their grounds as possible, to prosecute the ling fishery; but that they generally involve themselves in debt and large families. The writer further observes, that formerly there were some old regulations called country acts, by one of which it was enacted, that no pair should marry unless possessed of 40*l.* Scots of free gear. This regulation is not now enforced. It is said that these regulations were approved and confirmed by the parliament of Scotland in the reign of Queen Mary or James VI.

In the account of Bressay Burra and Quarff in Shetland,[776] it is observed that the farms are very small, and few have a plough. The object of the proprietors is to have as many

[774]One writer takes notice of this circumstance and observes, that formerly the births seem to have borne a greater proportion to the whole population than at present. Probably, he says, more were born, and there was a greater mortality. Parish of Montquitter, vol. vi. p. 121.

[775]Vol. i. p. 385.

[776]Vol. x. p. 194.

fishermen on their lands as possible—a great obstacle to improvements in agriculture. They fish for their masters, who either give them a fee totally inadequate, or take their fish at a low rate. The writer remarks, that "in most countries the increase of population is reckoned an advantage, and justly. It is however the reverse in the present state of Shetland. The farms are split. The young men are encouraged to marry without having any stock. The consequence is poverty and distress. It is believed that there is at present in these islands double the number of people that they can properly maintain."

The writer of the account of Auchterderran,[777] in the county of Fife, says, that the meagre food of the labouring man is unequal to oppose the effects of incessant hard labour upon his constitution, and by this means his frame is worn down before the time of nature's appointment; and adds, "That people continuing voluntarily to enter upon such a hard situation by marrying, shews how far the union of the sexes and the love of independence are principles of human nature." In this observation, perhaps the love of independence had better have been changed for the love of progeny.

The island of Jura[778] appears to be absolutely overflowing with inhabitants in spite of constant and numerous emigrations. There are sometimes 50 or 60 on a farm. The writer observes, that such a swarm of inhabitants, where manufactures and many other branches of industry are unknown, are a very great load upon the proprietors, and useless to the state.

Another writer[779] is astonished at the rapid increase of population, in spite of a considerable emigration to America in 1770, and a large drain of young men during the late war. He thinks it difficult to assign adequate causes for it; and observes, that, if the population continue to increase in this manner, unless some employment be found for the people, the country will soon be unable to support them. And in the account of the parish of Callander,[780] the writer says, that the villages of this place, and other villages in similar situations, are filled with naked and starving crowds of people, who are pouring down for shelter or for bread; and then observes, that whenever the population of a town or village exceeds the industry of its inhabitants, from that moment the place must decline.

A very extraordinary instance of a tendency to rapid increase occurs in the register of the parish of Duthil,[781] in the county of Elgin; and as errors of excess are not so probable as errors of omission, it seems to be worthy of attention. The proportion of annual births to the whole population is as 1 to 12, of marriages as 1 to 55, and of deaths the same. The births are to the deaths as 70 to 15, or 4 2/3 to 1. We may suppose some inaccuracy respecting the number of deaths, which seems to err on the side of defect; but the very extraordinary proportion of the annual births, amounting to 1/12 of the whole population, seems not to be easily liable to error; and the other circumstances respecting the parish tend to confirm the

[777] Vol. i. p. 449.

[778] Vol. xii. p. 317.

[779] Parish of Lochalsh, county of Ross, vol. xi. p. 422.

[780] Vol. xi. p. 574.

[781] Vol. iv. p. 308.

statement. Out of a population of 830, there were only three bachelors, and each marriage yielded seven children. Yet with all this, the population is supposed to have decreased considerably since 1745; and it appears that this excessive tendency to increase, had been occasioned by an excessive tendency to emigrate. The writer mentions very great emigrations; and observes, that whole tribes, who enjoyed the comforts of life in a reasonable degree, had of late years emigrated from different parts of Scotland, from mere humour, and a fantastical idea of becoming their own masters and freeholders.

Such an extraordinary proportion of births, caused evidently by habits of emigration, shews the extreme difficulty of depopulating a country merely by taking away a part of its people. Take but away its industry, and the sources of its subsistence, and it is done at once.

It may be observed that in this parish the average number of children to a marriage is said to be seven, though from the proportion of annual births to annual marriages it would appear to be only 4 2/3. This difference occurs in many other parishes, from which we may conclude that the writers of these accounts very judiciously adopted some other mode of calculation, than the mere uncorrected proportion of annual births to marriages; and probably founded the results they give, either on personal inquiries, or researches into their registers, to find the number of children, which had been born to each mother in the course of her marriage.

The women of Scotland appear to be prolific. The average of 6 children to a marriage is frequent; and of 7, and even 7½, not very uncommon. One instance is very curious, as it appears as if this number was actually living to each marriage, which would of course imply, that a much greater number had been and would be born. In the parish of Nigg,[782] in the county of Kincardine, the account says, that there are 57 land families, and 405 children, which gives nearly 7 1/9 each; 42 fisher families, and 314 children, nearly 7½ each. Of the land families which have had no children there were 7; of the fishers, none. If this statement be just, I should conceive that each marriage must have yielded, or would yield, in the course of its duration, as many as 9 or 10 births.

When from any actual survey it appears, that there are about 3 living children to each marriage, or 5 persons, or, only 4½ to a house, which are very common proportions, we must not infer that the average number of births to a marriage is not much above 3. We must recollect, that all the marriages or establishments of the present year are of course without children, all of the year before have only one, all of the year before that can hardly be expected to have as many as two, and all of the fourth year will certainly, in the natural course of things, have less than three. One out of five children is a very unusually small proportion to lose in the course of ten years; and after ten years, it may be supposed that the eldest begin to leave their parents; so that if each marriage be supposed accurately to yield 5 births in the course of its duration, the families which had increased to their full complement would only have four children; and a very large proportion of those which

[782] Vol. vii. p. 194.

were in the earlier stages of increase would have less than three;[783] and consequently, taking into consideration the number of families where one of the parents may be supposed to be dead, I much doubt whether in this case a survey would give 4½ to a family. In the parish of Duthil,[784] already noticed, the number of children to a marriage is mentioned as 7, and the number of persons to a house as only 5.

The poor of Scotland are in general supported by voluntary contributions, distributed under the inspection of the minister of the parish; and it appears, upon the whole, that they have been conducted with considerable judgment. Having no claim of right to relief,[785] and the supplies, from the mode of their collection, being necessarily uncertain, and never abundant, the poor have considered them merely as a last resource in cases of extreme distress, and not as a fund on which they might safely rely, and an adequate portion of which belonged to them by the laws of their country in all difficulties.

The consequence of this is, that the common people make very considerable exertions to avoid the necessity of applying for such a scanty and precarious relief. It is observed, in many of the accounts, that they seldom fail of making a provision for sickness and for age; and, in general, the grown-up children and relations of persons, who are in danger of falling upon the parish, step forward, if they are in any way able, to prevent such a degradation, which is universally considered as a disgrace to the family.

The writers of the accounts of the different parishes frequently reprobate in very strong terms the system of English assessments for the poor, and give a decided preference to the Scotch mode of relief. In the account of Paisley,[786] though a manufacturing town, and with a numerous poor, the author still reprobates the English system, and makes an observation on this subject, in which perhaps he goes too far. He says, that, though there are in no country such large contributions for the poor as in England, yet there is no where so great a number of them; and their condition, in comparison of *the poor of other countries, is truly most miserable.*

In the account of Caerlaverock,[787] in answer to the question, How ought the poor to be supplied? it is most judiciously remarked, "that distress and poverty multiply in proportion to the funds created to relieve them; that the measures of charity ought to remain invisible, till the moment when it is necessary that they should be distributed; that in the country parishes of Scotland in general, small occasional voluntary collections are sufficient; that the legislature has no occasion to interfere to augment the stream, which is already copious

---

[783] It has been calculated that, on an average, the difference of age in the children of the same family is about two years.

[784] Vol. iv. p. 308.

[785] It has lately been stated in Parliament, that the poor-laws of Scotland are not materially different from those of England, though they have been very differently understood and executed; but, whatever may be the laws on the subject, the practice is generally as here represented; and it is the practice alone that concerns the present question.

[786] Vol. vii. p. 74.

[787] Id. vi. p. 21.

enough; in fine, that the establishment of a poor's rate would not only be unnecessary but hurtful, as it would tend to oppress the landholder, without bringing relief on the poor."

These, upon the whole, appear to be the prevailing opinions of the clergy of Scotland. There are, however, some exceptions; and the system of assessments is sometimes approved, and the establishment of it recommended. But this is not to be wondered at. In many of these parishes the experiment had never been made; and without being thoroughly aware of the principle of population from theory, or having fully seen the evils of poor-laws in practice, nothing seems, on a first view of the subject, more natural than the proposal of an assessment, to which the uncharitable, as well as the charitable, should be made to contribute according to their abilities, and which might be increased or diminished, according to the wants of the moment.

The endemic and epidemic diseases in Scotland fall chiefly, as is usual, on the poor. The scurvy is in some places extremely troublesome and inveterate; and in others it arises to a contagious leprosy, the effects of which are always dreadful, and not unfrequently mortal. One writer calls it the scourge and bane of human nature.[788] It is generally attributed to cold and wet situations, meagre, and unwholesome food, impure air from damp and crowded houses, indolent habits, and the want of attention to cleanliness.

To the same causes, in a great measure, are attributed the rheumatisms which are general, and the consumptions which are frequent among the common people. Whenever, in any place, from particular circumstances, the condition of the poor has been rendered worse, these disorders, particularly the latter, have been observed to prevail with greater force.

Low nervous fevers, and others of a more violent and fatal nature, are frequently epidemic, and sometimes take off considerable numbers; but the most fatal epidemic, since the extinction of the plague which formerly visited Scotland, is the small-pox, the returns of which are, in many places, at regular intervals; in others, irregular, but seldom at a greater distance than 7 or 8 years. Its ravages are dreadful, though in some parishes not so fatal as they were some time ago. The prejudices against inoculation are still great; and as the mode of treatment must almost necessarily be bad in small and crowded houses, and the custom of visiting each other during the disorder still subsists in many places, it may be imagined that the mortality must be considerable, and the children of the poor the principal sufferers. In some parishes of the Western Isles and the Highlands, the number of persons to a house has increased from 4½ and 5, to 6½ and 7. It is evident, that if such a considerable increase, without the proper accommodations for it, cannot generate the disease, it must give to its devastations tenfold force when it arrives.

Scotland has at all times been subject to years of scarcity, and occasionally even to dreadful famines. The years 1635, 1680; 1688, the concluding years of the 16th century, the years 1740, 1756, 1766, 1778, 1782, and 1783, are all mentioned, in different places, as years of very great sufferings from want. In the year 1680, so many families perished from

[788] Parishes of Forbes and Kearn, County of Aberdeen, vol. xi. 189.

this cause, that for six miles, in a well-inhabited extent, there was not a smoke remaining.[789] The seven years at the end of the 16th century were called the ill years. The writer of the account of the parish, of Montquhitter[790] says, that of 16 families, on a farm in that neighbourhood, 13 were extinguished; and on another, out of 169 individuals, only 3 families (the proprietors included) survived. Extensive farms, now containing a hundred souls, being entirely desolated, were converted into a sheepwalk. The inhabitants of the parish in general were diminished by death to one-half, or, as some affirm, to one-fourth of the preceding number. Until 1709 many farms were waste. In 1740, another season of scarcity occurred; and the utmost misery was felt by the poor, though it fell short of death. Many offered in vain to serve for their bread. Stout men accepted thankfully two pence a-day in full for their work. Great distress was also suffered in 1782 and 1783, but none died. "If at this critical period," the author says, "the American war had not ceased; if the copious magazines, particularly of pease, provided for the navy, had not been brought to sale, what a scene of desolation and horror would have been exhibited in this country!"

Many similar descriptions occur in different parts of the Statistical Account; but these will be sufficient to shew the nature and intensity of the distress which has been occasionally felt from want.

The year 1783 depopulated some parts of the Highlands, and is mentioned as the reason why in these places the number of people was found to have diminished since Dr. Webster's survey. Most of the small farmers in general, as might be expected, were absolutely ruined by the scarcity; those of this description in the Highlands were obliged to emigrate to the Lowlands as common labourers,[791] in search of a precarious support. In some parishes, at the time of the last survey, the effect of the ruin of the farmers, during this bad year, was still visible in their depressed condition, and the increased poverty and misery of the common people, which is a necessary consequence of it.

In the account of the parish of Grange,[792] in the county of Banff, it is observed, that the year 1783 put a stop to all improvements by green crops, and made the farmers think of nothing but raising grain. Tenants were most of them ruined. Before this period, consumptions were not near so frequent as they have been since. This may be justly attributed to the effects of the scarcity and bad victual in the year 1783, to the long inclement harvests in 1782 and 1787, in both which seasons the labourers were exposed to much cold and wet during the three months that the harvests continued; but principally to the change that has taken place in the manner of living among the lower ranks. Formerly every householder could command a draught of small beer, and killed a sheep now and then out of his own little flock; but now the case is different. The frequent want of the necessaries of life among the poor, their damp and stinking houses, and dejection of mind

[789] Parish of Duthil, vol. iv. p. 305.

[790] Vol. vi. p. 121.

[791] Parish of Kincardine, County of Ross, vol. iii. p. 505.

[792] Vol. ix. P. 550.

among the middling classes, appear to be the principal causes of the prevailing distempers and mortality of this parish. Young people are cut off by consumptions, and the more advanced by dropsies and nervous fevers.

The state of this parish, which, though there are others like it, may be considered as an exception to the average state of Scotland, was, without doubt, occasioned by the ruin of the tenants; and the effect is not to be wondered at; as no greater evil can easily happen to a country, than the loss of agricultural stock and capital.

We may observe that the diseases of this parish are said to have increased, in consequence of the scarcity and bad victual of 1783. The same circumstance is noticed in many other parishes; and it is remarked, that though few people died of absolute famine, yet that mortal diseases almost universally followed.

It is remarked also, in some parishes, that the number of the births and marriages is affected by years of scarcity and plenty.

Of the parish of Dingwall,[793] in the county of Ross, it is observed that, after the scarcity of 1783, the births were 16 below the average, and 14 below the lowest number of late years. The year 1787 was a year of plenty; and the following year the births increased in a similar proportion, and were 17 above the average, and 11 above the highest of the other years.

In the account of Dunrossness,[794] in Orkney, the writer says that the annual number of marriages depends much on the seasons. In good years they may amount to thirty or upwards; but, when crops fail, will hardly come up to the half of that number.

The whole increase of Scotland, since the time of Dr. Webster's survey in 1755, is about 260,000,[795] for which a proportionate provision has been made in the improved state of agriculture and manufactures, and in the increased cultivation of potatoes, which in some places form two-thirds of the diet of the common people. It has been calculated that the half of the surplus of births in Scotland is drawn off in emigrations; and it cannot be doubted that this drain tends greatly to relieve the country, and to improve the condition of those which remain. Scotland is certainly still overpeopled, but not so much as it was a century or half a century ago, when it contained fewer inhabitants.

The details of the population of Ireland are but little known. I shall only observe therefore, that the extended use of potatoes has allowed of a very rapid increase of it during the last century. But the cheapness of this nourishing root, and the small piece of ground which, under this kind of cultivation, will in average years produce the food for a family, joined to the ignorance and depressed state of the people, which have prompted them to follow their inclinations with no other prospect than an immediate bare subsistence,

---

[793] Vol. iii. p. i.

[794] Vol. vii. p. 391.

[795] According to the returns in the enumeration of 1800, the whole population of Scotland was above 1,590,000, and therefore the increase up to that time was above 320,000. In 1810 the population was 1,805,688; and in 1820, 2,093,456.

have encouraged marriage to such a degree, that the population is pushed much beyond the industry and present resources of the country; and the consequence naturally is, that the lower classes of people are in the most impoverished and miserable state. The checks to the population are of course chiefly of the positive kind, and arise from the diseases occasioned by squalid poverty, by damp and wretched cabins, by bad and insufficient clothing, and occasional want. To these positive checks have, of late years, been added the vice and misery of intestine commotion, of civil war, and of martial law.

According to the late enumeration in 1821, the population of Ireland amounted to 6,801,827, and in 1695 it was estimated only at 1,034,000. If these numbers be correct it affords an example of continued increase for 125 years together, at such a rate as to double the population in about 45 years—a more rapid increase than has probably taken place in any other country of Europe, during the same length of time.

In the peculiar circumstances of Ireland, it would be very interesting to know the average mortality, and the proportions of births and marriages to the population. But unfortunately no correct parochial registers have been kept, and the information, however much to be desired, is unattainable.

3.5.11 Book II, Chapter XI: On the Fruitfulness of Marriages.

It would be extremely desirable to be able to deduce from the registers of births, deaths and marriages in different countries, and the actual population with the rate of increase, the real prolifickness of marriages, and the true proportion of the born which lives to marry. Perhaps the problem may not be capable of an accurate solution; but we shall make some approximation towards it, and be able to account for some of the difficulties which appear in many registers, if we attend to the following considerations.

It should be premised, however, that in the registers of most countries there is reason to believe that the omissions in the births and deaths are greater than in the marriages; and consequently, that the proportion of marriages is almost always given too great. In the enumerations which have lately taken place in this country, while it is supposed with reason that the registry of marriages is nearly correct, it is known with certainty that there are very great omissions in the births and deaths; and it is probable that similar omissions, though not perhaps to the same extent, prevail in other countries.

If we suppose a country where the population is stationary, where there are no emigrations, immigrations, or illegitimate children, and where the registers of births deaths and marriages are accurate, and continue always in the same proportion to the population, then the proportion of the annual births to the annual marriages will express the number of children born to each marriage, including second and third marriages, and when corrected for second and third marriages, it will also express the proportion of the born which lives to marry, once or oftener; while the annual mortality will accurately express the expectation of life.

But if the population be either increasing or decreasing, and the births, deaths and marriages increasing or decreasing in the same ratio, such a movement will necessarily disturb all the proportions, because the events which are contemporary in the registers are not contemporary in the order of nature, and an increase or decrease must have been taking place in the interval.

In the first place, the births of any year cannot in the order of nature have come from the contemporary marriages, but must have been derived principally from the marriages of preceding years.

To form a judgment then of the prolifickness of marriages taken as they occur, including second and third marriages, let us cut off a certain period of the registers of any country (30 years for instance) and inquire what is the number of births which has been produced by all the marriages included in the period cut off. It is evident, that with the marriages at the beginning of the period will be arranged a number of births proceeding from marriages not included in the period; and at the end, a number of births produced by the marriages included in the period will be found arranged with the marriages of a succeeding period. Now, if we could subtract the former number, and add the latter, we should obtain exactly all the births produced by the marriages of the period, and of course the real prolifickness of those marriages. If the population be stationary, the number of births to be added would exactly equal the number to be subtracted, and the proportion of births to marriages, as found in the registers, would exactly represent the real prolifickness of marriages. But if the population be either increasing or decreasing, the number to be added would never be equal to the number to be subtracted, and the proportion of births to marriages in the registers would never truly represent the prolifickness of marriages. In an increasing population the number to be added would evidently be greater than the number to be subtracted, and of course the proportion of births to marriages as found in the registers would always be too small to represent the true prolifickness of marriages. And the contrary effect would take place in a decreasing population. The question therefore is, what we are to add, and what to subtract, when the births and deaths are not equal.

The average proportion of births to marriages in Europe is about 4 to 1. Let us suppose, for the sake of illustration, that each marriage yields four children, one every other year.[796] In this case it is evident that, wherever we begin the period in the registers, the marriages of the preceding eight years will only have produced half of their births, and the other half will be arranged with the marriages included in the period, and ought to be subtracted from them. In the same spanner the marriages of the last eight years of the period will only have produced half of their births, and the other half ought to be added. But half of the births of any eight years may be considered as nearly equal to all the births of the succeeding 3¾ years. In instances of the most rapid increase it will rather exceed the births of the next 3½ years, and, in cases of slow increase, approach towards the births of the next 4 years. The

[796] In the statistical account of Scotland it is said, that the average distance between the children of the same family has been calculated to be about two years.

mean therefore may be taken at 3¾ years.[797] Consequently, if we subtract the births of the first 3¾ years of the period, and add the births of the 3¾ years subsequent to the period, we shall have a number of births nearly equal to the births produced by all the marriages included in the period, and of course the prolifickness of these marriages. But if the population of a country be increasing regularly, and the births, deaths and marriages continue always to bear the same proportion to each other, and to the whole population, it is evident that all the births of any period will bear the same proportion to all the births of any other period of the same extent, taken a certain number of years later, as the births of any single year, or an average of five years, to the births of a single year, or an average of five years, taken the same number of years later; and the same will be true with regard to the marriages. And consequently, to estimate the prolifickness of marriages, we have only to compare the marriages of the present year, or average of five years, with the births of a subsequent year, or average of five years, taken 3¾ years later.

We have supposed, in the present instance, that each marriage yields four births; but the average proportion of births to marriages in Europe is 4 to 1;[798] and as the population of Europe is known to be increasing at present, the prolifickness of marriages must be greater than 4. If, allowing for this circumstance, we take the distance of 4 years instead of 3¾ years, we may not be far from the truth. And though undoubtedly the period will differ in different countries, yet it will not differ so much as we might at first imagine; because in countries where the marriages are more prolific, the births generally follow at shorter intervals, and where they are less prolific, at longer intervals; and with different degrees of prolifickness, the length of the period might still remain the same.[799]

It will follow from these observations, that the more rapid is the increase of population, the more will the real prolifickness of marriages exceed the proportion of births to marriages in the registers.

The rule which has been here laid down attempts to estimate the prolifickness of marriages taken as they occur; but this prolifickness should be carefully distinguished from the prolifickness of first marriages or of married women, and still more from the natural prolifickness of women in general taken at the most favourable age. It is probable, that the natural prolifickness of women is nearly the same in most parts of the world; but the prolifickness of marriages is liable to be affected by a variety of circumstances peculiar to each country, and particularly by the number of late marriages. In all countries the second and third marriages alone form a most important consideration, and materially influence the

[797] According to the rate of increase which has lately been taking place in England (1802,) the period by calculation would be about 3¾ years.

[798] The true proportion will be greater, if, as before stated, there is reason to believe that in all registers the omissions in the births and deaths are more numerous than in the marriages.

[799] In places where there are many migrations of people, the calculations will of course be disturbed. In towns, particularly, where there is a frequent change of inhabitants, and where it often happens that the marriages of the people in the neighbouring country are celebrated, the inferences from the proportion of births to marriages are not to be depended on.

average proportions. According to Sussmilch, in all Pomerania, from 1748 to 1756 both included, the number of persons who married were 56,956, and of these 10,586 were widows and widowers.[800] According to Busching, in Prussia and Silesia, for the year 1781, out of 29,308 persons who married, 4,841, were widows and widowers,[801] and consequently the proportion of marriages will be given full one sixth too much. In estimating the prolifickness of married women, the number of illegitimate births[802] would tend, though in a slight degree, to counterbalance the overplus of marriages; and as it is found that the number of widowers who marry again, is greater than the number of widows, the whole of the correction should not on this account be applied; but in estimating the proportion of the born which lives to marry from a comparison of the marriages with the births or deaths, which is what we are now about to proceed to, the whole of this correction is always necessary.

It is obvious, in the second place, that the marriages of any year can never be contemporary with the births from which they have resulted, but must always be at such a distance from them as is equal to the average age of marriage. If the population be increasing, the marriages of the present year have resulted from a smaller number of births than the births of the present year, and of course the marriages, compared with the contemporary births, will always be too few to represent the proportion of the born which lives to marry; and the contrary will take place if the population be decreasing; and, to find this proportion, we must compare the marriages of any year with the births of a previous year at the distance of the average age of marriage.

But on account of the distance of this period, it may be often more convenient, though it is not essentially so correct, to compare the marriages with the contemporary deaths. The average age of marriage will almost always be much nearer to the average age of death than marriage is to birth; and consequently the annual marriages compared with the contemporary annual deaths will much more nearly represent the true proportion of the born living to marry, than the marriages compared with the births.[803] The marriages compared with the

[800] Göttliche Ordnung, vol. i. tables, p. 98.

[801] Sussmilch, vol. iii. tables, p. 95.

[802] In France, before the revolution, the proportion of illegitimate births was 1/47 of the whole number. Probably it is less in this country.

[803] Dr. Price very justly says (Observ. on Revers. Pay. vol. i. p. 269. 4th. edit.) "that the general effect of an increase while it is going on in a country is to render the proportion of persons marrying annually, to the annual deaths *greater* and to the annual births *less* than the true proportion marrying out of any given number born. This proportion generally lies between the other two proportions, but always nearest the first." In these observations I entirely agree with him, but in a note to this passage he appears to me to fall into an error. He says, that if the prolifickness of marriages be increased (the *probabilities of life* and *the encouragement to marriage* remaining the same) both the annual births and burials would increase in proportion to the annual weddings. That the proportion of annual births would increase is certainly true; and I here acknowledge my error in differing from Dr. Price on this point in my last edition; but I still think that the proportion of burials to weddings would not necessarily increase under the circumstances here supposed.

births, after a proper allowance has been made for second and third marriages, can never represent the true proportion of the born living to marry, unless when the population is absolutely stationary; but although the population be increasing or decreasing, the average age of marriage may still be equal to the average of death; and in this case the marriages in the registers compared with the contemporary deaths, (after the correction for second or third marriages,) will nearly represent the true proportion of the born living to marry.[804] Generally, however, when an increase of population is going forwards, the average age of marriage is less than the average of death, and then the proportion of marriages, compared with the contemporary deaths, will be too great to represent the true proportion of the born living to marry; and, to find this proportion, we must compare the marriages of any particular year with the deaths of a subsequent year at such a distance from it in the registers, as is equal to the difference between the average age of marriage and the average age of death.

There is no necessary connection between the average age of marriage and the average age of death. In a country, the resources of which will allow of a rapid increase of population, the expectation of life or the average age of death may be extremely high, and yet the age of marriage be very early; and the marriages then, compared with the contemporary deaths in the registers, would (even after the correction for second and third marriages) be very much too great to represent the true proportion of the born living to marry. In such a country we might suppose the average age of death to be 40, and the age of marriage only 20; and in this case, which however would be a rare one, the distance between marriage and death would be the same as between birth and marriage.

If we apply these observations to registers in general, though we shall seldom be able to obtain the true proportion of the born living to marry on account of the proportions of births, deaths, and marriages not remaining the same, and of our not knowing the average age of marriage, yet we may draw many useful inferences from the information which they contain, and reconcile some apparent contradictions; and it will generally be found that, in those countries where the marriages bear a very large proportion to the deaths, we shall see reason to believe that the age of marriage is much earlier than the average age of death.

In the Russian table for the year 1799, produced by Mr. Tooke, and referred to, p. 317, the proportion of marriages to deaths appeared to be as 100 to 210. When corrected for second and third marriages, by subtracting one sixth from the marriages, it will be as 100 to 252. From which it would seem to follow, that out of 252 births 200 of them had lived to marry; but we cannot conceive any country to be so healthy as that 200 out of 252 should live to marry. If however we suppose, what seems to be probable, that the age of marriage in Russia is 15 years earlier than the expectation of life or the average age of

[804]The reader will be aware that, as all the born must die, deaths, may in some cases be taken as synonymous with births. If we had the deaths registered of all the births which had taken place in a country during a certain period, distinguishing the married from the unmarried, it is evident that the number of those who died married, compared with the whole number of deaths, would accurately express the proportion of the births which had lived to marry.

death, then, in order to find the proportion which lives to marry, we must compare the marriages of the present year with the deaths 15 years later. Supposing the births to deaths to be (as stated p. 317) 183 to 100, and the mortality 1 in 50, the yearly increase will be about 1/60 of the population; and consequently in 15 years the deaths will have increased a little above .28; and the result will be, that the marriages, compared with the deaths 15 years later will be as 100 to 322. Out of 322 births it will appear that 200 live to marry, which, from the known healthiness of children in Russia, and the early age of marriage, is a possible proportion. The proportion of marriages to births, being as 100 to 385, the prolifickness of marriages, according to the rule laid down, will be as 100 to 411; or each marriage will on an average, including second and third marriages, produce 4.11 births.

The lists given in the earlier part of the chapter on Russia are probably not correct. It is suspected with reason, that there are considerable omissions both in the births and deaths, but particularly in the deaths; and consequently the proportion of marriages is given too great. There may also be a further reason for this large proportion of marriages in Russia. The Empress Catherine, in her instructions for a new code of laws, notices a custom prevalent among the peasants, of parents obliging their sons, while actually children, to marry full-grown women, in order to save the expense of buying female slaves. These women, it is said, generally become the mistresses of the father; and the custom is particularly reprobated by the Empress as prejudicial to population. This practice would naturally occasion a more than usual number of second and third marriages, and of course more than usually increase the proportion of marriages to births in the registers.

In the Transactions of the Society at Philadelphia (vol. iii. No. vii. p. 25,) there is a paper by Mr. Barton, entitled *Observations on the Probability of Life in the United States,* in which it appears, that the proportion of marriages to births is as 1 to 4½. He mentions indeed 6½, but his numbers give only 4½. As however this proportion was taken principally from towns, it is probable that the births are given too low; and I think we may very safely take as many as five for the average of towns and country. According to the same authority the mortality is about 1 in 45; and if the population doubles every 25 years, the births would be about 1 in 20. The proportion of marriages to deaths would on these suppositions be as 1 to 2 2/9; and, corrected for second and third marriages, as 1 to 2.7 nearly. But we cannot suppose, that out of 27 births 20 should live to marry. If however the age of marriage be ten years earlier than the mean age of death, which is highly probable, we must compare the marriages of the present year with the deaths ten years later, in order to obtain the true proportion of the born which lives to marry. According to the progress of population here stated, the increase of the deaths in ten years would be a little above .3, and the result will be, that 200 out of 351, or about 20 out of 35, instead of twenty out of 27, will live to marry.[805] The marriages compared with the births 4 years later, according to the rule laid down, will in this case give 5.58 for the prolifickness of marriages. The calculations of Mr. Barton respecting the age to which half of the born live, cannot possibly be applicable to America in general. The registers, on which they are founded, are taken from Philidelphia and one or two small towns and villages, which do not appear to be so healthy as the moderate towns of Europe, and therefore can form no criterion for the country in general.

In England the average proportion of marriages to births appears of late years to have been about 100 to 350. If we add 1/7 to the births instead of 1/6, which in the chapter on *the Checks to Population in England,* I conjectured might be nearly the amount of the omissions in the births and deaths, this will allow for the circumstance of illegitimate births; and the marriages will then be to the births as 1 to 4, to the deaths as 1 to 3.[806] Corrected for second and third marriages, the proportion of marriages to deaths will be as 1 to 3.6. Supposing the age of marriage in England about 7 years earlier than the mean age of death, the increase in these 7 years, according to the present progress of population of 1/120 yearly, would be .06, and the proportion living to marry would be 200 out of 381, or rather more than half.[807] The marriages compared with the births four years later will give 4.136 for the prolifickness of marriages.

These instances will be sufficient to shew the mode of applying the rules which have been given, in order to form a judgment, from registers, of the prolifickness of marriages, and the proportion of the born which lives to marry; but it must still be remembered that they are only approximations, and intended rather to explain apparent difficulties, than to obtain results which can be depended upon as correct.

---

[805] If the proportions mentioned by Mr. Barton be just, the expectation of life in America is considerably less than in Russia, which is the reason that I have taken only ten years for the difference between the age of marriage and the age of death, instead of fifteen years, as in Russia. According to the mode adopted by Dr. Price, (vol. i. p. 272,) of estimating the expectation of life in countries the population of which is increasing, this expectation in Russia would be about 38, (births 1/26, deaths 1/50, mean 1/38), and supposing the age of marriage to be 23, the difference would be 15.

Since this was written, I have seen reason to believe, from some calculations of Mr. Milne, actuary to the Sun Life Assurance Society, that Dr. Price's mode of estimating the expectation of life in countries that are increasing is by no means correct, and that the true expectation of life in such countries lies very much nearer the proportion of the annual mortality, than a mean between the annual mortality and the proportion of annual births; but I retain the mean proportion in the calculations of this chapter, because I find that this mean expresses more nearly the period when the deaths will equal the present births, or accord with the present marriages, than the distance of the expectation of life. In a progressive country, where the annual births considerably exceed the annual deaths, the period at which the annual deaths will equal the present annual births is less distant than the expectation of life.

[806] This applies to the state of population before 1800.

[807] Births 1/30, deaths 1/40, mean 1/35; and on the supposition that the age of marriage is 28, the difference would be 7.

It will be observed how very important the correction for second and third marriages is. Supposing each marriage to yield four births, and the births and deaths to be equal, it would at first appear necessary that, in order to produce this effect, exactly half of the born should live to marry; but if, on account of the second and third marriages, we subtract 1/6 from the marriages, and then compare them with the deaths, the proportion will be as 1 to 4 4/5; and it will appear that, instead of one half, it will only be necessary that 2 children out of 4 4/5 should live to marry. Upon the same principle, if the births were to the marriages as 4 to 1, and exactly half of the born live to marry, it might be supposed at first that the population would be stationary; but if we subtract 1/6 from the marriages; and then take the proportion of deaths to marriages as 4 to 1, we shall find that the deaths in the registers, compared with the marriages, would only be as 3 1/3 to 1; and the births would be to the deaths as 4 to 3 1/3, or 12 to 10, which is a tolerably fast rate of increase.

It should be further observed, that as a much greater number of widowers marry again than of widows, if we wish to know the proportion of males which lives to marry, we must subtract full 1/5 from the marriages instead of 1/6.[808] According to this correction, if each marriage yielded 4 births, it would only be necessary that two male children out of 5 should live to marry in order to keep up the population; and if each marriage yielded 5 births, less than one third would be necessary for this purpose; and so for the other calculations. In estimating the proportion of males living to marry, some allowance ought also to be made for the greater proportion of male births.

Three causes appear to operate in producing an excess of the births above the deaths: 1. the prolifickness of marriages; 2. the proportion of the born which lives to marry; and 3. the earliness of these marriages compared with the expectation of life, or the shortness of a generation by marriage and birth, compared with the passing away of a generation by death. This latter cause Dr. Price seems to have omitted to consider. For though he very justly says that the rate of increase, supposing the prolific powers the same, depends upon the encouragement to marriage, and the expectation of a child just born; yet in explaining himself, he seems to consider an increase in the expectation of life, merely as it affects the increase of the number of persons who reach maturity and marry, and not as it affects, besides, the distance between the age of marriage and the age of death. But it is evident that, if there be any principle of increase, that is, if one marriage in the present generation yields more than one in the next, including second and third marriages, the quicker these generations are repeated, compared with the passing away of a generation by death, the more rapid will be the increase.

A favourable change in either of these three causes, the other two remaining the same, will clearly produce an effect upon population, and occasion a greater excess of the births above the deaths in the registers. With regard to the two first causes, though an increase in

[808]Of 28,473 marriages in Pomerania, 5,964 of the men were widowers. Sussmilch, vol. i. tables, p. 98. And according to Busching, of 14,759 marriages in Prussia and Silesia, 3,071 of the men were widowers. Sussmilcb, vol iii. tables, p. 95. Muret calculates that 100 men generally marry 110 women. Mémoires par la Société Economique de Berne. Année 1766, premiere partie, page 30.

either of them will produce the same kind of effect on the proportion of births to deaths, yet their effects on the proportion of marriages to births will be in opposite directions. The greater is the prolifickness of marriages, the greater will be the proportion of births to marriages; and the greater is the number of the born which lives to be married, the less will be the proportion of births to marriages.[809] Consequently, if within certain limits, the prolificness of marriages and the number of the born living to marry increase at the same time, the proportion of births to marriages in the registers may still remain unaltered. And this is the reason why the registers of different countries, with respect to births and marriages, are often found the same under very different rates of increase.

The proportion of births to marriages, indeed, forms no criterion whatever, by which to judge of the rate of increase. The population of a country may be stationary or declining with a proportion of 5 to 1, and may be increasing with some rapidity with a proportion of 4 to 1. But given the rate of increase, which may be obtained from other sources, it is clearly desirable to find in the registers a small rather than a large proportion of births to marriages; because the smaller this proportion is, the greater must be the proportion of the born which lives to marry, and of course the more healthy must be the country.

Crome[810] observes that, when the marriages of a country yield less than 4 births, the population is in a very precarious state; and he estimates the prolifickness of marriages by the proportion of yearly births to marriages. If this observation were just, the population of many countries of Europe would be in a precarious state, as in many countries the proportion of births to marriages in the registers is rather below than above 4 to 1. It has been shown in what manner this proportion in the registers should be corrected, in order to make it a just representation of the prolifickness of marriages; and if a large part of the born live to marry, and the age of marriage be considerably earlier than the expectation of life, such a proportion in the registers is by no means inconsistent with a rapid increase. In Russia it has appeared that the proportion of births to marriages is less than 4 to 1; and yet its population increases faster than that of any other nation in Europe. In England the

[809] Dr. Price himself has insisted strongly upon this, (vol. i. p. 270. 4th edit.) and yet he says (p. 275) that healthfulness and prolifickness are probably causes of increase seldom separated, and refers to registers of births and weddings as a proof of it. But though these causes may undoubtedly exist together, yet if Dr. Price's reasoning be just, such co-existence cannot possibly be inferred from the lists of births and weddings. Indeed the two countries, Sweden and France, to the registers of which he refers as showing the prolifickness of their marriages, are known to be by no means remarkably healthy; and the registers of towns to which be alludes, though they may show, as be intends, a want of prolifickness, yet, according, to his previous reasoning, show at the same time great healthiness, and therefore ought not to be produced as a proof of the absence of both. The general fact that Dr. Price wishes to establish may still remain true, that country situations are both more healthy and more prolific than towns: but this fact certainly cannot be inferred merely from lists of births and marriages. With regard to the different countries of Europe, it will generally be found, that those are the most healthy which are the least prolific, and those the most prolific which are the least healthy. The earlier age of marriage in unhealthy countries is the obvious reason of this fact.

[810] Ueber die Bevölkerung der Europais. Staat. p. 91.

population increases more rapidly than in France; and yet in England the proportion of births to marriages, when allowance has been made for omissions, is about 4 to 1; in France 4 4/5 to 1. To occasion so rapid a progress as that which has taken place in America, it will indeed be necessary that all the causes of increase should be called into action; and if the prolifickness of marriages be very great, the proportion of births to marriages will certainly be above 4 to 1: but in all ordinary cases, where the whole power of procreation has not room to expand itself, it is surely better that the actual increase should arise from that degree of healthiness in the early stages of life which causes a great proportion of the born to live to maturity and to marry, than from a great degree of prolifickness accompanied by a great mortality. And consequently in all ordinary cases a proportion of births to marriages as 4, or less than 4, to 1 cannot be considered as an unfavourable sign.

It should be observed that it does not follow that the marriages of a country are early, or that the preventive check to population does not prevail, because the greater part of the born lives to marry. In such countries as Norway and Switzerland, where half of the born live to above 40, it is evident that, though rather more than half live to marry, a large portion of the people between the ages of 20 and 40 would be living in an unmarried state, and the preventive check would appear to prevail to a great degree. In England it is probable that half of the born live to above 35;[811] and though rather more than half live to marry, the preventive check might prevail considerably (as we know it does), though not to the same extent as in Norway and Switzerland.

The preventive check is perhaps best measured by the smallness of the proportion of yearly births to the whole population. The proportion of yearly marriages to the population is only a just criterion in countries similarly circumstanced, but is incorrect where there is a difference in the prolifickness of marriages or in the proportion of the population under the age of puberty, and in the rate of increase. If all the marriages of a country, be they few or many, take place young, and be consequently prolific, it is evident that, to produce the same proportion of births, a smaller proportion of marriages will be necessary; or with the same proportion of marriages a greater proportion of births will be produced. This latter case seems to be applicable to France, where both the births and deaths are greater than in Sweden, though the proportion of marriages is nearly the same, or rather less. And when, in two countries compared, one of them has a much greater part of its population under the age of puberty than the other, it is evident that any general proportion of the yearly marriages to the whole population will not imply the same operation of the preventive check among those of a marriageable age.

It is, in part, the small proportion of the population under the age of puberty, as well as the influx of strangers, that occasions in towns a greater proportion of marriages than in the country, although there can be little doubt that the preventive check prevails most in towns. The converse of this will also be true; and consequently in such a country as America,

[811] At present (1825), and for the last ten, or even twenty years, there is reason to believe that half of the born live to 45 years.

where half of the population is under sixteen, the proportion of yearly marriages will not accurately express how little the preventive check really operates.

But on the supposition of nearly the same natural prolifickness in the women of most countries, the smallness of the proportion of births will generally indicate, with tolerable exactness, the degree in which the preventive check prevails, whether arising principally from late, and consequently unprolific, marriages, or from a large proportion of the population above the age of puberty dying unmarried.

That the reader may see at once the rate of increase, and the period of doubling, which would result from any observed proportion of births to deaths, and of these to the whole population, I subjoin two tables from Sussmilch, calculated by Euler, which I believe are very correct. The first is confined to the supposition of a mortality of 1 in 36, and therefore can only be applied to countries where such a mortality is known to take place. The other is general, depending solely upon the proportion which the excess of the births above the burials bears to the whole population, and therefore may be applied universally to all countries, whatever may be the degree of their mortality. I have now also (1825) added a third table as convenient on account of the custom of decennial enumerations in this and some other countries. It is calculated by the Rev. B. Bridge, of Peter House, Cambridge, and shows the rate of increase, or period of doubling, from the observed per-centage increase of any ten years, supposing such rate of increase to continue.

It will be observed that, when the proportion between the births and burials is given, the period of doubling will be shorter, the greater the mortality; because the births as well as deaths are increased by this supposition, and they both bear a greater proportion to the whole population than if the mortality were smaller, and there were a greater number of people in advanced life.

The mortality of Russia, according to Mr. Tooke, is 1 in 58, and the proportion of births 1 in 26. Allowing for the omissions in the burials, if we assume the mortality to be 1 in 52, then the births will be to the deaths as 2 to 1, and the proportion which the excess of births bears to the whole population will be 1/52.[812]

According to Table II. the period of doubling will, in this case, be about 36 years. But if we were to keep the proportion of births to deaths as 2 to 1, and suppose a mortality of 1 in 36, as in Table I., the excess of births above the burials would be 1/36 of the whole population, and the period of doubling would be only 25 years.

[812]The proportions here mentioned are different from those which have been taken from the additional table in Mr. Tooke's second edition; but they are assumed here as more easily and clearly illustrating the subject.

TABLE I.

	If the proportion of deaths to births be as	Then the excess of the births will be	The proportion of the excess of the births, to the whole population, will be	And therefore the period of doubling will be
	11	277	1/360	250 years
	12	555	1/180	125
	13	833	1/120	83½
	14	1110	1/90	62¾
	15	1388	1/72	50¼
10:	16	1666	1/60	42
	17	1943	1/51	35¾
	18	2221	1/45	31⅔
	19	2499	1/40	28
	20	2777	1/36	25 3/10
	22	3332	1/30	21⅙
	25	4165	1/24	17
	30	5554	1/18	12⅘

TABLE II.

The proportion of the excess of births above the deaths to the whole of the living.	Periods of doubling in years and ten thousandth parts.	The proportion of the excess of births above the deaths to the whole of the living.	Periods of doubling in years and ten thousandth parts.
10	7.2722	110	76.5923
11	7.9659	120	83.5230
12	8.6595	130	90.4554
13	9.3530	140	97.3868
1: 14	10.0465	1: 150	104.3183
15	10.7400	160	111.2598
16	111.2598	170	118.1813
17	12.1266	180	125.1128
18	12.8200	190	138.9757
19	13.5133	200	138.9757
21	14.9000	210	145.9072
22	15.5932	220	152.8387
23	16.2864	230	159.7702
24	16.9797	240	166.7017
25	17.6729	250	173.6332
26	18.3662	260	180.5647
27	19.0594	270	187.4961
28	19.7527	280	194.4275
29	20.4458	290	201.3590
30	21.1391	300	208.2905
32	22.5255	310	215.2220
34	23.9119	320	222.1535
36	25.2983	330	229.0850
38	26.6847	340	236.0164
40	28.0711	350	242.9479
42	29.4574	360	249.8794
44	30.8438	370	256.8109
46	32.2302	380	263.7425
48	43.6161	390	270.6740
50	35.0029	400	277.6055
55	38.4687	410	284.5370
60	41.9345	420	291.4685

(continued)

TABLE II. (continued)

The proportion of the excess of births above the deaths to the whole of the living.	Periods of doubling in years and ten thousandth parts.	The proportion of the excess of births above the deaths to the whole of the living.	Periods of doubling in years and ten thousandth parts.
60	41.9345	420	291.4685
65	45.4003	430	298.4000
70	48.8661	440	305.3314
75	52.3318	450	312.2629
80	55.7977	460	319.1943
85	59.2634	470	326.1258
90	62.7292	480	333.0573
95	66.1950	490	339.9888
100	69.6607	500	346.9202
		1: 1000	693.49.

TABLE III.

I. Per centage increase in ten years.	II. Period of doubling.	I. Per centage increase in ten years.	II. Period of doubling.	I. Per centage increase in ten years.	II. Period of doubling.
	Years		Years		Years
1	696.60	16	46.70	30.5	26.03
1.5	465.55	16.5	45.38	31	25.67
2	350.02	17	44.14	31.5	25.31
2.5	280.70	17.5	42.98	32	24.96
3	234.49	18	41.87	32.5	24.63
3.5	201.48	18.5	40.83	33	24.30
4	176.73	19	39.84	33.5	23.99
4.5	157.47	19.5	38.91	34	23.68
5	142.06	20	38.01	34.5	23.38
5.5	129.46			35	23.09
6	118.95	20.5	37.17	35.5	22.81
6.5	110.06	21	36.36	36	22.54
7	102.44	21.5	35.59	36.5	22.27

(continued)

7.5	95.84	22	34.85	37	22.01
8	90.06	22.5	34.15	37.5	21.76
8.5	84.96	23	33.48	38	21.52
9	80.43	23.5	32.83	38.5	21.28
9.5	76.37	24	32.22	39	21.04
10	72.72	24.5	31.63	39.5	20.82
		25	31.06	40	20.61
10.5	69.42	25.5	30.51		
11	66.41	26	29.99	41	20.17
11.5	63.67	26.5	29.48	42	19.76
12	61.12	27	28.99	43	19.37
12.5	58.06	27.5	28.53	44	19.00
13	56.71	28	28.07	45	18.65
13.5	54.73	28.5	27.65	46	18.31
14	52.90	29	27.22	47	17.99
14.5	51.19	29.5	26.81	48	17.68
15	49.59	30	26.41	49	17.38
15.5	48.10			50	17.06

3.5.12 Book II, Chapter XII: Effects of Epidemics on Registers of Births, Deaths, and Marriages.

It appears clearly from the very valuable tables of mortality, which Sussmilch has collected, and which include periods of 50 or 60 years, that all the countries of Europe are subject to periodical sickly seasons, which check their increase; and very few are exempt from those great and wasting plagues which, once or twice perhaps in a century, sweep off the third or fourth part of their inhabitants. The way in which these periods of mortality affect all the general proportions of births, deaths, and marriages, is strikingly illustrated in the tables for Prussia and Lithuania, from the year 1692 to the year 1757.[813]

The table, from which this is copied, contains the marriages, births and deaths for every particular year during the whole period; but to bring it into a smaller compass, I have retained only the general average drawn from the shorter periods of five and four years, except where the numbers for the individual years presented any fact worthy of particular observation. The year 1711, immediately succeeding the great plague, is not included by Sussmilch in any general average; but he has given the particular numbers, and if they be accurate they shew the very sudden and prodigious effect of a great mortality on the number of marriages.

[813] Sussmilch, Göttliche Ordnung, vol. i. table xxi. p. 83 of the tables.

Sussmilch calculates that above one third of the people was destroyed by the plague; and yet, notwithstanding this great diminution of the population, it will appear by a reference to the table, that the number of marriages in the year 1711 was very nearly double the average of the six years preceding the plague.[814] To produce this effect, we may suppose that almost all who were at the age of puberty were induced, from the demand for labour and the number of vacant employments, immediately to marry. This immense number of marriages in the year could not possibly be accompanied by a great proportional number of births, because we cannot suppose that the new marriages could each yield more than one birth in the year, and the rest must come from the marriages which had continued unbroken through the plague. We cannot therefore be surprised that the proportion of births to marriages in this year should be only 2.7 to 1, or 27 to 10. But though the proportion of births to marriages could not be great; yet, on account of the extraordinary number of marriages, the absolute number of births must be great; and as the number of deaths would naturally be small, the proportion of births to deaths is prodigious, being 320 to 100; an excess of births as great, perhaps, as has ever been known in America.

In the next year, 1712, the number of marriages must of course diminish exceedingly; because, nearly all who were at the age of puberty having married the year before, the marriages of this year would be supplied principally by those who had arrived at this age, subsequent to the plague. Still, however, as all who were marriageable had not probably married the year before, the number of marriages in the year 1712 is great in proportion to the population; and, though not much more than half of the number which took place during the preceding year, is greater than the average number in the last period before the plague. The proportion of births to marriages in 1712, though greater than in the preceding year, on account of the smaller comparative number of marriages, is, with reference to other countries, not great, being as 3.6 to 1, or 3.6 to 10. But the proportion of births to deaths, though less than in the preceding year, when so very large a proportion of the people married, is, with reference to other countries, still unusually great, being as 220 to 100; an excess of births, which, calculated on a mortality of 1 in 36, would double the population of a country (according to Table I. page 496) in 21 1/8 years.

From this period the number of annual marriages begins to be regulated by the diminished population, and of course to sink considerably below the average number of marriages before the plague, depending principally on the number of persons rising annually to a marriageable state. In the year 1720, about nine or ten years after the plague,

---

[814]The number of people before the plague, according to Sussmilch's calculations, (vol. i. ch. ix. sect. 173,) was 570,000, from which if we subtract 247,733, the number dying in the plague, the remainder, 322,267, will be the population after the plague; which, divided by the number of marriages and the number of births for the year 1711, makes the marriages about one twenty-sixth part of the population, and the births about one tenth part. Such extraordinary proportions could only occur in any country in an individual year. If they were to continue, they would double the population in less than ten years. It is possible that there may be a mistake in the table, and that the births and marriages of the plague years are included in the year 1711; though as the deaths are carefully separated, it seems very strange that it should be so. It is however a matter of no great importance. The other years are sufficient to illustrate the general principle.

the number of annual marriages, either from accident, or the beginning operation of the preventive check, is the smallest; and it is at this time that the proportion of births to marriages rises very high. In the period from 1717 to 1721 the proportion, as appears by the table, is 49 to 10; and in the particular years 1719 and 1720, it is 50 to 10 and 55 to 10.

Sussmilch draws the attention of his readers to the fruitfulness of marriages in Prussia after the plague, and mentions the proportion of 50 annual births to 10 annual marriages as a proof of it. There are the best reasons from the general average for supposing that the marriages in Prussia at this time were very fruitful; but certainly the proportion of this individual year, or even period, is not a sufficient proof of it, being evidently caused by a smaller number of marriages taking place in the year, and not by a greater number of births.[815] In the two years immediately succeeding the plague, when the excess of births above the deaths was so astonishing, the births bore a small proportion to the marriages; and according to the usual mode of calculation, it would have followed that each marriage yielded only 2.7 or 3.6 children. In the last period of the table, (from 1752 to 1756,) the births are to the marriages as 5 to 1, and in the individual year 1756, as 6.1 to 1: and yet during this period the births are to the deaths only as 148 to 100, which could not have been the case, if the high proportion of births to marriages had indicated a much greater number of births than usual, instead of a smaller number of marriages.

The variations in the proportion of births to deaths, in the different periods of 64 years included in the table, deserve particular attention. If we were to take an average of the four years immediately succeeding the plague, the births would be to the deaths in the proportion of above 22 to 10, which, supposing the mortality to be 1 in 36, would double the population in twenty-one years. If we take the twenty years from 1711 to 1731, the average proportion of the births to deaths will appear to be about 17 to 10, a proportion which (according to Table I. page 496) would double the population in about thirty-five years. But if, instead of 20 years, we were to take the whole period of 64 years, the average proportion of births to deaths turns out to be but a little more than 12 to 10; a proportion which would not double the population in less than 125 years. If we were to include the mortality of the plague, or even of the epidemic years 1736 and 1737, in too short a period, the deaths might exceed the births, and the population would appear to be decreasing.

Sussmilch thinks that, instead of 1 in 36, the mortality in Prussia, after the plague, might be 1 in 38; and it may appear perhaps to some of my readers, that the plenty occasioned by such an event ought to make a still greater difference. Dr. Short has particularly remarked that an extraordinary healthiness generally succeeds any very great mortality;[816] and I have no doubt that the observation is just, comparing similar ages together. But, under the most favourable circumstances, infants under three years are more subject to death than at other ages; and the extraordinary proportion of children which usually follows a very great mortality, counterbalances at first the natural healthiness of the period, and prevents it from making much difference in the general mortality.

[815] Sussmilch, Göttliche Ordnung, vol. i. c. v. s. lxxxvi. p. 175.

[816] History of Air, Seasons, &c. vol. ii. p. 344.

If we divide the population of Prussia after the plague, by the number of deaths in the year 1711, it will appear, that the mortality was nearly 1 in 31, and was therefore increased rather than diminished, owing to the prodigious number of children born in that year. But this greater mortality would certainly cease, as soon as these children began to rise into the firmer stages of life, and then probably Sussmilch's observations would be just. In general, however, we shall observe that a great previous mortality produces a more sensible effect on the births than on the deaths. By referring to the table it will appear, that the number of annual deaths regularly increases with the increasing population, and nearly keeps up the same relative proportion all the way through. But the number of annual births is not very different during the whole period, though in this time the population had more than doubled itself; and therefore the *proportion* of births to the whole population, at first and at last, must have changed in an extraordinary degree.

It will appear therefore how liable we should be to err in assuming a given proportion of births, for the purpose of estimating the past population of any country. In the present instance, it would have led to the conclusion, that the population was scarcely diminished by the plague, although from the number of deaths it was known to be diminished one third.

Variations of the same kind, though not in the same degree, appear in the proportions of births, deaths and marriages, in all the tables which Sussmilch has collected; and as writers on these subjects have been too apt to form calculations for past and future times from the proportions of a few years, it may be useful to draw the attention of the reader to a few more instances of such variations.

In the Churmark of Brandenburgh,[817] during 15 years, ending with 1712, the proportion of births to deaths was nearly 17 to 10. For 6 years, ending with 1718, the proportion sunk to 13 to 10; for 4 years, ending with 1752, it was only 11 to 10; and for 4 years, ending with 1756, 12 to 10. For 3 years, ending with 1759, the deaths very greatly exceeded the births. The proportion of the births to the whole population is not given; but it is not probable that the great variations observable in the proportion of births to deaths should have arisen solely from the variations in the deaths. The proportion of births to marriages is tolerably uniform, the extremes being only 38 to 10 and 35 to 10, and the mean about 37 to 10. In this table no very great epidemics occur till the 3 years beginning with 1757, and beyond this period the lists are not continued.

In the dukedom of Pomerania,[818] the average proportion of births to deaths for 60 years (from 1694 to 1756 both included) was 138 to 100; but in some of the periods of six years it was as high as 177 to 100, and 155 to 100. In others it sunk as low as 124 to 100, and 130 to 100. The extremes in the proportions of births to marriages of the different periods of 5 and 6 years, were 36 to 10 and 43 to 10, and the mean of the 60 years about 38 to 10. Epidemic years appear to have occurred occasionally, in three of which the deaths exceeded the births; but this temporary diminution of population produced no corresponding diminution of births, and the two individual years which contain the greatest

[817] Sussmilch's Göttliche Ordnung, vol. i. tables, p. 88.

[818] Sussmilch, vol. i, tables, p. 91.

proportion of marriages in the whole table occur, one in the year after, and the other two years after epidemics. The excess of deaths however was not great till the three years ending with 1759, with which the table concludes.

In the Neumark of Brandenburgh,[819] for 60 years, from 1695 to 1756 both included, the average proportion of births to deaths in the first 30 years was 148 to 100, in the last 30 years 127 to 100, in the whole 60 years 136 to 100. In some periods of 5 years it was as high as 171 and 167 to 100. In others as low as 118 and 128 to 100. For 5 years ending with 1726, the yearly average of births was 7012; for 5 years ending with 1746, it was 6927, from which, judging by the births, we might infer that the population had decreased in this interval of 20 years; but it appears from the average proportion of births and deaths during this period, that it must have considerably increased, notwithstanding the intervention of some epidemic years. The proportion of births to the whole population must therefore have decidedly changed. Another interval of 20 years in the same tables gives a similar result, both with regard to the births and marriages. The extremes of the proportion of births to marriages are 34 to 10, and 42 to 10, the mean about 38 to 10. The 3 years beginning with 1757, were, as in the other tables, very fatal years.

In the dukedom of Magdeburgh,[820] during 64 years ending with 1756, the average proportion of births to deaths was 123 to 100; in the first 28 years of the period 142 to 100, and in the last 34 years only 112 to 100; during one period of 5 years it was as high as 170 to 100; and in two periods the deaths exceeded the births. Slight epidemics appear to be interspersed rather thickly throughout the table. In the two instances, where three or four occur in successive years and diminish the population, they are followed by an increase of marriages and births. The extremes of the proportions of births to marriages are 42 to 10 and 34 to 10, and the mean of the 64 years 39 to 10. On this table Sussmilch remarks, that though the average number of deaths shews an increased population of one third from 1715 or 1720, yet the births and marriages would prove it to be stationary, or even declining. In drawing this conclusion however, he adds the three epidemic years ending with 1759, during which both the marriages and births seem to have diminished.

In the principality of Halberstadt,[821] the average proportion of births to deaths for 68 years, ending with 1756, was 124 to 100; but in some periods of 5 years it was as high as 160 to 100, and in others as low as 110 to 100. The increase in the whole 68 years was considerable, and yet for 5 years ending with 1723, the average number of births was 2818; and for 4 years ending with 1750, 2628, from which it would appear that the population in 27 years had considerably diminished. A similar appearance occurs with regard to the marriages during a period of 32 years. In the 5 years ending with 1718, they were 727; in the 5 years ending with 1750, 689. During both these periods the proportion of deaths would have shewn a considerable increase. Epidemics seem to have occurred frequently; and in almost all the instances, in which they were such as for the deaths to

[819] Id. p. 99.

[820] Sussmilch, vol. i. tables, p. 103.

[821] Sussmilch, vol. i. tables, p. 108.

exceed the births, they were immediately succeeded by a more than usual proportion of marriages, and in a few years by an increased proportion of births. The greatest number of marriages in the whole table occurs in the year 1751, after an epidemic in the year 1750, in which the deaths had exceeded the births above one third, and the four or five following years contain the largest proportion of births. The extremes of the proportions of births to marriages are 42 to 10 and 34 to 10; the mean of the 68 years 38 to 10.

The remaining tables contain similar results; but these will be sufficient to shew the variations which are continually occurring in the proportions of the births and marriages, as well as of the deaths, to the whole population.

It will be observed that the least variable of the proportions is that which the births and marriages bear to each other; and the obvious reason is, that this proportion is principally influenced by the prolifickness of marriages, which will not of course be subject to great changes. We can hardly indeed suppose, that the prolifickness of marriages should vary so much as the different proportions of births to marriages in the tables. Nor is it necessary that it should, as another cause will contribute to produce the same effect. The births which are contemporary with the marriages of any particular year, belong principally to marriages which had taken place some years before; and therefore, if for four or five years a large proportion of marriages were to take place; and then accidentally for one or two years a small proportion, the effect would be a large proportion of births to marriages in the registers during these one or two years; and on the contrary, if for four or five years few marriages comparatively were to take place, and then for one or two years a great number, the effect would be a small proportion of births to marriages in the registers. This was strikingly illustrated in the table for Prussia and Lithuania, and would be confirmed by an inspection of all the other tables collected by Sussmilch; in which it appears that the extreme proportions of births to marriages are generally more affected by the number of marriages than the number of births, and consequently arise more from the variations in the disposition or encouragement to matrimony, than from the variations in the prolifickness of marriages.

The common epidemical years which are interspersed throughout these tables, will not of course have the same effects on the marriages and births as the great plague in the table for Prussia; but in proportion to their magnitude, their operation will in general be found to be similar. From the registers of many other countries, and particularly of towns, it appears that the visitations of the plague were frequent at the latter end of the 17th, and the beginning of the 18th centuries.

In contemplating the plagues and sickly seasons which occur in these tables after a period of rapid increase, it is impossible not to be impressed with the idea, that the number of inhabitants had in these instances exceeded the food and the accommodations necessary to preserve them in health. The mass of the people would, upon this supposition, be obliged to live worse, and a greater number of them would be crowded together in one house; and these natural causes would evidently contribute to produce sickness, even though the country, absolutely considered, might not be crowded and populous. In a country even thinly inhabited, if an increase of population take place before more food is raised, and more houses are built, the inhabitants must be distressed for room and subsistence. If in the Highlands of Scotland, for the next ten or twelve years, the marriages were to be either more frequent or more prolific, and no emigration were to take place, instead of five to a cottage, there might be seven; and this, added to the necessity of worse living, would evidently have a most unfavourable effect on the health of the common people.

TABLE IV.

Annual Average.	Marriages.	Births.	Deaths.	Proportion of marriages to births.	Proportion of deaths to births.
5 yrs to 1697	5747	19715	14862	10:34	100:132
5 yrs–1702	6070	24112	14474	10:39	100:165
6 yrs–1708	6082	26896	16430	10:44	100:163
In 1709 & 1710	a plague	number destroyed in 2 years.	247733		
In 1711	12028	32522	10131	10:27	100:320
In 1712	267	22970	10445	10:36	100:220
5 yrs to 1716	4968	21603	11984	10:43	100:180
5 yrs–1721	4324	21396	12039	10:49	100:177
5 yrs-1726	4719	21452	12863	10:45	100:166
5 yrs–1731	4808	29554	12825	10:42	100:160
4 yrs–1735	5424	22692	15475	10:41	100:146
In 1736	5280	21859	26371	Epidemic	
In 1737	5765	18930	24480	years.	
5 yrs to 1742	5582	22099	15255	10:39	100:144
4 yrs–1746	5469	25275	15117	10:46	100:167
5 yrs–1751	6423	28235	17272	10:43	100:163
5 yrs–1756	5599	28892	19154	10:50	100:148
In the 16 yrs before the plague	95585	380516	245763	10:39	100:154
In 46 yrs after the plague	248777	1083872	690324	10:43	100:157
In 62 good yrs	344361 936087	1464388	936087	10:43	100:156
More born than died		528301			
In the 2 plague years	5477	23977	247733		
In all the 64 years including the plague 1183820	340838	1488365	1183820	10:42	100:125
More born than died.		304545			

3.5.13 Book II, Chapter XIII: General Deductions from the Preceding View of Society.

That the checks which have been mentioned are the immediate causes of the slow increase of population, and that these checks result principally from an insufficiency of subsistence, will be evident from the comparatively rapid increase which has invariably taken place, whenever, by some sudden enlargement in the means of subsistence, these checks have in any considerable degree been removed.

It has been universally remarked that all new colonies settled in healthy countries, where room and food were abundant, have constantly made a rapid progress in population. Many of the colonies from ancient Greece, in the course of one or two centuries, appear to have rivalled, and even surpassed, their mother cities. Syracuse and Agrigentum in Sicily, Tarentum and Locri in Italy, Ephesus and Miletus in Lesser Asia, were, by all accounts, at least equal to any of the cities of ancient Greece. All these colonies had established themselves in countries inhabited by savage and barbarous nations, which easily gave place to the new settlers, who had of course plenty of good land. It is calculated that the Israelites, though they increased very slowly while they were wandering in the land of Canaan, on settling in a fertile district of Egypt, doubled their numbers every fifteen years during the whole period of their stay.[822] But not to dwell on remote instances, the European settlements in America bear ample testimony to the truth of a remark, that has never I believe been doubted. Plenty of rich land to be had for little or nothing, is so powerful a cause of population, as generally to overcome all obstacles.

No settlements could easily have been worse managed than those of Spain, in Mexico, Peru, and Quito. The tyranny, superstition, and vices of the mother country were introduced in ample quantities among her children. Exorbitant taxes were exacted by the crown; the most arbitrary restrictions were imposed on their trade; and the governors were not behind hand in rapacity and extortion for themselves as well as their masters. Yet under all these difficulties, the colonies made a quick progress in population. The city of Quito, which was but a hamlet of Indians, is represented by Ulloa as containing fifty or sixty thousand inhabitants above fifty years ago.[823] Lima, which was founded since the conquest, is mentioned by the same author as equally or more populous before the fatal earthquake in 1746. Mexico is said to contain a hundred thousand inhabitants; which, notwithstanding the exaggerations of the Spanish writers, is supposed to be five times greater than what it contained in the time of Montezuma.[824]

In the Portuguese colony of Brazil, governed with almost equal tyranny, there were supposed to be, above thirty years ago, six hundred thousand inhabitants of European extraction.[825]

---

[822] Short's New Observ. on Bills of Mortality, p. 259, 8vo. 1750.

[823] Voy. d'Ulloa, tom. i. liv. v. ch. v. p. 229, 4to. 1752.

[824] Smith's Wealth of Nations, vol. ii. b. iv. ch. viii. p. 363.

[825] Id. p. 365.

The Dutch and French colonies, though under the government of exclusive companies of merchants, still persisted in thriving under every disadvantage.[826]

But the English North-American colonies, now the powerful people of the United States of America, far outstripped all the others in the progress of their population. To the quantity of rich land which they possessed in common with the Spanish and Portuguese colonies, they added a greater degree of liberty of liberty and equality. Though not without some restrictions on their foreign commerce, they were allowed the liberty of managing their own internal affairs. The political institutions which prevailed were favourable to the alienation and division of property. Lands which were not cultivated by the proprietor within a limited time, were declared grantable to any other person. In Pennsylvania there was no right of primogeniture; and in the provinces of New England, the eldest son had only a double share. There were no tithes in any of the States, and scarcely any taxes.

And on account of the extreme cheapness of good land, and a situation favourable to the exportation of grain, a capital could not be more advantageously employed than in agriculture; which, at the same time that it affords the greatest quantity of healthy work, supplies the most valuable produce to the society.

The consequence of these favourable circumstances united, was a rapidity of increase almost without parallel in history. Throughout all the northern provinces the population was found to double itself in 25 years. The original number of persons which had settled in the four provinces of New England in 1643, was 21,200. Afterwards it was calculated that more left them than went to them. In the year 1760 they were increased to half a million. They had, therefore, all along doubled their number in 25 years. In New Jersey the period of doubling appeared to be 22 years, and in Rhode Island still less. In the back settlements, where the inhabitants applied themselves solely to agriculture, and luxury was not known, they were supposed to double their number in fifteen years. Along the sea-coast, which would naturally be first inhabited, the period of doubling was about 35 years, and in some of the maritime towns the population was absolutely at a stand.[827] From the late census made in America, it appears that, taking all the States together, they have still continued to double their numbers within 25 years;[828] and as the whole population is now so great as not

---

[826] Id. p. 368, 369.

[827] Price' s Observ. on Revers. Paym. vol. i. p. 282, 283, and vol. ii. p. 260. I have lately had an opportunity of seeing some extracts from the sermon of Dr. Styles, from which Dr. Price has taken these facts. Speaking of Rhode Island, Dr. Styles says that, though the period of doubling for the whole colony is 25 years, yet that it is different in different parts, and within land is 20 and 15 years. The population of the five towns of Gloucester, Situate, Coventry, West Greenwich and Exeter, was 5033, A. D. 1798, and 6986, A. D. 1755; which implies a period of doubling of 15 years only. He mentions afterwards, that the county of Kent doubles in 20 years, and the county of Providence in 18 years.

[828] See an article in the Supplement to the Encyclopædia Britannica on Population, p. 308; and a curious table, p. 310, calculated by Mr. Milne, Actuary to the Sun Life Assurance Office, which strikingly confirms and illustrates the computed rate of increase in the United States, and shews that it cannot be essentially affected by immigrations.

to be materially affected by the emigrations from Europe, and as it is known that, in some of the towns and districts near the sea-coast, the progress of population has been comparatively slow; it is evident, that in the interior of the country in general, the period of doubling from procreation only must have been considerably less than 25 years.

The population of the United States of America, according to the fourth census, in 1820, was 7,861,710. We have no reason to believe that Great Britain is less populous at present, for the emigration of the small parent stock which produced these numbers. On the contrary, a certain degree of emigration is known to be favourable to the population of the mother country. It has been particularly remarked that the two Spanish provinces, from which the greatest number of people emigrated to America, became in consequence more populous.

Whatever was the original number of British emigrants which increased so fast in North America, let us ask, Why does not an equal number produce an equal increase in the same time in Great Britain? The obvious reason to be assigned is the want of food; and that this want is the most efficient cause of the three immediate checks to population, which have been observed to prevail in all societies; is evident from the rapidity with which even old states recover the desolations of war, pestilence, famine, and the convulsions of nature. They are then for a short time placed a little in the situation of new colonies; and the effect is always answerable to what might be expected. If the industry of the inhabitants be not destroyed, subsistence will soon increase beyond the wants of the reduced numbers; and the invariable consequence will be, that population, which before perhaps was nearly stationary, will begin immediately to increase, and will continue its progress till the former population is recovered.

The fertile province of Flanders, which has been so often the seat of the most destructive wars, after a respite of a few years has always appeared as rich and populous as ever. The undiminished population of France, which has before been noticed, is an instance very strongly in point. The tables of Sussmilch afford continual proofs of a very rapid increase after great mortalities; and the table for Prussia and Lithuania, which I have inserted,[829] is particularly striking in this respect. The effects of the dreadful plague in London, in 1666, were not perceptible 15 or 20 years afterwards. It may even be doubted whether Turkey and Egypt are upon an average much less populous for the plagues which periodically lay them waste. If the number of people which they contain be considerably less now than formerly, it is rather to be attributed to the tyranny and oppression of the governments under which they groan, and the consequent discouragements to agriculture, than to the losses which they sustain by the plague. The traces of the most destructive famines in China, Indostan, Egypt, and other countries, are by all accounts very soon obliterated; and the most tremendous convulsions of nature, such as volcanic eruptions and earthquakes, if they do not happen so frequently as to drive away the inhabitants or destroy their spirit of industry, have been found to produce but a trifling effect on the average population of any state.

---

[829] See p. 500.

It has appeared from the registers of different countries, which have already been produced, that the progress of their population is checked by the periodical, though irregular, returns of plagues and sickly seasons. Dr. Short, in his curious researches into bills of mortality, often uses the expression—"terrible correctives of the redundance of mankind;"[830] and in a table of all the plagues, pestilences and famines, of which he could collect accounts, shews the constancy and universality of their operation.

The epidemical years in his table, or the years in which the plague or some great and wasting epidemic prevailed, (for smaller sickly seasons seem not to be included;) are 431,[831] of which 32 were before the Christian æra.[832] If we divide therefore the years of the present æra by 399, it will appear, that the periodical returns of such epidemics, to some countries that we are acquainted with, have been on an average only at the interval of about 4½ years.

Of the 254 great famines and dearths enumerated in the table, 15 were before the Christian æra,[833] beginning with that which occurred in Palestine, in the time of Abraham. If, subtracting these 15, we divide the years of the present æra by the remainder, it will appear that the average interval between the visits of this dreadful scourge has been only about 7½ years.

How far these "terrible correctives to the redundance of mankind" have been occasioned by the too rapid increase of population, is a point which it would be very difficult to determine with any degree of precision. The causes of most of our diseases appear to us to be so mysterious, and probably are really so various, that it would be rashness to lay too much stress on any single one; but it will not perhaps be too much to say, that *among* these causes we ought certainly to rank crowded houses and insufficient or unwholesome food, which are the natural consequences of an increase of population faster than the accommodations of a country with respect to habitations and food will allow.

Almost all the histories of epidemics, which we possess, tend to confirm this supposition, by describing them in general as making their principal ravages among the lower classes of people. In Dr. Short's tables this circumstance is frequently mentioned;[834] and it further appears that a very considerable proportion of the epidemic years either followed or were accompanied by seasons of dearth and bad food.[835] In other places he also mentions great plagues as diminishing particularly the numbers of the lower or servile sort of people;[836] and in speaking of different diseases he observes that those which are occasioned by bad and unwholesome food generally last the longest.[837]

---

[830] New Observ. on Bills of Mortality, p. 96.

[831] Hist. of Air, Seasons, &c. vol. ii. p. 366.

[832] Id. vol. ii. p. 202.

[833] Hist. of Air, Seasons, &c, vol. ii. p. 206.

[834] Hist. of Air; Seasons, &c. vol. ii. p. 206. et seq.

[835] Id. vol. ii. p. 206. et seq. and 336.

[836] New Observ. p. 125.

[837] New Observ. p. 108.

We know from constant experience, that fevers are generated in our jails, our manufactories, our crowded workhouses and in the narrow and close streets of our large towns; all which situations appear to be similar in their effects to squalid poverty; and we cannot doubt that causes of this kind, aggravated in degree, contributed to the production and prevalence of those great and wasting plagues formerly so common in Europe, but which now, from the mitigation of these causes, are every where considerably abated, and in many places appear to be completely extirpated.

Of the other great scourge of mankind, famine, it may be observed that it is not in the nature of things, that the increase of population should absolutely produce one. This increase, though rapid, is necessarily gradual; and as the human frame cannot be supported, even for a very short time, without food, it is evident, that no more human beings can grow up than there is provision to maintain. But though the principle of population cannot absolutely produce a famine, it prepares the way for one; and by frequently obliging the lower classes of people to subsist nearly on the smallest quantity of food that will support life, turns even a slight deficiency from the failure of the seasons into a severe dearth; and may be fairly said, therefore, to be one of the principal causes of famine. Among the signs of an approaching dearth, Dr. Short mentions one or more years of luxuriant crops together;[838] and this observation is probably just, as we know that the general effect of years of cheapness and abundance is to dispose a great number of persons to marry; and under such circumstances the return to a year merely of an average crop might produce a scarcity.

The small-pox, which may be considered as the most prevalent and fatal epidemic in Europe, is of all others, perhaps, the most difficult to account for, though the periods of its returns are in many places regular.[839] Dr. Short observes, that from the histories of this disorder it seems to have very little dependence upon the past or present constitution of the weather or seasons, and that it appears epidemically at all times and in all states of the air, though not so frequently in a hard frost. We know of no instances, I believe, of its being clearly generated under any circumstances of situation. I do not mean therefore to insinuate that poverty and crowded houses ever absolutely produced it; but I may be allowed to remark, that in those places where its returns are regular, and its ravages among children, particularly among those of the lower class, are considerable, it necessarily follows that these circumstances, in a greater degree than usual, must always precede and accompany its appearance; that is, from the time of its last visit, the average number of children will be increasing, the people will, in consequence, be growing poorer, and the houses will be more crowded till another visit removes this superabundant population.

In all these cases, how little soever force we maybe disposed to attribute to the effects of the principle of population in the actual production of disorders, we cannot avoid allowing

[838] Hist. of Air, Seasons, &c: vol. ii. p. 367.

[839] Id. vol. ii. p. 411.

their force as predisposing causes to the reception of contagion, and as giving very great additional force to the extensiveness and fatality of its ravages.

It is observed by Dr. Short that a severe mortal epidemic is generally succeeded by an uncommon healthiness, from the late distemper having carried off most of the declining and worn out constitutions.[840] It is probable, also, that another cause of it may be the greater plenty of room and food, and the consequently meliorated condition of the lower classes of the people. Sometimes, according to Dr. Short, a very fruitful year is followed by a very mortal and sickly one, and mortal ones often succeeded by very fruitful, as if Nature sought either to prevent or quickly repair the loss by death. In general the next year after sickly and mortal ones is prolific in proportion to the breeders left.[841]

This last effect we have seen most strikingly exemplified in the table for Prussia and Lithuania.[842] And from this and other tables of Sussmilch, it also appears that, when the increasing produce of a country and the increasing demand for labour, so far meliorate the condition of the labourer as greatly to encourage marriage, the custom of early marriages is generally continued, till the population has gone beyond the increased produce, and sickly seasons appear to be the natural and necessary consequence. The continental registers exhibit many instances of rapid increase, interrupted in this manner by mortal diseases; and the inference seems to be, that those countries where subsistence is increasing sufficiently to encourage population, but not to answer all its demands, will be more subject to periodical epidemics, than those where the increase of population is more nearly accommodated to the average produce.

The converse of this will of course be true. In those countries which are subject to periodical sicknesses, the increase of population, or the excess of births above the deaths, will be greater in the intervals of these periods than is usual in countries not so much subject to these diseases. If Turkey and Egypt have been nearly stationary in their average population for the last century, in the intervals of their periodical plagues, the births must have exceeded the deaths in a much greater proportion than in such countries as France and England.

It is for these reasons that no estimates of future population or depopulation, formed from any existing rate of increase or decrease, can be depended upon. Sir William Petty calculated that in the year 1800 the city of London would contain 5,359,000[843] inhabitants, instead of which it does not now contain a fifth part of that number. Mr. Eaton has lately prophesied the extinction of the population of the Turkish empire in another century,[844] an event which will certainly fail of taking place. If America were to continue increasing at the same rate as at present for the next 150 years, her population would exceed the population

[840] Hist. of Air, Seasons, &c. vol. ii. p. 344.

[841] New Observ. p. 191.

[842] Id. p. 500.

[843] Political Arithmetic, p. 17.

[844] Survey of the Turkish Empire, c. vii. p. 281.

of China; but though prophecies are dangerous, I will venture to say that such an increase will not take place in that time, though it may perhaps in five or six hundred years.

Europe was without doubt formerly more subject to plagues and wasting epidemics than at present; and this will account, in a great measure, for the greater proportion of births to deaths in former times, mentioned by many authors; as it has always been a common practice to estimate these proportions from too short periods, and generally to reject the years of plague as accidental.

The average proportion of births to deaths in England during the last century may be considered as about 12 to 10, or 120 to 100. The proportion in France for ten years, ending in 1780, was about 115 to 100.[845] Though these proportions undoubtedly varied at different periods during the century, yet we have reason to think that they did not vary in any very considerable degree; and it will appear therefore, that the population of France and England had accommodated itself more nearly to the average produce of each country than many other states. The operation of the preventive check—wars—the silent though certain destruction of life in large towns and manufactories—and the close habitations and insufficient food of many of the poor—prevent population from outrunning the means of subsistence; and, if I may use an expression which certainly at first appears strange, supersede the necessity of great and ravaging epidemics to destroy what is redundant. If a wasting plague were to sweep off two millions in England, and six millions in France, it cannot be doubted that, after the inhabitants had recovered from the dreadful shock, the proportion of births to deaths would rise much above the usual average in either country during the last century.

In New Jersey the proportion of births to deaths, on an average of 7 years, ending with 1743, was 300 to 100. In France and England the average proportion cannot be reckoned at more than 120 to 100. Great and astonishing as this difference is, we ought not to be so wonder-struck at it, as to attribute it to the miraculous interposition of Heaven. The causes of it are not remote, latent and mysterious, but near us, round about us, and open to the investigation of every inquiring mind. It accords with the most liberal spirit of philosophy to believe that no stone can fall, or plant rise, without the immediate agency of divine power. But we know from experience, that these operations of what we call nature have been conducted almost invariably according to fixed laws. And since the world began, the causes of population and depopulation have been probably as constant as any of the laws of nature with which we are acquainted.

The passion between the sexes has appeared in every age to be so nearly the same, that it may always be considered, in algebraic language, as a given quantity. The great law of necessity, which prevents population from increasing in any country beyond the food which it can either produce or acquire, is a law so open to our view, so obvious and evident to our understandings, that we cannot for a moment doubt it. The different modes which nature takes to repress a redundant population, do not indeed appear to us so certain and

[845] Necker de l'Administration des Finances, tom. i. c. ix. p. 255.

regular; but though we cannot always predict the mode, we may with certainty predict the fact. If the proportion of the births to the deaths for a few years indicates an increase of numbers much beyond the proportional increased or acquired food of the country, we may be perfectly certain that, unless an emigration take place, the deaths will shortly exceed the births, and that the increase which had been observed for a few years cannot be the real average increase of the population of the country. If there were no other depopulating causes, and if the preventive check did not operate very strongly, every country would without doubt be subject to periodical plagues and famines.

The only true criterion of a real and permanent increase in the population of any country, is the increase of the means of subsistence. But even this criterion is subject to some slight variations, which however are completely open to our observation. In some countries population seems to have been forced; that is, the people have been habituated by degrees to live almost upon the smallest possible quantity of food. There must have been periods in such countries, when population increased permanently without an increase in the means of subsistence. China, India and the countries possessed by the Bedoween Arabs, as we have seen in the former part of this work, appear to answer to this description. The average produce of these countries seems to be but barely sufficient to support the lives of the inhabitants, and of course any deficiency from the badness of the seasons must be fatal. Nations in this state must necessarily be subject to famines.

In America, where the reward of labour is at present so liberal, the lower classes might retrench very considerably in a year of scarcity, without materially distressing themselves. A famine therefore seems to be almost impossible. It may be expected, that in the progress of the population of America, the labourers will in time be much less liberally rewarded. The numbers will in this case permanently increase, without a proportional increase in the means of subsistence.

In the different countries of Europe there must be some variations in the proportion of the number of inhabitants, and the quantity of food consumed, arising from the different habits of living which prevail in each state. The labourers in the south of England are so accustomed to eat fine wheaten bread, that they will suffer themselves to be half starved before they will submit to live like the Scotch peasants.

They might perhaps, in time, by the constant operation of the hard law of necessity, be reduced to live even like the lower classes of the Chinese, and the country would then with the same quantity of food support a greater population. But to effect this must always be a difficult, and every friend to humanity will hope, an abortive attempt.

I have mentioned some cases where population may permanently increase without a proportional increase in the means of subsistence. But it is evident that the variation in different states between the food and the numbers supported by it is restricted to a limit beyond which it cannot pass. In every country, the population of which is not absolutely decreasing, the food must be necessarily sufficient to support and continue the race of labourers.

Other circumstances being the same, it may be affirmed that countries are populous according to the quantity of human food which they produce or can acquire; and happy, according to the liberality with which this food is divided, or the quantity which a day's labour will purchase. Corn countries are more populous than pasture countries, and rice countries more populous than corn countries. But their happiness does not depend either upon their being thinly or fully inhabited, upon their poverty or their riches, their youth or their age; but on the proportion which the population and the food bear to each other.

This proportion is generally the most favourable in new colonies, where the knowledge and industry of an old state operate on the fertile unappropriated land of a new one. In other cases the youth or the age of a state is not, in this respect, of great importance. It is probable that the food of Great Britain is divided in more liberal shares to her inhabitants at the present period, than it was two thousand, three thousand, or four thousand years ago. And it has appeared that the poor and thinly-inhabited tracts of the Scotch Highlands are more distressed by a redundant population than the most populous parts of Europe.

If a country were never to be overrun by a people more advanced in arts, but left to its own natural progress in civilization; from the time that its produce might be considered as an unit, to the time that it might be considered as a million, during the lapse of many thousand years, there might not be a single period when the mass of the people could be said to be free from distress, either directly or indirectly, for want of food. In every state in Europe, since we have first had accounts of it, millions and millions of human existences have been repressed from this simple cause, though perhaps in some of these states an absolute famine may never have been known.

Must it not then be acknowledged by an attentive examiner of the histories of mankind, that, in every age and in every state in which man has existed or does now exist,

The increase of population is necessarily limited by the means of subsistence:

Population invariably increases when the means of subsistence increase,[846] unless prevented by powerful and obvious checks:

These checks, and the checks which keep the population down to the level of the means of subsistence, are moral restraint, vice, and misery?

In comparing the state of society which has been considered in this second book with that which formed the subject of the first, I think it appears that in modern Europe the positive checks to population prevail less, and the preventive checks more than in past times, and in the more uncivilized parts of the world.

War, the predominant check to the population of savage nations, has certainly abated, even including the late unhappy revolutionary contests; and since the prevalence of a greater degree of personal cleanliness, of better modes of clearing and building towns, and of a more equable distribution of the products of the soil from improving knowledge in political economy, plagues, violent diseases and famines have been certainly mitigated, and have become less frequent.

[846] By an increase in the means of subsistence, as the expression is used here, is always meant such an increase as the mass of the population can command; otherwise it can be of no avail in encouraging an increase of people.

With regard to the preventive check to population, though it must be acknowledged that that branch of it which comes under the head of moral restraint,[847] does not at present prevail much among the male part of society; yet I am strongly disposed to believe that it prevails more than in those states which were first considered; and it can scarcely be doubted that in modern Europe a much larger proportion of women pass a considerable part of their lives in the exercise of this virtue, than in past times and among uncivilized nations. But however this may be, if we consider only the general term which implies principally a delay of the marriage union from prudential considerations, without reference to consequences, it may be considered in this light as the most powerful of the checks, which in modern Europe keep down the population to the level of the means of subsistence.

[847] The reader will recollect the confined sense in which I use this term.

An Essay on the Principle of Population, vol. 2 [1826, 6th ed.]

4

Thomas Robert Malthus

4.1 Book III: Of the Different Systems or Expedients Which Have Been Proposed or Have Prevailed in Society, as They Affect the Evils Arising from the Principle of Population

4.1.1 Book III, Chapter I: Of Systems of Equality. Wallace, Condorcet.

To a person who views the past and present states of mankind in the light in which they have appeared in the two preceding books, it cannot but be a matter of astonishment, that all the writers on the perfectibility of man and of society, who have noticed the argument of the principle of population, treat it always very lightly, and invariably represent the difficulties arising from it as at a great and almost immeasurable distance. Even Mr. Wallace, who thought the argument itself of so much weight as to destroy his whole system of equality, did not seem to be aware that any difficulty would arise from this cause, till the whole earth had been cultivated like a garden, and was incapable of any further increase of produce. If this were really the case, and a beautiful system of equality were in other respects practicable, I cannot think that our ardour in the pursuit of such a scheme ought to be damped by the contemplation of so remote a difficulty. An event at such a distance might fairly be left to Providence. But the truth is, that, if the view of the argument given in this essay be just, the difficulty, so far from being remote, is imminent and immediate. At every period during the progress of cultivation, from the present moment to the time when the whole earth was become like a garden, the distress for want of food would be constantly pressing on all mankind, if they were equal. Though the produce of the earth would be

T. R. Malthus (✉)
University of Oxford, Oxford, UK

D. Coleman (ed.), *Thomas Robert Malthus – An Essay on the Principle of Population*, Classic Texts in the Sciences,
https://doi.org/10.1007/978-3-031-94503-8_4

increasing every year, population would have the power of increasing much faster, and this superior power must necessarily be checked by the periodical or constant action of moral restraint, vice, or misery.

M. Condorcet's *Esquisse d'un Tableau Historique des Progrès de l'Esprit Humain* was written, it is said, under the pressure of that cruel proscription which terminated in his death. If he had no hopes of its being seen during his life, and of its interesting France in his favour, it is a singular instance of the attachment of a man to principles, which every day's experience was, so fatally for himself, contradicting. To see the human mind in one of the most enlightened nations of the world, debased by such a fermentation of disgusting passions, of fear, cruelty, malice, revenge, ambition, madness and folly, as would have disgraced the most savage nations in the most barbarous age, must have been such a tremendous shock to his ideas of the necessary and inevitable progress of the human mind, as nothing but the firmest conviction of the truth of his principles, in spite of all appearances, could have withstood.

This posthumous publication is only a sketch of a much larger work, which he proposed should be executed. It necessarily wants therefore that detail and application, which can alone prove the truth of any theory. A few observations will be sufficient to show how completely this theory is contradicted, when it is applied to the real, and not to an imaginary, state of things.

In the last division of the work, which treats of the future progress of man towards perfection, M. Condorcet says that, comparing in the different civilized nations of Europe the actual population with the extent of territory, and observing their cultivation, their industry, their divisions of labour, and their means of subsistence, we shall see that it would be impossible to preserve the same means of subsistence, and consequently the same population, without a number of individuals who have no other means of supplying their wants than their industry.

Having allowed the necessity of such a class of men, and adverting afterwards to the precarious revenue of those families that would depend so entirely on the life and health of their chief,[1] he says very justly, "There exists then a necessary cause of inequality, of dependence, and even of misery, which menaces without ceasing the most numerous and active class of our societies." The difficulty is just and well stated; but his mode of removing it will, I fear, be found totally inefficacious.

By the application of calculations to the probabilities of life, and the interest of money, he proposes that a fund should be established, which should assure to the old an assistance produced in part by their own former savings, and in part by the savings of individuals, who in making the same sacrifice die before they reap the benefit of it. The same or a similar fund should give assistance to women and children who lose their husbands and fathers;

[1] To save time and long quotations, I shall here give the substance of M. Condorcet's sentiments, and I hope that I shall not misrepresent them; but I refer the reader to the work itself, which will amuse, if it do not convince him.

and afford a capital to those who were of an age to found a new family, sufficient for the development of their industry. These establishments, he observes, might be made in the name and under the protection of the society. Going still further, he says, that by the just application of calculations, means might be found of more completely preserving a state of equality, by preventing credit from being the exclusive privilege of great fortunes, and yet giving it a basis equally solid, and by rendering the progress of industry and the activity of commerce less dependent on great capitalists.

Such establishments and calculations may appear very promising upon paper; but when applied to real life, they will be found to be absolutely nugatory. M. Condorcet allows that a class of people which maintains itself entirely by industry, is necessary to every state. Why does he allow this? No other reason can well be assigned, than because he conceives, that the labour necessary to procure subsistence for an extended population will not be performed without the goad of necessity. If by establishments upon the plans that have been mentioned, this spur to industry be removed; if the idle and negligent be placed upon the same footing with regard to their credit and the future support of their wives and families, as the active and industrious; can we expect to see men exert that animated activity in bettering their condition, which now forms the master-spring of public prosperity? If an inquisition were to be established to examine the claims of each individual, and to determine whether he had or had not exerted himself to the utmost, and to grant or refuse assistance accordingly, this would be little else than a repetition upon a larger scale of the English poor-laws, and would be completely destructive of the true principles of liberty and equality.

But independently of this great objection to these establishments, and supposing for a moment that they would give no check to production, the greatest difficulty remains yet behind.

If every man were sure of a comfortable provision for a family, almost every man would have one; and if the rising generation were free from the fear of poverty, population must increase with unusual rapidity. Of this M. Condorcet seems to be fully aware himself; and after having described further improvements, he says,

"But in this progress of industry and happiness, each generation will be called to more extended enjoyments, and in consequence, by the physical constitution of the human frame, to an increase in the number of individuals. Must not a period then arrive when these laws, equally necessary, shall counteract each other; when, the increase of the number of men surpassing their means of subsistence, the necessary result must be, either a continual diminution of happiness and population—a movement truly retrograde; or at least a kind of oscillation between good and evil? In societies arrived at this term, will not this oscillation be a constantly subsisting cause of periodical misery? Will it not mark the limit, when all further melioration will become impossible, and point out that term to the perfectibility of the human race, which it may reach in the course of ages, but can never pass?"

He then adds,

"There is no person who does not see how very distant such a period is from us. But shall we ever arrive at it? It is equally impossible to pronounce for or against the future realization of an event, which cannot take place but at in æra when the human race will have attained improvements, of which we can at present scarcely form a conception."

M. Condorcet's picture of what may be expected to happen, when the number of men shall surpass their means of subsistence, is justly drawn. The oscillation which he describes will certainly take place, and will without doubt be a constantly subsisting cause of periodical misery. The only point in which I differ from M. Condorcet in this description is with regard to the period when it may be applied to the human race. M. Condorcet thinks that it cannot possibly be applicable but at an æra extremely distant. If the proportion between the natural increase of population and of food in a limited territory, which was stated in the beginning of this essay, and which has received considerable confirmation from the poverty that has been found to prevail in every stage of human society, be in any degree near the truth; it will appear, on the contrary, that the period when the number of men surpasses their means of easy subsistence has long since arrived; and that this necessary oscillation, this constantly subsisting cause of periodical misery, has existed in most countries ever since we have had any histories of mankind, and continues to exist at the present moment.

M. Condorcet, however, goes on to say, that should the period which he conceives to be so distant, ever arrive, the human race, and the advocates of the perfectibility of man, need not be alarmed at it. He then proceeds to remove the difficulty in a manner which I profess not to understand. Having observed that the ridiculous prejudices of superstition would by that time have ceased to throw over morals a corrupt and degrading austerity, he alludes either to a promiscuous concubinage, which would prevent breeding, or to something else as unnatural. To remove the difficulty in this way will surely, in the opinion of most men, be to destroy that virtue and purity of manners, which the advocates of equality and of the perfectibility of man profess to be the end and object of their views.

The last question which M. Condorcet proposes for examination is the organic perfectibility of man. He observes, if the proofs which have been already given, and which, in their developement, will receive greater force in the work itself, are sufficient to establish the indefinite perfectibility of man, upon the supposition of the same natural faculties and the same organization which he has at present; what will be the certainty, what the extent of our hopes, if this organization, these natural faculties themselves, be susceptible of melioration?

From the improvement of medicine; from the use of more wholesome food and habitations; from a manner of living, which will improve the strength of the body by exercise, without impairing it by excess; from the destruction of the two great causes of the degradation of man, misery and too great riches; from the gradual removal of transmissible and contagious disorders by the improvement of physical knowledge, rendered more efficacious by the progress of reason and of social order; he infers, that though man will not absolutely become immortal, yet the duration between his birth and natural death will increase without ceasing, will have no assignable term, and may properly be expressed by the word indefinite. He then defines this word to mean either a constant approach to an unlimited extent without ever reaching it; or an increase in the immensity of ages to an extent greater than any assignable quantity.

But surely the application of this term in either of these senses to the duration of human life is in the highest degree unphilosophical, and totally unwarranted by any appearances in the laws of nature. Variations from different causes are essentially distinct from a regular and unretrograde increase. The average duration of human life will, to a certain degree, vary from healthy or unhealthy climates, from wholesome or unwholesome food, from virtuous or vicious manners, and other causes; but it may be fairly doubted whether there has been really the smallest perceptible advance in the natural duration of human life, since first we had any authentic history of man. The prejudices of all ages have indeed been directly contrary to this supposition; and though I would not lay much stress upon these prejudices, they must have some tendency to prove that there has been no marked advance in an opposite direction.

It may perhaps be said, that the world is yet so young, so completely in its infancy, that it ought not to be expected that any difference should appear so soon.

If this be the case, there is at once an end of all human science. The whole train of reasonings from effects to causes will be destroyed. We may shut our eyes to the book of nature, as it will no longer be of any use to read it. The wildest and most improbable conjectures may be advanced with as much certainty, as the most just and sublime theories, founded on careful and reiterated experiments. We may return again to the old mode of philosophizing, and make facts bend to systems, instead of establishing systems upon facts. The grand and consistent theory of Newton will be placed upon the same footing as the wild and eccentric hypotheses of Descartes. In short, if the laws of nature be thus fickle and inconstant; if it can be affirmed, and be believed, that they will change, when for ages and ages they have appeared immutable; the human mind will no longer have any incitements to inquiry, but must remain sunk in inactive torpor, or amuse itself only in bewildering dreams and extravagant fancies.

The constancy of the laws of nature, and of effects and causes, is the foundation of all human knowledge; and if, without any previous observable symptoms or indications of a change, we can infer that a change will take place, we may as well make any assertion whatever; and think it as unreasonable to be contradicted, in affirming that the moon will come in contact with the earth to-morrow, as in saying that the sun will rise at its expected time.

With regard to the duration of human life, there does not appear to have existed, from the earliest ages of the world to the present moment, the smallest permanent symptom or indication of increasing prolongation. The observable effects of climate, habit, diet, and other causes, on length of life, have furnished the pretext for asserting its indefinite extension; and the sandy foundation on which the argument rests is, that because the limit of human life is undefined, because you cannot mark its precise term, and say so far exactly shall it go, and no farther, therefore its extent may increase for ever, and be properly termed indefinite or unlimited. But the fallacy and absurdity of this argument will sufficiently appear from a slight examination of what M. Condorcet calls the organic perfectibility or degeneration of the race of plants and animals, which, he says, may be regarded as one of the general laws of nature.

I have been told that it is a maxim among some of the improvers of cattle, that you may breed to any degree of nicety you please; and they found this maxim upon another, which is, that some of the offspring will possess the desirable qualities of the parents in a greater degree. In the famous Leicestershire breed of sheep, the object is to procure them with small heads and small legs. Proceeding upon these breeding maxims, it is evident that we might go on till the heads and legs were evanescent quantities; but this is so palpable an absurdity, that we may be quite sure the premises are not just, and that there really is a limit, though we cannot see it, or say exactly where it is. In this case, the point of the greatest degree of improvement, or the smallest size of the head and legs, may be said to be undefined; but this is very different from unlimited, or from indefinite, in M. Condorcet's acceptation of the term. Though I may not be able in the present instance to mark the limit, at which further improvement will stop, I can very easily mention a point, at which it will not arrive. I should not scruple to assert, that were the breeding to continue for ever, the heads and legs of these sheep would never be so small as the head and legs of a rat.

It cannot be true therefore, that, among animals, some of the offspring will possess the desirable qualities of the parents in a greater degree; or that animals are indefinitely perfectible.

The progress of a wild plant to a beautiful garden-flower is perhaps more marked and striking than any thing that takes place among animals; yet, even here, it would be the height of absurdity to assert that the progress was unlimited or indefinite. One of the most obvious features of the improvement is the increase of size. The flower has grown gradually larger by cultivation. If the progress were really unlimited, it might be increased ad infinitum; but this is so gross an absurdity, that we may be quite sure that, among plants as well as among animals, there is a limit to improvement, though we do not exactly know where it is. It is probable that the gardeners who contend for flower-prizes have often applied stronger dressing without success. At the same time it would be highly presumptuous in any man to say, that he had seen the finest carnation or anemone that could ever be made to grow. He might, however, assert, without the smallest chance of being contradicted by a future fact, that no carnation or anemone could ever, by cultivation, be increased to the size of a large cabbage; and yet there are assignable quantities greater than a cabbage. No man can say that he has seen the largest ear of wheat, or the largest oak, that could ever grow; but he might easily, and with perfect certainty, name a point of magnitude at which they would not arrive. In all these cases, therefore, a careful distinction should be made between an unlimited progress, and a progress where the limit is merely undefined.

It will be said, perhaps, that the reason why plants and animals cannot increase indefinitely in size is, that they would fall by their own weight. I answer, how do we know this but from experience? from experience of the degree of strength, with which these bodies are formed. I know, that a carnation long before it reached the size of a cabbage would not be supported by its stalk; but I only know this from my experience of the weakness and want of tenacity in the materials of a carnation-stalk. There might be substances of the same size that would support as large a head as a cabbage.

The reasons of the mortality of plants are at present perfectly unknown to us. No man can say why such a plant is annual, another biennial, and another endures for ages. The whole affair in all these cases, in plants, animals, and in the human race, is an affair of experience; and I only conclude, that man is mortal, because the invariable experience of all ages has proved the mortality of that organized substance, of which his visible body is made.

"What can we reason but from what we know?"

Sound philosophy will not authorize me to alter this opinion of the mortality of man on earth, till it can be clearly proved that the human race has made, and is making, a decided progress toward an illimitable extent of life. And the chief reason why I adduce the two particular instances from animals and plants was to expose and illustrate, if I could, the fallacy of that argument, which infers an unlimited progress merely because some partial improvement has taken place, and that the limit of this improvement cannot be precisely ascertained.

The capacity of improvement in plants and animals, to a certain degree, no person can possibly doubt. A clear and decided progress has already been made; and yet I think it appears that it would be highly absurd to say, that this progress has no limits. In human life, though there are great variations from different causes, it may be doubted whether, since the world began, any organic improvement whatever of the human frame can be clearly ascertained. The foundations, therefore, on which the arguments for the organic perfectibility of man rest are unusually weak, and can only be considered as mere conjectures. It does not, however, by any means, seem impossible that, by an attention to breed, a certain degree of improvement similar to that among animals might take place among men. Whether intellect could be communicated may be a matter of doubt; but size, strength, beauty, complexion, and, perhaps, even longevity, are in a degree transmissible. The error does not lie in supposing a small degree of improvement possible, but in not discriminating between a small improvement, the limit of which is undefined, and an improvement really unlimited. As the human race, however, could not be improved in this way, without condemning all the bad specimens to celibacy, it is not probable that an attention to breed should ever become general; indeed I know of no well-directed attempts of this kind, except in the ancient family of the Bickerstaffs, who are said to have been very successful in whitening the skins and increasing the height of their race by prudent marriages, particularly by that very judicious cross with Maud the milkmaid, by which some capital defects in the constitutions of the family were corrected.

It will not be necessary, I think, in order more completely to shew the improbability of any approach in man towards immortality on earth, to urge the very great additional weight, that an increase in the duration of life would give to the argument of population.

M. Condorcet's book may be considered not only as a sketch of the opinions of a celebrated individual, but of many of the literary men in France at the beginning of the revolution. As such, though merely a sketch, it seems worthy of attention.

Many, I doubt not, will think that the attempting gravely to controvert so absurd a paradox, as the immortality of man on earth, or, indeed, even the perfectibility of man and society, is a waste of time and words; and that such unfounded conjectures are best answered by neglect. I profess, however, to be of a different opinion. When paradoxes of this kind are advanced by ingenious and able men, neglect has no tendency to convince them of their mistakes. Priding themselves on what they conceive to be a mark of the reach and size of their own understandings, of the extent and comprehensiveness of their views, they will look upon this neglect merely as an indication of poverty and narrowness in the mental exertions of their contemporaries, and only think that the world is not yet prepared to receive their sublime truths.

On the contrary, a candid investigation of these subjects, accompanied with a perfect readiness to adopt any theory warranted by sound philosophy, may have a tendency to convince them that, in forming improbable and unfounded hypotheses, so far from enlarging the bounds of human science, they are contracting it; so far from promoting the improvement of the human mind, they are obstructing it: they are throwing us back again almost into the infancy of knowledge; and weakening the foundations of that mode of philosophizing, under the auspices of which science has of late made such rapid advances. The late rage for wide and unrestrained speculation seems to have been a kind of mental intoxication, arising perhaps from the great and unexpected discoveries, which had been made in various branches of science. To men elate and giddy with such successes, every thing appeared to be within the grasp of human powers; and under this illusion they confounded subjects where no real progress could be proved, with those where the progress had been marked, certain and acknowledged. Could they be persuaded to sober themselves with a little severe and chastised thinking, they would see that the cause of truth and of sound philosophy cannot but suffer, by substituting wild flights and unsupported assertions for patient investigation and well-supported proofs.

4.1.2 Book III, Chapter II: Of Systems of Equality. Godwin.

In reading Mr. Godwin's ingenious work on political justice, it is impossible not to be struck with the spirit and energy of his style, the force and precision of some of his reasonings, the ardent tone of his thoughts, and particularly with that impressive earnestness of manner which gives an air of truth to the whole. At the same time it must be confessed that he has not proceeded in his inquiries with the caution that sound philosophy requires; his conclusions are often unwarranted by his premises; he fails sometimes in removing objections which he himself brings forward; he relies too much on general and abstract propositions, which will not admit of application; and his conjectures certainly far outstrip the modesty of nature. The system of equality, which Mr. Godwin proposes, is, on a first view of it, the most beautiful and engaging of any that has yet appeared. A melioration of society to be produced merely by reason and conviction gives more promise of permanence than any change effected and maintained by force. The unlimited exercise of private judgment is a doctrine grand and captivating, and has a vast superiority over those systems, where every individual is in a manner the slave of the public. The substitution of benevolence, as the master-spring and moving principle of society, instead of self-love, appears at first sight to be a consummation devoutly to be wished. In short, it is impossible to contemplate the whole of this fair picture, without emotions of delight and admiration, accompanied with an ardent longing for the period of its accomplishment. But alas! that moment can never arrive. The whole is little better than a dream—a phantom of the imagination. These "gorgeous palaces" of happiness and immortality, these "solemn temples" of truth and virtue, will dissolve, "like the baseless fabric of a vision," when we awaken to real life, and contemplate the genuine situation of man on earth.

Mr. Godwin, at the conclusion of the third chapter of his eighth book, speaking of population, says, "there is a principle in human society, by which population is perpetually kept down to the level of the means of subsistence. Thus among the wandering tribes of America and Asia we never find, through the lapse of ages, that population has so increased, as to render necessary the cultivation of the earth."[2] This principle, which Mr. Godwin thus mentions as some mysterious and occult cause, and which he does not attempt to investigate, has appeared to be the law of necessity—misery, and the fear of misery.

The great error under which Mr. Godwin labours throughout his whole work is, the attributing of almost all the vices and misery that prevail in civil society to human institutions, political regulations and the established administration of property are, with him, the fruitful sources of all evil, the hotbeds of all the clinics that degrade mankind. Were this really a true state of the case, it would not seem an absolutely hopeless task, to remove evil completely from the world; and reason seems to be the proper and adequate instrument for effecting so great a purpose. But the truth is, that though human institutions appear to be, and indeed often are, the obvious and obtrusive causes of much mischief to society, they are, in reality, light and superficial, in comparison with those deeper-seated causes of evil, which result from the laws of nature and the passions of mankind.

In a chapter on the benefits attendant upon a system of equality, Mr. Godwin says, "The spirit of oppression, the spirit of servility, and the spirit of fraud, these are the immediate growth of the established administration of property. They are alike hostile to intellectual improvement. The other vices of envy, malice and revenge, are their inseparable companions. In a state of society where men lived in the midst of plenty, and where all shared alike the bounties of nature these sentiments would inevitably expire. The narrow principle of selfishness would vanish. No man being obliged to guard his little store, or provide with anxiety and pain for his restless wants, each would lose his individual

[2] P. 460, 8vo. 2d edit.

existence in the thought of the general good. No man would be an enemy to his neighbours, for they would have no subject of contention; and of consequence philanthropy would resume the empire which reason assigns her. Mind would be delivered from her perpetual anxiety about corporal support; and be free to expatiate in the field of thought which is congenial to her. Each would assist the inquiries all."[3]

This would indeed be a happy state. But that it is merely an imaginary picture with scarcely a feature near the truth, the reader, I am afraid, is already too well convinced.,

Man cannot live in the midst of plenty. All cannot share alike the bounties of nature. Were there no established administration of property, every man would be obliged to guard with force his little store. Selfishness would be triumphant. The subjects of contention would be perpetual. Every individual would be under a constant anxiety about corporal support, and not a single intellect would be left free to expatiate in the field of thought.

How little Mr. Godwin has turned his attention to the real state of human society, will sufficiently appear from the manner in which he endeavours to remove the difficulty of a superabundant population. He says, "The obvious answer to this objection is, that to reason thus is to foresee difficulties at a great distance. Three-fourths of the habitable globe are now uncultivated. The parts already cultivated are capable of immeasurable improvement. Myriads of centuries of still increasing population may pass away, and the earth be still found sufficient for the subsistence of its inhabitants."[4]

I have already pointed out the error of supposing that no distress or difficulty would arise from a redundant population, before the earth absolutely refused to produce any more. But let us imagine for a moment Mr. Godwin's system of equality realized, and see how soon this difficulty might be expected to press, under so perfect a form of society. A theory that will not admit of application cannot possibly be just.

Let us suppose all the causes of vice and misery in this island removed. War and contention cease. Unwholesome trades and manufactories do not exist. Crowds no longer collect together in great and pestilent cities for purposes of court intrigue, of commerce, and of vicious gratification. Simple, healthy and rational amusements take place of drinking, gaming and debauchery. There are no towns sufficiently large to have any prejudicial effects on the human constitution. The greater part of the happy inhabitants of this terrestrial Paradise live in hamlets and farm-houses scattered over the face of the country. All men are equal. The labours of luxury are at an end; and the necessary labours of agriculture are shared amicably among all. The number of persons and the produce of the island we suppose to be the same as at present. The spirit of benevolence, guided by impartial justice, will divide this produce among all the members of society according to their wants. Though it would be impossible that they should all have animal food every day, yet vegetable food, with meat occasionally, would satisfy the desires of a frugal people, and would be sufficient to preserve them in health, strength and spirits.

---

[3]Political Justice, b. viii. c. iii. p. 458.

[4]Polit. Justice, b. viii, c. x. p. 510.

Mr. Godwin considers marriage as a fraud and a monopoloy.[5] Let us suppose the commerce of the sexes established upon principles of the most perfect freedom. Mr. Godwin does not think himself, that this freedom would lead to a promiscuous intercourse; and in this I perfectly agree with him. The love of variety is a vicious, corrupt and unnatural taste, and could not prevail in any great degree in a simple and virtuous state of society. Each man would probably select for himself a partner, to whom he would adhere, as long as that adherence continued to be the choice of both parties. It would be of little consequence, according to Mr. Godwin, how many children a woman had, or to whom they belonged. Provisions and assistance would spontaneously flow from the quarter in which they abounded to the quarter in which they were deficient.[6] And every man, according to his capacity, would be ready to furnish instruction to the rising generation.

I cannot conceive a form of society so favourable upon the whole to population. The irremediableness to marriage, as it is at present constituted, undoubtedly deters many from entering into this state. An unshackled intercourse on the contrary would be a most powerful incitement to early attachments; and as we are supposing no anxiety about the future support of children to exist, I do not conceive that there would be one woman in a hundred, of twenty-three years of age, without a family.

With these extraordinary encouragements to population, and every cause of depopulation, as we have supposed, removed, the numbers would necessarily increase faster than in any society that has ever yet been known. I have before mentioned that the inhabitants of the back settlements of America appear to double their numbers in fifteen years. England is certainly a more healthy country than the back settlements of America; and as we have supposed every house in the island to be airy and wholesome, and the encouragements to have a family greater even than in America, no probable reason can be assigned, why the population should not double itself in less, if possible, than fifteen years. But to be quite sure, that we do not go beyond the truth, we will only suppose the period of doubling to be twenty-five years; a ratio of increase, which is slower than is known to have taken place throughout all the United States of America.

There can be little doubt that the equalization of property which we have supposed, added to the circumstance of the labour of the whole community being directed chiefly to agriculture, would tend greatly to augment the produce of the country. But to answer the demands of a population increasing so rapidly, Mr. Godwin's calculation of half an hour a day would certainly not be sufficient. It is probable that the half of every man's time must be employed for this purpose. Yet with such or much greater exertions, a person who is acquainted with the nature of the soil in this country, and who reflects on the fertility of the lands already in cultivation, and the barrenness of those that are not cultivated, will be very much disposed to doubt, whether the whole average produce could possibly be doubled in twenty-five years from the present period. The only chance of success would be from the

[5] Polit. Justice, b. viii. c. viii. p. 498, et seq.

[6] Political Justice, b. vii. c. viii. p. 504.

ploughing up of most of the grazing countries, and putting an end almost entirely to animal food. Yet this scheme would probably defeat itself. The soil of England will not produce much without dressing; and cattle seem to be necessary to make that species of manure, which best suits the land.

Difficult however as it might be to double the average produce of the island in twenty-five years, let us suppose it effected. At the expiration of the first period therefore, the food, though almost entirely vegetable, would be sufficient to support in health the population increased from 11 to 22 millions.[7]

During the next period, where will the food be found, to satisfy the importunate demands of the increasing numbers? Where is the fresh land to turn up? Where is the dressing necessary to improve that which is already in cultivation? There is no person with the smallest knowledge of land but would say that it was impossible that the average produce of the country could be increased during the second twenty-five years by a quantity equal to what it at present yields. Yet we will suppose this increase, however improbable, to take place. The exuberant strength of the argument allows of almost any concession. Even with this concession, however, there would be 11 millions at the expiration of the second term unprovided for. A quantity equal to the frugal support of 33 millions would be to be divided among 44 millions.

Alas! what becomes of the picture, where men lived in the midst of plenty, where no man was obliged to provide with anxiety and pain for his restless wants; where the narrow principle of selfishness did not exist; where the mind was delivered from her perpetual anxiety about corporal support, and free to expatiate in the field of thought which is congenial to her? This beautiful fabric of the imagination vanishes at the severe touch of truth. The spirit of benevolence, cherished and invigorated by plenty, is repressed by the chilling breath of want. The hateful passions that had vanished reappear. The mighty law of self-preservation expels all the softer and more exalted emotions of the soul. The temptations to evil are too strong for human nature to resist. The corn is plucked up before it is ripe, or secreted in unfair proportions; and the whole black train of vices that belong to falsehood are immediately generated. Provisions no longer flow in for the support of a mother with a large family. The children are sickly from insufficient food. The rosy flush of health gives place to the pallid cheek and hollow eye of misery. Benevolence, yet lingering in a few bosoms, makes some faint expiring struggles, till at length self-love resumes his wonted empire, and lords it triumphant over the world.

No human institutions here existed, to the perverseness of which Mr. Godwin ascribes the original sin of the worst men.[8] No opposition had been produced by them between public and private good. No monopoly had been created of those advantages which reason directs to be left in common. No man had been goaded to the breach of order by unjust laws. Benevolence had established her reign in all hearts. And yet in so short a period as

[7]The numbers here mentioned refer to the enumeration of 1800.

[8]Polit. Justice, b. viii, c. iii, p. 340.

fifty years, violence, oppression, falsehood, misery, every hateful vice and every form of distress, which degrade and sadden the present state of society, seem to have been generated by the most imperious circumstances, by laws inherent in the nature of man, and absolutely independent of all human regulations.

If we be not yet too well convinced of the reality of this melancholy picture, let us but look for a moment into the next period of twenty-five years, and we shall see that, according to the natural increase of population, 44 millions of human beings would be without the means of support; and at the conclusion of the first century, the population would have had the power of increasing to 176 millions, while the food was only sufficient for 55 millions, leaving 121 millions unprovided for: and yet all this time we are supposing the produce of the earth absolutely unlimited, and the yearly increase greater than the boldest speculator can imagine.

This is undoubtedly a very different view of the difficulty arising from the principle of population from that which Mr. Godwin gives, when he says, "Myriads of centuries of still increasing population may pass away, and the earth be still found sufficient for the subsistence of its inhabitants."

I am sufficiently aware that the redundant millions which I have mentioned could never have existed. It is a perfectly just observation of Mr. Godwin, that "there is a principle in human society by which population is perpetually kept down to the level of the means of subsistence." The sole question is, what is this principle? Is it some obscure and occult cause? Is it some mysterious interference of Heaven, which at a certain period strikes the men with impotence, and the women with barrenness? Or is it a cause open to our researches, within our view; a cause which has constantly been observed to operate, though with varied force, in every state in which man has been placed? Is it not misery and the fear of misery, the necessary and inevitable results of the laws of nature in the present stage of man's existence, which human institutions, so far from aggravating, have tended considerably to mitigate, though they can never remove?

It may be curious to observe, in the case that we have been supposing, how some of the principal laws, which at present govern civilized society, would be successively dictated by the most imperious necessity. As man, according to Mr. Godwin, is the creature of the impressions to which he is subject, the goadings of want could not continue long, before some violations of public or private stock would necessarily take place. As these violations increased in number and extent, the more active and comprehensive intellects of the society would soon perceive that, while the population was fast increasing, the yearly produce of the country would shortly begin to diminish. The urgency of the case would suggest the necessity of some immediate measures being taken for the general safety. Some kind of convention would be then called, and the dangerous situation of the country stated in the stronger terms. It would be observed that while they lived in the midst of plenty, it was of little consequence who laboured the least, or who possessed the least, as every man was perfectly willing and ready to supply the wants of his neighbour. But that the question was no longer, whether one man should give to another that which he did not use himself; but whether he should give to his neighbour the food which was absolutely necessary to his own existence. It would be represented that the number of those who were in want very greatly exceeded the number and means of those who should supply them; that these pressing wants, which from the state of the produce of the country could not all be gratified, had occasioned some flagrant violations of justice; that these violations had already checked the increase of food, and would, if they were not by some means or other prevented, throw the whole community into confusion; that imperious necessity seemed to dictate that a yearly increase of produce should, if possible, be obtained at all events; that in order to effect this first great and indispensable purpose, it would be advisable to make a more complete division of land, and to secure every man's property against violation by the most powerful sanctions.

It might be urged perhaps by some objectors, that as the fertility of the land increased, and various accidents occurred, the shares of some men might be much more than sufficient for their support; and that when the reign of self-love was once established, they would not distribute their surplus produce without some compensation in return. It would be observed in answer, that this was an inconvenience greatly to be lamented; but that it was an evil which would bear no comparison to the black train of distresses inevitably occasioned by the insecurity of property; that the quantity of food, which one man could consume, was necessarily limited by the narrow capacity of the human stomach; that it was certainly not probable that he should throw away the rest; and if he exchanged his surplus produce for the labour of others, this would be better than that these others should absolutely starve.

It seems highly probable therefore that an administration of property, not very different from that which prevails in civilized states at present, would be established as the best (though inadequate) remedy for the evils which were pressing on the society.

The next subject which would come under discussion, intimately connected with the preceding, is the commerce of the sexes. It would be urged by those who had turned their attention to the true cause of the difficulties under which the community laboured, that while every man felt secure that all his children would be well provided for by general benevolence, the powers of the earth would be absolutely inadequate to produce food for the population which would ensue; that even if the whole attention and labour of the society were directed to this sole point, and if by the most perfect security of properly, and every other encouragement that could be thought of, the greatest possible increase of produce were yearly obtained, yet still the increase of food would by no means keep pace with the much more rapid increase of population; that some check to population therefore was imperiously called for; that the most natural and obvious check seemed to be, to make every man provide for his own children; that this would operate in some respect as a measure and a guide in the increase of population, as it might be expected that no man would bring beings into the world for whom he could not find the means of support; that, where this notwithstanding was the case, it seemed necessary for the example of others, that the disgrace and inconvenience attending such a conduct should fall upon that individual, who had thus inconsiderately plunged himself and his innocent children into want and misery.

The institution of marriage, or at least of some express or implied obligation on every man to support his own children, seems to be the natural result of these reasonings in a community under the difficulties that we have supposed.

The view of these difficulties presents us with a very natural reason, why the disgrace which attends a breach of chastity should be greater in a woman than in a man. It could not be expected that women should have resources sufficient to support their own children. When therefore a woman had lived with a man who had entered into no compact to maintain her children, and aware of the inconveniences that he might bring upon himself, had deserted her, these children must necessarily fall upon the society for support, or starve. And to prevent the frequent recurrence of such an inconvenience, as it would be highly unjust to punish so natural a fault by personal restraint or infliction, the men might agree to punish it with disgrace. The offence is besides more obvious and conspicuous in the woman, and less liable to any mistake. The father of a child may not always be known; but the same uncertainty cannot easily exist with regard to the mother. Where the evidence of the offence, was most complete, and the inconvenience to the society, at the same time, the greatest, there it was agreed that the largest share of blame should fall. The obligation on every man to support his children, the society would enforce by positive laws; and the greater degree of inconvenience or labour, to which a family would necessarily subject him, added to some portion of disgrace which every human being must incur who leads another into unhappiness, might be considered as a sufficient punishment for the man.

That a woman should at present be almost driven from society for an offence, which men commit nearly with impunity, seems undoubtedly to be a breach of natural justice. But the origin of the custom, as the most obvious and effectual method of preventing the frequent recurrence of a serious inconvenience to the community, appears to be natural, though not perhaps perfectly justifiable. This origin is now lost in the new train of ideas that the custom has since generated. What at first might be dictated by state necessity is now supported by female delicacy; and operates with the greatest force on that part of the society, where, if the original intention of the custom were preserved, there is the least real occasion for it.

When these two fundamental laws of society, the security of property, and the institution of marriage, were once established, inequality of conditions must necessarily follow. Those who were born after the division of property would come into a world already possessed. If their parents, from having too large a family, were unable to give them sufficient for their support, what could they do in a world where every thing was appropriated? We have seen the fatal effects that would result to society, if every man had a valid claim to an equal share of the produce of the earth. The members of a family, which was grown too large for the original division of land appropriated to it, could not then demand a part of the surplus produce of others as a debt of justice. It has appeared that from the inevitable laws of human nature some human beings will be exposed to want. These are the unhappy persons, who in the great lottery of life have drawn a blank. The number of these persons would soon exceed the ability of the surplus produce to supply. Moral merit is a very difficult criterion, except in extreme cases. The owners of surplus produce would in general seek some more obvious mark of distinction; and it seems to be both natural and just, that, except upon particular occasions, their choice should fall upon those who were able, and professed themselves willing, to exert their strength in procuring a further surplus produce, which would at once benefit the community and enable the proprietors to afford assistance to greater numbers. All who were in want of food would be urged by necessity to offer their labour in exchange for this article so absolutely necessary to existence. The fund appropriated to the maintenance of labour would be the aggregate quantity of food possessed by the owners of land beyond their own consumption. When the demands upon this fund were great and numerous it would naturally be divided into very small shares. Labour would be ill paid. Men would offer to work for a bare subsistence; and the rearing of families would be checked by sickness and misery. On the contrary, when this fund was increasing fast; when it was great in proportion to the number of claimants, it would be divided in much larger shares. No man would exchange his labour without receiving an ample quantity of food in return. Labourers would live in ease and comfort, and would consequently be able to rear a numerous and vigorous offspring.

On the state of this fund, the happiness, or the degree of misery, prevailing among the lower classes of people in every known state at present, chiefly depends; and on this happiness or degree of misery, depends principally the increase, stationariness, or decrease of population.

And thus it appears that a society constituted according to the most beautiful form that imagination can conceive, with benevolence for its moving principle instead of self-love, and with every evil disposition in all its members corrected by reason, not force, would, from the inevitable laws of nature, and not from any fault in human institutions, degenerate in a very short period into a society constructed upon a plan not essentially different from that which prevails in every known state at present; a society, divided into a class of proprietors and a class of labourers, and with self-love for the main-spring of the great machine.

In the supposition which I have made, I have undoubtedly taken the increase of population smaller, and the increase of produce greater, than they really would be. No reason can be assigned why, under the circumstances supposed, population should not increase faster than in any known instance. If then we were to take the period of doubling at fifteen years instead of twenty-five years, and reflect upon the labour necessary to double the produce in so short a time, even if we allow it possible; we may venture to pronounce with certainty, that, if Mr. Godwin's system of society were established, instead of myriads of centuries, not thirty years could elapse before its utter destruction from the simple principle of population.

I have taken no notice of emigration in this place, for obvious reasons. If such societies were instituted in other parts of Europe, these countries would be under the same difficulties with regard to population, and could admit no fresh members into their bosoms. If this beautiful society were confined to our island, it must have degenerated strangely from its original purity, and administer but a very small portion of the happiness it proposed, before any of its members would voluntarily consent to leave it, and live under such governments as at present exist in Europe, or submit to the extreme hardships of first settlers in new regions.

4.1.3 Book III, Chapter III: Of Systems of Equality (continued).

It was suggested to me same years since by persons for whose judgment I have a high respect, that it might be advisable, in a new edition, to throw out the matter relative to systems of equality, to Wallace, Condorcet and Godwin, as having in a considerable degree lost its interest, and as not being strictly connected with the main subject of the Essay, which is an explanation and illustration of the theory of population. But independently of its being natural for me to have some little partiality for that part of the work which led to those inquiries on which the main subject rests; I really think that there should be somewhere on record an answer to systems of equality founded on the principle of population; and perhaps such an answer is as appropriately placed, and is likely to have as much effect, among the illustrations and applications of the principle of population, as in any other situation to which it could be assigned.

The appearances in all human societies, particularly in all those which are the furthest advanced in civilization and improvement, will ever be such, as to inspire superficial observers with a belief that a prodigious change for the better might be effected by the introduction of a system of equality and of common property. They see abundance in some quarters, and want in others; and the natural and obvious remedy seems to be an equal division of the produce. They see a prodigious quantity of human exertion wasted upon trivial, useless, and sometimes pernicious objects, which might either be wholly saved or more effectively employed. They see invention after invention in machinery brought forward, which is seemingly calculated, in the most marked manner, to abate the sum of human toil. Yet with these apparent means of giving plenty, leisure and happiness to all, they still see the labours of the great mass of society undiminished, and their condition, if not deteriorated, in no very striking and palpable manner improved.

Under these circumstances, it cannot be a matter of wonder that proposals for systems of equality should be continually reviving. After periods when the subject has undergone a thorough discussion, or when some great experiment in improvement has failed, it is likely that the question should be dormant for a time, and that the opinions of the advocates of equality should be ranked among those errors which had passed away to be heard of no more. But it is probable that if the world were to last for any number of thousand years, systems of equality would be among those errors, which like the tunes of a barrel organ, to use the illustration of Dugald Stewart,[9] will never cease to return at certain intervals.

[9] Preliminary Dissertation to Supplement to the Encyclopædia Britannica, p. 121.

I am induced to make these remarks, and to add a little to what I have already said on systems of equality, instead of leaving out the whole discussion, by a tendency to a revival of this kind at the present moment.[10]

A gentleman, for whom I have a very sincere respect, Mr. Owen, of Lanark, has lately published a work entitled *A New View of Society,* which is intended to prepare the public mind for the introduction of a system involving a community of labour and of goods. It is also generally known that an idea has lately prevailed among some of the lower classes of society, that the land is the people's farm, the rent of which ought to be equally divided among them; and that they have been deprived of the benefits which belong to them from this their natural inheritance, by the injustice and oppression of their stewards, the landlords.

Mr. Owen is, I believe, a man of real benevolence, who has done much good; and every friend of humanity must heartily wish him success in his endeavours to procure an Act of Parliament for limiting the hours of working among the children in the cotton manufactories, and preventing them from being employed at too early an age. He is further entitled to great attention on all subjects relating to education, from the experience and knowledge which he must have gained in an intercourse of many years with two thousand manufacturers, and from the success which is said to have resulted from his modes of management. A theory professed to be founded on such experience is, no doubt, worthy of much more consideration than one formed in a closet.

The claims to attention possessed by the author of the new doctrines relating to land are certainly very slender; and the doctrines themselves indicate a very great degree of ignorance; but the errors of the labouring classes of society are always entitled to great indulgence and consideration. They are the natural and pardonable result of their liability to be deceived by first appearances, and by the arts of designing men, owing to the nature of their situation, and the scanty knowledge which in general falls to their share. And, except in extreme cases, it must always be the wish of those who are better informed, that they should be brought to a sense of the truth, rather by patience and the gradual diffusion of education and knowledge, than by any harsher methods.

After what I have already said on systems of equality in the preceding chapters, I shall not think it necessary to enter into a long and elaborate refutation of these doctrines. I merely mean to give an additional reason for leaving on record an answer to systems of equality, founded on the principle of population, together with a concise restatement of this answer for practical application.

Of the two decisive arguments against such systems, one is, the unsuitableness of a state of equality, both according to experience and theory, to the production of those stimulants to exertion which can alone overcome the natural indolence of man, and prompt him to the proper cultivation of the earth and the fabrication of those conveniences and comforts which are necessary to his happiness.

---

[10] Written in 1817.

And the other, the inevitable and necessary poverty and misery in which every system of equality must shortly terminate from the acknowledged tendency of the human race to increase faster than the means of subsistence, unless such increase be prevented by means infinitely more cruel than those which result from the laws of private property, and the moral obligation imposed on every man by the commands of God and nature to support his own children.

The first of these arguments has, I confess, always appeared to my own mind sufficiently conclusive. A state, in which an inequality of conditions offers the natural rewards of good conduct, and inspires widely and generally the hopes of rising and the fears of falling in society, is unquestionably the best calculated to develope the energies and faculties of man, and the best suited to the exercise and improvement of human virtue;[11] and history, in every case of equality that has yet occurred, has uniformly borne witness to the depressing and deadening effects which arise from the want of this stimulus. But still, perhaps, it may be true that neither experience nor theory on this subject is quite so decisive as to preclude all plausible arguments on the other side. It may be said that the instances which history records of systems of equality really carried into execution are so few, and those in societies so little advanced from a state of barbarism, as to afford no fair conclusions relative to periods of great civilization and improvement; that, in other instances, in ancient times, where approaches were made toward a tolerable equality of conditions, examples of considerable energy of character in some lines of exertion are not unfrequent; and that in modern times some societies, particularly of Moravians, are known to have had much of their property in common without occasioning the destruction of their industry. It may be said that, allowing the stimulus of inequality of conditions to have been necessary, in order to raise man from the indolence and apathy of the savage to the activity and intelligence of civilized life, it does not follow that the continuance of the same stimulus should be necessary when this activity and energy of mind has been once gained. It may *then* be allowable quietly to enjoy the benefit of a regimen which, like many other stimulants, having produced its proper effect at a certain point, must be left off, or exhaustion, disease, and death will follow.

These observations are certainly not of a nature to produce conviction in those who have studied the human character; but they are, to a certain degree, plausible, and do not admit of so definite and decisive an answer as to make the proposal for an experiment in modern times utterly absurd and unreasonable.

The peculiar advantage of the other argument against systems of equality, that which is founded on the principle of population, is, that it is not only still more generally and uniformly confirmed by experience, in every age and in every part of the world, but it is so pre-eminently clear in theory, that no tolerably plausible answer can be given to it; and,

[11] See this subject very ably treated in a work on the Records of the Creation, and the Moral Attributes of the Creator, by the Rev. John Bird Sumner, not long since published; a work of very great merit, which I hope soon to see in as extensive circulation as it deserves.

consequently, no decent pretence can be brought forward for an experiment. The affair is a matter of the most simple calculation applied to the known properties of land, and the proportion of births to deaths which takes place in almost every country village. There are many parishes in England, where, notwithstanding the actual difficulties attending the support of a family which must *necessarily* occur in every well-peopled country, and making no allowances for omissions in the registers, the births are to the deaths in the proportion of 2 to 1. This proportion, with the usual rate of mortality in country places, of about 1 in 50, would continue doubling the population in 41 years, if there were no emigrations from the parish. But in any system of equality, either such as that proposed by Mr. Owen, or in parochial partnerships in land, not only would there be no means of emigration to other parishes with any prospect of relief; but the rate of increase at first would, of course, be much greater than in the present state of society. What then, I would ask, is to prevent the division of the produce of the soil to each individual from becoming every year less and less, till the whole society and every individual member of it are pressed down by want and misery?[12]

This is a very simple and intelligible question. And surely no man ought to propose or support a system of equality, who is not able to give a rational answer to it, at least in theory. But, even in theory, I have never yet heard any thing approaching to a rational answer to it.

It is a very superficial observation which has sometimes been made, that it is a contradiction to lay great stress upon the efficacy of moral restraint in an improved and improving state of society, according to the present structure of it, and yet to suppose that it would not act with sufficient force in a system of equality, which almost always presupposes a great diffusion of information, and a great improvement of the human mind. Those who have made this observation do not see that the encouragement and motive to moral restraint are at once destroyed in a system of equality, and community of goods.

Let us suppose that in a system of equality, in spite of the best exertions to procure more food, the population is pressing hard against the limits of subsistence, and all are becoming very poor. It is evidently necessary under these circumstances, in order to prevent the society from starving, that the rate at which the population increases should be retarded.

[12] In the Spencean system, as published by the secretary of the Society of Spencean Philanthropists, it unfortunately happens, that after the *proposed* allowances have been made for the expenses of the government, and of the other bodies in the state which are intended to be supported, there would be absolutely no remainder; and the people would not derive a single sixpence from their estate, even at first, and on the supposition of the national debt being entirely abolished, without the slightest compensation to the national creditors.

The annual rent of the land, houses, mines, and fisheries, is estimated at 150 millions, about three times its real amount; yet, even upon this extravagant estimate, it is calculated that the division would only come to about four pounds a head, not more than is sometimes given to individuals from the poor's rates; a miserable provision! and yet constantly diminishing.

But who are the persons that are to exercise the restraint thus called for, and either to marry late or not at all? It does not seem to be a necessary consequence of a system of equality that all the human passions should be at once extinguished by it; but if not, those who might wish to marry would feel it hard that they should be among the number forced to restrain their inclinations. As all would be equal, and in similar circumstances, there would be no reason whatever why one individual should think himself obliged to practise the duty of restraint more than another. The thing however must be done, with any hope of avoiding universal misery; and in a state of equality, the necessary restraint could only be effected by some general law. But how is this law to be supported, and how are the violations of it to be punished? Is the man who marries early to be pointed at with the finger of scorn? is he to be whipped at the cart's tail? is he to be confined for years in a prison? is he to have his children exposed? Are not all direct punishments for an offence of this kind shocking and unnatural to the last degree? And yet, if it be absolutely necessary, in order to prevent the most overwhelming wretchedness, that there should be some restraint on the tendency to early marriages, when the resources of the country are only sufficient to support a slow rate of increase, can the most fertile imagination conceive one at once so natural, so just, so consonant to the laws of God and to the best laws framed by the most enlightened men, as that each individual should be responsible for the maintenance of his own children; that is, that he should be subjected to the natural inconveniences and difficulties arising from the indulgence of his inclinations, and to no other whatever?

That this natural check to early marriages arising from a view of the difficulty attending the support of a large family operates very widely throughout all classes of society in every civilized state, and may be expected to be still more effective, as the lower classes of people continue to improve in knowledge and prudence, cannot admit of the slightest doubt. But the operation of this natural check depends exclusively upon the existence of the laws of property, and succession; and in a state of equality and community of property could only be replaced by some artificial regulation of a very different stamp, and a much more unnatural character. Of this Mr. Owen is fully sensible, and has in consequence taxed his ingenuity to the utmost to invent some mode, by which the difficulties arising from the progress of population could be got rid of, in the state of society to which he looks forward. His absolute inability to suggest any mode of accomplishing this object that is not unnatural, immoral, or cruel in a high degree, together with the same want of success in every other person, ancient[13] or modern, who has made a similar attempt, seem to shew that the argument against systems of equality founded on the principle of population does not admit of a plausible answer, even in theory. The fact of the tendency of population to increase beyond the means of subsistence may be seen in almost every register of a country parish in the kingdom. The unavoidable effect of this tendency to depress the whole body of the people in want and misery, unless the progress of the population be somehow or

[13]The reader has already seen in ch. xiii. bk. i. the detestable means of checking population proposed by some ancient lawgivers in order to support their systems of equality.

other retarded, is equally obvious; and the impossibility of checking the rate of increase in a state of equality, without resorting to regulations that are unnatural, immoral or cruel, forms all argument at once conclusive against every such system.

4.1.4 Book III, Chapter IV: Of Emigration.

Although the resource of emigration seems to be excluded from such perfect societies as the advocates of equality generally contemplate, yet in that imperfect state of improvement, which alone can rationally be expected, it may fairly enter into our consideration. And as it is not probable that human industry should begin to receive its best direction throughout all the nations of the earth at the same time, it may be said that, in the case of a redundant population in the more cultivated parts of the world, the natural and obvious remedy which presents itself is emigration to those parts that are uncultivated. As these parts are of great extent, and very thinly peopled, this resource might appear, on a first view of the subject, an adequate remedy, or at least of a nature calculated to remove the evil to a distant period: but when we advert to experience and the actual state of the uncivilized parts of the globe, instead of any thing like an adequate remedy, it will appear but a slight palliative.

In the accounts which we have received of the peopling of new countries, the dangers, difficulties and hardships, with which the first settlers have had to struggle, appear to be even greater than we can well imagine they could be exposed to in their parent state. The endeavour to avoid that degree of unhappiness which arises from the difficulty of supporting a family might long have left the new world of America unpeopled by Europeans, if those more powerful passions, the thirst of gain, the spirit of adventure, and religious enthusiasm, had not directed and animated the enterprise. These passions enabled the first adventurers to triumph over every obstacle; but in many instances, in a way to make humanity shudder, and to defeat the true end of emigration. Whatever may be the character of the Spanish inhabitants of Mexico and Peru at the present moment, we cannot read the accounts of the first conquests of these countries, without feeling strongly, that the race destroyed was, in moral worth as well as numbers, superior to the race of their destroyers,

The parts of America settled by the English, from being thinly peopled, were better adapted to the establishment of new colonies; yet even here, the most formidable difficulties presented themselves. In the settlement of Virginia, begun by Sir Walter Raleigh and established by Lord Delaware, three attempts completely failed. Nearly half of the first colony was destroyed by the savages, and the rest, consumed and worn down by fatigue and famine, deserted the country, and returned home in despair. The second colony was cut off to a man in a manner unknown; but they were supposed to be destroyed by the Indians. The third experienced the same dismal fate; and the remains of the fourth, after it had been reduced by famine and disease in the course of six months from 500 to 60 persons, were returning in a famishing and desperate condition to England, when they were met in the mouth of the Chesapeak bay by Lord Delaware, with a squadron loaded with provisions, and every thing for their relief and defence.[14]

The first puritan settlers in New England were few in number. They landed in a bad season, and were only supported by their private funds. The winter was premature and terribly cold; the country was covered with wood, and afforded very little for the refreshment of persons sickly with such a voyage, or for the sustenance of an infant people. Nearly half of them perished by the scurvy, by want, and the severity of the climate; yet those who survived were not dispirited by their hardships, but, supported by their energy of character, and the satisfaction of finding themselves out of the reach of the spiritual arm, reduced this savage country by degrees to yield a comfortable subsistence.[15]

Even the plantation of Barbadoes, which increased afterwards with such extraordinary rapidity, had at first to contend with a country utterly desolate, an extreme want of provisions, a difficulty in clearing the ground unusually great from the uncommon size and hardness of the trees, a most disheartening scantiness and poverty in their first crops, and a slow and precarious supply of provisions from England.[16]

The attempt of the French in 1663, to form at once a powerful colony in Guiana, was attended with the most disastrous consequences. Twelve thousand men were landed in the rainy season, and placed under tents and miserable sheds. In this situation, inactive, weary of existence, and in want of all necessaries; exposed to contagious distempers, which are always occasioned by bad provisions, and to all the irregularities which idleness produces among the lower classes of society, almost the whole of them ended their lives in all the horrors of despair. The attempt was completely abortive. Two thousand men, whose robust constitutions had enabled them to resist the inclemency of the climate, and the miseries to which they had been exposed, were brought back to France; and the 26,000,000 of livres, which had been expended in the expedition, were totally lost.[17]

In the late settlements at Port Jackson in New Holland, a melancholy and affecting picture is drawn by Collins of the extreme hardships, with which, for some years, the infant colony had to struggle, before the produce was equal to its support. These distresses were undoubtedly aggravated by the character of the settlers; but those which were caused by the unhealthiness of a newly cleared country, the failure of first crops, and the uncertainty of supplies from so distant a mother-country, were of themselves sufficiently disheartening, to place in a strong point of view the necessity of great resources, as well as unconquerable perseverance, in the colonization of savage countries.

The establishment of colonies in the more thinly peopled regions of Europe and Asia would evidently require still greater resources. From the power and warlike character of the inhabitants of these countries, a considerable military force would be necessary, to prevent

[14] Burke's America, vol. ii. p. 219. Robertson, b. ix. p. 83, 86.

[15] Burke's America, vol. ii. p. 144.

[16] Burke's America, vol. ii. p. 85.

[17] Raynal, Hist. des Indes, tom. vii. liv. xiii. p. 43. 10 vols. 8vo. 1795.

their utter and immediate destruction. Even the frontier provinces of the most powerful states are defended with considerable difficulty from such restless neighbours; and the peaceful labours of the cultivator are continually interrupted by their predatory incursions. The late Empress Catherine of Russia found it necessary to protect by regular fortresses the colonies which she had established in the districts near the Wolga; and the calamities which her subjects suffered by the incursions of the Crim Tartars furnished a pretext, and perhaps a just one, for taking possession of the whole of the Crimea, and expelling the greatest part of these turbulent neighbours, and reducing the rest to a more tranquil mode of life.

The difficulties attending a first establishment from soil, climate and the want of proper conveniences, are of course nearly the same in these regions as in America. Mr. Eton, in his Account of the Turkish Empire, says that 75,000 Christians were obliged by Russia to emigrate from the Crimea, and sent to inhabit the country abandoned by the Nogai Tartars; but the winter coming on before the houses built for them were ready, a great part of them had no other shelter from the cold than what was afforded them by holes dull in the ground, covered with what they could procure, and the greatest part of them perished. Only seven thousand remained a few years afterwards. Another colony from Italy to the banks of the Borysthenes had, he says, no better fate, owing to the bad management of those, who were commissioned to provide for them.

It is needless to add to these instances, as the accounts given of the difficulties experienced in new settlements are all nearly similar. It has been justly observed by a correspondent of Dr. Franklin, that one of the reasons why we have seen so many fruitless attempts to settle colonies at an immense public and private expense by several of the powers of Europe is, that the moral and mechanical habits adapted to the mother-country are frequently not so to the new-settled one, and to external events, many of which are unforeseen; and that it is to be remarked that none of the English colonies became any way considerable, till the necessary manners were born and grew up in the country. Pallas particularly notices the want of proper habits in the colonies established by Russia, as one of the causes why they did not increase so fast as might have been expected.

In addition to this it may be observed, that the first establishment of a new colony generally presents an instance of a country peopled considerably beyond its actual produce; and the natural consequence seems to be, that this population, if not amply supplied by the mother-country, should at the commencement be diminished to the level of the first scanty productions, and not begin permanently to increase, till the remaining numbers had so far cultivated the soil, as to make it yield a quantity of food more than sufficient for their own support; and which consequently they could divide with a family. The frequent failures in the establishment of new colonies tend strongly to shew the order of precedence between food and population.

It must be acknowledged then, that the class of people, on whom the distress arising from a too rapidly increasing population would principally fall, could not possibly begin a new colony in a distant country. From the nature of their situation, they must necessarily be deficient in those resources, which alone could ensure success; and unless they could find leaders among the higher classes urged by the spirit of avarice or enterprise, or of religious or political discontent; or were furnished with means and support by government; whatever degree of misery they might suffer in their own country from the scarcity of subsistence, they would be absolutely unable to take possession of any of those uncultivated regions, of which there is such an extent on the earth.

When new colonies have been once securely established, the difficulty of emigration is indeed very considerably diminished; yet, even then, some resources are necessary to provide vessels for the voyage, and support and assistance till the emigrants can settle themselves, and find employment in their adopted country. How far it is incumbent upon a government to furnish these resources may be a question; but whatever be its duty in this particular, perhaps it is too much to expect that, except where any particular colonial advantages are proposed, emigration should be actively assisted.

The necessary resources for transport and maintenance are however frequently furnished by individuals or private companies. For many years before the American war, and for some few since, the facilities of emigration to this new world, and the probable advantages in view, were unusually great; and it must be considered undoubtedly as a very happy circumstance for any country, to have so comfortable an asylum for its redundant population. But I would ask, whether, even during these periods, the distress among the common people in this country was little or nothing; and whether every man felt secure before he ventured on marriage, that, however large his family might be, he should find no difficulty to supporting it without parish assistance. The answer, I fear, could not be in the affirmative.

It will be said that, when an opportunity of advantageous emigration is offered, it is the fault of the people themselves, if, instead of accepting it, they prefer a life of celibacy or extreme poverty in their own country. Is it then a fault for a man to feel an attachment to his native soil, to love the parents that nurtured him, his kindred, his friends, and the companions of his early years? Or is it no evil that he suffers, because he consents to bear it rather than snap these cords which nature has wound in close and intricate folds round the human heart? The great plan of Providence seems to require, indeed, that these ties should sometimes be broken; but the separation does not, on that account, give less pain; and though the general good may be promoted by it, it does not cease to be an individual evil. Besides, doubts and uncertainty must ever attend all distant emigrations, particularly in the apprehensions of the lower classes of people. They cannot feel quite secure, that the representations made to them of the high price of labour or the cheapness of land, are accurately true. They are placing themselves in the power of the persons who are to furnish them with the means of transport and maintenance, who may perhaps have an interest in deceiving them; and the sea which they are to pass, appears to them like the separation of death from all their former connections, and in a manner to preclude the possibility of return in case of failure, as they cannot expect the offer of the same means to bring them back. We cannot be surprised then, that, except where a spirit of enterprise is added to the uneasiness of poverty, the consideration of these circumstances should frequently

"Make them rather bear the ills they suffer,

Than fly to others which they know not of."

If a tract of rich land as large as this island were suddenly annexed to it, and sold in small lots, or let out in small farms, the case would be very different, and the melioration of the state of the common people would be sudden and striking; though the rich would be continually complaining of the high price of labour, the pride of the lower classes, and the difficulty of getting work done. These, I understand, are not unfrequent complaints among the men of property in America.

Every resource however from emigration, if used effectually, as this would be, must be of short duration. There is scarcely a state in Europe, except perhaps Russia, the inhabitants of which do not often endeavour to better their condition by removing to other countries. As these states therefore have nearly all rather a redundant than deficient population in proportion to their product, they cannot be supposed to afford any effectual resources of emigration to each other. Let us suppose for a moment, that in this more enlightened part of the globe, the internal economy of each state were so admirably regulated, that no checks existed to population, and that the different governments provided every facility for emigration. Taking the population of Europe, excluding Russia, at a hundred millions, and allowing a greater increase of produce than is probable, or even possible, in the mother-countries, the redundancy of parent stock in a single century would be eleven hundred millions, which, added to the natural increase of the colonies during the same time, would more than double what has been supposed to be the present population of the whole earth.

Can we imagine, that in the uncultivated parts of Asia, Africa, or America, the greatest exertions and the best-directed endeavours could, in so short a period, prepare a quantity of land sufficient for the support of such a population? If any sanguine person should feel a doubt upon the subject, let him only add 25 or 50 years more, and every doubt must be crushed in overwhelming conviction.

It is evident, therefore, that the reason why the resource of emigration has so long continued to be held out as a remedy to redundant population is, because, from the natural unwillingness of people to desert their native country, and the difficulty of clearing and cultivating fresh soil, it never is or can be adequately adopted. If this remedy were indeed really effectual, and had power so far to relieve the disorders of vice and misery in old states, as to place them in the condition of the most prosperous new colonies, we should soon see the phial exhausted; and when the disorders returned with increased virulence, every hope from this quarter would be for ever closed.

It is clear, therefore, that with any view of making room for an unrestricted increase of population, emigration is perfectly inadequate; but as a partial and temporary expedient, and with a view to the more general cultivation of the earth, and the wider extension of civilization, it seems to be both useful and proper; and if it cannot be proved that governments are bound actively to encourage it, it is not only strikingly unjust, but in the highest degree impolitic in them to prevent it. There are no fears so totally ill-grounded as the fears of depopulation from emigration. The *vis inertiæ* of the great body of the people, and their attachment to their homes, are qualities so strong and general, that we may rest assured they will not emigrate unless, from political discontents or extreme poverty, they are in such a state as will make it as much for the advantage of their country as of themselves that they should go out of it. The complaints of high wages in consequence of emigrations are of all others the most unreasonable, and ought the least to be attended to. If the wages of labour in any country be such as to enable the lower classes of people to live with tolerable comfort, we may be quite certain that they will not emigrate; and if they be not such, it is cruelty and injustice to detain them.

In all countries the progress of wealth must depend mainly upon the industry, skill, and success of individuals, and upon the state and demands of other countries. Consequently, in all countries great variations may take place at different times in the rate at which wealth increases, and in the demand for labour. But though the progress of population is mainly regulated by the effective demand for labour, it is obvious that the number of people cannot conform itself immediately to the state of this demand. Some time is required to bring more labour into the market when it is wanted; and some time to check the supply when it is flowing in with too great rapidity. If these variations amount to no more than that natural sort of oscillation noticed in as early part of this work, which seems almost always to accompany the progress of population and food, they should be submitted to as a part of the usual course of things. But circumstances may occasionally give them great force, and then, during the period that the supply of labour is increasing faster than the demand, the labouring classes are subject to the most severe distress. If, for instance, from a combination of external and internal causes, a very great stimulus should be given to the population of a country for ten or twelve years together, and it should then comparatively cease, it is clear that labour will continue flowing into the market with almost undiminished rapidity, while the means of employing and paying it have been essentially contracted. It is precisely under these circumstances that emigration is most useful as a temporary relief; and it is in these circumstances that Great Britain finds herself placed at present.[18] Though no emigration should take place, the population will by degrees conform itself to the state of the demand for labour; but the interval must be marked by the most severe distress, the amount of which can scarcely be reduced by any human efforts; because, though it may be mitigated at particular periods, and as it affects particular classes, it will be proportionably extended over a larger space of time, and a greater number of people. The only real relief in such a case is emigration; and the subject at the present moment is well worthy the attention of the government, both as a matter of humanity and policy.

4.1.5 Book III, Chapter V: Of Poor Laws.

To remedy the frequent distresses of the poor, laws to enforce their relief have been instituted; and in the establishment of a general system of this kind England has particularly

[18] 1816 and 1817.

distinguished herself. But it is to be feared, that, though it may have alleviated a little the intensity of individual misfortune, it has spread the evil over a much larger surface.

It is a subject often started in conversation, and mentioned always as a matter of great surprise, that, notwithstanding the immense sum which is annually collected for the poor in this country, there is still so much distress among them. Some think that the money must be embezzled for private use; others, that the churchwardens and overseers consume the greatest part of it in feasting. All agree that somehow or other it must be very ill managed. In short, the fact, that even before the late scarcities three millions were collected annually for the poor, and yet that their distresses were not removed, is the subject of continual astonishment. But a man who looks a little below the surface of things would be much more astonished, if the fact were otherwise than it is observed to be; or even if a collection universally of eighteen shillings in the pound, instead of four, were materially to alter it.

Suppose, that by a subscription of the rich the eighteen pence or two shillings, which men earn now, were made up five shillings: it might be imagined, perhaps, that they would then be able to live comfortably, and have a piece of meat every day for their dinner. But this would be a very false conclusion. The transfer of three additional shillings a day to each labourer would not increase the quantity of meat in the country. There is not at present enough for all to have a moderate share. What would then be the consequence? the competition among the buyers in the market of meat would rapidly raise the price from eight pence or nine pence to two or three shillings in the pound, and the commodity would not be divided among many more than it is at present. When an article is scarce, and cannot be distributed to all, he that can shew the most valid patent, that is, he that offers the most money, becomes the possessor. If we can suppose the competition among the buyers of meat to continue long enough for a greater number of cattle to be reared annually, this could only be done at the expense of the corn, which would be a very disadvantageous exchange; for it is well known, that the country could not then support the same population; and when subsistence is scarce in proportion to the number of people, it is of little consequence, whether the lowest members of the society possess two shillings or five. They must, at all events, be reduced to live upon the hardest fare, and in the smallest quantity.

It might be said, perhaps, that the increased number of purchasers in every article would give a spur to productive industry, and that the whole produce of the island would be increased. But the spur that these fancied riches would give to population would more than counterbalance it; and the increased produce would be to be divided among a more than proportionably increased number of people.

A collection from the rich of eighteen shillings in the pound, even if distributed in the most judicious manner, would have an effect similar to that resulting from the supposition which I have just made; and no possible sacrifices of the rich, particularly in money, could for any time prevent the recurrence of distress among the lower members of society, whoever they were. Great changes might indeed be made. The rich might become poor, and some of the poor rich: but while the present proportion between population and food continues, a part of the society must necessarily find it difficult to support a family, and this difficulty will naturally fall on the least fortunate members.

It may at first appear strange, but I believe it is true, that I cannot by means of money raise the condition of a poor man, and enable him to live much better than he did before, without proportionably depressing others in the same class. If I retrench the quantity of food consumed in my house, and give him what I have cut off, I then benefit him without depressing any but myself and family, who perhaps may be well able to bear it. If I turn up a piece of uncultivated land, and give him the produce, I then benefit both him and all the members of society, because what he before consumed is thrown into the common stock, and probably some of the new produce with it. But if I only give him money, supposing the produce of the country to remain the same, I give him a title to a larger share of that produce than formerly, which share he cannot receive without diminishing the shares of others. It is evident that this effect in individual instances must be so small as to be totally imperceptible; but still it must exist, as many other effects do which, like some of the insects that people the air, elude our grosser perceptions.

Supposing the quantity of food in any country to remain the same for many years together, it is evident that this food must be divided according to the value of each man's patent, or the sum of money which he can afford to spend in this commodity so universally in request. It is a demonstrative truth, therefore, that the patents of one set of men could not be increased in value without diminishing the value of the patents of some other set of men. If the rich were to subscribe and give five shillings a day to five hundred thousand men, without retrenching their own tables, no doubt can exist, that as these men would live more at their ease, and consume a greater quantity of provisions, there would be less food remaining to divide among the rest; and consequently each man's patent would be diminished in value, or the same member of pieces of silver would purchase a smaller quantity of subsistence, and the price of provisions would universally rise.

These general reasonings have been strikingly confirmed during the late scarcities.[19] The supposition which I have made of a collection from the rich of eighteen shillings in the pound has been newly realized; and the effect has been such as might have been expected. If the same distribution had been made when no scarcity existed, a considerable advance in the price of provisions would have been a necessary consequence; but following as it did a scarcity, its effect must have been doubly powerful. No person, I believe, will venture to doubt, that if we were to give three additional shillings a day to every labouring man in the kingdom, as I before supposed, in order that he might have meat for his dinner, the price of meat would rise in the most rapid and unexampled manner. But surely, in a deficiency of corn, which renders it impossible for every man to have his usual share, if we still continue to furnish each person with the means of purchasing the same quantity as before, the effect must be in every respect similar.

It seems in great measure to have escaped observation, that the price of corn in a scarcity will depend much more upon the obstinacy with which the same degree of consumption is persevered in, than on the degree of the actual deficiency. A deficiency of one half of a

[19]The scarcities referred to in this chapter were those of 1800 and 1801.

crop, if the people could immediately consent to consume only one half of what they did before, would produce little or no effect on the price of corn. A deficiency of one-twelfth, if exactly the same consumption were to continue for ten or eleven months, might raise the price of corn to almost any height. The more is given in parish assistance, the more power is furnished of persevering in the same consumption; and, of course, the higher will the price rise before the necessary diminution of consumption is effected.

It has been asserted by some people, that high prices do not diminish consumption. If this were really true, we should see the price of a bushel of corn at a hundred pounds or more, in every deficiency, which could not be fully and completely remedied by importation. But the fact is, that high prices do ultimately diminish consumption; but, on account of the riches of the country, the unwillingness of the people to resort to substitutes, and the immense sums which are distributed by parishes, this object cannot be attained till the prices become excessive, and force even the middle classes of society, or, at least, those immediately above the poor, to save in the article of bread from the actual inability of purchasing it in the usual quantity. The poor who were assisted by their parishes, had no reason whatever to complain of the high price of grain; because it was the excessiveness of this price, and this alone, which by enforcing such a saving left a greater quantity of corn for the consumption of the lowest classes, which corn the parish allowances enabled them to command. The greatest sufferers in the scarcity were, undoubtedly, the classes immediately above the poor; and these were in the most marked manner depressed by the excessive bounties given to those below them. Almost all poverty is relative; and I much doubt whether these people would have been rendered so poor, if a sum equal to half of these bounties had been taken directly out of their pockets, as they were, by the new distribution of the money of the society which actually took place.[20] This distribution, by giving to the poorer classes a command of food so much greater than that to which their degree of skill and industry entitle them, in the actual circumstances of the country, diminished exactly in the came proportion that command over the necessaries of life, which the classes above them, by their superior skill and industry, would naturally possess; and it may be a question, whether the degree of assistance which the poor received, and which prevented them from resorting to the use of those substitutes which, in every other country on such occasions the great law of necessity teaches, was not more than overbalanced by the severity of the pressure on so large a body of the people from the extreme high prices,

[20] Supposing the lower classes to earn on an average ten shillings a week, and the classes just above them twenty, it is not to be doubted, that in a scarcity these latter would be more straightened in their power of commanding the necessaries of life, by a donation of ten shillings a week to those below them, than by the subtraction of five shillings a week from their own earnings. In the one case, they would be all reduced to a level; the price of provisions would rise in an extraordinary manner from the greatness of the competition; and all would be straightened for subsistence. In the other case, the losses above the poor would still maintain a considerable part of their relative superiority; the price of provisions would by no means rise in the same degree; and their remaining fifteen shillings would purchase much more than their twenty shillings in the former case.

and the permanent evil which must result from forcing sees many persons on the parish, who before thought themselves almost out of the reach of want.

If we were to double the fortunes of all those who possess above a hundred a year, the effect on the price of grain would be slow and inconsiderable; but if we were to double the price of labour throughout the kingdom, the effect in raising the price of grain would be rapid and great. The general principles on this subject will not admit of dispute; and that, in the particular case which we have been considering, the bounties to the poor were of a magnitude to operate very powerfully in this manner will sufficiently appear, if we recollect that before the late scarcities the sum collected for the poor was estimated at three millions, and that during the year 1801 it was said to be ten millions. An additional seven millions acting at the bottom of the scale,[21] and employed exclusively in the purchase of provisions, joined to a considerable advance in the price of wages in many parts of the kingdom, and increased by a prodigious sum expended in voluntary charity, must have had a most powerful effect in raising the price of the necessaries of life, if any reliance can be placed on the clearest general principles confirmed as much as possible by appearances. A man with a family has received, to my knowledge, fourteen shillings a week from the parish. His common earnings were ten shillings a week, and his weekly revenue, therefore, twenty-four. Before the scarcity he had been in the habit of purchasing a bushel of flour a week with eight shillings perhaps, and, consequently, had two shillings out of his ten, to spare for other necessaries. During the scarcity he was enabled to purchase the same quantity at nearly three times the price. He paid twenty-two shillings for his bushel of flour, and had, as before, two shillings remaining for other wants. Such instances could not possibly have been universal, without raising the price of wheat very much higher than it really was during any part of the dearth. But similar instances were by no means unfrequent; and the system itself of measuring the relief given by the price of grain was general.

If the circulation of the country had consisted entirely of specie, which could not have been immediately increased, it would have been impossible to have given such an additional sum as seven millions to the poor without embarrassing, to a great degree, the operations of commerce. On the commencement, therefore, of this extensive relief, which would necessarily occasion a proportionate expenditure in provisions throughout all the ranks of society, a great demand would be felt for an increased circulating medium. The nature of the medium then principally in use was such, that it could be created immediately on demand. From the accounts of the Bank of England, as laid before Parliament, it appeared, that no very great additional issues of paper took place from this quarter. The

[21] See a small pamphlet published in November, 1800, entitled, *An Investigation of the Cause of the present high Price of Provisions.* This pamphlet was mistaken by some for an inquiry into the cause of the scarcity, and as each it would naturally appear to be incomplete, adverting, as it does, principally to a single cause. But the sole object of the pamphlet was to give the principal reason for the extreme high price of provisions, in proportion to the degree of the scarcity, admitting the deficiency of one-fourth, as stated in the Duke of Portland's letter; which, I am much inclined to think, was very near the truth.

three millions and a half added to its former average issues were not probably much above what was sufficient to supply the quantity of specie that had been withdrawn from the circulation., If this supposition be true, (and the small quantity of gold which made its appearance at that time furnishes the strongest reason for believing that nearly as much as this must have been withdrawn,) it would follow that the part of the circulation originating in the Bank of England, though changed in its nature, had not been much increased in its quantity; and with regard to the effect of the circulating medium on the prices of all commodities, it cannot be doubted that it would be precisely the same, whether this medium were made up principally of guineas, or of pound-notes and shillings which would pass current for guineas.

The demand, therefore, for an increased circulating medium was left to be supplied principally by the country banks, and it could not be expected that they should hesitate in taking advantage of so profitable an opportunity. The paper issues of a country bank are, as I conceive, measured by the quantity of its notes which will remain in circulation; and this quantity is again measured, supposing a confidence to be established, by the sum of what is wanted to carry on all the money transactions of the neighbourhood. From the high price of provisions, all these transactions became more expensive. In the single article of the weekly payment of labourers' wages, including the parish allowances, it is evident that a very great addition to the circulating medium of the neighbourhood would be wanted. Had the country banks attempted to issue the same quantity of paper without such a particular demand for it, they would quickly have been admonished of their error by its rapid and pressing return upon them; but at this time it was wanted for immediate and daily use, and was, therefore, eagerly absorbed into the circulation. It may even admit of a question, whether, under similar circumstances, the country banks would not have issued nearly the same quantity of paper, if the Bank of England had not been restricted from payment in specie. Before this event the issues of the country banks in paper were regulated by the quantity which the circulation would take up; and after, as well as before, they were obliged to pay the notes which returned upon them in Bank of England circulation. The difference in the two cases would arise principally from the pernicious custom, adopted since the restriction of the bank, of issuing one and two pound notes, and from the little preference that many people might feel, if they could not get gold, between country bank paper and Bank of England paper.

This very great issue of country bank paper during the years 1800 and 1801 was evidently, therefore, in its origin, rather a consequence than a cause of the high price of provisions; but being once absorbed into the circulation, it must necessarily affect the price of all commodities, and throw very great obstacles in the way of returning cheapness, This is the great mischief of the system. During the scarcity, it is not to be doubted that the increased circulation, by preventing the embarrassments which commerce and speculation must otherwise have felt, enabled the country to continue all the branches of its trade with less interruption, and to import a much greater quantity of grain, than it could have done otherwise; but to overbalance these temporary advantages, a lasting evil might be entailed upon the community, and the prices of a time of scarcity might become permanent, from the difficulty of reabsorbing this increased circulation.

In this respect, however, it is much better that the great issue of paper should have come from the country banks than from the Bank of England. During the restriction of payment in specie, there is no possibility of forcing the bank to retake its notes when too abundant; but with regard to the country banks, as soon as their notes are not wanted in the circulation, they will be returned; and if the Bank of England notes be not increased, the whole circulating medium will thus be diminished.

We may consider ourselves as peculiarly fortunate, that the two years of scarcity were succeeded by two events the best calculated to restore plenty and cheapness—an abundant harvest, and a peace; which together produced a general conviction of plenty, in the minds both of buyers and sellers; and by rendering the first slow to purchase, and the others eager to sell, occasioned a glut in the market, and a consequent rapid fall of price, which has enabled parishes to take off their allowances to the poor, and thus to present a return of high prices, when the alarm among the sellers was over.

If the two years of scarcity had been succeeded merely by years of average crops, I am strongly disposed to believe, that, as no glut would have taken place in the market, the price of grain would have fallen only in a comparatively inconsiderable degree, the parish allowances could not leave been resumed, the increased quantity of paper would still have been wanted, and the price of all commodities might by degrees have been regulated permanently according to the increased circulating medium.

If instead of giving the temporary assistance of parish allowances, which might be withdrawn on the first fall of price, we had raised universally the wages of labour, it is evident, that the obstacles to a diminution of the circulation and to returning cheapness would have been still farther increased; and the high price of labour would have become permanent, without any advantage whatever to the labourer.

There is no one that more ardently desires to see a real advance in the price of labour than myself; but the attempt to effect this object by forcibly raising the nominal price, which was practised to a certain degree, and recommended almost universally during the late scarcities, every thinking man must reprobate as puerile and ineffectual.

The price of labour, when left to find its natural level, is a most important political barometer, expressing the relation between the supply of provisions, and the demand for them; between the quantity to be consumed and the number of consumers; and taken on the average, independently of accidental circumstances, it further expresses clearly the wants of the society respecting population; that is, whatever may be the number of children to a marriage necessary to maintain exactly the present population, the price of labour will be just sufficient to support this number, or be above it, or below it, according to the state of the real funds for the maintenance of labour, whether stationary, progressive or retrograde. Instead, however, of considering it in this light, we consider it as something which we may raise or depress at pleasure, something which depends principally upon his Majesty's justices of the peace. When an advance in the price of provisions already expresses that the demand is too great for the supply, in order to put the labourer in the same condition as before, we raise the price of labour, that is, we increase the demand, and are then much surprised that the price of provisions continues rising. In this we act much in the same manner as if, when the quicksilver in the common weather-glass stood at *stormy,* we were to raise it by some mechanical pressure to *settled fair,* and then be greatly astonished that it continued raining.

Dr. Smith has clearly shewn, that the natural tendency of a year of scarcity is either to throw a number of labourers out of employment, or to oblige them to work for less than they did before, from the inability of masters to employ the same number at the same price. The raising of the price of wages tends necessarily to throw more out of employment, and completely to prevent the good effects, which, he says, sometimes arise from a year of moderate scarcity, that of making the lower classes of people do more work, and become more careful and industrious. The number of servants out of place, and of manufacturers wanting employment during the late scarcities, were melancholy proofs of the truth of these reasonings. If a general rise in the wages of labour had taken place proportioned to the price of provisions, none but farmers and a few gentlemen could have afforded to employ the same number of workmen as before. Additional crowds of servants and manufacturers would have been turned off; and those who were thus thrown out of employment would of course have no other refuge than the parish. In the natural order of things a scarcity must tend to lower, instead of to raise, the price of labour.

After the publication and general circulation of such a work as Adam Smith's, I confess it appears to me strange, that so many men, who would yet aspire to be thought political economists, should still think that it is in the power of the justices of the peace or even of the omnipotence of parliament to alter by a *fiat* the whole circumstances of the country; and when the demand for provisions is greater than the supply, by publishing a particular edict, to make the supply at once equal to or greater than the demand. Many men who would shrink at the proposal of a maximum, would propose themselves, that the price of labour should be proportioned to the price of provisions, and do not seem to be aware that the two proposals are very nearly of the same nature, and that both tend directly to famine. It matters not whether we enable the labourer to purchase the same quantity of provisions which he did before, by fixing their price, or by raising in proportion the price of labour. The only advantage on the side of raising the price of labour is, that the rise in the price of provisions, which necessarily follows it, encourages importation: but putting importation out of the question, which might possibly be prevented by war, or other circumstances, a universal rise of wages in proportion to the price of provisions, aided by adequate parish allowances to those who were thrown out of work, would, by preventing any kind of saving, in the same manner as a maximum, cause the whole crop to be consumed in nine months, which ought to have lasted twelve, and thus produce a famine. At the same time we must not forget, that both humanity and true policy imperiously require, that we should give every assistance to the poor on these occasions, that the nature of the case will admit. If provisions were to continue at the price of scarcity, the wages of labour must necessarily rise, or sickness and famine would quickly diminish the number of labourers; and the supply of labour being unequal to the demand, its price would soon rise in a still greater proportion than the price of provisions. But even one or two years of scarcity, if the poor were left entirely to shift for themselves, might produce some effect of this kind, and consequently it is our interest, as well as our duty, to give them temporary aid in such seasons of distress. It is on such occasions that every cheap substitute for bread, and every mode of economizing food should be resorted to. Nor should we be too ready to complain of that high price of corn, which by encouraging importation increases the supply.

As the inefficacy of poor-laws, and of attempts forcibly to raise the price of labour, is most conspicuous in a scarcity, I have thought myself justified in considering them under this view; and as these causes of increased price received great additional force during the late scarcity from the increase of the circulating medium, I trust, that the few observations which I have made on this subject will be considered as an allowable digression.

4.1.6 Book III, Chapter VI: Of Poor Laws, continued.

Independently of any considerations respecting a year of deficient crops, it is evident, that an increase of population, without a proportional increase of food, must lower the value of each man's earnings. The food must necessarily be distributed in smaller quantities, and consequently a day's labour will purchase a smaller quantity of provisions. An increase in the price of provisions will arise either from an increase of population faster than the means of subsistence, or from a different distribution of the money of the society. The food of a country which has been long peopled, if it be increasing, increases slowly and regularly, and cannot be made to answer any sudden demands; but variations in the distribution of the money of the society are not unfrequently occurring, and are undoubtedly among the causes which, occasion the continual variations in the prices of provisions.

The poor-laws of England tend to depress the general condition of the poor in these two ways. Their first obvious tendency is to increase population without increasing the food for its support. A poor man may marry with little or no prospect of being able to support a family without parish assistance. They may be said, therefore, to create the poor which they maintain; and as the provisions of the country must, in consequence of the increased population, be distributed to every man in smaller proportions, it is evident that the labour of those who are not supported by parish assistance, will purchase a smaller quantity of provisions than before, and consequently more of them must be driven to apply for assistance.

Secondly, the quantity of provisions consumed in workhouses, upon a part of the society that cannot in general be considered as the most valuable part, diminishes the shares that would otherwise belong to more industrious and more worthy members, and thus, in the same manner, forces more to become dependent. If the poor in the workhouses were to live better than they do now, this new distribution of the money of the society would tend more conspicuously to depress the condition of those out of the workhouses by occasioning an advance in the price of provisions.

Fortunately for England, a spirit of independence, still remains among the peasantry. The poor-laws are strongly calculated to eradicate this spirit. They have succeeded in part; but had they succeeded as completely as might have been expected, their pernicious tendency would not have been so long concealed.

Hard as it may appear in individual instances, dependent poverty ought to be held disgraceful. Such a stimulus seems to be absolutely necessary to promote the happiness of the great mass of mankind; and every general attempt to weaken this stimulus, however benevolent its intention, will always defeat its own purpose. If men be induced to marry from the mere prospect of parish provision, they are not only unjustly tempted to bring unhappiness and dependence upon themselves and children, but they are tempted, without knowing it, to injure all in the same class with themselves.

The poor-laws of England appear to have contributed to raise the price of provisions, and to lower the real price of labour. They have therefore contributed to impoverish that class of people whose only possession is their labour. It is also difficult to suppose that they have not powerfully contributed to generate that carelessness and want of frugality observable among the poor, so contrary to the disposition generally to be remarked among petty tradesmen and small farmers. The labouring poor, to use a vulgar expression, seem always to live from hand to mouth. Their present wants employ their whole attention; and they seldom think of the future. Even when they have an opportunity of saving, they seldom exercise it; but all that they earn beyond their present necessities goes, generally speaking, to the ale-house. The poor-laws may therefore be said to diminish both the power and the will to save, among the common people; and thus to weaken one of the strongest incentives to sobriety and industry, and consequently to happiness.

It is a general complaint among master manufacturers, that high wages ruin all their workmen; but it is difficult to conceive that these men would not save a part of their high wages for the future support of their families, instead of spending it in drunkenness and dissipation, if they did not rely on parish assistance for support in case of accidents. And that the poor employed in manufactures consider this assistance as a reason why they may spend all the wages which they earn, and enjoy themselves while they can, appears to be evident, from the number of families that, upon the failure of any great manufactory, immediately fall upon the parish; when perhaps the wages earned in this manufactory while it flourished, were sufficiently above the price of common country labour, to have allowed them to save enough for their support till they could find some other channel for their industry.

A man who might not be deterred from going to the ale-house from the consideration that on his death or sickness he should leave his wife and family upon the parish, might yet hesitate in thus dissipating his earnings, if he were assured, that in either of these cases his family must starve, or be left to the support of casual bounty.

The mass of happiness among the common people cannot but be diminished, when one of the strongest checks to idleness and dissipation is thus removed; and positive institutions, which render dependent poverty so general, weaken that disgrace which, for the best and most humane reasons, ought to be attached to it.

The poor-laws of England were undoubtedly instituted for the most benevolent purpose; but it is evident they have failed in attaining it. They certainly mitigate some cases of severe distress, which might otherwise occur; though the state of the poor who are supported by parishes, considered in all its circumstances, is very miserable. But one of the principal objections to the system is, that for the assistance which some of the poor receive, in itself almost a doubtful blessing, the whole class of the common people of England is subjected to a set of grating, inconvenient, and tyrannical laws, totally inconsistent with the genuine spirit of the constitution. The whole business of settlements, even in its present amended state, is contradictory to all ideas of freedom. The parish persecution of men whose families are likely to become chargeable, and of poor women who are near lying-in, is a most disgraceful and disgusting tyranny. And the obstructions continually occasioned in the market of labour by these laws, have a constant tendency to add to the difficulties of those who are struggling to support themselves without assistance.

These evils attendant on the poor-laws seem to be irremediable. If assistance be to be distributed to a certain class of people, a power must be lodged somewhere of discriminating the proper objects, and of managing the concerns of the institutions that are necessary; but any great interference with the affairs of other people is a species of tyranny, and in the common course of things, the exercise of this power may be expected to become grating to those who are driven to ask for support. The tyranny of churchwardens and overseers is a common complaint among the poor; but the fault does not lie so much in these persons, who probably before they were in power were not worse than other people, but in the nature of all such institutions.

I feel persuaded that if the poor-laws had never existed in this country, though there might have been a few more instances of very severe distress, the aggregate mass of happiness among the common people would have been much greater than it is at present.

The radical defect of all systems of the kind is that of tending to depress the condition of those that are not relieved by parishes, and to create more poor. If, indeed, we examine some of our statutes strictly with reference to the principle of population, we shall find that they attempt an absolute impossibility; and we cannot be surprised, therefore, that they should constantly fail in the attainment of their object.

The famous 43d of Elizabeth, which has been so often referred to and admired, enacts, that the overseers of the poor "shall take order from time to time, by and with the consent of two or more justices, for setting to work the children of all such, whose parents shall not by the said persons be thought able to keep and maintain their children; and also such persons married or unmarried, as, having no means to maintain them, use no ordinary and daily trade of life to get their living by; and also to raise, weekly or otherwise, by taxation of every inhabitant, and every occupier of lands in the said parish, (in such competent sums as they shall think fit,) a convenient stock of flax, hemp, wool, thread, iron, and other necessary ware and stuff, to set the poor to work."

What is this but saying, that the funds for the maintenance of labour in this country may be increased at, will, and without limit, by a *fiat* of government, or an assessment of the overseers? Strictly speaking, this clause is as arrogant and as absurd, as if it had enacted that two ears of wheat should in future grow where one only had grown before. Canute, when he commanded the waves not to wet his princely foot, did not in reality assume a greater power over the laws of nature. No directions are given to the overseers how to increase the funds for the maintenance of labour; the necessity of industry, economy and enlightened exertion, in the management of agricultural and commercial capital, is not insisted on for this purpose; but it is expected that a miraculous increase of these funds should immediately follow an edict of the government used at the discretion of some ignorant parish officers.

If this clause were really and *bonâ fide* put in execution, and the shame attending the receiving of parish assistance worn off, every labouring man might marry as early as he pleased, under the certain prospect of having all his children properly provided for; and as, according to the supposition, there would be no check to population from the consequences of poverty after marriage, the increase of people would be rapid beyond example in old states. After what has been said in the former parts of this work, it is submitted to the reader, whether the utmost exertions of the most enlightened government could, in this case, make the food keep pace with the population; much less a mere arbitrary edict, the tendency of which is certainly rather to diminish than to increase the funds for the maintenance of productive labour.

In the actual circumstances of every country, the prolific power of nature seems to be always ready to exert nearly its full force; but within the limit of possibility, there is nothing perhaps more improbable, or more out of the reach of any government to effect, than the direction of the industry of its subjects in such a manner, as to produce the greatest quantity of human sustenance that the earth could bear. It evidently could not be done without the most complete violation of the law of property, from which every thing that is valuable to man has hitherto arisen. Such is the disposition to marry, particularly in very young people, that, if the difficulties of providing for a family were entirely removed, very few would remain single at twenty-two. But what statesman or rational government could propose that all animal food should be prohibited, that no horses should be used for business or pleasure, that all the people should live upon potatoes, and that the whole industry of the nation should be exerted in the production of them, except what was required for the mere necessaries of clothing and houses? Could such a revolution be effected, would it be desirable? particularly as in a few years, notwithstanding all these exertions, want, with less resource than ever, would inevitably recur.

After a country has once ceased to be in the peculiar situation of a new colony, we shall always find that in the actual state of its cultivation, or in that state which may rationally be expected from the most enlightened government, the increase of its food can never allow for any length of time an unrestricted increase of population; and therefore the due execution of the clause in the 43d of Elizabeth, as a permanent law, is a physical impossibility.

It will be said, perhaps, that the fact contradicts the theory; and that the clause in question has remained in force, and has been executed, during the last two hundred years. In answer to this, I should say without hesitation, that it has not really been executed;

and that it is merely owing to its incomplete execution, that it remains on our statute-book at present.

The scanty relief granted to persons in distress, the capricious and insulting manner in which it is sometimes distributed by the overseers, and the natural and becoming pride, not yet quite extinct among the peasantry of England, have deterred the more thinking and virtuous part of them from venturing on marriage, without some better prospect of maintaining their families than mere parish assistance. The desire of bettering our condition, and the fear of making it worse, like the *vis medicatrix naturæ* in physics, is the *vis medicatrix reipublicæ* in politics, and is continually counteracting the disorders arising from narrow human institutions. In spite of the prejudices in favour of population, and the direct encouragements to marriage from the poor-laws, it operates as a preventive check to increase; and happy for this country is it, that it does so. But besides that spirit of independence and prudence, which checks the frequency of marriage, notwithstanding the encouragements of the poor-laws, these laws themselves occasion a check of no inconsiderable magnitude, and thus counteract with one hand what they encourage with the other. As each parish is obliged to maintain its own poor, it is naturally fearful of increasing their number; and every landholder is in consequence more inclined to pull down than to build cottages, except when the demand for labourers is really urgent. This deficiency of cottages operates necessarily as a strong check to marriage; and this check is probably the principal reason why we have been able to continue the system of the poor-laws so long.

Those who are not prevented for a time from marrying by these causes, are either relieved very scantily at their own homes, where they suffer all the consequences arising from squalid poverty; or they are crowded together in close and unwholesome workhouses, where a great mortality almost universally takes place, particularly among the young children. The dreadful account given by Jonas Hanway of the treatment of parish children in London is well known; and it appears from Mr. Howlett and other writers, that in some parts of the country their situation is not very much better. A great part of the redundant population occasioned by the poor-laws is thus taken off by the operation of the laws themselves, or at least by their ill execution. The remaining part which survives, by causing the funds for the maintenance of labour to be divided among a greater number than can be properly maintained by them, and by turning a considerable share from the support of the diligent and careful workman to the support of the idle and negligent, depresses the condition of all those who are out of the workhouses, forces more into them every year, and has ultimately produced the enormous evil, which we all so justly deplore; that of the great and unnatural proportion of the people which is now become dependent upon charity.

If this be a just representation of the manner in which the clause in question has been executed, and of the effects which it has produced, it must be allowed that we have practised an unpardonable deceit upon the poor, and have promised what we have been very far from performing.

The attempts to employ the poor on any great scale in manufactures have almost invariably failed, and the stock and materials have been wasted. In those few parishes which, by better management or larger foods, have been enabled to persevere in this system, the effect of these new manufactures in the market must have been to throw out of employment many independent workmen, who were before engaged in fabrications of a similar nature. This effect has been placed in a strong point of view by Daniel de Foe, in an address to parliament, entitled, *Giving Alms no Charity*. Speaking of the employment of parish children in manufactures, he says, "For every skein of worsted these poor children spin, there must be a skein the less spun by some poor family that spun it before; and for every piece of baize so made in London, there must be a piece the less made at Colchester, or somewhere else."[22] Sir F. M. Eden, on the same subject, observes, that "whether mops and brooms are made by parish children or by private workmen, no more can be sold than the public is in want of."[23]

It will be said, perhaps, that the same reasoning might be applied to any new capital brought into competition in a particular trade or manufacture, which can rarely be done without injuring, in some degree, those that were engaged in it before. But there is a material difference in the two cases. In this the competition is perfectly fair, and what every man on entering into business must lay his account to. He may rest secure that he will not be supplanted, unless his competitor possess superior skill and industry. In the other case the competition is supported by a great bounty; by which means, notwithstanding very inferior skill and industry on the part of his competitors, the independent workman may be undersold, and unjustly excluded from the market. He himself perhaps is made to contribute to this competition against his own earnings; and the funds for the maintenance of labour are thus turned from the support of a trade which yields a proper profit, to one which cannot maintain itself without a bounty. It should be observed in general, that when a fund for the maintenance of labour is raised by assessment, the greatest part of it is not a new capital brought into trade, but an old one, which before was much more profitably employed, turned into a new channel. The farmer pays to the poor's rates, for the encouragement of a bad and unprofitable manufacture, what he would have employed on his land with infinitely more advantage to his country. In the one case, the foods for the maintenance of labour are daily diminished; in the other, daily increased. And this obvious tendency of assessments for the employment of the poor, to decrease the real funds for the maintenance of labour in any country, aggravates the absurdity of supposing that it is in the

---

[22] See Extracts from Daniel de Foe, in Sir F. M. Eden's valuable Work on the poor, vol. i. p. 261.

[23] Sir F. M. Eden, speaking of the supposed right of the poor to be supplied with employment while able to work, and with a maintenance when incapacitated from labour, very justly remarks, "It may however be doubted, whether any right, the gratification of which seems to be impracticable, can be said to exist," vol. i. p.447. No man has collected so many materials for forming a judgement on the effects of the poor-laws as Sir F. M. Eden, and the result he thus expresses: "Upon the whole therefore there seems to be just grounds for concluding, that the sum of good to be expected from a compulsory maintenance of the poor will be far outbalanced by the sum of evil which it will inevitably create," vol. i. p. 467—I am happy to have the sanction of so practical an inquirer to my opinion of the poor-laws.

power of a government to find employment for all its subjects, however fast they may increase.

It is not intended that these reasonings should be applied against every mode of employing the poor on a limited scale, and with such restrictions as may not encourage at the same time their increase. I would never wish to push general principles too far; though I think that they ought always to be kept in view. In particular cases the individual good to be obtained may be so great, and the general evil so slight, that the former may clearly overbalance the latter.

My intention is merely to shew that the poor-laws as a general system are founded on a gross error: and that the common declamation on the subject of the poor, which we see so often in print, and hear continually in conversation, namely, that the market price of labour ought always to be sufficient decently to support a family, and that employment ought to be found for all those who are willing to work, is in effect to say, that the funds for the maintenance of labour in this country are not only infinite, but not subject to variation; and that, whether the resources of a country be rapidly progressive, slowly progressive, stationary or declining, the power of giving full employment and good wages to the labouring classes must always remain exactly the same,—a conclusion which contradicts the plainest and most obvious principles of supply and demand, and involves the absurd position that a definite quantity of territory can maintain an infinite population.

4.1.7 Book III, Chapter VII: Of Poor Laws, continued.

The remarks made in the last chapter on the nature and effects of the poor-laws have been in the most striking manner confirmed by the experience of the years 1815, 1816 and 1817.[24] During these years, two points of the very highest importance have been established, so as no longer to admit of a doubt in the mind of any rational man.

The first is, that the country does not in point of fact fulfil the promise which it makes to the poor in the poor-laws, to maintain and find in employment, by means of parish assessments, those who are unable to support themselves or their families, either from want of work or any other cause.

And secondly, that with a very great increase of legal parish assessments, aided by the most liberal and praiseworthy contributions of voluntary charity, the country has been wholly unable to find adequate employment for the numerous labourers and artificers who were able as well as willing to work.

It can no longer surely be contended that the poor-laws really perform what they promise, when it is known that many almost starving families have been found in London and other great towns, who are deterred from going on the parish by the crowded, unhealthy and horrible state of the workhouses into which they would be received, if

[24]This chapter was written in 1817.

indeed they could be received at all; when it is known that many parishes have been absolutely unable to raise the necessary assessments, the increase of which, according to the existing laws, have tended only to bring more and more persons upon the parish, and to make what was collected less and less effectual; and when it is known that there has been an almost universal cry from one end of the kingdom to the other for voluntary charity to come in aid of the parochial assessments.

These strong indications of the inefficiency of the poor-laws may be considered not only as incontrovertible proofs of the fact that they do not perform what they promise, but as affording the strongest presumption that they cannot do it. The best of all reasons for the breach of a promise, is, the absolute impossibility of executing it; indeed it is the only plea that can ever be considered as valid. But though it may be fairly pardonable not to execute an impossibility, it is unpardonable knowingly to promise one. And if it be still thought advisable to act upon these statutes as far as is practicable, it would surely be wise so to alter the terms in which they are expressed, and the general interpretation given to them, as not to convey to the poor a false notion of what really is within the range of practicability.

It has appeared further as a matter of fact, that very large voluntary contributions, combined with greatly increased parochial assessments, and aided by the most able and incessant exertions of individuals, have failed to give the necessary employment to those who have been thrown out of work by the sudden falling off of demand which has occurred during the last two or three years.

It might perhaps have been foreseen that, as the great movements of society, the great causes which render a nation progressive, stationary or declining, for longer or shorter periods, cannot be supposed to depend much upon parochial assessments or the contributions of charity, it could not be expected that any efforts of this kind should have power to create, in a stationary or declining state of things, that effective demand for labour which only belongs to a progressive state. But to those who did not see this truth before, the melancholy experience of the last two years[25] must have brought it home with an overpowering conviction.

It does not however by any means follow that the exertions which have been made to relieve the present distresses have been ill directed. On the contrary, they have not only been prompted by the most praiseworthy motives; they have not only fulfilled the great moral duty of assisting our fellow-creatures in distress; but they have in point of fact done great good, or at least prevented great evil. Their partial failure does not necessarily indicate either a want of energy or a want of skill in those who have taken the lead in these efforts, but merely that a part only of what has been attempted is practicable.

It is practicable to mitigate the violence and relieve the severe pressure of the present distress, so as to carry the sufferers through to better times, though even this can only be done at the expense of some sacrifices, not merely of the rich, but of other classes of the poor. But it is impracticable by any exertions, either individual or national, to restore at

[25]The years 1816 and 1817.

once that brisk demand for commodities and labour which has been lost by events, that, however they may have originated, are now beyond the power of control.

The whole subject is surrounded on all sides by the most formidable difficulties, and in no state of things is it so necessary to recollect the saying of Daniel de Foe quoted in the last chapter. The manufacturers all over the country, and the Spitalfields weavers in particular, are in a state of the deepest distress, occasioned immediately and directly by the want of demand for the produce of their industry, and the consequent necessity felt by the masters of turning off many of their workmen, in order to proportion the supply to the contracted demand. It is proposed, however, by some well-meaning people, to raise by subscription a fund for the express purpose of setting to work again those who have been turned off by their masters, the effect of which can only be to continue glutting a market, already much too fully supplied. This is most naturally and justly objected to by the masters, as it prevents them from withdrawing the supply, and taking the only course which can prevent the total destruction of their capitals, and the necessity of turning off all their men instead of a part.

On the other hand, some classes of merchants and manufacturers clamour very loudly for the prohibition of all foreign commodities which may enter into competition with domestic products, and interfere, as they intimate, with the employment of British industry. But this is most naturally and most justly deprecated by other classes of British subjects, who are employed to a very great extent in preparing and manufacturing those commodities which are to purchase our imports from foreign countries. And it must be allowed to be perfectly true that a court-ball, at which only British stuffs are admitted, may be the means of throwing out of employment in one quarter of the country just as many persons as it furnishes with employment in another.

Still, it would be desirable if possible to employ those that were out of work, if it were merely to avoid the bad moral effects of idleness, and of the evil habits which might be generated by depending for a considerable time on mere alms. But the difficulties just stated will show, that we ought to proceed in this part of the attempt with great caution, and that the kinds of employment which ought to be chosen are those, the results of which will not interfere with existing capitals. Such are public works of all descriptions, the making and repairing of roads, bridges, railways, canals, &c.; and now perhaps, since the great loss of agricultural capital, almost every sort of labour upon the land, which could be carried on by public subscription.

Yet even in this way of employing labour, the benefit to some must bring with it disadvantages to others. That portion of each person's revenue which might go in subscriptions of this kind, must of course be lost to the various sorts of labour which its expenditure in the usual channels would have supported; and the want of demand thus occasioned in these channels must cause the pressure of distress to be felt in quarters which might otherwise have escaped it. But this is an effect which, in such cases, it is impossible to avoid; and, as a temporary measure, it is not only charitable but just, to spread the evil over a larger surface, in order that its violence on particular parts may be so mitigated as to be made bearable by all.

The great object to be kept in view, is to support the people through their present distresses, in the hope (and I trust a just one) of better times. The difficulty is without doubt considerably aggravated by the prodigious stimulus which has been given to the population of the country of late years, the effects of which cannot suddenly subside. But it will be seen probably, when the next returns of the population are made, that the marriages and births have diminished, and the deaths increased in a still greater degree than in 1800 and 1801; and the continuance of this effect to a certain degree for a few years will retard the progress of the population, and combined with the increasing wants of Europe and America from their increasing riches, and the adaptation of the supply of commodities at home to the new distribution of wealth occasioned by the alteration of the circulating medium, will again give life and energy to all our mercantile and agricultural transactions, and restore the labouring classes to full employment and good wages.[26]

On the subject of the distresses of the poor, and particularly the increase of pauperism of late years, the most erroneous opinions have been circulated. During the progress of the war, the increase in the proportion of persons requiring parish assistance was attributed chiefly to the high price of the necessaries of life. We have seen these necessaries of life experience a great and sudden fall, and yet at the same time a still larger proportion of the population requiring parish assistance.

It is now said that taxation is the sole cause of their distresses, and of the extraordinary stagnation in the demand for labour; yet I feel the firmest conviction, that if the whole of the taxes were removed to-morrow, this stagnation, instead of being at an end, would be considerably aggravated. Such an event would cause another great and general rise in the value of the circulating medium, and bring with it that discouragement to industry with which such a convulsion in society must ever be attended. If, as has been represented, the labouring classes now pay more than half of what they receive in taxes, he must know very little indeed of the principles on which the wages of labour are regulated, who can for a moment suppose that, when the commodities on which they are expended have fallen one half by the removal of taxes, these wages themselves would still continue of the same nominal value. Were they to remain but for a short time the same, while all commodities had fallen, and the circulating medium had been reduced in proportion, it would be quickly seen that multitudes of them would be at once thrown out of employment.

The effects of taxation are no doubt in many cases pernicious in a very high degree; but it may be laid down as a rule which has few exceptions, that the relief obtained by taking off a tax, is in no respect equal to the injury inflicted in laying it on; and generally it may be

[26] 1825. This has, in a considerable degree, taken place; but it has been owing rather to the latter causes noticed than to the former. It appeared, by the returns of 1821, that the scarce years of 1817 end 1818 had but a slight effect in diminishing the number of marriages and births, compared with the effect of the great proportion of plentiful years in increasing them; so that the population proceeded with great rapidity during the ten years ending with 1820. But this great increase of the population has prevented the labouring classes from being so fully employed as might have been expected from the prosperity of commerce and agriculture during the last two or three years.

said that the specific evil of taxation consists in the check which it gives to production, rather than the diminution which it occasions in demand. With regard to all commodities indeed of home production and home demand, it is quite certain that the conversion of capital into revenue, which is the effect of loans, must necessarily increase the proportion of demand to the supply; and the conversion of the revenue of individuals into the revenue of the government, which is the effect of taxes properly imposed, however hard upon the individuals so taxed, can have no tendency to diminish the general amount of demand. It will of course diminish the demands of the persons taxed by diminishing their powers of purchasing; but to the exact amount that the powers of these persons are diminished, will the powers of the government and of those employed by it be increased. If an estate of five thousand a year has a mortgage upon it of two thousand, two families, both in very good circumstances, may be living upon the rents of it, and both have considerable demands for houses, furniture, carriages, broad cloth, silks, cottons, &c. The man who owns the estate is certainly much worse off than if the mortgage-deed was burnt, but the manufacturers and labourers who supply the silks, broad-cloth, cottons, &c., are so far from being likely to be benefited by such burning, that it would be a considerable time before the new wants and tastes of the enriched owner had restored the former demand; and if he were to take a fancy to spend his additional income in horses, hounds and menial servants, which is probable, not only would the manufacturers and labourers who had before supplied their silks, cloths and cottons, be thrown out of employment, but the substituted demand would be very much less favourable to the increase of the capital and general resources of the country.

The foregoing illustration represents more nearly than may generally be imagined the effects of a national debt on the labouring classes of society, and the very great mistake of supposing that, because the demands of a considerable portion of the community would be increased by the extinction of the debt, these increased demands would not be balanced, and often more than balanced, by the loss of the demand from the fund-holders and government.

It is by no means intended by these observations to intimate that a national debt may not be so heavy as to be extremely prejudicial to a state. The division and distribution of property, which is so beneficial when carried only to a certain extent, is fatal to production when pushed to extremity. The division of an estate of five thousand a year will generally tend to increase demand, stimulate production, and improve the structure of society; but the division of an estate of eighty pounds a year will generally be attended with effects directly the reverse.

But, besides the probability that the division of property occasioned by a national debt may in many cases be pushed too far, the process of the division is effected by means which sometimes greatly embarrass production. This embarrassment must necessarily take place to a certain extent in almost every species of taxation; but under favourable circumstances it is overcome by the stimulus given to demand compared with supply. During the late war, from the prodigious increase of produce and population, it may fairly be presumed that the power of production was not essentially impeded, notwithstanding the enormous amount of taxation; but in the state of things which has occurred since the peace, and under a most extraordinary fall of the exchangeable value of the raw produce of the land, and a great consequent diminution of the circulating medium, the very sudden increase of the weight and pressure of taxation must greatly aggravate the other causes which discourage production. This effect has been felt to a considerable extent on the land; but the distress in this quarter is already much mitigated;[27] and among the mercantile and manufacturing classes, where the greatest numbers are without employment, the evil obviously arises, not so much from the want of capital and the means of production, as the want of a market for the commodity when produced—a want, for which the removal of taxes, however proper, and indeed absolutely necessary as a permanent measure, is certainly not the immediate and specific remedy.

The principal causes of the increase of pauperism, independently of the present crisis, are, first, the general increase of the manufacturing system and the unavoidable variations of manufacturing labour; and secondly, and more particularly, the practice which has been adopted in some counties, and is now spreading pretty generally all over the kingdom, of paying a considerable portion of what ought to be the wages of labour out of the parish rates. During the war, when the demand for labour was great and increasing, it is quite certain that nothing but a practice of this kind could for any time have prevented the wages of labour from rising fully in proportion to the necessaries of life, in whatever degree these necessaries might have been raised by taxation. It was seen, consequently, that in these parts of Great Britain where this practice prevailed the least; the wages of labour rose the most. This was the case in Scotland, and some parts of the North of England, where the improvement in the condition of the labouring classes, and their increased command over the necessaries and conveniences of life, were particularly remarkable. And if, in some other parts of the country, where the practice did not greatly prevail, and especially in the towns, wages did not rise in the same degree, it was owing to the influx and competition of the cheaply raised population of the surrounding counties.

It is a just remark of Adam Smith, that the attempts of the legislature to raise the pay of curates had always been ineffectual, on account of the cheap and abundant supply of them, occasioned by the bounties given to young persons educated for the church at the universities. And it is equally true that no human efforts can keep up the price of day-labour so as to enable a man to support on his earnings a family of a moderate size, so long as those who have more than two children are considered as having a valid claim to parish assistance.

If this system were to become universal, and I own it appears to me that the poor-laws naturally lead to it, there is no reason whatever why parish assistance should not by degrees begin earlier and earlier; and I do not hesitate to assert that, if the government and constitution of the country were in all other respects as perfect as the wildest visionary thinks he could make them; if parliaments were annual, suffrage universal, wars, taxes and

[27] Written in 1817. It increased again afterwards from another great fall in the price of corn, subsequent to 1818.

pensions unknown, and the civil list fifteen hundred a year, the great body of the community might still be a collection of paupers.

I have been accused of proposing a law to prohibit the poor from marrying. This is not true. So far from proposing such a law, I have distinctly said that, if any person chooses to marry without having a prospect of being able to maintain a family, he ought to have the most perfect liberty so to do; and whenever any prohibitory propositions have been suggested to me as advisable by persons who have drawn wrong inferences from what I have said, I have steadily and uniformly reprobated them. I am indeed most decidedly of opinion that any positive law to limit the age of marriage would be both unjust and immoral; and my greatest objection to a system of equality and the system of the poor-laws (two systems which, however different in their outset, are of a nature calculated to produce the same results) is, that the society in which they are effectively carried into execution, will ultimately be reduced to the miserable alternative of choosing between universal want and the enactment of *direct* laws against marriage.

What I have really proposed is a very different measure. It is the *gradual* and *very gradual* abolition of the poor-laws.[28] And the reason why I have ventured to suggest a proposition of this kind for consideration is my firm conviction, that they have lowered very decidedly the wages of the labouring classes, and made their general condition essentially worse than it would have been if these laws had never existed. Their operation is every where depressing; but it falls peculiarly hard upon the labouring classes in great towns. In country parishes the poor do really receive some compensation for their low wages; their children, beyond a certain number, are really supported by the parish; and though it must be a most grating reflection to a labouring man, that it is scarcely possible for him to marry without becoming the father of paupers; yet if he can reconcile himself to this prospect, the compensation, such as it is, is, no doubt, made to him. But in London and all the great towns of the kingdom, the evil is suffered without the compensation. The population raised by bounties in the country naturally and necessarily flows into the towns, and as naturally and necessarily tends to lower wages in them; while, in, point of fact, those who marry in towns, and have large families, receive no assistance from their parishes, unless they are actually starving; and altogether the assistance which the manufacturing classes obtain for the support of their families, in aid of their lowered wages, is perfectly inconsiderable.

To remedy the effects of this competition from the country, the artificers and manufacturers in towns have been apt to combine, with a view to keep up the price of labour, and to prevent persons from working below a certain rate. But such combinations are not only illegal,[29] but irrational and ineffectual; and if the supply of workmen in any

---

[28] So gradual as not to affect any individuals at present alive, or who will be born within the next two years.

[29] This has since been altered; but the subsequent part of the passage is particularly applicable to the present time—the end of the year 1825. The workmen are beginning to find that, if they could raise their wages above what the state of the demand and the prices of goods will warrant, it is absolutely

particular branch of trade be such as would naturally lower wages, the keeping them up forcibly must have the effect of throwing so many out of employment, as to make the expense of their support fully equal to the gain acquired by the higher wages, and thus render these higher wages in reference to the whole body perfectly futile.

It may be distinctly stated to be an *absolute impossibility* that all the different classes of society should be both well paid and fully employed, if the supply of labour on the whole exceed the demand; and as the poor-laws tend in the most marked manner to make the supply of labour exceed the demand for it, their effect must be, either to lower universally all wages, or, if some are kept up artificially, to throw great numbers of workmen out of employment, and thus constantly to increase the poverty and distress of the labouring classes of society.

If these things be so (and I am firmly convinced that they are) it cannot but be a subject of the deepest regret to those who are anxious for the happiness of the great mass of the community, that the writers which are now most extensively read among the common people should have selected for the subject of reprobation exactly that line of conduct which can alone generally improve their condition, and for the subject of approbation that system which must inevitably depress them in poverty and wretchedness.

They are taught that there is no occasion whatever for them to put any sort of restraint upon their inclinations, or exercise any degree of prudence in the affair of marriage; because the parish is bound to provide for all that are born. They are taught that there is as little occasion to cultivate habits of economy, and make use of the means afforded them by saving banks, to lay by their earnings while they are single, in order to furnish a cottage when they marry, and enable them to set out in life with decency and comfort; because, I suppose, the parish is bound to cover their nakedness, and to find them a bed and a chair in a workhouse.

They are taught that any endeavour on the part of the higher classes of society to inculcate the duties of prudence and economy can only arise from a desire to save the money which they pay in poor-rates; although it is absolutely certain that the *only* mode, consistent with the laws of morality and religion, of giving to the poor the largest share of the property of the rich, without sinking the whole community in misery, is the exercise on the part of the poor of prudence in marriage, and of economy both before and after it.

They are taught that the command of the Creator to increase and multiply is meant to contradict those laws which he has himself appointed for the increase and multiplication of the human race; and that it is equally the duty of a person to marry early, when, from the impossibility of adding to the food of the country in which he lives, the greater part of his offspring must die prematurely, and consequently no multiplication follow from it, as when the children of such marriages can all be well maintained, and there is room and food for a great and rapid increase of population.

---

impossible that all, or nearly all, should be employed. The masters could not employ the same number as before, without inevitable ruin.

They are taught that, in relation to the condition of the labouring classes, there is no other difference between such a country as England, which has been long well peopled, and where the land, which is not yet taken into cultivation, is comparatively barren, and such a country as America, where millions and millions of acres of fine land are yet to be had for a trifle, except what arises from taxation.

And they are taught, O monstrous absurdity! that the only reason why the American labourer earns a dollar a day, and the English labourer earns two shillings, is that the English labourer pays a great part of these two shillings in taxes.

Some of these doctrines are so grossly absurd that I have no doubt they are rejected at once by the common sense of many of the labouring classes. It cannot but strike them that, if their main dependence for the support of their children is to be on the parish, they can only expect parish fare, parish clothing, parish furniture, a parish house, and parish government, and they must know that persons living in this way cannot possibly be in a happy and prosperous state.

It can scarcely escape the notice of the common mechanic, that the scarcer workmen are upon any occasion the greater share do they retain of the value of what they produce for their masters; and it is a most natural inference, that prudence in marriage, which is the only moral means of preventing an excess of workmen above the demand, can be the only mode of giving to the poor permanently a large share of all that is produced in the country.

A common man, who has read his Bible, must be convinced that a command given to a rational being by a merciful God cannot be intended so to be interpreted as to produce only disease and death instead of multiplication; and a plain sound understanding would make him see that, if, in a country in which little or no increase of food is to be obtained, every man were to marry at eighteen or twenty, when he generally feels most inclined to it, the consequence must be increased poverty, increased disease, and increased mortality, and not increased numbers, as long, at least, as it continues to be true (which he will hardly be disposed to doubt) that additional numbers cannot live without additional food.

A moderately shrewd judgment would prompt any labourer acquainted with the nature of land to suspect that there must be some great difference, quite independent of taxation, between a country such as America, which might easily be made to support fifty times as many inhabitants as it contains at present, and a country such as England, which could not, without extraordinary exertions, be made to support two or three times as many. He would, at least, see that there would be a prodigious difference in the power of maintaining an additional number of cattle, between a small farm already well stocked, and a very large one which had not the fiftieth part of what it might be made to maintain; and as he would know that both rich and poor must live upon the produce of the earth as well as all other animals, he would be disposed to conclude that what was so obviously true in one case, could not be false in the other. These considerations might make him think it natural and probable that in those countries where there was a great want of people, the wages of labour would be such as to encourage early marriages and large families, for the best of all possible reasons, because all that are born may be very easily and comfortably supported; but that in those countries which were already nearly full, the wages of labour cannot be such as to give the same encouragement to early marriages, for a reason surely not much worse, because the persons so brought into the world cannot be properly supported.

There are few of our mechanics and labourers who have not heard of the high prices of bread, meat and labour in this country compared with the nations of the continent, and they have generally heard at the same time that these high prices were chiefly occasioned by taxation, which, though it had raised among other things the money wages of labour, had done harm rather than good to the labourer, because it had before raised the price of the bread and beer and other articles in which he spent his earnings. With this amount of information, the meanest understanding would revolt at the idea that the very same cause which had kept the money price of labour in all the nations of Europe much lower than in England, namely, the absence of taxation, had been the means of raising it to more than double in America. He would feel quite convinced that, whatever might be the cause of the high money wages of labour in America, which he might not perhaps readily understand, it must be something very different indeed from the mere absence of taxation, which could only have an effect exactly opposite.

With regard to the improved condition of the lower classes of people in France since the revolution, which has also been much insisted upon; if the circumstances accompanying it were told at the same time, it would afford the strongest presumption against the doctrines which have been lately promulgated. The improved condition of the labouring classes in France since the revolution has been accompanied by a greatly diminished proportion of births, which has had its natural and necessary effect in giving to these classes a greater share of the produce of the country, and has kept up the advantage arising from the sale of the church lands and other national domains, which would otherwise have been lost in a short time. The effect of the revolution in France has been, to make every person depend more upon himself and less upon others. The labouring classes are therefore become more industrious, more saving and more prudent in marriage than formerly; and it is quite certain that without these effects the revolution would have done nothing for them. An improved government has, no doubt, a natural tendency to produce these effects, and thus to improve the condition of the poor. But if an extensive system of parochial relief, and such doctrines as have lately been inculcated, counteract them, and prevent the labouring classes from depending upon their own prudence and industry, then any change for the better in other respects becomes comparatively a matter of very little importance; and, under the best form of government imaginable, there may be thousands on thousands out of employment and half starved.

If it be taught that all who are born have a *right* to support on the land, whatever be their number, and that there is no occasion to exercise any prudence in the affair of marriage so as to check this number, the temptations, according to all the known principles of human nature, will inevitably be yielded to, and more and more will gradually become dependent on parish assistance. There cannot therefore be a greater inconsistency and contradiction than that those who maintain these doctrines respecting the poor, should still complain of the number of paupers. Such doctrines and a crowd of paupers are unavoidably united; and it is utterly beyond the power of any revolution or change of government to separate them.

4.1.8 Book III, Chapter VIII: Of the Agricultural System.

As it is the nature of agriculture to produce subsistence for a greater number of families than can be employed in the business of cultivation, it might perhaps be supposed that a nation which strictly pursued an agricultural system would always have more food than was necessary for its inhabitants, and that its population could never be checked from the want of the means of subsistence.

It is indeed obviously true that the increase of such a country is not immediately checked, either by the want of power to produce, or even by the deficiency of the actual produce of the soil compared with the population. Yet if we examine the condition of its labouring classes, we shall find that the real wages of their labour are such as essentially to check and regulate their increase, by checking and regulating their command over the means of subsistence.

A country under certain circumstances of soil and situation, and with a deficient capital, may find it advantageous to purchase foreign commodities with its raw produce rather than manufacture them at home: and in this case it will necessarily grow more raw produce than it consumes. But this state of things is very little connected either with the permanent condition of the labouring classes of the society or the rate of their increase; and in a country where the agricultural system entirely predominates, and the great mass of its industry is directed towards the land, the condition of the people is subject to almost every degree of variation.

Under the agricultural system perhaps are to be found the two extremes in the condition of the poor; instances where they are in the best state, and instances where they are in the worst state of any of which we have accounts.

In a country where there is an abundance of good land, where there are no difficulties in the way of its purchase and distribution, and where there is an easy foreign vent for raw produce, both the profits of stock and the wages of labour will be high. These high profits and high wages, if habits of economy pretty generally prevail, will furnish the means of a rapid accumulation of capital and a great and continued demand for labour, while the rapid increase of population which will ensue will maintain undiminished the demand for produce, and check the fall of profits. If the extent of territory be considerable, and the population comparatively inconsiderable, the land may remain understocked both with capital and people for some length of time, notwithstanding a rapid increase of both; and it is under these circumstances of the agricultural system that labour is able to command the greatest portion of the necessaries of life, and that the condition of the labouring classes of society is the best.

The only drawback to the wealth of the labouring classes under these circumstances is the relatively low value of the raw produce.

If a considerable part of the manufactured commodities used in such a country be purchased by the export of its raw produce, it follows as a necessary consequence that the relative value of its raw produce will be lower, and of its manufactured produce higher, than in the countries with which such a trade is carried on. But where a given portion of raw produce will not command so much of manufactured and foreign commodities as in other countries, the condition of the labourer cannot be exactly measured by the quantity of raw produce which falls to his share. If, for instance, in one country the yearly earnings of a labourer amount in money value to fifteen quarters of wheat and in another to nine, it would be incorrect to infer that their relative condition, and the comforts which they enjoy, were in the same proportion, because the whole of a labourer's earnings are not spent in food; and if that part which is not so spent will, in the country where the value of fifteen quarters is earned, not go near so far in the purchase of clothes and other conveniences as in the countries where the value of nine quarters is earned, it is clear that altogether the situation of the labourer in the latter country may approach nearer to that of the labourer in the former than might at first be supposed.

At the same time it should be recollected that *quantity* always tends powerfully to counterbalance any deficiency of value; and the labourer who earns the greatest number of quarters may still command the greatest quantity of necessaries and conveniences combined, though not to the extent implied by the proportions of the raw produce.

America affords a practical instance of the agricultural system in a state the most favourable to the condition of the labouring classes. The nature of the country is such as to make it answer to employ a very large proportion of its capital in agriculture; and the consequence has been a very rapid increase of it. This rapid increase both of the quantity and value of capital has kept up a steady and continued demand for labour. The labouring classes have in consequence been peculiarly well paid. They have been able to command an unusual quantity of the necessaries of life, and the progress of population has been unusually rapid.

Yet even here, some little drawback has been felt from the relative cheapness of corn. As America till the late war imported the greatest part of its manufactures from England, and as England imported flour and wheat from America, the value of food in America compared with manufactures must have been decidedly less than in England. Nor would this effect take place merely with relation to the foreign commodities imported into America, but also to those of its home manufactures, in which it has no particular advantage. In agriculture, the abundance of good land would counterbalance the high wages of labour and high profits of stock, and keep the price of corn moderate, notwithstanding the great expense of these two elements of price. But in the production of manufactured commodities they must necessarily tell, without any particular advantage to counterbalance them, and must in general occasion in home goods, as well as foreign, a high price compared with food.

Under these circumstances, the condition of the labouring classes of society cannot in point of conveniences and comforts be so much better than that of the labourers of other countries as the relative quantity of food which they earn might seem to indicate; and this conclusion is sufficiently confirmed by experience. In some very intelligent Travels through a great part of England, written in 1810 and 1811 by Mr. Simond, a French gentleman, who had resided above twenty years in America, the author seems to have been evidently much struck with the air of convenience and comfort in the houses of our peasantry, and the neatness and cleanliness of their dress. In some parts of his tour he saw so many neat cottages, so much good clothing, and so little appearance of poverty and distress, that he could not help wondering where the poor of England and their dwellings were concealed. These observations, coming from an able, accurate and apparently most impartial observer, just landed from America and visiting England for the first time, are curious and instructive; and the facts which they notice, though they may arise in part from the different habits and modes of life prevailing in the two countries, must be occasioned in a considerable degree by the causes above mentioned.

A very striking instance of the disadvantageous effect of a low relative price of food on the condition of the poor may be observed in Ireland. The food of Ireland has increased so rapidly during the last century, and so large a portion of that which forms the principal support of the lower classes of society has been obtained by them, that the increase of population has been more rapid than in almost any known country, except America. The Irish labourer paid in potatoes has earned perhaps the means of subsistence for double the number of persons that could be supported by the earnings of an English labourer paid in wheat; and the increase of population in the two countries during the last century has been nearly in proportion to the relative quantity of the customary food awarded to the labourers in each. But their general condition with respect to conveniences and comforts is very far indeed from being in a similar proportion. The great quantity of food which land will bear when planted with potatoes, and the consequent cheapness of the labour supported by them; tends rather to raise than to lower the rents of land, and as far as rent goes, to keep up the price of the materials of manufactures and all other sorts of raw produce, except potatoes. The indolence and want of skill which usually accompany such a state of things tend further to render all wrought commodities comparatively dear. In home manufactures, therefore, a great relative disadvantage will be suffered, and a still greater both in the raw and manufactured produce of foreign countries. The value of the food which the Irish labourer earns above what he and his family consume will go but a very little way in the purchase of clouting, lodging and other conveniences; and the consequence is that his condition in these respects is extremely miserable, at the same time that his means of subsistence, such as they are, may be comparatively abundant.

In Ireland the money price of labour is not much more than the half of what it is in England. The quantity of food earned by no means makes up for its very low price. A certain portion therefore of the Irish labourer's wages (a fourth or a fifth for instance) will go but a very little way in the purchase of manufactures and foreign produce. In the United States, on the other hand, even the money wages of labour are nearly double those of England. Though the American labourer therefore cannot purchase manufactures and foreign produce with the food that he earns so cheap as the English labourer, yet the greater quantity of this food more than makes up for its lower price. His condition, compared with the labouring classes of England, though it may not be so much superior as their relative means of subsistence might indicate, must still on the whole have decidedly the advantage; and altogether, perhaps, the United States may be produced as an instance of the agricultural system in which the condition of the labouring classes is the best of any that we know.

The instances where, under the agricultural system, the condition of the lower classes of society is very wretched, are more frequent. When the accumulation of capital stops, whatever may be the cause, the population, before it comes to a stand, will always be pressed on as near to the limits of the actual means of subsistence, as the habits of the lower classes of the society will allow; that is, the real wages of labour will sink, till they are only just sufficient to maintain a stationary population. Should this happen, as it frequently does, while land is still in abundance and capital scarce, the profits of stock will naturally be high; but corn will be very cheap, owing to the goodness and plenty of the land, and the stationary demand for it, notwithstanding the high profits of stock; while these high profits, together with the want of skill and proper division of labour, which usually attend a scanty capital, will render all domestic manufactured commodities comparatively very dear. This state of things will naturally be unfavourable to the generation of those habits of prudential restraint which most frequently arise from the custom of enjoying conveniences and comforts, and it is to be expected that the population will not stop till the wages of labour, estimated even in food, are very low. But in a country where the wages of labour estimated in food are low, and that food is relatively of a very low value, both with regard to domestic and foreign manufactures, the condition of the labouring classes of society must be the worst possible.

Poland, and some parts of Russia, Siberia and European Turkey, afford instances of this kind. In Poland the population seems to be almost stationary or very slowly progressive; and as both the population and produce are scanty, compared with the extent of territory, we may infer with certainty that its capital is scanty, and yet slowly progressive. It follows, therefore, that the demand for labour increases very slowly, and that the real wages of labour, or the command of the labouring classes over the necessaries and conveniences of life, are such as to keep the population down to the level of the slowly increasing quantity that is awarded to them. And as from the state of the country the peasantry cannot have been much accustomed to conveniences and comforts, the checks to its population are more likely to be of the positive than of the preventive kind.

Yet here corn is in abundance, and great quantities of it are yearly exported. Hence it appears that it is not either the power of the country to produce food, or even what it actually produces, that limits and regulates the progress of population, but the quantity and value of the food which in the actual state of things is awarded to the labourer, and the rate at which these funds appropriated increase.

In the present case the demand for labour is very small, and though the population is inconsiderable, it is greater than the scanty capital of the country can fully employ; the condition of the labourer therefore is depressed by his being able to command only such a quantity of food as will maintain a stationary or very slowly increasing population. It is further depressed by the low relative value of the food that he earns, which gives to any surplus he may possess a very small power in the purchase of manufactured commodities or foreign produce.

Under these circumstances, we cannot be surprised that all accounts of Poland should represent the condition of the lower classes of society as extremely miserable; and the other parts of Europe which resemble Poland in the state of their land and capital, resemble it in the condition of their people.

In justice, however, to the agricultural system, it should be observed that the premature check to the capital and the demand for labour, which occurs in some of the countries of Europe, while land continues in considerable plenty, is not occasioned by the particular direction of their industry, but by the vices of the government and the structure of the society, which prevent its full and fair developement in that direction.

Poland is continually brought forward as an example of the miserable effects of the agricultural system. But nothing surely can be less fair. The misery of Poland does not arise from its directing its industry chiefly to agriculture, but from the little encouragement given to industry of any kind, owing to the state of property and the servile condition of the people. While the land is cultivated by boors, the produce of whose exertions belongs entirely to their masters, and the whole society consists mainly of these degraded beings and the lords and owners of great tracts of territory, there will evidently be no class of persons possessed of the means either of furnishing an adequate demand at home for the surplus produce of the soil, or of accumulating fresh capital and increasing the demand for labour. In this miserable state of things, the best remedy would unquestionably be the introduction of manufactures and commerce; because the introduction of manufactures and commerce could alone liberate the mass of the people from slavery and give the necessary stimulus to industry and accumulation. But were the people already free and industrious, and landed property easily divisible and alienable, it might still answer to such a country as Poland to purchase its finer manufactures from foreign countries by means of its raw products, and thus to continue essentially agricultural for many years. Under these new circumstances, however, it would present a totally different picture from that which it exhibits at present; and the condition of the people would more resemble that of the inhabitants of the United States of America, than of the inhabitants of the unimproved countries of Europe. Indeed America is perhaps the only modern instance of the fair operation of the agricultural system. In every country of Europe, and in most of its colonies in other parts of the world, formidable obstacles still exist to the employment of capital upon the land, arising from the remains of the feudal system. But these obstacles which have essentially impeded cultivation have been very far indeed from proportionably encouraging other branches of industry. Commerce and manufactures are necessary to agriculture; but agriculture is still more necessary to commerce and manufactures. It must ever be true that the surplus produce of the cultivators, taken in its most enlarged sense, measures and limits the growth of that part of the society which is not employed upon the land. Throughout the whole world the number of manufacturers, of merchants, of proprietors, and of persons engaged in the various civil and military professions, must be exactly proportioned to this surplus produce, and cannot in the nature of things increase beyond it. If the earth had been so niggardly of her produce as to oblige all her inhabitants to labour for it, no manufactures or idle persons could ever have existed. But her first intercourse with man was a voluntary present, not very large indeed, but sufficient as a fund for his subsistence till he could procure a greater. And the power to procure a greater was given to him in that quality of the earth by which it may be made to yield a much larger quantity of food, and of the materials of clothing and lodging, than is necessary to feed, clothe and lodge the persons employed in the cultivation of the soil. This quality is the foundation of that surplus produce which peculiarly distinguishes the industry employed upon the land. In proportion as the labour and ingenuity of man exercised upon the land have increased this surplus produce, leisure has been given to a greater number of persons to employ themselves in all the inventions which embellish civilized life; while the desire to profit by these inventions has continued to stimulate the cultivators to increase their surplus produce. This desire indeed may be considered as almost absolutely necessary to give it its proper value, and to encourage its further extension; but still the order of precedence is, strictly speaking, the surplus produce; because the foods for the subsistence of the manufacturer must be advanced to him before he can complete his work; and no step can be taken in any other sort of industry unless the cultivators obtain from the soil more than they themselves consume.

If in asserting the peculiar productiveness of the labour employed upon the land, we look only to the clear monied rent yielded to a certain number of proprietors, we undoubtedly consider the subject in a very contracted point of view. In the advanced stages of society, this rent forms indeed the most prominent portion of the surplus produce here meant; but it may exist equally in the shape of high wages and profits during the earlier periods of cultivation, when there is little or no rent. The labourer who earns a value equal to fifteen or twenty quarters of corn in the year may have only a family of three or four children, and not consume in kind above five or six quarters; and the owner of the farming stock, which yields high profits, may consume but a very moderate proportion of them in food and raw materials. All the rest, whether in the shape of wages and profits, or of rents, may be considered as a surplus produce from the soil, which affords the means of subsistence and the materials of clothing and lodging to a certain number of people according to its extent, some of whom may live without manual exertions, and others employ themselves in modifying the raw materials obtained from the earth into the forms best suited to the gratification of man.

It will depend of course entirely upon its answering to a country to exchange a part of the surplus produce for foreign commodities, instead of consuming it at home, whether it is to be considered as mainly agricultural or otherwise. And such an exchange of raw produce for manufactures, or peculiar foreign products, may for a period of some extent suit a state, which might resemble Poland in scarcely any other feature but that of exporting corn.

It appears then, that countries in which the industry of the inhabitants is principally directed towards the land, and in which corn continues to be exported, may enjoy great abundance or experience great want, according to the particular circumstances in which they are placed. They will in general not be much exposed to the temporary evils of scarcity arising from the variations of the seasons; but the quantity of food permanently awarded to the labourer may be such as not to allow of an increase of population; and their state, in respect to their being progressive, stationary or declining, will depend upon other causes than that of directing their attention principally to agriculture.

4.1.9 Book III, Chapter IX: Of the Commercial System.

A country which excels in commerce and manufactures, may purchase corn from a great variety of others; and it may be supposed, perhaps, that, proceeding upon this system, it may continue to purchase an increasing quantity, and to maintain a rapidly increasing population, till the lands of all the nations with which it trades are fully cultivated. As this is an event necessarily at a great distance, it may appear that the population of such a country will not be checked from the difficulty of procuring subsistence till after the lapse of a great number of ages.

There are, however, causes constantly in operation, which will occasion the pressure of this difficulty, long before the event here contemplated has taken place, and while the means of raising food in the surrounding countries may still be comparatively abundant.

In the first place, advantages which depend exclusively upon capital and skill, and the present possession of particular channels of commerce, cannot in their nature be permanent. We know how difficult it is to confine improvements in machinery to a single spot; we know that it is the constant object, both of individuals and countries, to increase their capital; and we know, from the past history of commercial states, that the channels of trade are not unfrequently taking a different direction. It is unreasonable therefore to expect that any one country, merely by the force of skill and capital, should remain in possession of markets uninterrupted by foreign competition. But, when a powerful foreign competition takes place, the exportable commodities of the country in question must soon fall to prices which will essentially reduce profits; and the fall of profits will diminish both the power and the will to save. Under these circumstances the accumulation of capital will be slow, and the demand for labour proportionably slow, till it comes nearly to a stand; while, perhaps, the new competitors either by raising their own raw materials or by some other advantages, may still be increasing their capitals and population with some degree of rapidity.

But, secondly, even if it were possible for a considerable time to exclude any formidable foreign competition, it is found that domestic competition produces almost unavoidably the same effects. If a machine be invented in a particular country, by the aid of which one man can do the work of ten, the possessors of it will of course at first make very unusual profits; but, as soon as the invention is generally known, so much capital and industry will be brought into this new and profitable employment, as to make its products greatly exceed both the foreign and domestic demand at the old prices. These prices, therefore, will continue to fall, till the stock and labour employed in this direction cease to yield unusual profits. In this case it is evident that, though in an early period of such a manufacture, the product of the industry of one man for a day might have been exchanged for such a portion of food as would support forty or fifty persons; yet, at a subsequent period, the product of the same industry might not purchase the support of ten.

In the cotton trade of this country, which has extended itself so wonderfully during the last twenty-five years, very little effect has hitherto been produced by foreign competition.[30] The very great fall which has taken place in the prices of cotton goods has been almost exclusively owing to domestic competition; and this competition has so glutted both the home and foreign markets, that the present capitals employed in the trade, notwithstanding the very peculiar advantages which they possess from the saving of labour, have ceased to possess any advantage whatever in the general rate of their profits. Although, by means of the admirable machinery used in the spinning of cotton, one boy or girl can now do as much as many grown persons could do formerly; yet neither the wages of the labourer, nor the profits of his master, are higher than in those employments where no machinery is used, and no saving of labour accomplished.

The country has, however, in the mean time, been very greatly benefited. Not only have all its inhabitants been enabled to obtain a superior fabric for clothing, at a less expense of labour and property, which must be considered as a great and permanent advantage; but the high temporary profits of the trade have occasioned a great accumulation of capital, and consequently a great demand for labour; while the extending markets abroad and the new values thrown into the market at home, have created such a demand for the products of every species of industry, agricultural and colonial, as well as commercial and manufacturing, as to prevent a fall of profits.

This country, from the extent of its lands, and its rich colonial possessions, has a large *arena* for the employment of an increasing capital; and the general rate of its profits are not, as it appears, very easily and rapidly reduced by accumulation. But a country, such as we are considering, engaged principally in manufactures, and unable to direct its industry to the same variety of pursuits, would sooner find its rate of profits diminished by an increase of capital, and no ingenuity in machinery which was not continually progressive could save it, after a certain period, from low profits and low wages, and their natural consequences, a check to population.

Thirdly. A country which is obliged to purchase both the raw materials of its manufactures and the means of subsistence for its population from foreign countries, is almost entirely dependent for the increase of its wealth and population on the increasing wealth and demands of the countries with which it trades.

It has been sometimes said, that a manufacturing country is no more dependent upon the country which supplies it with food and raw materials, than the agricultural country is on that which manufactures for it; but this is really an abuse of terms. A country with great resources in land may find it decidedly for its advantage to employ the main part of its capital in cultivation, and to import its manufactures. In so doing, it will often employ the whole of its industry most productively, and most rapidly increase its stock. But, if the slackness of its neighbours in manufacturing, or any other cause, should either considerably check or altogether prevent the importation of manufactures, a country with food and raw materials provided at home cannot be, long at a loss. For a time it would not certainly be so well supplied; but manufacturers and artizans would soon be found, and would soon

[30] 1816.

acquire tolerable skill;[31] and though the capital and population of the country might not, under the new circumstances in which it was placed, increase so rapidly as before, it would still have the power of increasing in both to a great extent.

On the other hand, if food and raw materials were denied to a nation merely manufacturing, it is obvious that it could not longer exist. But not only does the absolute existence of such a nation, on an extreme supposition, depend upon its foreign commerce, but its progress in wealth must be almost entirely measured by the progress and demand of the countries which deal with it. However skilful, industrious and saving such a nation might be, if its customers, from indolence and want of accumulation, would not or could not take off a yearly increasing value of its commodities, the effects of its skill and machinery would be but of very short duration.

That the cheapness of manufactured commodities, occasioned by skill and machinery in one country, is calculated to encourage an increase of raw produce in others, no person can doubt; but we know at the same time that high profits may continue for a considerable period in an indolent and ill-governed state, without producing an increase of wealth; yet, unless such an increase of wealth and demand were produced in the surrounding countries, the increasing ingenuity and exertions of the manufacturing and commercial state would be lost in continually falling prices. It would not only be obliged, as its skill and capital increased, to give a larger quantity of manufactured produce for the raw produce which it received in return; but might be unable, even with the temptation of reduced prices, to stimulate its customers to such purchases as would allow of an increasing importation of food and raw materials; and without such an increasing importation, it is quite obvious that the population must become stationary.

It would come to the same thing, whether this inability to obtain an increasing quantity of food were occasioned by the advancing money price of corn, or the falling money price of manufactures. In either case the effect would be the same; and it is certain that this effect might take place in either way, from increasing competition and accumulation in the manufacturing nation, and the want of them in the agricultural, long before any essential increase of difficulty had occurred in the production of corn.

Fourthly. A nation which is obliged to purchase from others nearly the whole of its raw materials, and the means of its subsistence, is not only dependent entirely upon the demands of its customers, as they may be variously affected by indolence, industry or caprice, but it is subjected to a necessary and unavoidable diminution of demand in the natural progress of these countries towards that proportion of skill and capital which they may reasonably be expected after a certain time to possess. It is generally an accidental and temporary, not a natural and permanent division of labour, which constitutes one state the manufacturer and the carrier of others. While, in these landed nations, agricultural profits continue very high, it may fully answer to them to pay others as their manufacturers and carriers; but when the profits on land fall, or the tenures on which it can be held are not such

[31] This has been fully exemplified in America (1816).

as to encourage the investment of an accumulating capital, the owner of this capital will naturally look towards commerce and manufactures for its employment; and, according to the just reasoning of Adam Smith and the Economists, finding at home both the materials of manufactures, the means of subsistence, and the power of carrying on their own trade with foreign countries, they will probably be able to conduct the business of manufacturing and carrying for themselves at a cheaper rate than if they allowed it to continue in the hands of others. As long as the agricultural nations continued to apply their increasing capital principally to the land, this increase of capital would be of the greatest possible advantage to the manufacturing and commercial nation. It would be indeed the main cause and great regulator of its progress in wealth and population. But after they had turned their attention to manufactures and commerce, their further increase of capital would be the signal of decay and destruction to the manufactures and commerce, which they had before supported. And thus, in the natural progress of national improvement, and without the competition of superior skill and capital, a purely commercial state must be undersold and driven out of the markets by those who possess the advantage of land.

In the distribution of wealth during the progress of improvement, the interests of an independent state in relation to others are essentially different from those of a particular province, in relation to the kingdom to which it belongs, a point which has not been sufficiently attended to. If agricultural capital increases, and agricultural profits diminish in Sussex, the overflowing stock will go to London, Manchester, Liverpool, or some other place where it can probably be engaged in manufactures or commerce more advantageously than at home. But if Sussex were an independent kingdom, this could not take place; and the corn which is now sent to London must be withdrawn to support manufacturers and traders living within its confines. If England, therefore, had continued to be separated into the seven kingdoms of the Heptarchy, London could not possibly have been what it is; and that distribution of wealth and population which takes place at present, and which we may fairly presume is the most beneficial to the whole of the realm, would have been essentially changed, if the object had been to accumulate the greatest quantity of wealth and population in particular districts, instead of the whole island. But at all times the interest of each independent state is to accumulate the greatest quantity of wealth within its limits. Consequently, the interest of an independent state, with regard to the countries with which it trades, can rarely be the same as the interest of a province with regard to the empire to which it belongs; and the accumulation of capital which would occasion the withdrawing of the exports of corn in the one case, would leave them perfectly undisturbed in the other.

If, from the operation of one or more of the causes above enumerated, the importation of corn into a manufacturing and commercial country should be essentially checked, and should either actually decrease, or be prevented from increasing, it is quite evident that its population must be checked nearly in the same proportion.

Venice presents a striking instance of a commercial state, at once stopped in its progress to wealth and population by foreign competition. The discovery made by the Portuguese of a passage to India by the Cape of Good Hope completely turned the channel of the Indian trade. The high profits of the Venetians, which had been the foundation of their rapidly increasing wealth, and of their extraordinary preponderance as a naval and commercial power, were not only suddenly reduced; but the trade itself, on which these high profits had been made, was almost annihilated, and their power and wealth were shortly contracted to those more confined limits which suited their natural resources.

In the middle of the 15th century, Bruges in Flanders was the great *entrepôt* of the trade between the north and the south of Europe. Early in the 16th century its commerce began to decline under the competition of Antwerp. Many English and foreign merchants in consequence left the declining city, to settle in that which was rapidly increasing in commerce and wealth. About the middle of the 16th century Antwerp was at the zenith of its power. It contained above a hundred thousand inhabitants, and was universally allowed to be the most illustrious mercantile city, and to carry on the most extensive and richest commerce of any in the north of Europe.

The rising greatness of Amsterdam was favoured by the unfortunate siege and capture of Antwerp by the duke of Parma; and the competition of the extraordinary industry and persevering exertions of the Hollanders not only prevented Antwerp from recovering her commerce, but gave a severe blow to the foreign trade of almost all the other Hanse Towns.

The subsequent decline of the trade of Amsterdam itself was caused partly by the low profits arising from home competition and abundance of capital; partly by excessive taxation, which raised the price of the necessaries of life; but more than either, perhaps, by the progress of other nations possessing greater natural advantages, and being able, even with inferior skill, industry and capital, beneficially to carry on much of that trade which had before fallen almost exclusively into the hands of the Dutch.

As early as 1669 and 1670, when Sir William Temple was in Holland, the effects of abundance of capital and domestic competition were such, that most of the foreign trades were losing ones, except the Indian, and that none of them gave a profit of more than two or three per cent.[32] In such a state of things both the power and the will to save must be greatly diminished. The capital must have been either stationary or declining, or at the best very slowly progressive. In fact, Sir William Temple gives it as his opinion that the trade of Holland had for some years passed its meridian, and begun sensibly to decay.[33] Subsequently, when the progress of other nations was still more marked, it appeared from undoubted documents that most of the trades of Holland, as well as its fisheries, had decidedly fallen off, and that no branch of its commerce had retained its former vigour, except the American and African trades, and that of the Rhine and Maese, which are independent of foreign power and competition.

In 1669, the whole population of Holland and West Friezeland was estimated by John de Witt at 2,400,000.[34] In 1778, the population of the seven provinces was estimated only at

[32] Temple's Works, vol. i. p. 60, fol.

[33] Temple's Works, vol. i. p. 67.

[34] Interest of Holland, vol. i. p. 9.

2,000,000;[35] and thus, in the course of above a hundred years, the population, instead of increasing, as is usual, had greatly diminished.

In all these cases of commercial states, the progress of wealth and population seems to have been checked by one or more of the causes above mentioned, which must necessarily affect more or less the power of commanding the means of subsistence.

Universally it may be observed, that if, from any cause or causes whatever, the funds for the maintenance of labour in any country cease to be progressive, the effective demand for labour will also cease to be progressive; and wages will be reduced to that sum, which, under the existing prices of provisions, and the existing habits of the people, will just keep up, and no more than keep up, a stationary population. A state so circumstanced is under a moral impossibility of increasing, whatever may be the plenty of corn, or however high may be the profits of stock in other countries.[36] It may indeed at a subsequent period, and under new circumstances, begin to increase again. If by some happy invention in mechanics, the discovery of some new channel of trade, or an unusual increase of agricultural wealth and population in the surrounding countries, its exports, of whatever kind, were to become unusually in demand, it might again import an increasing quantity of corn, and might again increase its population. But as long as it is unable to make yearly additions to its imports of food, it will evidently be unable to furnish the means of support to an increasing population; and it will necessarily experience this inability, when, from the state of its commercial transactions, the funds for the maintenance of its labour become stationary, or begin to decline.

4.1.10 Book III, Chapter X: Of Systems of Agriculture and Commerce combined.

In a country the most exclusively confined to agriculture, some of its raw materials will always be worked up for domestic use. In the most commercial state, not absolutely confined to the walls of a town, some part of the food of its inhabitants, or of its cattle, will be drawn from the small territory in its neighbourhood. But, in speaking of systems of agriculture and commerce combined, something much further than this kind of combination is intended; and it is meant to refer to countries, where the resources in land, and the capitals employed in commerce and manufactures, are both considerable, and neither preponderating greatly over the other.

A country so circumstanced possesses the advantages of both systems, while at the same time it is free from the peculiar evils which belong to each, taken separately.

---

[35] Richesse de la Hollande, vol. ii. p. 349.

[36] It is a curious fact, that among the causes of the decline of the Dutch trade, Sir William Temple reckons the cheapness of corn, which, he says, "has been for these dozen years, or more, general in these parts of Europe." (Vol. i. p. 69.) This cheapness, he says, impeded the vent of spices and other Indian commodities among the Baltic nations, by diminishing their power of purchasing.

The prosperity of manufactures and commerce in any state implies at once that it has freed itself from the worst parts of the feudal system. It shews that the great body of the people are not in a state of servitude; that they have both the power and the will to save; that when capital accumulates it can find the means of secure employment, and consequently that the government is such as to afford the necessary protection to property. Under these circumstances, it is scarcely possible that it should ever experience that premature stagnation in the demand for labour, and the produce of the soil, which at times has marked the history of most of the nations of Europe. In a country in which manufactures and commerce flourish, the produce of the soil will always find a ready market at home; and such a market is peculiarly favourable to the progressive increase of capital. But the progressive increase of capital, and particularly of the quantity and value of the funds for the maintenance of labour, is the great cause of a demand for labour, and of good corn wages, while the high relative price of corn, occasioned by the improved machinery and extended capital employed in manufactures, together with the prosperity of foreign commerce, enables the labourer to exchange any given portion of his earnings in corn for a large proportion both of domestic and foreign conveniences and luxuries. Even when the effective demand for labour begins to slacken, and the corn wages to be reduced, still the high relative value of corn keeps up comparatively the condition of the labouring classes; and though their increase is checked, yet a very considerable body of them may still be well lodged and well clothed, and able to indulge themselves in the conveniences and luxuries of foreign produce. Nor can they ever be reduced to the miserable condition of the people in those countries, where, at the same time that the demand for labour is stationary, the value of corn, compared with manufactures and foreign commodities, is extremely low.

All the peculiar disadvantages therefore of a purely agricultural country are avoided by the growth and prosperity of manufactures and commerce.

In the same manner it will be found that the peculiar disadvantages attending states merely manufacturing and commercial will be avoided by the possession of resources in land.

A country which raises its own food cannot by any sort of foreign competition be reduced at once to a necessarily declining population. If the exports of a merely commercial country be essentially diminished by foreign competition, it may lose, in a very short time, its power of supporting the same number of people; but if the exports of a country which has resources in land be diminished, it will merely lose some of its foreign conveniencies and luxuries; and the great and most important of all trades, the domestic trade carried on between the towns and the country, will remain comparatively undisturbed. It may indeed be checked in the rate of its progress for a time by the want of the same stimulus; but there is no reason for its becoming retrograde; and there is no doubt that the capital thrown out of employment by the loss of foreign trade will not lie idle. It will find some channel in which it can be employed with advantage, though not with the same advantage as before; and will be able to maintain an increasing population, though not increasing at the same rate as under the stimulus of a prosperous foreign trade.

The effects of home competition will in like manner be very different in the two states we are comparing.

In a state merely manufacturing and commercial, home competition and abundance of capital may so reduce the price of manufactured compared with raw produce, that the increased capital employed in manufactures may not procure in exchange an increased quantity of food. In a country where there are resources in land this cannot happen; and though from improvements in machinery and the decreasing fertility of the new land taken into cultivation, a greater quantity of manufactures will be given for raw produce, yet the mass of manufactures can never fall in value, owing to a competition of capital in this species of industry, unaccompanied by a correspondent competition of capital on land.

It should also be observed that in a state, the revenue of which consists solely in profits and wages, the diminution of profits and wages may greatly impair its disposable income. The increase in the amount of capital and in the number of labourers may in many cases not be sufficient to make up for the diminished rate of profits and wages. But where the revenue of the country consists of rents as well as profits and wages, a great part of what is lost in profits and wages is gained in rents, and the disposable income remains comparatively unimpaired.

Another eminent advantage possessed by a nation which is rich in land, as well as in commerce and manufactures, is, that the progress of its wealth and population is in a comparatively slight degree dependent upon the state and progress of other countries. A nation, whose wealth depends exclusively on manufactures and commerce, cannot increase without an increase in the raw products of the countries with which it trades; or taking away a share of what they have been in the habit of actually consuming, which will rarely be parted with; and thus the ignorance and indolence of others may not only be prejudicial, but fatal to its progress.

A country with resources in land can never be exposed to these inconveniences; and if its industry, ingenuity, and economy increase, its wealth and population will increase, whatever may be the situation and conduct of the nations with which it trades. When its manufacturing capital becomes redundant, and manufactured commodities are too cheap, it will have no occasion to wait for the increasing raw products of its neighbours. The transfer of its own redundant capital to its own land will raise fresh products, against which its manufactures may be exchanged, and by the double operation of diminishing comparatively the supply, and increasing the demand, enhance their price. A similar operation, when raw produce is too abundant, will restore the level between the profits of agriculture and manufactures. And upon the same principle the stock of the country will be distributed through its various and distant provinces, according to the advantages presented by each situation for the employment, either of agricultural or manufacturing capital.

A country, in which in this manner agriculture, manufactures, and commerce, and all the different parts of a large territory, act and re-act upon each other in turn, might evidently go on increasing in riches and strength, although surrounded by Bishop Berkely's wall of brass. Such a country would naturally make the most of its foreign commerce, whatever might be the actual state of it; and its increase or decrease would be the addition or removal of a powerful stimulus to its own produce; but still the increase of this produce, to a very considerable extent, would be independent of foreign countries; and though it might be retarded by a failure of foreign commerce, it could not either be stopped or be made retrograde.

A fourth advantage derived from the union of agriculture and manufactures, particularly when they are nearly balanced, is, that the capital and population of such a country can never be forced to make a retrograde movement, merely by the natural progress of other countries to that state of improvement to which they are all constantly tending.

According to all general principles, it will finally answer to most landed nations, both to manufacture for themselves, and to conduct their own commerce. That raw cottons should be shipped in America, carried some thousands of miles to another country, unshipped there, to be manufactured and shipped again for the American market, is a state of things which cannot be permanent. That it may last for some time, there can be no doubt; and I am very far from meaning to insinuate that an advantage, while it lasts, should not be used, merely because it will not continue for ever. But if the advantage be in its nature temporary, it is surely prudent to have this in view, and to use it in such a way, that when it ceases, it may not have been productive, on the whole, of more evil than good.

If a country, owing to temporary advantages of this kind, should have its commerce and manufactures so greatly preponderate as to make it necessary to support a large portion of its people on foreign corn, it is certain that the progressive improvement of foreign countries in manufactures and commerce might, after a time, subject it to a period of poverty and of retrograde movements in capital and population, which might more than counterbalance the temporary benefits before enjoyed; while a nation in which the commercial and manufacturing population continued to be supported by its agriculture, might receive a very considerable stimulus to both, from such temporary advantages, without being exposed to any essential evil on their ceasing.

The countries which thus unite great landed resources with a prosperous state of commerce and manufactures, and in which the commercial part of the population never essentially exceeds the agricultural part, are eminently secure from sudden reverses. Their increasing wealth seems to be out of the reach of all common accidents; and there is no reason to say that they might not go on increasing in riches and population for hundreds, nay almost thousands of years.

We must not, however, imagine that there is no limit to this progress though it is distant, and has certainly not been attained by any large landed nation yet known.

We have already seen that the limit to the population of commercial nations is the period when, from the actual state of foreign markets, they are unable regularly to import an increasing quantity of food. And the limit to the population of a nation which raises the whole of its food on its own territory is, when the land has been so fully occupied and worked, that the employment of another labourer on it will not, on an average, raise an additional quantity of food sufficient to support a family of such a size as will admit of an increase of population.

This is evidently the extreme practical limit to the progress of population, which no nation has ever yet reached, nor, indeed, ever will; since no allowance has been here made either for other necessaries besides food, or for the profits of stock, both of which, however low, must always be something not inconsiderable.

Yet, even this limit is very far short of what the earth is capable of producing, if all were employed upon it who were not employed in the production of other necessaries; that is, if soldiers, sailors, menial servants, and all the artificers of luxuries, were made to labour upon the land. They would not, indeed, produce the support of a family, and ultimately not even of themselves; but, till the earth absolutely refused to yield any more, they would continue to add something to the common stock; and, by increasing the means of subsistence, would afford the means of supporting an increasing population. The whole people of a country might thus be employed during their whole time in the production of mere necessaries, and no leisure be left for other pursuits of any kind. But this state of things could only be effected by the forced direction of the national industry into one channel by public authority. Upon the principal of private property, which it may be fairly presumed will always prevail in society, it could never happen. With a view to the individual interest, either of a landlord or farmer, no labourer can ever be employed on the soil, who does not produce more than the value of his wages; and if these wages be not on an average sufficient to maintain a wife, and rear two children to the age of marriage, it is evident that both the population and produce must come to a stand. Consequently, at the most extreme practical limit of population, the state of the land must be such as to enable the last employed labourers to produce the maintenance of as many, probably, as four persons.

And it is happy for mankind that such are the laws of nature. If the competition for the necessaries of life, in the progress of population, could reduce the whole human race to the necessity of incessant labour for them, man would be continually tending to a state of degradation; and all the improvements which had marked the middle stages of his career would be completely lost at the end of it; but, in reality, and according to the universal principle of private property, at the period when it will cease to answer to employ more labour upon the land, the excess of raw produce, not actually consumed by the cultivators, will, in the shape of rents, profits, and wages, particularly the first, bear nearly as great a proportion to the whole as at any previous period, and, at all events, sufficient to support a large part of the society living, either without manual labour, or employing themselves in modifying the raw materials of the land into the forms best suited to the gratification of man.

When we refer, therefore, to the practical limits of population, it is of great importance to recollect that they must be always very far short of the utmost power of the earth to produce food.

It is also of great importance to recollect that, long before this practical limit is attained in any country, the rate of the increase of population will gradually diminish. When the capital of a country becomes stationary from bad government, indolence, extravagance, or a sudden shock to commerce, it is just possible that the check to population may, in some degree, be sudden, though, in that case, it cannot take place without a considerable convulsion. But when the capital of a country comes to a stop from the continued progress of accumulation and the exhaustion of the cultivable land, both the profits of stock and the wages of labour must have been gradually diminishing for a long period, till they are both ultimately so low as to afford no further encouragement to an increase of stock, and no further means for the support of an increasing population. If we could suppose that the capital employed upon the land was, at all times, as great as could possibly be applied with the same profit, and there were no agricultural improvements to save labour, it is obvious that, as accumulation proceeded, profits and wages would regularly fall, and the diminished rate in the progress of population would be quite regular. But practically this can never happen; and various causes, both natural and artificial, will concur to prevent this regularity, and occasion great variations at different times in the rate at which the population proceeds towards its final limit.

In the first place, land is practically almost always understocked with capital. This arises partly from the usual tenures on which farms are held, which, by discouraging the transfer of capital from commerce and manufactures, leaves it principally to be generated on the land; and partly from the very nature of much of the soil of almost all large countries, which is such that the employment of a small capital upon it may be little productive, while the employment of a large capital in draining, or in changing the character of the soil by a sufficient quantity of natural and artificial manures, may be productive in a high degree; and partly also, from the circumstance that after every fall of profits and wages there will often be room for the employment of a much greater capital upon the land than is at the command of those, who, by being in the actual occupation of farms, can alone so employ it.

Secondly; improvements in agriculture. If new and superior modes of cultivation be invented, by which not only the land is better managed, but is worked with less labour, it is obvious that inferior land may be cultivated at higher profits than could be obtained from richer land before; and an improved system of culture, with the use of better instruments, may, for a long period, more than counterbalance the tendency of an extended cultivation and a great increase of capital to yield smaller proportionate returns.

Thirdly; improvements in manufactures. When by increased skill and the invention of unproved machinery in manufactures one man becomes capable of doing as much as eight or ten could before, it is well known that, from the principle of home competition and the consequent great increase of quantity, the prices of such manufactures will greatly fall; and, as far as they include the necessaries and accustomed conveniences of labourers and farmers, they must tend to diminish that portion of the value of the whole produce which is consumed necessarily on the land, and leave a larger remainder. From this larger remainder may be drawn a higher rate of profits, notwithstanding the increase of capital and extension of cultivation.

Fourthly; the prosperity of foreign commerce. If from a prosperous foreign commerce our labour and domestic commodities rise considerably in price, while foreign commodities are advanced comparatively very little, an event which is very common, it is evident that the farmer or labourer will be able to obtain the tea, sugar, cottons, linens, leather, tallow, timber, &c. which he stands in need of, for a smaller quantity of corn or labour than before; and this increased power of purchasing foreign commodities will have precisely the same effect, in allowing the means of an extended cultivation without a fall of profits, as the improvements in manufactures just referred to.

Fifthly; a temporary increase in the relative price of raw produce from increased demand. Allowing, what is certainly not true, that a rise in the price of raw produce, will, after a certain number of years, occasion a proportionate rise in labour[37] and other commodities; yet, during the time that the price of raw produce takes the lead, it is obvious that the profits of cultivation may increase under an extended agriculture, and a continued accumulation of capital. And these intervals, it should be observed, must be of infinite importance in the progress of the wealth of a landed nation, particularly with reference to the causes of deficient capital upon the land before mentioned. If the land, for the most part, generates the new capital which is employed in extending its cultivation; and if the employment of a considerable capital for a certain period will often put land in such a state, that it can be cultivated afterwards at comparatively little expense; a period of high agricultural profits, though it may last only eight or ten years, may often be the means of giving to a country what is equivalent to a fresh quantity of land.

Though it is unquestionably and necessarily true, therefore, that the *tendency* of a continually increasing capital and extending cultivation is to occasion a progressive fall both of profits and wages; yet the causes above enumerated are evidently sufficient to account for great and long irregularities in this progress.

We see, in consequence, in all the states of Europe, great variations at different periods in the progress of their capital and population. After slumbering for years in a state almost stationary, some countries have made a sudden start, and have begun increasing at a rate almost approaching to that of new colonies. Russia and parts of Prussia have afforded instances of this kind, and have continued this rate of progress after the accumulation of capital and the extension of cultivation had been proceeding with great rapidity for many years.

From the operation of the same causes we have seen similar variations in our own country. About the middle of last century the interest of money was at 3 per cent.; and we may conclude that the profits of stock were nearly in proportion. At that time, as far as can be collected from the births and marriages, the population was increasing but slowly. From 1720 to 1750, a period of 30 years, the increase is calculated to have been only about 900,000 on a population of 5,565,000.[38] Since this period it cannot be doubted that the capital of the country has been prodigiously enlarged, and its cultivation very greatly extended; yet, during the last twenty years, we have seen the interest of money at above 5 per cent., with profits in proportion; and, from 1800 to 1811, an increase of population

[37] A rise, which is occasioned exclusively by the increased quantity of labour which may be required in the progress of society to raise a given quantity of corn on the last land taken into cultivation, must, of course, be peculiar to raw produce, and will not be communicated to those commodities, in the production of which there is no increase of labour.

[38] Population Abstracts, Preliminary Observations, table, p. xxv.

equal to 1,200,000 on 9,287,000, a rate of increase about two and a half times as great as at the former period.

But, notwithstanding these causes of irregularity in the progress of capital and population, it is quite certain that they cannot reach their necessary practical limit but by a very gradual process. Before the accumulation of capital comes to a stop from *necessity,* the profits of stock must, for a long time, have been so low as to afford scarcely any encouragement to an excess of saving above expenditure; and before the progress of population is finally stopped, the real wages of labour most have been gradually diminishing, till, under the existing habits of the people, they could only support such families as would just keep up, and no more than keep up, the actual population.

It appears then, that it is the union of the agricultural and commercial systems, and not either of them taken separately, that is calculated to produce the greatest national prosperity; that a country with an extensive and rich territory, the cultivation of which is stimulated by improvements in agriculture, manufactures, and foreign commerce, has such various and abundant resources, that it is extremely difficult to say when they will reach their limits. That there is, however, a limit, which, if the capital and population of a country continue increasing, they must ultimately reach, and cannot pass; and that this limit, upon the principle of private property, must be far short of the utmost power of the earth to produce food.

4.1.11 Book III, Chapter XI: Of Corn-Laws. Bounties upon Exportation.

It has been observed that some countries, with great resources in land, and an evident power of supporting a greatly increased population from their own soil, have yet been in the habit of importing large quantities of foreign corn, and have become dependent upon other states for a great part of their supplies.

The causes which may lend to this state of things seem to be chiefly the following:

First; any obstacles which the laws, constitution and customs of a country present to the accumulation of capital on the land, which do not apply with equal force to the increasing employment of capital in commerce and manufactures.

In every state in which the feudal system has prevailed, there are laws and customs of this kind, which prevent the free division and alienation of land like other property, and render the preparations for an extension of cultivation often both very difficult and very expensive. Improvements in such countries are chiefly carried on by tenants, a large part of whom have not leases, or at least leases of any length; and though their wealth and respectability have of late years very greatly increased, yet it is not possible to put them on a footing with enterprising owners, and to give them the same independence, and the same encouragement to employ their capitals with spirit, as merchants and manufacturers.

Secondly; a system of direct or indirect taxation, of such a nature as to throw a weight upon the agriculture of a country, which is either unequal, or, from peculiar circumstances, can be better borne by commerce and manufactures.

It is universally allowed that a direct tax on corn grown at home, if not counterbalanced by a corresponding tax on the importation of it, might be such as to destroy at once the cultivation of grain, and make a country import the whole of its consumption; and a partial effect of the same kind would follow, if, by a system of indirect taxation, the general price of labour were raised and yet by means of drawbacks on home and foreign commodities, by an abundance of colonial produce, and by those peculiar articles,[39] the demand for which abroad would not be much affected by the increase of price, the value of the whole of the exports, though not the quantity, might admit of increase.

Thirdly; improved machinery, combined with extensive capital and a very advantageous division of labour.

If in any country, by means of capital and machinery, one man be enabled to do the work of ten, it is quite obvious that before the same advantages are extended to other countries, a rise in the price of labour will but very little interfere with the power of selling those sorts of commodities, in the production of which the capital and machinery are so effectively applied. It is quite true that an advance in the necessary wages of labour, which increases the expense of raising corn, may have the same effect upon many commodities besides corn; and if there were no others, no encouragement would be given to the importation of foreign grain, as there might be no means by which it could be purchased cheaper abroad. But a large class of the exportable commodities of a commercial country are of a different description. They are either articles in a considerable degree peculiar to the country and its dependencies, or such as have been produced by superior capital and machinery, the prices of which are determined rather by domestic than foreign competition. All commodities of this kind will evidently be able to support without essential injury an advance in the price of labour, some permanently, and others for a considerable time. The rise in the price of the commodity so occasioned, or rather the prevention of that fall which would otherwise have taken place, may always indeed have the effect of decreasing in some degree the *quantity* of the commodity exported; but it by no means follows that it will diminish the whole of its bullion value in the foreign country, which is precisely what determines the bullion value, and generally the quantity of the returns. If cottons in this country were now to fall to half their present price, we should undoubtedly export a greater quantity than we do at present; but I very much doubt whether we should export double the quantity, at least for many years, and yet we must do this to enable us to command as much foreign produce as before. In this case, as in numerous others of the same kind, quantity and value go together to a certain point, though not at an equal pace; but, beyond this point, a further increase of quantity only diminishes the whole value produced, and the amount of the returns that can be obtained for it.

It is obvious then that a country, notwithstanding a high comparative price of labour and of materials, may easily stand a competition with foreigners in those commodities to which

[39] A rise in the price of labour in China would certainly increase the returns which it receives for its teas.

it can apply a superior capital and machinery with great effect; although such a price of labour and materials might give an undisputed advantage to foreigners in agriculture and some other sorts of produce, where the same saving of labour cannot take place. Consequently such a country may find it cheaper to purchase a considerable part of its supplies of grain from abroad with its manufactures and peculiar products, than to grow the whole at home.

If, from all or any of these causes, a nation becomes habitually dependent on foreign countries for the support of a considerable portion of its population, it must evidently be subjected, while such dependence lasts, to some of those evils which belong to a nation purely manufacturing and commercial. In one respect, indeed, it will still continue to have a great superiority. It will possess resources in land, which may be resorted to when its manufactures and commerce, either from foreign competition, or any other causes, begin to fail. But, to balance this advantage, it will be subjected, during the time that large importations are necessary, to much greater fluctuations in its supplies of corn, than countries wholly manufacturing and commercial. The demands of Holland and Hamburgh may be known with considerable accuracy by the merchants who supply them. If they increase, they increase gradually; and, not being subject from year to year to any great and sudden variations, it might be safe and practicable to make regular contracts for the average quantity wanted. But it is otherwise with such countries as England and Spain. Their wants are necessarily very variable, from the variableness of the seasons; and if the merchants were to contract with exporting countries for the quantity required in average years, two or three abundant seasons might ruin them. They must necessarily wait to see the state of the crops in each year, in order safely to regulate their proceedings; and though it is certainly true that it is only the deficiency from the average crop, and not the whole deficiency, which may be considered altogether in the light of a new demand in Europe; yet the largeness and previous uncertainty of this whole deficiency, the danger of making contracts for a stated quantity annually, and the greater chance of hostile combinations against large and warlike states, must greatly aggravate the difficulties of procuring a steady supply; and if it be true that unfavourable seasons are not unfrequently general, it is impossible to conceive that they should not occasionally be subject to great variations of price.

It has been sometimes stated that scarcities are partial, not general, and that a deficiency in one country is always compensated by a plentiful supply in others. But this seems to be quite an unfounded supposition. In the evidence brought before the Committee of the House of Commons in 1814, relating to the corn-laws, one of the corn merchants being asked whether it frequently happened that crops in the country bordering upon the Baltic failed, when they failed here, replied, "When crops are unfavourable in one part of Europe, it generally happens that they are more or less so in another."[40] If any person will take the trouble to examine the contemporaneous prices of corn in the different countries of Europe for some length of time, he will be convinced that the answer here given is perfectly just. In

[40]Report, p. 93.

the last hundred and fifty years, above twenty will be found in which the rise of prices is common to France and England, although there was seldom much intercourse between them in the trade of corn: and Spain and the Baltic nations, as far as their prices have been collected, appear frequently to have shared in the same general deficiency. Even within the last five years, two have occurred, the years 1811-12, and 1816-1817, in which, with extraordinary high prices in this country, the imports have been comparatively inconsiderable; which can only have arisen from those scarcities having been general over the greatest part of Europe.

Under these circumstances let us suppose that two million quarters of foreign grain were the average quantity annually wanted in this country, and suppose, at the same time, that a million quarters were deficient from a bad season; the whole deficiency to be supplied would then be three millions.

If the scarcity were general in Europe, it may fairly be concluded, that some states would prohibit the export of their corn entirely, and others tax it very highly; and if we could obtain a million or fifteen hundred thousand quarters, it is probably as much as we could reasonably expect. We should then, however, be two millions or fifteen hundred thousand quarters deficient. On the other hand, if we had habitually grown our own consumption, and were deficient a million of quarters from a bad season, it is scarcely probable that, notwithstanding a general scarcity, we should not be able to obtain three or four hundred thousand quarters in consequence of our advanced prices; particularly if the usual prices of our corn and labour were higher than in the rest of Europe. And in this case the sum of our whole deficiency would only be six or seven hundred thousand quarters, instead of fifteen hundred thousand or two millions of quarters. If the present year (1816-17) had found us in a state in which our growth of corn had been habitually far short of our consumption, the distresses of the country would have been dreadfully aggravated.

To provide against accidents of this kind, and to secure a more abundant and, at the time, a more steady supply of grain, a system of corn-laws has been recommended, the object of which is to discourage by duties or prohibitions the importation of foreign corn, and encourage by bounties the exportion of corn of home growth.

A system of this kind was completed in our own country in 1688,[41] the policy of which has been treated of at some length by Adam Smith.

In whatever way the general question may be finally decided, it must be allowed by all those who acknowledge the efficacy of the great principle of supply and demand that the line of argument taken by the auther of the *Wealth of Nations* against the system is essentially erroneous.

He first states that, whatever extension of the foreign market can be occasioned by the bounty, must in every particular year be altogether at the expense of the home market, as every bushel of corn which is exported by means of the bounty, and which would not have

[41]Though the object here stated may not have been the specific object of the law of 1688, it is certainly the object for which the system has been subsequently recommended.

been exported without the bounty, would have remained in the home market to increase the consumption, and to lower the price of that commodity.[42]

In this observation he evidently misapplies the term market. Because, by selling a commodity lower, it is easy to get rid of a greater quantity of it, in any particular market, than would have gone off otherwise, it cannot justly be said that by this process such a market is proportionally extended. Though the removal of the two taxes mentioned by Adam Smith as paid on account of the bounty would certainly increase the power of the lower classes to purchase, yet in each particular year the consumption must ultimately be limited by the population, and the increase of consumption from the removal of these taxes would by no means be sufficient to give the same encouragement to cultivation as the addition of the foreign demand. If the price of British corn in the home market rise in consequence of the bounty, before the price of production is increased, (and an immediate rise is distinctly acknowledged by Adam Smith,) it is an unanswerable proof that the effectual demand for British corn is extended by it; and that the diminution of demand at home, whatever it may be, is more than counterbalanced by the extension of demand abroad.

Adam Smith goes on to say that the two taxes paid by the people on account of the bounty, namely, the one to the government to pay this bounty, and the other paid in the advanced price of the commodity, must either reduce the subsistence of the labouring poor, or occasion an augmentation in their pecuniary wages proportioned to that in the pecuniary price of their subsistence. So far as it operates in the one way it must reduce the ability of the labouring poor to educate and bring up their children, and must so far tend to restrain the population of the country. So far as it operates in the other, it must reduce the ability of the employers of the poor to employ so great a number as they otherwise might do, and must so far tend to restrain the industry of the country.

It will be readily allowed that the tax occasioned by the bounty will have the one or the other of the effects here contemplated; but it cannot be allowed that it will have both. Yet it is observed, that though the tax, which that institution imposes upon the whole body of the people, be very burdensome to those who pay it, it is of very little advantage to those who receive it. This is surely a contradiction. If the price of labour rise in proportion to the price of wheat, as it subsequently asserted, how is the labourer rendered less competent to support a family? If the price of labour do not rise in proportion to the price of wheat, how is it possible to maintain that the landlords and farmers are not able to employ more labourers on their land? Yet in this contradiction the author of the *Wealth of Nations* has had respectable followers; and some of those who have agreed with him in his opinion that corn regulates the prices of labour, and of all other commodities, still insist on the injury done to the labouring classes of society by a rise in the price of corn, and the benefit they would derive from a fall.

[42] Vol. ii. b. iv. c. 5.

The main argument however which Adam Smith adduces against the bounty is, that as the money price of corn regulates that of all other home-made commodities, the advantage to the proprietor from the increase of money price is merely apparent, and not real; since what he gains in his sales he must lose in his purchases.

This position, though true to a certain extent, is by no means true to the extent of preventing the movement of capital to or from the land, which is the precise point in question. The money price of corn in a particular country is undoubtedly by far the most powerful ingredient in regulating the price of labour, and of all other commodities; but it is not enough for Adam Smith's position that it should be the most powerful ingredient; it must be shewn that, other causes remaining the same, the price of every article will rise and fall exactly in proportion to the price of corn, and this is very far from being the case. Adam Smith himself excepts all foreign commodities; but when we reflect upon the vast amount of our imports, and the quantity of foreign articles used in our manufactures, this exception alone is of the greatest importance. Wool and raw hides, two most important materials of home growth, do not, according to Adam Smith's own reasonings, (Book I. c. xi. p. 363, et seq. [pars. 209-222—Ed.] depend much upon the price of corn and the rent of land; and the prices of flax, tallow, and leather, are of course greatly influenced by the quantity we import. But woollen cloths, cotton and linen goods, leather, soap, candles, tea, sugar, &c. which are comprehended in the above-named articles, form almost the whole of the clothing and luxuries of the industrious classes of society.

It should be further observed that in all countries, the industry of which is greatly assisted by fixed capital, the part of the price of the wrought commodity which pays the profits of such capital will not necessarily rise in consequence of an advance in the price of corn, except as it requires gradual renovation; and the advantage derived from machinery which has been constructed before the advance in the price of labour will naturally last for some years.

In the case also of great and numerous taxes on consumption, a rise or fall in the price of corn, though it would increase or decrease that part of the wages of labour which resolves itself into food, evidently would not increase or decrease that part which is destined for the payment of taxes.

It cannot then be admitted as a general position that the money price of corn in any country is a just measure of the real value of silver in that country. But all these considerations, though of great weight to the owners of land, will not influence the farmers beyond the present leases. At the expiration of a lease, any particular advantage which a farmer had received from a favourable proportion between the price of corn and of labour would be taken from him, and any disadvantage from an unfavourable proportion be made up to him. The sole cause which would determine the proportion of capital employed in agriculture, would be the extent of the effectual demand for corn; and if the bounty had really enlarged this demand, which it certainly would have done, it is impossible to suppose that more capital would not be employed than upon the land.

When Adam Smith says that the nature of things has stamped upon corn a real value, which cannot be altered by merely altering the money price, and that no bounty upon exportation, no monopoly of the home market, can raise that value, nor the freest competition lower it, it is obvious that he changes the question from the profits of the growers of corn, or of the proprietors of the land, to the physical value of corn itself. I certainly do not mean to say that the bounty alters the physical value of corn, and makes a bushel of it support equally well a greater number of labourers than it did before; but I certainly do mean to say that the bounty to the British cultivator does, in the actual state of things, really increase the demand for British corn, and thus encourage him to sow more that he otherwise would do, and enables him in consequence to employ more bushels of corn in the maintenance of a greater number of labourers.

If Adam Smith's theory were true, and what he calls the real price of corn were unchangeable, or not capable of experiencing a relative increase or decrease of value compared with labour and other commodities, agriculture would indeed be in an unfortunate situation. It would be at once excluded from the operation of that principle so beautifully explained in the *Wealth of Nations,* by which capital flows from one employment to another, according to the various and necessarily fluctuating wants of society. But surely we cannot doubt that the real price of corn varies, though it may not vary so much as the real price of other commodities; and that there are periods when all wrought commodities are cheaper, and periods when they are dearer, in proportion to the price of corn; and in the one case capital flows from manufactures to agriculture, and in the other from agriculture to manufactures. To overlook these periods or consider them of slight importance, is not allowable; because in every branch of trade these periods form the grand encouragement to an increase of supply. Undoubtedly the profits of trade in any particular branch of industry can never long remain higher than in others; but how are they lowered except by the influx of capital occasioned by these high profits? It never can be a national object permanently to increase the profits of any particular set of dealers. The national object is the increase of supply; but this object cannot be attained except by previously increasing the profits of these dealers, and thus determining a greater quantity of capital to this particular employment. The ship-owners and sailors of Great Britain do not make greater profits now than they did before the Navigation Act; but the object of the nation was not to increase the profits of ship-owners and sailors, but the quantity of shipping and seamen; and this could not be done but by a law, which, by increasing the demand for them, raised the profits of the capital before employed in this way, and determined a greater quantity to flow into the same channel. The object of a nation in the establishment of a bounty is, not to increase the profits of the farmers, or the rents of the landlords, but to determine a greater quantity of the national capital to the land, and consequently to increase supply; and though, in the case of an advance in the price of corn from an increased demand, the rise of wages, the rise of rents, and the fall of silver tend, in some degree, to obscure our view of the subject; yet we cannot refuse to acknowledge that the real price of corn varies during periods sufficiently long to affect the determination of capital, or we shall be reduced to the dilemma of owning that no possible degree of demand can encourage the growth of corn.

It must be allowed then that the peculiar argument relating to the nature of corn brought forward by Adam Smith upon this occasion cannot be maintained, and that a bounty upon the exportation of corn must enlarge the demand for it and encourage its production in the same manner, if not in the same degree, as a bounty upon the exportation of any other commodity.

But it has been urged further, that this increased production of corn must necessarily occasion permanent cheapness; and a period of considerable length, during the first 64 years of the last century, while a bounty was in full operation in this country, has been advanced as a proof of it. In this conclusion, however, it may be reasonably suspected, that an effect, in its nature temporary, though it may be of some duration, has been mistaken for one which is permanent.

According to the theory of demand and supply, the bounty might be expected to operate, in the following manner:

It is frequently stated in the *Wealth of Nations* that a great demand is followed by a great supply; a great scarcity by a great plenty; an unusual dearness by an unusual cheapness. A great and indefinite demand is indeed generally found to produce a supply more than proportioned to it. This supply as naturally occasions unusual cheapness; but this cheapness, when it comes, must in its turn check the production of the commodity; and this check, upon the same principle, is apt to continue longer than necessary, and again to occasion a return to high prices.

This appears to be the manner in which a bounty upon the exportation of corn, if granted under circumstances favourable to its efficiency, might be expected to operate, and this seems to have been the manner in which it really did operate in the only instance where it has been fairly tried.

Without meaning to deny the concurrence of other causes, or attempting to estimate the relative efficiency of the bounty, it is impossible not to acknowledge that, when the growing price of corn was, according to Adam Smith, only 28 shillings a quarter, and the corn-markets of England were as low as those of the continent, a premium of five shillings a quarter upon exportation must have occasioned an increase of real price, and given encouragement to the cultivation of grain. But the changes produced in the direction of capital to or from the land will always be slow. Those who have been in the habit of employing their stock in mercantile concerns do not readily turn it into the channel of agriculture; and it is a still more difficult and slower operation to withdraw capital from the soil, to employ it in commerce. For the first 25 years after the establishment of the bounty in this country the price of corn rose 2 or 3 shillings in the quarter; but owing probably to the wars of William and Anne, to bad seasons, and a scarcity of money, capital seems to have accumulated slowly on the land, and no great surplus growth was effected. It was not till after the peace of Utrecht that the capital of the country began in a marked manner to increase; and it is impossible that the bounty should not gradually have directed a larger portion of this accumulation to the land than would otherwise have gone to it. A surplus growth, and a fall of price for thirty or forty years, followed.

It will be said that this period of low prices was too long to be occasioned by a bounty, even according to the theory just laid down. This is perhaps true, and in all probability the period would have been shorter if the bounty alone had operated; but in this case other causes powerfully combined with it.

The fall in the price of British corn was accompanied by a fall of prices on the continent. Whatever were the general causes which produced this effect in foreign countries, it is probable that they were not wholly inoperative in England. At all events nothing could be so powerfully calculated to produce cheapness, and to occasion a slow return to high prices, as a considerable surplus growth, which was unwillingly received, and only at low prices, by other nations. When such a surplus growth had been obtained, some time would necessarily be required to destroy it by cheapness, particularly as the moral stimulus of the bounty would probably continue to act long after the fall of prices had commenced. If to these causes we add that a marked fall in the rate of interest, about the same time, evinced an abundance of capital, and a consequent difficulty of finding a profitable employment for it; and consider further the natural obstacles to the moving of capital from the land; we shall see sufficient reason why even a long period might elapse without any essential alteration in the comparative abundance and cheapness of corn.

Adam Smith attributes this cheapness to a rise in the value of silver. The fall in the price of corn which took place in France and some other countries about the same time might give some countenance to the conjecture. But the accounts we have lately had of the produce of the mines during the period in question does not sufficiently support it; and it is much more probable that it arose from the comparative state of peace in which Europe was placed after the termination of the wars of Louis XIV., which facilitated the accumulation of capital on the land, and encouraged agricultural improvements.

With regard to this country, indeed, it is observed by Adam Smith himself, that labour[43] and other articles were rising; a fact very unfavourable to the supposition of an increased value of the precious metals. Not only the money price of corn fell, but its value relative to other articles was lowered, and this fall of relative value, together with great exportations, clearly pointed to a relative abundance of corn, in whatever way it might be occasioned, as the main cause of the facts observed, rather than a scarcity of silver. This great fall in the British corn-market, particularly during the ten years from 1740 to 1750, accompanied by a great fall in the continental markets, owing in some degree perhaps to the great exportations of British corn, especially during the years 1748, 1749, and 1750, must necessarily have given some check to its cultivation, while the increase of the real price of labour must at the same time have given a stimulus to the increase of population. The united operation of these two causes is exactly calculated first to diminish, and ultimately to destroy a surplus of corn; and as, after 1764, the wealth and manufacturing population of Great Britain

[43] It is certainly a very remarkable fact, that although Adam Smith repeatedly states, in the most distinct manner, that labour alone is the true measure of the value of silver and of all other commodities, he should suppose that silver was rising at the very time when he says the money price of labour was rising. There cannot be a more decided contradiction.

increased more rapidly than those of her neighbours, the returning stimulus to agriculture, considerable as it was, arising almost exclusively from a home demand, was incapable of producing a surplus; and not being confined, as before, to British cultivation, owing to the alteration in the corn-laws, was inadequate even to effect an independent supply. Had the old corn-laws remained in full force, we should still probably have lost our surplus growth, owing to the causes above mentioned, although, from their restrictive clauses, we should certainly have been nearer the growth of an independent supply immediately previous to the scarcity of 1800.

It is not therefore necessary, in order to object to the bounty, to say with Adam Smith that the fall in the price of corn which took place during the first half of the last century must have happened in spite of the bounty, and could not possibly have happened in consequence of it. We may allow, on the contrary, what I think we ought to allow according to all general principles, that the bounty, when granted under favourable circumstances, is really calculated, after going through a period of dearness, to produce the surplus and the cheapness which its advocates promise;[44] but according to the same general principles we must allow that this surplus and cheapness, from their operating at once as a check to produce, and an encouragement to population, cannot be for any great length of time maintained.

The objection then to a bounty on corn, independently of the objections to bounties in general, is, that when imposed under the most favourable circumstances it cannot produce permanent cheapness: and if it be imposed under unfavourable circumstances; that is, if an attempt be made to force exportation by an adequate bounty at a time when the country does not fully grow its own consumption; it is obvious not only that the tax necessary for the purpose must be a very heavy one, but that the effect will be absolutely prejudicial to the population, and the surplus growth will be purchased by a sacrifice very far beyond its worth.

But notwithstanding the strong objections to bounties on general grounds, and their inapplicability in cases which are not unfrequent, it must be acknowledged that, while they are operative, that is, while they produce an exportation which would not otherwise have taken place, they unquestionably encourage an increased growth of corn in the countries in which they are established, or maintain it at a point to which it would not otherwise have attained.

Under peculiar and favourable circumstances a country might maintain a considerable surplus growth for a great length of time, with an inconsiderable increase of the growing price of corn; and perhaps little or no increase of the average price, including years of

[44] As far as the bounty might tend to force the cultivation of poorer land, so far no doubt it would have a tendency to raise the price of corn; but we know from experience that the rise of price naturally occasioned in this way is continually counteracted by improvements in agriculture. A, a matter of fact it must be allowed, that, during the period of the last century when corn was falling, more land must have been taken into cultivation.

scarcity.[45] If from any period during the last century, when an average excess of growth for exportation had been obtained by the stimulus of a bounty, the foreign demand for our corn had increased at the same rate as the domestic demand, our surplus growth might have become permanent. After the bounty had ceased to stimulate to fresh exertions, its influence would by no means be lost. For some years it would have given the British grower an absolute advantage over the foreign grower. This advantage would of course gradually diminish; because it is the nature of all effectual demand to be ultimately supplied, and oblige the producers to sell at the lowest price they can afford consistently with the general rate of profits. But, after having experienced a period of decided encouragement, the British grower would find himself in the habit of supplying a larger market than his own upon equal terms with his competitors. And if the foreign and British markets continued to extend themselves equally, he would continue to proportion his supplies to both; because, unless a particular increase of demand were to take place at home, he could never withdraw his foreign supply without lowering the price of his whole crop; and the nation would thus be in possession of a constant store for years of scarcity.

But even supposing that, by a bounty, combined with the most favourable state of prices in other countries, a particular state could maintain permanently an average excess of growth for exportation, it must not of course be imagined that its population would not still be checked by the difficulty of procuring subsistence. It would indeed be less exposed to the particular pressure arising from years of scarcity; but in other respects it would be subject to the same checks as those already described in the preceding chapters; and whether there was an habitual exportation or not, the population would be regulated by the real wages of labour, and would come to a stand when the necessaries which these wages could command were not sufficient, under the actual habits of the people, to encourage an increase of numbers.

4.1.12 Book III, Chapter XII: Of Corn-Laws. Restrictions upon Importation.

The laws which prohibit the importation of foreign grain, though by no means unobjectionable, are not open to the same objections as bounties, and must be allowed to be adequate to the object they have in view,—the maintenance of an independent supply. A country, with landed resources, which determines never to import corn but when the price indicates an approach towards a scarcity, will necessarily, in average years, supply its own wants. Though we may reasonably therefore object to restrictions upon the importation of foreign corn, on the grounds of their tending to present the most profitable employment of the national capital and industry, to check population, and to discourage the export of our

[45] The average price is different from the growing price. Years of scarcity, which must occasionally occur, essentially affect the average price; and the growth of a surplus quantity of corn, which tends to present scarcity, will tend to lower this average, and make it approach nearer to the growing price.

manufactures; yet we cannot deny their tendency to encourage the growth of corn at home, and to procure and maintain an independent supply. A bounty, it has appeared, sufficient to make it answer its purpose in forcing a surplus growth, would, in many cases, require so very heavy a direct tax, and would bear so large a proportion to the whole price of the corn, as to make it in some countries next to impracticable. Restrictions upon importation impose no direct tax upon the people. On the contrary, they might be made, if it were thought advisable, sources of revenue to the government, and they can always, without difficulty, be put in execution, and be made infallibly to answer their express purpose of securing, in average years, a sufficient growth of corn for the actual population.

We have considered, in the preceding chapters, the peculiar disadvantages which attend a system either almost exclusively agricultural or exclusively commercial, and the peculiar advantages which attend a system in which they are united, and flourish together. It has further appeared that, in a country with great landed resources, the commercial population may, from particular causes, so far predominate, as to subject it to some of the evils which belong to a state purely commercial and manufacturing, and to a degree of fluctuation in the price of corn greater than is found to take place in such a state. It is obviously possible, by restrictions upon the importation of foreign corn, to maintain a balance between the agricultural and commercial classes. The question is not a question of the effieciency or inefficiency of the measure proposed, but of its policy or impolicy. The object can certainly be accomplished, but it may be purchased too dear; and to those who do not at once reject all inquiries on points of this kind, as impeaching a principle which they hold sacred, the question, whether a balance between the agricultural and commercial classes of society, which would not take place naturally, ought, under certain circumstances, to be maintained artificially, must appear to be a most important practical question.

One of the objections to the admission of the doctrine that restrictions upon importation are advantageous is, that it cannot possibly be laid down as a general rule that every state ought to raise its own corn. There are some states so circumstanced that the rule is clearly and obviously inapplicable to them.

In the first place, there are many states which have made some figure in history, the territories of which have been perfectly inconsiderable compared with their main town or towns, and utterly incompetent to supply the actual population with food. In such communities, what is called the principal internal trade of a large state, the trade which is carried on between the towns and the country, must necessarily be a foreign trade, and the importation of foreign corn is absolutely necessary to their existence. They may be said to be born without the advantage of land, and, to whatever risks and disadvantages a system merely commercial and manufacturing may be exposed, they have no power of choosing any other. All that they can do is to make the most of their own situation, compared with the situation of their neighbours, and to endeavour by superior industry, skill, and capital, to make up for so important a deficiency. In these efforts, some states of which we have accounts have been wonderfully successful; but the reverses to which they have been subject have been almost as conspicuous as the degree of their prosperity compared with the scantiness of their natural resources.

Secondly, restrictions upon the importation of foreign corn are evidently not applicable to a country which, from its soil and climate. is subject to very great and sudden variations in its home supplies, from the variations of the seasons. A country so circumstanced will unquestionably increase its chance of a steady supply of grain by opening as many markets for importation and exportation as possible, and this will probably be true, even though other countries occasionally prohibit or tax the exports of their grain. The peculiar evil to which such a country is subject can only be mitigated by encouraging the freest possible foreign trade in corn.

Thirdly, restrictions upon importation are not applicable to a country which has a very barren territory, although it may be of some extent. An attempt fully to cultivate and improve such a territory by forcibly directing capital to it would probably, under any circumstances, fail; and the actual produce obtained in this way might be purchased by sacrifices which the capital and industry of the nation could not possibly continue to support. Whatever advantages those countries may enjoy, which possess the means of supporting a considerable population from their own soil, such advantages are not within the reach of a state so circumstanced. It must either consent to be a poor and inconsiderable community, or it must place its chief dependence on other resources than those of land. It resembles in many respects those states which have a very small territory; and its policy, with regard to the importation of corn, must of course be nearly the same.

In all these cases there can be no doubt of the impolicy of attempting to maintain a balance between the agricultural and commercial classes of society which would not take place naturally.

Under other and opposite circumstances, however, this impolicy is by no means so clear.

If a nation possesses a large territory consisting of land of an average quality, it may without difficulty support from its own soil a population fully sufficient to maintain its rank in wealth and power among the countries with which it has relations either of commerce or of war. Territories of a certain extent must ultimately in the main support their own population. As each exporting country approaches towards that complement of wealth and population to which it is naturally tending, it will gradually withdraw the corn which for a time it had spared to its more manufacturing and commercial neighbours, and leave them to subsist on their own resources. The peculiar products of each soil and climate are objects of foreign trade, which can never, under any circumstances, fail. But food is not a peculiar product; and the country which produces it in the greatest abundance may, according to the laws which govern the progress of population, have nothing to spare for others. An extensive foreign trade in corn beyond what arises from the variableness of the seasons in different countries is rather a temporary and incidental trade, depending chiefly upon the different stages of improvement which different countries may have reached, and on other accidental circumstances, than a trade which is in its nature permanent, and the stimulus to which will remain in the progress of society unabated. In the wildness of speculation it has been suggested (of course more in jest than in earnest), that Europe ought to grow its corn in America, and devote itself solely to manufactures and commerce, as the best sort of division of the labour of the globe. But even on the extravagant supposition that the natural course of things might lead to such a division of labour for a time, and that by such means Europe could raise a population greater than its lands could possibly support, the consequences ought justly to be dreaded. It is an unquestionable truth that it must answer to every territorial state, in its natural progress to wealth, to manufacture for itself, unless the countries from which it had purchased its manufactures possess some advantages peculiar to them besides capital and skill. But when upon this principle America began to withdraw its corn from Europe, and the agricultural exertions of Europe were inadequate to make up for the deficiency, it would certainly be felt that the temporary advantages of a greater degree of wealth and population (supposing them to have been really attained) had been very dearly purchased by a long period of retrograde movements and misery.

If then a country be of such a size that it may fairly be expected finally to supply its own population with food; if the population which it can thus support from its own resources in land be such as to enable it to maintain its rank and power among other nations; and further, if there be reason to fear not only the final withdrawing of foreign corn used for a certain time, which might be a distant event, but the immediate effects that attend a great predominance of a manufacturing population, such as increased unhealthiness, increased turbulence, increased fluctuations in the price of corn, and increased variableness in the wages of labour; it may not appear impolitic artificially to maintain a more equal balance between the agricultural and commercial classes by restricting the importation of foreign corn, and making agriculture keep pace with manufactures.

Thirdly, if a country be possessed of such a soil and climate, that the variations in its annual growth of corn are less than in most other countries, this may be an additional reason for admitting the policy of restricting the importation of foreign corn. Countries are very different in the degree of variableness to which their annual supplies are subject; and though it is unquestionably true that if all were nearly equal in this respect, and the trade in corn *really* free, the steadiness of price in a particular state would increase with an increase in the number of the nations connected with it by the commerce of grain; yet it by no means follows that the same conclusion will hold good when the premises are essentially different; that is, when some of the countries taken into the circle of trade are subject to very great comparative variations in their supplies of grain, and when this defect is aggravated by the acknowledged want of real freedom in the foreign trade of corn.

Suppose, for instance, that the extreme variations above and below the average quantity of corn grown, were in England ¼ and in France 1/3, a free intercourse between the two countries would probably increase the variabless of the English markets. And if, in addition to England and France, such a country as Bengal could be brought near, and admitted into the circle—a country in which, according to Sir George Colebrook, rice is sometimes sold four times as cheap in one year as in the succeeding without famine or scarcity;[46] and

[46] Husbandry of Bengal, p. 108. Note. He observes in the text of the same page that the price of corn fluctuates much more than in Europe.

where, notwithstanding the frequency of abundant harvests, deficiencies sometimes occur of such extent as necessarily to destroy a considerable portion of the population; it is quite certain that the supplies both of England and France would become very much more variable than before the accession.

In point of fact, there is reason to believe that the British isles, owing to the nature of their soil and climate, are peculiarly free from great variations in their annual produce of grain. If the compare the prices of corn in England and France from the period of the commencement of the Eton tables to the beginning of the revolutionary war, we shall find that in England the highest price of the quarter of wheat of 8 bushels during the whole of that time was 3*l.* 15*s.* 6¾*d.* (in 1648), and the lowest price 1*l.* 2*s.* 1*d.* (in 1743), while in France the highest price of the septier was 62 francs 78 centuries (in 1662), and the lowest price 8 francs 89 centimes (in 1718).[47] In the one case the difference is a little above 3¼ times, and in the other very nearly 7 times. In the English tables, during periods of ten or twelve years, only two instances occur of a variation amounting to as much as 3 times; in the French tables, during periods of the same length, one instance occurs of a variation of 4 times or above. These variations may, perhaps, have been aggravated by a want of freedom in the internal trade of corn, but they are strongly confirmed by the calculations of Turgot, which relate solely to variations of produce, without reference to any difficulties or obstructions in its free transport from one part of the country to another.

On land of an average quality he estimates the produce at seven septiers the arpent in years of great abundance; and three septiers the arpent in years of great scarcity; while the medium produce he estimates at five septiers the arpent.[48] These calculations he conceives are not far removed from the truth; and proceeding on these grounds he observes that, in a very abundant year, the produce will be five months above its ordinary consumption, and in a very scarce year as much below. These variations are, I should think, much greater than those which take place in this country, at least if we may judge from prices, particularly as in a given degree of scarcity in the two countries there is little doubt that, from the superior riches of England, and the extensive parish relief which it affords to the poorer classes in times of dearth, its prices would rise more above the usual average than those of France.

If we look to the prices of wheat in Spain during the same period, we shall find, in like manner, much greater variations than in England. In a table of the prices of the fanega of wheat in the market of Seville from 1675 to 1764 inclusive, published in the Appendix to the Bullion Report,[49] the highest price is 48 reals vellon (in 1667), and the lowest price 7 reals vellon (in 1720), a difference of nearly seven times; and in periods of ten or twelve years the difference is, in two or three instances, as much as four times. In another table, from 1753 to 1792 inclusive, relating to the towns of Old Castille, the highest price in 1790 was 109 reals vellon the fanega, and in 1792 the lowest was only 16 reals vellone the

[47] Garnier's edition of the Wealth of Nations, vol. ii. table, p. 188.

[48] Œuvres de Turgot, tom. vi. p. 143. Edit. 1803.

[49] Appendix, p. 182.

fanega. In the market of Medina del Rio Seco, a town of the kingdom of Leon, surrounded by a very fine corn country, the price of the load of four fanegas of wheat was, in May, 1800, 100 reals vellon, and in May, 1804, 600 reals vellon, and these were both what are called *low prices,* as compared with the highest prices of the year. The difference would be greater if the high prices were compared with the low prices. Thus, in 1799, the low price of the four fanegas was 88 reals vellon, and in 1804 the high price of the four fanegas was 640 reals vellon—a difference of above seven times in so short a period as six years.[50]

In Spain, foreign corn is freely admitted; yet the variation of price, in the towns of Andalusia, a province adjoining the sea, and penetrated by the river Guadalquiver, though not so great as those just mentioned, seem to shew that the coasts of the Mediterranean by no means furnish very steady supplies. It is known, indeed, that Spain is the principal competitor of England in the purchase of grain in the Baltic; and as it is quite certain that what maybe called the growing or usual price of corn in Spain is much lower than in England, it follows, that the difference between the prices of plentiful and scarce years must be very considerable.

I have not the means of ascertaining the variations in the supplies and prices of the northern nations. They are, however, occasionally great, as it is well known that some of these countries are at times subject to very severe scarcities. But the instances already produced are sufficient to shew, that a country which is advantageously circumstanced with regard to the steadiness of its home supplies may rather diminish than increase this steadiness by uniting its interests with a country less favourably circumstanced in this respect; and this steadiness will unquestionably be still further diminished, if the country which is the most variable in its supplies is allowed to inundate the other with its crops when they are abundant, while it reserves to itself the privilege of retaining them in a period of slight scarcity, when its commercial neighbour happens to be in the greatest want.[51]

3dly, if a nation be possessed of a territory, not only of sufficient extent to maintain under its actual cultivation a population adequate to a state of the first rank, but of sufficient unexhausted fertility to allow of a very great increase of population, such a circumstance would of course make the measure of restricting the importation of foreign corn more applicable to it.

A country which, though fertile and populous, had been cultivated nearly to the utmost, would have no other means of increasing its population than by the admission of foreign corn. But the British isles shew at present no symptoms whatever of this species of exhaustion. The necessary accompaniments of a territory worked to the utmost are very low profits and interest, a very slack demand for labour, low wages, and a stationary population. Some of these symptoms may indeed take place without an exhausted territory; but an exhausted territory cannot take place without all these symptoms. Instead, however,

---

[50] Bullion Report. Appendix, p. 185.

[51] These two circumstances essentially change the premises on which the question of a free importation, as applicable to a particular state, must rest.

of such symptoms, we have seen in this country, during the twenty years previous to 1814, a high rate of profits and interest, a very great demand for labour, good wages, and an increase of population more rapid, perhaps, than during any period of our previous history. The capitals which were laid out in bringing new land into cultivation, or improving the old, must necessarily have yielded good returns, or, under the actual rate of general profits, they would not have been so employed: and although it is strictly true that, as capital accumulates upon the land, its profits must ultimately diminish; yet owing to the increase of agricultural skill, and other causes noticed in a former chapter, these two events do not by any means always keep pace with each other. Though they must, finally unite and terminate the career of their progress together, they are often, during the course of their progress, separated for a considerable time, and at a considerable distance. In some countries, and some soils, the quantity of capital which can be absorbed before any essential diminution of profits necessarily takes place is so great, that its limit is not easily calculated; and certainly, when we consider what has been actually done in some districts of England and Scotland, and compare it with what remains to be done in other districts, we mast allow that no near approach to this limit has yet been made. On account of the high money price of labour, and of the materials of agricultural capital, occasioned partly by direct and indirect taxation, and partly, or perhaps chiefly by the great prosperity of our foreign commerce,[52] new lands cannot be brought into cultivation, nor great improvements made on the old, without a high money price of grain; but these lands, when they have been so brought into cultivation, or improved, have by no means turned out unproductive. The quantity and value of their produce have borne a full and fair proportion to the quantity of capital and labour employed upon them; and they were cultivated with great advantage both to individuals and the state, as long as the same, or nearly the same, relations between the value of produce and the cost of production, which prompted this cultivation, continued to exist.

In such a state of the soil, the British empire might unquestionably be able not only to support from its own agricultural resources its present population, but double, and in time, perhaps, even treble the number; and consequently a restriction upon the importation of foreign corn, which might be thought greatly objectionable in a country which had reached nearly the end of its resources, might appear in a very different light in a country capable of supporting from its own lands a very great increase of population.

But it will be said, that although a country may be allowed to be capable of maintaining from its own soil not only a great, but an increasing population, yet, if it be acknowledged that, by opening its ports for the free admission of foreign corn, it may be made to support a

[52]No restrictions upon the importation of grain, however absurdly severe, could permanently maintain our corn and labour at a much higher price than in the rest of Europe, if such restrictions were essentially to interfere with the prosperity of our foreign commerce. When the money price of labour is high in any country, or, what is the same thing, when the value of money is low, nothing can prevent it from going out to find its level, but some comparative advantages, either natural or acquired, which enable such country to maintain the abundance of its exports, notwithstanding the high money price of its labour.

greater and more rapidly increasing population, it is unjustifiable to go out of our way to check this tendency, and to prevent that degree of wealth and population which would naturally take place.

This is unquestionably a powerful argument; and granting fully the premises, it cannot be answered upon the principles of political economy solely. I should say, however, that if it could be clearly ascertained that the addition of wealth and population so acquired would subject the society to a greater degree of uncertainty in its supplies corn, greater fluctuations in the wages of labour, greater unhealthiness and immorality owing to a larger proportion of the population being employed in manufactories, and a greater chance of long and depressing retrograde movements occasioned by the natural progress of those countries from which corn had been imported; I should have no hesitation in considering such wealth and population as much too dearly purchased. The happiness of a society is, after all, the legitimate end even of its wealth, power, and population. It is certainly true that with a view to the structure of society most favourable to this happiness, and an adequate stimulus to the production of wealth from the soil, a very considerable admixture of commercial and manufacturing population with the agricultural is absolutely necessary; but there is no argument so frequently and obviously fallacious as that which infers that what is good to a certain extent is good to any extent; and though it will be most readily admitted that, in a large landed nation, the evils which belong to the manufacturing and commercial system are much more than counterbalanced by its advantages, as long as it is supported by agriculture; yet, in reference to the effect of the excess which is not so supported, it may fairly be doubted whether the evils do not decidedly predominate.

It is observed by Adam Smith, that the "capital which is acquired to any country by commerce and manufactures is all a very uncertain and precarious possession, till some part of it has been secured and realized in the cultivation and improvement of its lands.[53]

It is remarked in another place, that the monopoly of the colony trade, by raising the rate of mercantile profit, discourages the improvement of the soil, and retards the natural increase of that great original source of revenue—the rent of land.[54]

Now it is certain that, at no period, have the manufactures, commerce and colony trade of the country been in a state to absorb so much capital as during the twenty years ending with 1814. From the year 1764 to the peace of Amiens, it is generally allowed that the commerce and manufactures of the country increased faster than its agriculture, and that it became gradually more and more dependent on foreign corn for its support. Since the peace of Amiens the state of its colonial monopoly and its manufactures has been such as to demand an unusual quantity of capital; and if the peculiar circumstances of the subsequent war, the high freights and insurance, and the decrees of Buonaparte, had not rendered the importation of foreign corn extremely difficult and expensive, we should at this moment, according to all general principles, have been in the habit of supporting a much larger

[53] Vol. ii. b. iii c. 4. p. 137.

[54] Vol. ii. b. iv. c. 8. p. 495.

portion of our population upon it, than at any former period of our history. The cultivation of the country would be in a very different state from what it is at present. Very few or none of those great improvements would have taken place which may be said to have purchased fresh land for the state that no fall of price can destroy. And the peace, or accidents of different kinds, might have curtailed essentially both our colonial and manufacturing advantages, and destroyed or driven away our capital before it had spread itself on the soil, and become national property.

As it is, the practical restrictions thrown in the way of importing foreign corn during the war have forced our steam-engines and our colonial monopoly to cultivate our lands; and those very causes which, according to Adam Smith, tend to draw capital from agriculture, and would certainly have so drawn it if we could have continued to purchase foreign corn at the market prices of France and Holland, have been the means of giving such a spur to our agriculture, that it has not only kept pace with a very rapid increase of commerce and manufactures, but has recovered the distance at which it had for many years been left behind, and now marches with them abreast.

But restrictions upon the importation of foreign corn in a country which has great landed resources, not only tend to spread every commercial and manufacturing advantage possessed, whether permanent or temporary, on the soil, and thus, in the language of Adam Smith, secure and realize it; but also tend to prevent those great oscillations in the progress of agriculture and commerce, which are seldom unattended with evil.

It is to be recollected, and it is a point of great importance to keep constantly in our minds, that the distress which has been experienced among almost all classes of society from the sudden fall of prices, except as far as it has been aggravated by the state of the currency, has been occasioned by *natural,* not *artificial* causes.

There is a tendency to an alternation in the rate of the progress of agriculture and manufactures in the same manner as there is a tendency to an alternation in the rate of the progress of food and population. In periods of peace and uninterrupted trade, these alternations, though not favourable to the happiness and quiet of society, may take place without producing material evil; but the intervention of war is always liable to give them a force and rapidity that must unavoidably produce a convulsion in the state of property.

The war that succeeded to the peace of Amiens found us dependent upon foreign countries for a very considerable portion of our supplies of corn; and we now grow our own consumption, notwithstanding an unusual increase of population in the interval. This great and sudden change in the state of our agriculture could only have been effected by very high prices occasioned by an inadequate home supply and the great expense and difficulty of importing foreign corn. But the rapidity with which this change has been effected must necessarily create a glut in the market as soon as the home growth of corn became fully equal or a little in excess above the home consumption; and, aided only by a small foreign importation, must inevitably occasion a very sudden fall of prices. If the ports had continued open for the free importation of foreign corn, there can be little doubt that the price of corn in 1815 would have been still considerably lower. This low price of corn, even if by means of lowered rents our present state of cultivation could be in a great degree preserved, must give such a check to future improvement, that if the ports were to continue open, we should certainly not grow a sufficiency at home to keep pace with our increasing population; and at the end of ten or twelve years we might be found by a new war in the same state that we were at the commencement of the present. We should then have the same career of high prices to pass through, the same excessive stimulus to agriculture[55] followed by the same sudden and depressing check to it, and the same enormous loans borrowed with the price of wheat at 90 or 100 shillings a quarter, and the monied incomes of the landholders and industrious classes of society nearly in proportion, to be paid when wheat is at 50 or 60 shillings a quarter, and the incomes of the landlords and industrious classes of society greatly reduced—a state of things which cannot take place without an excessive aggravation of the difficulty of paying taxes, and particularly that invariable monied amount which pays the interest of the national debt.

On the other hand a country which so restricts the importations of foreign corn as on an average to grow its own supplies, and to import merely in periods of scarcity, is not only certain of spreading every invention in manufactures and every peculiar advantage it may possess from its colonies or general commerce on the land, and thus of fixing them to the spot and rescuing them from accidents; but is necessarily exempt from those violent and distressing convulsions of property which almost unavoidably arise from the coincidence of a general war and an insufficient home supply of corn.

If the late war had found us independent of foreigners for our average consumption, not even our paper currency could have made the prices of our corn approach to the prices which were at one time experienced.[56] And if we had continued, during the course of the contest, independent of foreign supplies, except in an occasional scarcity, it is impossible that the growth of our own consumption, or a little above it, should have produced at the end of the war so universal a feeling of distress.

The chief practical objection to which restrictions on the importation of corn are exposed is a glut from an abundant harvest, which cannot be relieved by exportation. And in the consideration of that part of the question which relates to the fluctuations of prices this objection ought to have its full and fair weight. But the fluctuation of prices arising from this cause has sometimes been very greatly exaggerated. A glut which might essentially distress the farmers of a poor country, might be comparatively little felt by the farmers of a rich one; and it is difficult to conceive that a nation with an ample capital, and not under the influence of a great shock to commercial confidence, as this country was in

[55] According to the evidence before the House of Lords (Reports, p. 49), the freight and insurance alone on a quarter of corn were greater by 48 shillings in 1811 than in 1814. Without any artificial interference then, it appears that war alone may occasion unavoidably a prodigious increase of price.

[56] According to Mr. Tooke (High and Low Prices, p. 215), if the last war had found us with a growth beyond our consumption, we should have witnessed a totally different set of phenomena connected with prices. It will be found upon examination, that the prices of our corn led the way to the excess and diminution of our paper currency, rather than followed, although the prices of corn could never have been either so high or so low if this excess and diminution had not taken place.

1815, would find much difficulty in reserving the surplus of one year to supply the wants of the next or some future year. It may fairly indeed be doubted whether, in such a country as our own, the fall of price arising from this cause would be so great as that which would be occasioned by the sudden pouring in of the supplies from an abundant crop in Europe, particularly from those states which do not regularly export corn. If our ports were always open, the existing laws of France would still prevent such a supply as would equalize prices; and French corn would only come in to us in considerable quantities in years of great abundance, when we were the least likely to want it, and when it was most likely to occasion a glut.[57]

But if the fall of price occasioned in these two ways would not be essentially different, as it is quite certain that the rise of price in years of general scarcity would be less in those countries which habitually grow their own supplies; it must be allowed that the range of variation will be the least under such a system of restrictions as, without preventing importation when prices are high, will secure in ordinary years a growth equal to the consumption.[58]

---

[57] Almost all the corn merchants who gave their evidence before the committees of the two houses in 1814 seemed fully aware of the low prices likely to be occasioned by an abundant crop in Europe, if our ports were open to receive it.

[58] [1825.] In the sixth number of the Westminster Review, in which prodigious stress is laid upon the necessary effect of the corn laws in occasioning great fluctuation, in the prices of corn, a table, said to be from the very highest mercantile authority, is given of the average prices of wheat at Rotterdam for each of the ten years ending with 1824. The purpose for which the table is produced, is to shew the average price of wheat in Holland during these ten years; but it incidentally shews that, even in Holland, which in many respects must be peculiarly favourable to steady prices, a free trade in corn can by no means secure them.

In the year 1817, the price per last of 86 Winchester bushels was 574 guilders; and in 1824, it was only 147 guilders; a difference of nearly four times. During the same period of ten years the greatest variation in the average price of each year in England, was between 94*s.* 9*d.* which was the price in 1817, and 43*s.* 9*d.* which was the price in 1822, (Appendix to Mr. Tooke's work on High and Low Prices. Table xii. p. 31.)—a difference short of 2 1/5! ! [Note: There are in fact two exclamation points in the original text of the 6th edition (p. 207). The arithmetic change does work out to being just short of 2.2 times: there is no missing unit of measurement as a potential typographical error. Additionally, there are very few typographical errors in this edition. Quite possibly this is one of the first instances in print of a doubling of exclamation points for extra emphasis.—Econlib Editor.]

It is repeated over and over again, apparently without the slightest reference to facts, that the freedom of the trade in corn would infallibly secure us from the possibility of a scarcity. The writer of the article *Corn Laws* in the supplement to the Encyclopædia Britannica goes so far as to say, "it is constantly found that when the crops of one country fail, plenty reigns in some other quarter**** There is always abundance of food in the world. To enjoy a constant plenty, we have only to lay aside our prohibitions and restrictions, and cease to counteract the benevolent wisdom of Providence." The same kind of language is repeated in the Review above adverted to: "If there be a bad harvest," it is said, "in one country, there is a good one in another, and the surplus produce of the latter supplies the deficiency of the former, &c. &c:." Now there are the best reasons for believing that these statements are decidedly contradicted by the most enlarged experience. In the first place, if they were true, and if

One objection however to systems of restriction must always remain. They are essentially unsocial. I certainly think that, in reference to the interests of a particular state, a restriction upon the importation of foreign corn may sometimes be advantageous; but I feel still more certain that in reference to the interests of Europe in general the most perfect freedom of trade in corn, as well as in every other commodity, would be the most advantageous. Such a perfect freedom, however, could hardly fail to be followed by a

the general plenty alluded to were only prevented by the want of a free trade in corn, we should necessarily see a great rise of prices in one country, contemporaneous with a great fall in others; but a slight glance at the prices of corn in the countries of the commercial world for the last one or two centuries will be sufficient to convince any impartial person that, on the contrary, there is a very remarkable sympathy of prices at the same periods, which is absolutely inconsistent the truth of the above statements. Secondly, all travellers who have paid any attention to the seasons, agree in stating that the same sort of weather often prevails in different countries at the same time. The peculiar and excessive heats of the very last summer not only prevailed generally over the greatest part of Europe, but extended even to America. Mr. Tooke, On High and Low Prices, (p. 247. 2d Edit.) quotes a passage from Mr. Lowe's work on the Present State of England, in which he observes, that "The public, particularly the untravelled part of the public, are hardly of the similarity of temperature prevailing throughout what may be called the corn-country of Europe, we mean Great Britain, Ireland, the North of France, the Netherlands, Denmark, the northwest of Germany, and in some measure Poland and the northeast of Germany." He then goes on to state instances of scarcity in different countries of Europe at the same time. And in the justness of these remarks, on the prevalence of a general similarity of seasons in Europe within certain latitudes, Mr. Tooke says he perfectly concurs. Many of the corn-merchants examined before the Committees of the two Houses, both in 1814 and 1821, expressed similar opinions; and I do not recollect a single instance of the opinion, that good and bad harvests generally balance each other in different countries, being stated by any person who had been in a situation to observe the facts. Such statements, therefore, must be considered as mere assertions quite unsupported by the least shadow of proof.

I am very far however from meaning to say that the circumstance of different countries having often an abundance or deficiency of corn at the same time, though it must prevent the possibility of steady prices, is a decisive reason against the abolition or alteration of the corn-laws. The most powerful of all the arguments against restrictions is their unsocial tendency, and the acknowledged injury which they must do to the interests of the commercial world in general. The weight of this argument is increased rather than diminished by the numbers which may suffer from scarcity at the same time. And at a period when our ministers are the most laudably setting an example of a more liberal system of commercial policy, it would be greatly desirable that foreign nations should not have so marked an exception as our present corn-laws to cast in our teeth. A duty on importation not too high, and a bounty nearly such as was recommended by Mr. Ricardo, would probably be best suited to our present situation, and best secure steady prices. A duty on foreign corn would resemble the duties laid by other countries on our manufactures as objects of taxation, and would not in the same manner impeach the principles of free trade.

But whatever system we may adopt, it is essential to a sound determination, and highly useful in preventing disappointments, that all the arguments both for and against corn-laws should be thoroughly and impartially considered; and it is because on a calm, and, as far as I can judge, an impartial review of the arguments of this chapter, they still appear to me of weight sufficient to deserve such consideration, and not as a kind of protest against the abolition or change of the corn-laws, that I republish them in another edition.

more free and equal distribution of capital, which, though it would greatly advance the riches and happiness of Europe, would unquestionably render some parts of it poorer and less populous than they are at present; and there is little reason to expect that individual states will ever consent to sacrifice the wealth within their own confines to the wealth of the world.

It is further to be observed, that, independently of more direct regulations, taxation alone produces a system of discouragements and encouragements which essentially interferes with the natural relations of commodities to each other; and as there is no hope of abolishing taxation, it may sometimes be only by a further interference that these natural relations can be restored.

A perfect freedom of trade therefore is a vision which it is to be feared can never be realized. But still it should be our object to make as near approaches to it as we can. It should always be considered as the great general rule. And when any deviations from it are proposed, those who propose them are bound clearly to make out the exception.

4.1.13 Book III, Chapter XIII: Of Increasing Wealth, As It Affects the Condition of the Poor.

The professed object of Adam Smith's *Inquiry* is the *Nature and Causes of the Wealth of Nations.* There is another, however, still more interesting, which he occasionally mixes with it—the causes which affect the happiness and comfort of the lower orders of society, which in every nation form the most numerous class. These two subjects are, no doubt, nearly connected; but the nature and extent of this connexion, and the mode in which increasing wealth operates on the condition of the poor, have not been stated with sufficient correctness and precision.

Adam Smith, in his chapter on the wages of labour, considers every increase in the stock or revenue of the society as an increase in the funds for the maintenance of labour; and having before laid down the position that the demand for those who live by wages can only increase in proportion to the increase of the funds for the payment of wages, the conclusion naturally follows, that every increase of wealth tends to increase the demand for labour and to improve the condition of the lower classes of society.[59]

Upon a nearer examination, however, it will be found that the funds for the maintenance of labour do not necessarily increase with the increase of wealth, and very rarely increase in *proportion* to it; and that the condition of the lower classes of society does not depend exclusively upon the increase of the funds for the maintenance of labour, or the power of supporting a greater number of labourers.

Adam Smith defines the wealth of a state to be the annual produce of its land and labour. This definition evidently includes manufactured produce as well as the produce of the land.

[59] Vol i. book i. c. 8.

Now, upon the supposition that a nation, from peculiar situation and circumstances, was unable to procure an additional quantity of food, it is obvious that the produce of its labour would not necessarily come to a stand, although the produce of its land or its power of importing corn were incapable of further increase. If the materials of manufactures could be obtained either at home or from abroad, improved skill and machinery might work them up to a greatly increased amount with the same number of hands, and even the number of hands might be considerably increased by an increased taste for manufactures, compared with war and menial service, and by the employment consequently of a greater proportion of the whole population in manufacturing and commercial labour.

That such a case does not frequently occur will be most readily allowed. It is not only however possible, but forms the specific limit to the increase of population in the natural progress of cultivation, with which limit, the limit to the further progress of wealth is obviously not contemporary. But though cases of this kind do not often occur, because these limits are seldom reached; yet approximations to them are constantly taking place, and in the usual progress of improvement the increase of wealth and capital is rarely accompanied with a proportionately increased power of supporting an additional number of labourers.

Some ancient nations, which, according to the accounts we have received of them, possessed but an inconsiderable quantity of manufacturing and commercial capital, appear to have cultivated their lands highly by means of an agrarian division of property, and were unquestionably very populous. In such countries, though full of people already, there would evidently be room for a very great increase of capital and riches; but, allowing all the weight that is in any degree probable to the increased production or importation of food occasioned by the stimulus of additional capital, there would evidently not be room for a proportionate increase of the means of subsistence.

If we compare the early state of our most flourishing European kingdoms with their present state, we shall find this conclusion confirmed almost universally by experience.

Adam Smith, in treating of the different progress of opulence in different nations, says, that England, since the time of Elizabeth, has been continually advancing in commerce and manufactures. He than adds, "The cultivation and improvement of the country has no doubt been gradually advancing. But it seems to have followed slowly and at a distance the more rapid progress of commerce and manufactures. The greater part of the country must probably have been cultivated before the reign of Elizabeth, and a very great part of it still remains uncultivated, and the cultivation of the far greater part is much inferior to what it might be."[60] The same observation is applicable to most of the other countries of Europe. The best land would naturally be the first occupied. This land, even with that sort of indolent cultivation and great waste of labour which particularly marked the feudal times, would be capable of supporting a considerable population; and on the increase of capital, the increasing taste for conveniences and luxuries, combined with the decreasing power of

[60] Vol. ii. book iv. c. 4. p. 133.

production in the new land to be taken into cultivation, would naturally and necessarily direct the greatest part of this new capital to commerce and manufactures, and occasion a more rapid increase of wealth than of population.

The population of England accordingly in the reign of Elizabeth appears to have been nearly five millions, which would not be very far short of the half of what it is at present (1811); but when we consider the very great proportion which the products of commercial and manufacturing industry now bear to the quantity of food raised for human consumption, it is probably a very low estimate to say that the mass of wealth or the stock and revenue of the country must, independently of any change in the value of the circulating medium, have increased above four times. Few of the other countries in Europe have increased to the same extent in commercial and manufacturing wealth as England; but as far as they have proceeded in this career, all appearances clearly indicate that the progress of their general wealth has been greater than the progress of their means of supporting an additional population.

That every increase of the stock or revenue of a nation cannot be considered as an increase of the real funds for the maintenance of labour will appear in a striking light in the case of China.

Adam Smith observes, that China has probably long been as rich as the nature of her laws and institutions will admit; but intimates that, with other laws and institutions, and if foreign commerce were held in honour, she might still be much richer.

If trade and foreign commerce were held in great honour in China, it is evident that, from the great number of her labourers and the cheapness of her labour, she might work up manufactures for foreign sale to a great amount. It is equally evident that, from the great bulk of provisions and the prodigious extent of her inland territory, she could not in return import such a quantity as would be any sensible addition to her means of subsistence. Her immense amount of manufactures, therefore, she would either consume at home, or exchange for luxuries collected from all parts of the world. At present the country appears to be over-peopled compared with what its stock can employ, and no labour is spared in the production of food. An immense capital could not be employed in China in preparing manufactures for foreign trade, without altering this state of things, and taking off some labourers from agriculture, which might have a tendency to diminish the produce of the country. Allowing, however, that this would be made up, and, indeed, more than made up, by the beneficial effects of improved skill and economy of labour in the cultivation of the poorest lands, yet, as the quantity of subsistence could be but little increased, the demand for manufactures which would raise the price of labour, would necessarily be followed by a proportionate rise in the price of provisions, and the labourer would be able to command but little more food than before. The country would, however, obviously be advancing in wealth; the exchangeable value of the annual produce of its land and labour would be annually augmented; yet the real funds for the maintenance of labour would be nearly stationary. The argument perhaps appears clearer when applied to China, because it is generally allowed that its wealth has been long stationary, and its soil cultivated nearly to the utmost.[61]

In all those cases, it is not on account of any undue preference given to commerce and manufactures, compared with agriculture, that the effect just described takes place, but merely because the powers of the earth in the production of food have narrower limits than the skill and tastes of mankind in giving value to raw materials, and consequently in the approach towards the limits of subsistence there is naturally more room, and consequently more encouragement, for the increase of the one species of wealth than of the other.

It must be allowed then, that the funds for the maintenance of labour do not *necessarily* increase with the increase of wealth, and very *rarely* increase in *proportion* to it.

But the condition of the lower classes of society certainly does not depend exclusively upon the increase of the funds for the maintenance of labour, or the means of supporting more labourers. That these means form always a very powerful ingredient in the condition of the labouring classes, and the main ingredient in the increase of population, is unquestionable. But, in the first place, the comforts of the lower classes of society do not depend solely upon food, nor even upon strict necessaries; and they cannot be considered as in a good state unless they have the command of some conveniences and even luxuries. Secondly, the tendency in population fully to keep pace with the means of subsistence must in general prevent the increase of these means from having a great and permanent effect in improving the condition of the poor. And, thirdly, the cause which has the most lasting effect in improving the situation of the lower classes of society depends chiefly upon the conduct and prudence of the individuals themselves, and is, therefore, not immediately and necessarily connected with an increase in the means of subsistence.

With a view, therefore, to the other causes which affect the condition of the labouring classes, as well as the increase of the means of subsistence, it may be desirable to trace more particularly the mode in which increasing wealth operates, and to state both the disadvantages as well as the advantages with which it is accompanied.

In the natural and regular progress of a country to a state of great wealth and population, there are two disadvantages to which the lower classes of society seem necessarily to be subjected. The first is, a diminished power of supporting children under the existing habits of the society with respect to the necessaries of life. And the second, the employment of a larger proportion of the population in occupations less favourable to health, and more exposed to fluctuations of demand and unsteadiness of wages.

A diminished power of supporting children is an absolutely unavoidable consequence of the progress of a country towards the utmost limits of its population. If we allow that the power of a given quantity of territory to produce food has some limit, we must allow that as this limit is approached, and the increase of population becomes slower and slower, the

[61] How far this latter opinion is to be depended upon it is not very easy to say. Improved skill and a saving of labour would certainly enable the Chinese to cultivate some lands with advantage which they cannot cultivate now, but the more general use of horses, instead of men, might prevent this extended cultivation from giving any encouragement to an increase of people.

power of supporting children will be less and less, till finally, when the increase of produce stops, it becomes only sufficient to maintain, on an average, families of such a size as will not allow of a further addition of numbers. This state of things is generally accompanied by a fall in the *corn* price of labour; but should this effect be prevented by the prevalence of prudential habits among the lower classes of society, still the result just described must take place; and though, from the powerful operation of the preventive check to increase, the wages of labour estimated even in corn might not be low, yet it is obvious that, in this case, the power of supporting children would rather be nominal than real; and the moment this power began to be exercised to its apparent extent, it would cease to exist.

The second disadvantage to which the lower classes of society are subjected in the progressive increase of wealth is, that a larger portion of them is engaged in unhealthy occupations, and in employments in which the wages of labour are exposed to much greater fluctuations than in agriculture, and the simpler kinds of domestic trade.

On the state of the poor employed in manufactories with respect to health, and the fluctuations of wages, I will beg leave to quote a passage from Dr. Aikin's Description of the Country round Manchester:—

"The invention and improvements of machines to shorten labour have had a surprising influence to extend our trade, and also to call in hands from all parts, particularly children for the cotton-mills. It is the wise plan of Providence, that in this life there shall be no good without its attendant inconvenience. There are many which are too obvious in these cotton-mills, and similar factories, which counteract that increase of population usually consequent on the improved facility of labour. In these, children of a very tender age are employed, many of them collected from the workhouses in London and Westminster, and transported in crowds as apprentices to masters resident many hundred miles distant, where they serve unknown, unprotected and forgotten by those to whose care nature or the laws had consigned them. These children are usually too long confined to work in close rooms, often during the whole night. The air they breathe from the oil, &c. employed in the machinery, and other circumstances, is injurious; little attention is paid to their cleanliness; and frequent changes from a warm and dense to a cold and thin atmosphere are pre-disposing causes to sickness and debility, and particularly to the epidemic fever which is so generally to be met with in these factories. It is also much to be questioned if society does not receive detriment from the manner in which children are thus employed during their early years. They are not generally strong to labour, or capable of pursuing any other branch of business when the term of their apprenticeship expires. The females are wholly uninstructed in sewing, knitting, and other domestic affairs requisite to make them notable and frugal wives and mothers. This is a very great misfortune to them and to the public, as is sadly proved by a comparison of the families of labourers in husbandry and those of manufacturers in general. In the former we meet with neatness, cleanliness, and comfort; in the latter with filth, rags, and poverty, although their wages may be nearly double to those of the husbandman. It must be added, that the want of early religious instruction and example, and the numerous and indiscriminate association in these buildings, are very unfavourable to their future conduct in life."[62]

In the same work it appears that the register for the collegiate church of Manchester, from Christmas, 1793, to Christmas 1794, shewed a decrease of 168 marriages, 538 christenings, and 250 burials. In the parish of Rochdale, in the neighbourhood, a still more melancholy reduction in proportion to the number of people took place. In 1792 the births were 746, the burials 646, and the marriages 339. In 1794 the births were 373, the burials 671, and the marriages 199. The cause of this sudden check to population was the failure of demand and of commercial credit which occurred at the commencement of the war, and such a check could not have taken place in so sudden a manner without the most severe distress, occasioned by the sudden reduction of wages.

In addition to the fluctuations arising from the changes from peace to war and from war to peace, it is well known how subject particular manufactures are to fail from the caprices of taste. The weavers of Spitalfields were plunged into the most severe distress by the fashion of muslins instead of silks; and great numbers of workmen in Sheffield and Birmingham were for a time thrown out of employment owing to the adoption of shoe strings and covered buttons, instead of buckles and metal buttons. Our manufactures, taken in the mass, have increased with prodigious rapidity, but in particular places they have failed; and the parishes where this has happened are invariably loaded with a crowd of poor in the most distressed and miserable condition.

In the evidence brought before the House of Lords during the inquiries which preceded the Corn-Bill of 1815, various accounts are produced from different manufactories, intended to shew that the high price of corn has rather the effect of lowering than of raising the price of manufacturing labour.[63] Adam Smith has clearly and correctly stated that the money price of labour depends upon the money price of provisions, and the state of the demand and the supply of labour. And he shews how much he thinks it is occasionally affected by the latter cause, by explaining in what manner it may vary in an opposite direction from the price of provisions during the pressure of a scarcity. The accounts brought before the House of Lords are a striking illustration of this part of his proposition; but they certainly do not prove the incorrectness of the other part of it, as it is quite obvious that, whatever may take place for a few years, the supply of manufacturing labour cannot possibly be continued in the market unless the natural or necessary price, that is, the price necessary to continue it in the market, be paid, and this of course is not done unless the money price be so proportioned to the price of provisions that labourers are enabled to bring up families of such a size as will supply the number of hands required.

---

[62]P. 219. Dr. Aikin says that endeavours have been made to remedy these evils, which in some factories have been attended with success. And it is very satisfactory to be able to add, that, since this account was written, the situation of the children employed in the cotton-mills has been further very essentially improved, partly by the interference of the legislature, and partly by the humane and liberal exertions of individuals.

[63]Reports, p. 51.

But though these accounts do not in any degree invalidate the usual doctrines respecting labour, or the statements of Adam Smith, they shew very clearly the great fluctuations to which the condition of the manufacturing labourer is subjected.

In looking over these accounts it will be found that in some cases the price of weaving has fallen a third, or nearly one-half, at the same time that the price of wheat has risen a third, or nearly one half; and yet these proportions do not always express the full amount of the fluctuations, as it sometimes happens that when the price is low, the state of the demand will not allow of the usual number of hours of working; and when the price is high, it will admit of extra hours.

That from the same causes there are sometimes variations of a similar kind in the price of task-work in agriculture will be readily admitted; but, in the first place, they do not appear to be nearly so considerable; and secondly, the great mass of agricultural labourers is employed by the day, and a sudden and general fall in the money price of agricultural day-labour is an event of extremely rare occurrence.[64]

It must be allowed then, that in the natural and usual progress of wealth, the means of marrying early and supporting a family are diminished, and a greater proportion of the population is engaged in employments less favourable to health and morals, and more subject to fluctuations in the price of labour, than the population employed in agriculture.

These are no doubt considerable disadvantages, and they would be sufficient to render the progress of riches decidedly unfavourable to the condition of the poor, if they were not counteracted by advantages which nearly, if not fully, counterbalance them.

And, first, it is obvious that the profits of stock are that source of revenue from which the middle classes are chiefly maintained; and the increase of capital, which is both the cause and effect of increasing riches, may be said to be the efficient cause of the emancipation of the great body of society from a dependence on the landlords. In a country of limited extent, consisting of fertile land divided into large properties, as long as the capital remains inconsiderable, the structure of society is most unfavourable to liberty and good government. This was exactly the state of Europe in the feudal times. The landlords could in no other way spend their incomes than by maintaining a great number of idle followers; and it was by the growth of capital in all the employments to which it is directed, that the pernicious power of the landlords was destroyed, and their dependent followers were turned into merchants, manufacturers, tradesmen, farmers, and independent labourers;—a change of prodigious advantage to the great body of society, including the labouring classes.

Secondly; in the natural progress of cultivation and wealth, the production of all additional quantity of corn will require more labour, while, at the same time, from the accumulation and better distribution of capital, the continual improvements made in

[64] Almost the only instance on record in this country is that which has lately taken place (1815 and 1816), occasioned by an unparalleled fall in the exchangeable value of the raw produce, which has necessarily disabled the holders of it from employing the same quantity of labour at the same price.

machinery, and the facilities opened to foreign commerce, manufactures and foreign commodities will be produced or purchased with less labour; and consequently a given quantity of corn will command a much greater quantity of manufactures and foreign commodities than while the country was poor. Although, therefore, the labourer may earn less corn than before, the superior value which every portion which he does not consume in kind will have in the purchase of conveniences, may more than counterbalance this diminution. He will not indeed have the same power of maintaining a large family; but with a small family he may be better lodged and clothed, and better able to command the decencies and comforts of life.

Thirdly; it seems to be proved by experience, that the labouring classes of society seldom acquire a decided taste for conveniences and comforts till they become plentiful compared with food, which they never do till food has become in some degree scarce. If the labourer can obtain the bill support of himself and family by two or three days' labour; and if, to furnish himself with conveniences and comforts, he must work three or four days more, he will generally think the sacrifice too great compared with the objects to be obtained, which are not strictly necessary to him, and will therefore often prefer the luxury of idleness to the luxury of improved lodging and clothing. This is said by Humboldt to be particularly the case in some parts of South America, and to a certain extent prevails in Ireland, India, and all countries where food is plentiful compared with capital and manufactured commodities. On the other hand, if the main part of the labourer's time be occupied in procuring food, habits of industry are necessarily generated, and the remaining time, which is but inconsiderable compared with the commodities it will purchase, is seldom grudged. It is under these circumstances, particularly when combined with a good government, that the labouring classes of society are most likely to acquire a decided taste for the conveniences and comforts of life; and this taste may be such as even to prevent, after a certain period, a further fall in the corn price of labour. But if the corn price of labour continues tolerably high while the relative value of commodities compared with corn falls very considerably, the labourer is placed in a most favourable situation. Owing to his decided taste for conveniences and comforts, the good corn wages of labour will not generally lead to early marriages; yet in individual cases, where large families occur, there will be the means of supporting them independently, by the sacrifice of the accustomed conveniences and comforts; and thus the poorest of the lower classes will rarely be stinted in food, while the great mass of them will not only have sufficient means of subsistence, but be able to command no inconsiderable quantity of those conveniences and comforts, which, at the same time that they gratify a natural or acquired want, tend unquestionably to improve the mind and elevate the character.

On an attentive review, then, of the effects of increasing wealth on the condition of the poor, it appears that, although such an increase does not imply a proportionate increase of the funds for the maintenance of labour, yet it brings with it advantages to the lower classes of society which may fully counterbalance the disadvantages with which it is attended; and, strictly speaking, the good or bad condition of the poor is not *necessarily* connected with any particular stage in the progress of society to its full complement of wealth. A rapid increase of wealth indeed, whether it consists principally in additions to the means of subsistence or to the stock of conveniences and comforts, will always, *cæteris paribus,* have a favourable effect on the poor; but the influence even of this cause is greatly modified and altered by other circumstances, and nothing but the union of individual prudence with the skill and industry which produce wealth can permanently secure to the lower classes of society that share of it which it is, on every account, so desirable that they should possess.

4.1.14 Book III, Chapter XIV: General Observations.

It has been observed, that many countries at the period of their greatest degree of populousness have lived in the greatest plenty, and have been able to export corn; but at other periods, when their population was very low, have lived in continual poverty and want, and have been obliged to import corn. Egypt, Palestine, Rome, Sicily and Spain are cited as particular exemplifications of this fact; and it has been inferred that an increase of population in any state, not cultivated to the utmost, will tend rather to augment than diminish the relative plenty of the whole society: and that, as Lord Kaimes observes, a country cannot easily become too populous for agriculture; because agriculture has the signal property of producing food in proportion to the number of consumers.[65]

The general facts, from which these inferences are drawn, there is no reason to doubt; but the inferences by no means follow from the premises. It is the nature of agriculture, (as it has before been observed,) particularly when well conducted, to produce support for a considerable number above that which it employs; and consequently if these members of the society, or, as Sir James Steuart calls them, the free hands, do not increase so as to reach the limit of the number which can be supported by the surplus produce, the whole population of the country may continue for ages increasing with the improving state of agriculture, and yet always be able to export corn. But this increase, after a certain period, will be very different from the natural and unrestricted increase of population; it will merely follow the slow augmentation of produce from the gradual improvement of agriculture; and population will still be checked by the difficulty of procuring subsistence. The precise measure of the population in a country thus circumstanced will not indeed be the quantity of food, because part of it is exported, but the quantity of employment. The state of this employment however will necessarily regulate the wages of labour, on which the power of the lower classes of people to procure food depends; and according as the employment of the country is increasing, whether slowly or rapidly, these wages will be such, as either to check or encourage early marriages; such, as to enable a labourer to support only two or three, or as many as five or six children.

In stating that in this, and all the other cases and systems which have been considered, the progress of population will be mainly regulated and limited by the real wages of labour,

[65] Sketches of the History of Man, b. i. sketch i. p. 106, 107. 8vo. 1788.

it is necessary to remark that, practically, the current wages of day-labour estimated in the necessaries of life do not always correctly represent the quantity of these necessaries which it is in the power of the lower classes to consume; and that sometimes the error is in excess and sometimes in defect.

In a state of things when the prices of corn and of all sorts of commodities are rising, the money wages of labour do not always rise in proportion; but this apparent disadvantage to the labouring classes is sometimes more than counterbalanced by the plenty of employment, the quantity of task-work that can be obtained, and the opportunity given to women and children to add considerably to the earnings of the family. In this case, the power of the labouring classes to command the necessaries of life is much greater than is implied by the current rate of their wages, and will of course have a proportionably greater effect on the population.

On the other hand, when prices are generally falling, it often happens that the current rate of wages does not fall in proportion; but this apparent advantage is in the same manner often more than counterbalanced by the scarcity of work, and the impossibility of finding employment for all the members of a labourer's family who are able and willing to be industrious. In this case, the powers of the labouring classes to command the necessaries of life will evidently be less than is implied by the current vale of their wages.

In the same manner parish allowances distributed to families, the habitual practice of task-work, and the frequent employment of women and children, will affect population like a rise in the real wages of labour. And, on the other hand, the paying of every sort of labour by the day, the absence of employment for women and children, and the practice among labourers of not working more than three or four days in the week, either from inveterate indolence, or any other cause, will affect population like a low price of labour.

In all these cases the real earnings of the labouring classes throughout the year, estimated in food, are different from the apparent wages; but it will evidently be the average earnings of the families of the labouring classes throughout the year on which the encouragement to marriage, and the power of supporting children, will depend, and not merely the wages of day-labour estimated in food.

An attention to this very essential point will explain the reason why, in many instances, the progress of population does not appear to be regulated by what are usually called the real wages of labour; and why this progress may occasionally be greater, when the price of a day's labour will purchase rather less than the medium quantity of corn, than when it will purchase rather more.

In our own country, for instance, about the middle of the last century, the price of corn was very low; and, for twenty years together, from 1735 to 1755 a day's labour would, on an average, purchase a peck of wheat. During this period, population increased at a moderate rate; but not by any means with the same rapidity as from 1790 to 1811, when the average wages of day-labour would not in general purchase so much as a peck of wheat. In the latter case; however, there was a more rapid accumulation of capital, and a greater demand for labour; and though the continued rise of provisions still kept them rather ahead of wages, yet the fuller employment for every body that would work, the greater quantity of task-work done, the higher relative value of corn compared with manufactures, the increased use of potatoes, and the greater sums distributed in parish allowances, unquestionably gave to the lower classes of society the power of commanding a greater quantity of food, and will account for the more rapid increase of population in the latter period, in perfect consistency with the general principle.

On similar grounds, if, in some warm climates and rich soils, where corn is cheap, the quantity of food earned by a day's labour be such as to promise a more rapid progress in population than is really known to take place, the fact will be fully accounted for, if it be found that inveterate habits of indolence fostered by a vicious government, and a slack demand for labour, prevent any thing like constant employment.[66] It would of course require high corn wages of day-labour even to keep up the supply of a stationary population, where the days of working would only amount to half of the year.

In the case also of the prevalence of prudential habits, and a decided taste for the conveniences and comforts of life, as, according to the supposition, these habits and tastes do not operate as an encouragement to early marriages, and are not in fact spent almost entirely in the purchase of corn, it is quite consistent with the general principles laid down, that the population should not proceed at the same rate as is usual, *cæteris paribus,* in other countries, where the corn wages of labour are equally high.

The quantity of employment in any country will not of course vary from year to year, in the same manner as the quantity of produce must necessarily do, from the variation of the seasons; and consequently the check from want of employment will be much more steady in its operation, and much more favourable to the lower classes of people, than the check from the immediate want of food. The first will be the preventive check; the second the positive check. When the demand for labour is either stationary, or increasing very slowly, people not seeing any employment open by which they can support a family, or the wages of common labour being inadequate to this purpose, will of course be deterred from marrying. But if a demand for labour continue increasing with some rapidity, although the supply of food be uncertain, on account of variable seasons and a dependence on other countries, the population will evidently go on, till it is positively checked by famine or the diseases arising from severe want.

Scarcity and extreme poverty, therefore, may or may not accompany an increasing population, according to circumstances: but they must necessarily accompany a permanently declining population; because there never has been, nor probably ever will be, any other cause than want of food, which makes the population of a country permanently decline. In the numerous instances of depopulation which occur in history, the causes may always be traced to the want of industry or the ill direction of that industry, arising from violence, bad government, ignorance, &c. which first occasion a want of food, and of course depopulation follows. When Rome adopted the custom of importing all her corn,

[66]This observation is exemplified in the slow progress of population in some parts of the Spanish dominions in America, compared with its progress in the United States.

and laying all Italy into pasture, she soon declined in population. The causes of the depopulation of Egypt and Turkey have already been adverted to; and in the case of Spain, it was certainly not the numerical loss of people occasioned by the expulsion of the Moors, but the industry and capital thus expelled, which permanently injured her population. When a country has been depopulated by violent causes, if a bad government with its usual concomitant insecurity of property ensue, (which has generally been the case in all those countries which are now less peopled than formerly,) neither the food nor the population can recover itself; and the inhabitants will probably live in severe want. But when an accidental depopulation takes place in a country which was before populous and industrious, and in the habit of exporting corn, if the remaining inhabitants be left at liberty to exert, and do exert, their industry in the same direction as before, it is a strange idea to entertain, that they would then be unable to supply themselves with corn in the same plenty; particularly as the diminished numbers would of course cultivate principally the more fertile parts of their territory, and not be obliged, as in their more populous state, to apply to ungrateful soils. Countries in this situation would evidently have the same chance of recovering their former number, as they had originally of reaching this number; and indeed if absolute populousness were necessary to relative plenty, as some agriculturists have supposed,[67] it would be impossible for new colonies to increase with the same rapidity as old states.

The prejudices on the subject of population bear a very striking resemblance to the old prejudices about specie; and we know how slowly and with what difficulty these last have yielded to juster conceptions. Politicians, observing that states which were powerful and prosperous were, almost invariably populous, have mistaken an effect for a cause, and have concluded, that their population was the cause of their prosperity, instead of their prosperity being the cause of their population; as the old political economists concluded that the

[67] Among others, I allude more particularly to Mr. Anderson, who, in a *Calm Investigation into the Circumstances which have led to the present Scarcity of Grain in Britain* (published in 1801,) has laboured with extraordinary earnestness, and I believe with the best intentions, to impress this curious truth on the minds of his countrymen. The particular position which he attempts to prove is, *that an increase of population in any state, whose fields have not been made to attain their highest possible degree of productiveness (a thing that probably has never yet been seen on this globe), will necessarily have its means of subsistence rather augmented than diminished by that augmentation of its population; and the reverse.* The proposition is, to be sure, expressed rather obscurely; but from the context his meaning evidently is, that every increase of population tends to increase relative plenty, and *vice versâ.* He concludes his proofs by observing that, if the facts which he has thus brought forward and connected do not serve to remove the fears of those, who doubt the possibility of this country producing abundance to sustain its increasing population, (were it to augment in a ratio greatly more progressive than it has yet done,) he should doubt whether they could be convinced of it, were one even to rise from the dead to tell them so. I agree with Mr. A. entirely, respecting the importance of directing a greater part of the national industry to agriculture; but from the circumstance of its being possible for a country, with a certain direction of its industry, always to grow corn sufficient for its own supplies, although it may be very populous, he has been led into the strange error of supposing, that an agricultural country could support an unchecked population.

abundance of specie was the cause of national wealth, instead of being the effect of it. The annual produce of the land and labour, in both these instances, became in consequence a secondary consideration; and its increase, it was conceived, would naturally follow the increase of specie in the one case, or of population in the other. The folly of endeavouring by forcible means to increase the quantity of specie in any country, and the absolute impossibility of accumulating it beyond a certain level by any human laws that can be devised, are now fully established, and have been completely exemplified in the instances of Spain and Portugal. But the illusion still remains respecting population; and under this impression almost every political treatise has abounded in proposals to encourage population, with little or no comparative reference to the means of its support. Yet surely the folly of endeavouring to increase the quantity of specie in any country, without an increase of the commodities which it is to circulate, is not greater than that of endeavouring to increase the number of people, without an increase of the food which is to maintain them; and it will be found that the level, above which no human laws can raise the population of a country, is a limit more fixed and impassable than the limit to the accumulation of specie. However improbable in fact, it is possible to conceive that means might be invented of retaining a quantity of specie in a state, greatly beyond what was demanded by the produce of its land and labour, and the relative state of other countries. But when by great encouragements population has been raised to such a height, that this produce is meted out to each individual in the smallest portions that can support life, no stretch of ingenuity can even conceive the possibility of going farther.

It has appeared, I think, clearly, in the review of different societies given in the former part of this work, that those countries, the inhabitants of which were sunk in the most barbarous ignorance, or oppressed by the most cruel tyranny, however low they might be in actual population, were very populous in proportion to their means of subsistence; and upon the slightest failure of the seasons generally suffered the severities of want. Ignorance and despotism seem to have no tendency to destroy the passion which prompts to increase; but they effectually destroy the checks to it from reason and foresight. The improvident barbarian who thinks only of his present wants, or the miserable peasant who, from his political situation, feels little security of reaping what he has sown, will seldom be deterred from gratifying his passions by the prospect of inconveniences, which cannot be expected to press on him under three or four years. But though this want of foresight, which is fostered by ignorance and despotism, tends thus rather to encourage the procreation of children, it is absolutely fatal to the industry which is to support them. Industry cannot exist without foresight and security. The indolence of the savage is well known; and the poor Egyptian or Abyssinian farmer without capital, who rents land which is let out yearly to the highest bidder, and who is constantly subject to the demands of his tyrannical masters, to the casual plunder of an enemy, and not unfrequently to the violation of his miserable contract, can have no heart to be industrious, and, if he had, could not exercise that industry with success. Even poverty itself, which appears to be the great spur to industry, when it has once passed certain limits, almost ceases to operate. The indigence which is hopeless destroys all vigorous exertion, and confines the efforts to what is sufficient for bare existence. It is the hope of bettering our condition, and the fear of want, rather than want itself, that is the best stimulus to industry; and its most constant and best directed efforts will almost invariably be found among a class of people above the class of the wretchedly poor.

The effect of ignorance and oppression will therefore always be to destroy the springs of industry, and consequently to diminish the annual produce of the land and labour in any country; and this diminution will inevitably be followed by a decrease of the population, in spite of the birth of any number of children whatever annually. The desire of immediate gratification, and the removal of the restraints to it from prudence, may perhaps, in such countries, prompt universally to early marriages; but when these habits have once reduced the people to the lowest possible state of poverty, they can evidently have no further effect upon the population. Their only effect must be on the degree of mortality; and there is no doubt, that, if we could obtain accurate bills of mortality in those southern countries, where very few women remain unmarried, and all marry young, the proportion of the annual deaths would be 1 in 17, 18, or 20, instead of 1 in 34, 36, or 40, as in European states where the preventive check operates.

That an increase of population, when it follows in its natural order, is both a great positive good in itself, and absolutely necessary to a further increase in the annual produce of the land and labour of any country, I should be the last to deny. The only question is, what is the order of its progress? In this point Sir James Steuart, who has in general explained this subject so well, appears to me to have fallen into an error. He determines, that multiplication is the efficient cause of agriculture, and not agriculture of multiplication.[68] But though it may be allowed, that the increase of people, beyond what could easily subsist on the natural fruits of the earth, first prompted man to till the ground; and that the view of maintaining a family, or of obtaining some valuable consideration in exchange for the products of agriculture, still operates as the principal stimulus to cultivation; yet it is clear that these products, in their actual state, must be beyond the lowest wants of the existing population, before any permanent increase can possibly be supported. We know, that a multiplication of births has in numberless instances taken place, which has produced no effect upon agriculture, and has merely been followed by an increase of diseases; but perhaps there is no instance, where a permanent increase of agriculture has not effected a permanent increase of population somewhere or other. Consequently, agriculture may with more propriety be termed the efficient cause of population, than population of agriculture;[69] though they certainly re-act upon each other, and are mutually necessary to each other's support. This indeed seems to be the hinge on which the subject turns; and all the prejudices respecting population have, perhaps, arisen from a mistake about the order of precedence.

[68] Polit. Econ. vol. i. b. i. c. xviii. p. 114.

[69] Sir James Steuart explains himself afterwards by saying, that he means principally the multiplication of those persons, who have some valuable consideration to give for the products of agriculture: but this is evidently not mere increase of population, and such an explanation seems to admit the incorrectness of the general proposition.

The author of *L'Ami des Hommes,* in a chapter on the effects of a decay of agriculture upon population, acknowledges that he had fallen into a fundamental error in considering population as the source of revenue; and that he was afterwards fully convinced that revenue was the source of population.[70] From a want of attention to this most important distinction, statesmen, in pursuit of the desirable object of population, have been led to encourage early marriages, to reward the fathers of families, and to disgrace celibacy; but this, as the same author justly observes, is to dress and water a piece of land without sowing it, and yet to expect a crop.

What is here said of the order of precedence with respect to agriculture and population, does not invalidate what was said in an earlier part of this work on the tendency to an oscillation or alternation in the increase of population and food in the natural course of their progress. In this progress nothing is more usual than for the population to increase at certain periods faster than food; indeed it is a part of the general principle that it should do so; and when the money wages of labour are prevented from falling by the employment of the increasing population in manufactures, the rise in the price of corn which the increased competition for it occasions is practically the most natural and frequent stimulus to agriculture. But then it must be recollected that the greater relative increase of population absolutely implies a previous increase of food at some time or other greater than the lowest wants of the people. Without this, the population could not possibly have gone forward.[71]

Universally, when the population of a country is for a longer or shorter time stationary, owing to the low corn wages of labour, a case which is not unfrequent, it is obvious that nothing but a previous increase of food, or at least an increase of the portion awarded to the labourer, can enable the population again to proceed forwards.

And, in the same manner, with a view to any essential improvement in the condition of the labourer, which is to give him a greater command over the means of comfortable subsistence, it is absolutely necessary that, setting out from the lowest point, the increase of food must precede and be greater than the increase of population.

Strictly speaking then, as man cannot live without food, there can be no doubt that in the order of precedence food must take the lead; although when, from the state of cultivation and other causes, the average quantity of food awarded to the labourer is considerably more than sufficient to maintain a stationary population, it is quite natural that the diminution of this quantity, from the tendency of population to increase, should be one of the most powerful and constant stimulants to agriculture.

It is worthy also of remark that on this account a stimulus to the increase of agriculture is much more easy when, from the prevalence of prudential restraint, or any other cause, the

[70]Tom. viii. p. 84, 12mo. 9 vols. 1762.

[71]According to the principle of population, the human race has a tendency to increase faster than food. It has therefore a constant tendency to people a country fully up to the limits of subsistence, but by the laws of nature it can never go beyond them, meaning, of course, by these limits, the lowest quantity of food which will maintain a stationary population. Population, therefore, can never, strictly speaking, precede food.

labourer is well paid; as in this case a rise in the price of corn, occasioned either by the increase of population or a foreign demand, will increase for a time the profits of the farmer, and often enable him to make permanent improvements; whereas, when the labourer is paid so scantily that his wages will not allow even of any temporary diminution without a diminution of population, the increase of cultivation and population must from the first be accompanied with a fall of profits. The prevalence of the preventive check to population and the good average wages of the labourer will rather promote than prevent that occasional increase and decrease of them, which as a stimulus seems to be favourable to the increase both of food and population.

Among the other prejudices which have prevailed on the subject of populations it has been generally thought that, while there is either waste among the rich, or land remaining uncultivated in any country, the complaints for want of food cannot be justly founded; or at least that the pressure of distress upon the poor is to be attributed to the ill conduct of the higher classes of society and the bad management of the land. The real effect however of these two circumstances is merely to narrow the limit of the actual population; but they have little or no influence on what may be called the average pressure of distress on the poorer members of society. If our ancestors had been so frugal and industrious, and had transmitted such habits to their posterity, that nothing superfluous was now consumed by the higher classes, no horses were used for pleasure, and no land was left uncultivated, a striking difference would appear in the state of the actual population; but probably none whatever in the state of the lower classes of people, with respect to the price of labour, and the facility of supporting a family. The waste among the rich, and the horses kept for pleasure, have indeed a little the effect of the consumption of grain in distilleries, noticed before with regard to China. On the supposition, that the food consumed in this manner may be withdrawn on the occasion of a scarcity, and be applied to the relief of the poor, they operate certainly, as far as they go, like granaries, which are only opened at the time that they are most wanted, and must therefore tend rather to benefit than to injure the lower classes of society.

With regard to uncultivated land, it is evident that its effect upon the poor is neither to injure nor to benefit them. The sudden cultivation of it will indeed tend to improve their condition for a time, and the neglect of lands before cultivated will certainly make their situation worse for a certain period; but when no changes of this kind are going forward, the effect of uncultivated land on the lower classes operates merely like the possession of a smaller territory. It may indeed be a point of some importance to the poor, whether a country be in the habit of exporting or importing corn; but this point is not necessarily connected with the complete or incomplete cultivation of the whole territory, but depends upon the proportion of the surplus produce to those who are supported by it; and in fact this proportion is generally the greatest, in countries which have not yet completed the cultivation of all their territory. If every inch of land in this country were well cultivated, there would be no reason to expect, merely from this circumstance, that we should be able to export corn. Our power in this respect would depend entirely on the proportion of the surplus produce to the commercial population; and this of course would in its turn depend on the direction of capital to agriculture or commerce.

It is not probable that any country with a large territory should ever be completely cultivated: and I am inclined to think that we often draw very inconsiderate conclusions against the industry and government of states from the appearance of uncultivated lands in them. It seems to be the clear and express duty of every government to remove all obstacles and give every facility to the inclosure and cultivation of land; but when this has been done, the rest must be left to the operation of individual interest; and upon this principle it cannot be expected that any new land should be brought into cultivation, the manure and the labour necessary for which might be employed to greater advantage on the improvement of land already in cultivation; and this is a case which will very frequently occur. In countries possessed of a large territory, there will always be a great quantity of land of a middling quality, which requires constant dressing to prevent it from growing worse, but which would admit of very great improvement, if a greater quantity of manure and labour could be employed upon it. The great obstacle to the melioration of land is the difficulty, the expense, and sometimes the impossibility, of procuring a sufficient quantity of dressing. As this instrument of improvement, therefore, is in practice limited, whatever it may be in theory, the question will always be, how it may be most profitably employed? And in any instance, where a certain quantity of dressing and labour, employed to bring new land into cultivation, would have yielded a permanently greater produce, if employed upon old land, both the individual and the nation are losers. Upon this principle, it is not uncommon for farmers in some situations never to dress their poorest land, but to get from it merely a scanty crop every three or four years, and to employ the whole of their manure, which they practically feel is limited, on those parts of their farms where it will produce a greater proportional effect.

The case will be different, of course, in a small territory with a great population, supported on funds not derived from their own soil. In this case there will be a little or no choice of land, and a comparative superabundance of manure; and under such circumstances the poorest soils may be brought under cultivation. But for this purpose, it is not mere population that is wanted, but a population which can obtain the produce of other countries, while it is gradually improving its own; otherwise it would be immediately reduced in proportion to the limited produce of this small and barren territory; and the melioration of the land might perhaps never take place; or, if it did, it would take place very slowly indeed, and the population would always be exactly measured by this tardy rate, and could not possibly increase beyond it.

This subject is illustrated in the cultivation of the Campine in Brabant, which, according to the Abbé Mann,[72] consisted originally of the most barren and arid sand. Many attempts were made by private individuals to bring it under cultivation, but without success; which

[72] Memoir on the Agriculture of the Netherlands, published in vol. i. of Communications to the Board of Agriculture, p. 225.

proves, that, as a farming project, and considered as a sole dependence, the cultivation of it would not answer. Some religious houses, however, at last settled there; and being supported by other funds, and improving the land merely as a secondary object, they by degrees, in the course of some centuries, brought nearly the whole under cultivation, letting it out to farmers as soon as it was sufficiently improved.

There is no spot, however barren, which might not be made rich this way, or by the concentrated population of a manufacturing town; but this is no proof whatever that, with respect to population and food, population has the precedence; because this concentrated population could not possibly exist without the preceding existence of an adequate quantity of food in the surplus produce of some other district.

In a country like Brabant or Holland, where territory is the principal want, and not manure, such a district as the Campine is described to be may perhaps be cultivated with advantage. But in countries possessed of a large territory, and with a considerable quantity of land of a middling quality, the attempt to cultivate such a spot would be a palpable misdirection and waste both of individual and national resources.

The French have already found their error in bringing under cultivation too great a quantity of poor land. They are now sensible, that they have employed in this way a portion of labour and dressing, which would have produced a permanently better effect, if it had been applied to the further improvement of better land. Even in China, which is so fully cultivated and so fully peopled, barren heaths have been noticed in some districts, which proves that, distressed as the people appear to be for subsistence, it does not answer to them to employ any of their manure on such spots. These remarks will be still further confirmed, if we recollect that, in the cultivation of a large surface of bad land, there must necessarily be a great waste of seed corn.

We should not therefore be too ready to make inferences against the internal economy of a country from the appearance of uncultivated heaths, without other evidence. But the fact is, that, as no country has ever reached, or probably ever will reach, its highest possible acme of produce, it appears always as if the want of industry, or the ill direction of that industry, was the actual limit to a further increase of produce and population, and not the absolute refusal of nature to yield any more: but a man who is locked up in a room may be fairly said to be confined by the walls of it, though he may never touch them; and with regard to the principle of population, it is never the question whether a country will produce *any more,* but whether it may be made to produce a sufficiency to keep pace with a nearly unchecked increase of people. In China, the question is not, whether a certain additional quantity of rice might be raised by improved culture; but whether such an addition could be expected during the next twenty-five years, as would be sufficient to support an additional three hundred millions of people. And in this country, it is not the question whether, by cultivating all our commons, we could raise considerably more corn than at present; but whether we could raise sufficient for a population of twenty millions in the next twenty-five years, and forty millions in the next fifty years.[73]

The allowing of the produce of the earth to be absolutely unlimited, scarcely removes the weight of a hair from the argument, which depends entirely upon the differently increasing ratios of population and food: and all that the most enlightened governments and the most persevering and best guided efforts of industry can do is to make the necessary checks to population operate more equably, and in a direction to produce the least evil; but to remove them is a task absolutely hopeless.

---

[73] It may be thought that the effects here referred to as resulting from greatly increased resources, could not take place in a country where there were towns and manufactories; and that they are not consistent with what was said in a former part of this work, namely, that the ultimate check to population (the want of food) is never the immediate check, except in cases of actual famine.

If the expressions are unguardedly strong, they will certainly allow of considerable mitigation, without any sensible diminution in the practical force and application of the argument. But I am inclined to think that, though they are unquestionably strong, they are not very far from the truth. The great cause which fills towns and manufactories is an insufficiency of employment, and consequently the means of support in the country; and if each labourer, in the parish where he was born, could command food, clothing, and lodging for ten children, the population of the towns would soon bear but a small proportion to the population of the country. And if to this consideration we add that, in the case supposed, the proportion of births and marriages in towns would be greatly increased, and all the mortality arising from poverty almost entirely removed, I should by no means be surprised (after a short interval for the change of habits) at an increase of population, even in China, equal to that which is referred to in the text.

With regard to this country, as it is positively known that the rate of increase has changed from that which would double the population in 120 years, or more, to that which would double it in 55 years, under a great increase of towns and manufactures, I feel very little doubt that, if the resources of the country were so augmented and distributed, as that every man could marry at 18 or 20, with the certainty of being able to support the largest family, the population of the British Isles would go on increasing at a rate which would double the population in 25 years, It appears, from our registers, that England is a healthier country than America. At the time that America was increasing with extraordinary rapidity, in some of the towns the deaths exceeded the births. In the English towns, with their present improvements, I do not think this would ever be the case, if all the lower classes could marry as soon as they pleased, and there was little or no premature mortality from the consequences of poverty.

But whether the habits and customs of an old state could be so changed by an abundance of food, as to make it increase nearly like a new colony, is a question of mere curiosity. The argument only requires that a change from scanty to abundant means of supporting a family should occasion, in old states, a marked increase of population; and this, it is conceived, cannot possibly be denied.

Book IV, Chapter I.

4.2 Book IV: Of Our Future Prospects Respecting the Removal or Mitigation of the Evils Arising from the Principle of Population.

4.2.1 Book IV, Chapter I: Of Moral Restraint, and Our Obligation to Practise This Virtue.

As it appears that, in the actual state of every society which has come within our review, the natural progress of population has been constantly and powerfully checked; and as it seems evident that no improved form of government, no plans of emigration, no benevolent institutions, and no degree or direction of national industry, can prevent the continued action of a great check to population in some form or other; it follows that we must submit to it as an inevitable law of nature; and the only inquiry that remains is, how it may take place with the least possible prejudice to the virtue and happiness of human society.

All the immediate checks to population, which have been observed to prevail in the same and different countries, seem to be resolvable into moral restraint, vice and misery; and if our choice be confined to these three, we cannot long hesitate in our decision respecting which it would be most eligible to encourage.

In the first edition of this essay I observed, that as from the laws of nature it appeared, that some check to population must exist, it was better that this check should arise from a foresight of the difficulties attending a family and the fear of dependent poverty, than from the actual presence of want and sickness. This idea will admit of being pursued farther; and I am inclined to think that, from the prevailing opinions respecting population, which undoubtedly originated in barbarous ages, and have been continued and circulated by that part of every community which may be supposed to be interested in their support, we have been prevented from attending to the clear dictates of reason and nature on this subject.

Natural and moral evil seem to be the instruments employed by the Deity in admonishing us to avoid any mode of conduct which is not suited to our being, and will consequently injure our happiness. If we are intemperate in eating and drinking, our health is disordered; if we indulge the transports of anger, we seldom fail to commit acts of which we afterwards repent; if we multiply too fast, we die miserably of poverty and contagious diseases. The laws of nature in all these cases are similar and uniform. They indicate to us that we have followed these impulses too far, so as to trench upon some other law, which equally demands attention. The uneasiness we feel from repletion, the injuries that we inflict on ourselves or others in anger, and the inconveniencies we suffer on the approach of poverty, are all admonitions to us to regulate these impulses better; and if we heed not this admonition, we justly incur the penalty of our disobedience, and our sufferings operate as a warning to others.

From the inattention of mankind hitherto to the consequences of increasing too fast, it must be presumed, that these consequences are not so immediately and powerfully connected with the conduct which leads to them, as in the other instances: but the delayed knowledge of particular effects does not alter their nature, or our obligation to regulate our conduct accordingly, as soon as we are satisfied of what this conduct ought to be. In many other instances it has not been till after long and painful experience, that the conduct most favourable to the happiness of man has been forced upon his attention. The kind of food, and the mode of preparing it, best suited to the purposes of nutrition and the gratification of the palate; the treatment and remedies of different disorders; the bad effects on the human frame of low and marshy situations; the invention of the most convenient and comfortable clothing; the construction of good houses; and all the advantages and extended enjoyments, which distinguish civilized life, were not pointed out to the attention of man at once; but were the slow and late result of experience, and of the admonitions received by repeated failures.

Diseases have been generally considered as the inevitable inflictions of Providence; but, perhaps, a great part of them may more justly be considered as indications that we have offended against some of the laws of nature. The plague at Constantinople, and in other towns of the East, is a constant admonition of this kind to the inhabitants. The human constitution cannot support such a state of filth and torpor; and as dirt, squalid poverty, and indolence are, in the highest degree, unfavourable to happiness and virtue, it seems a benevolent dispensation, that such a state should, by the laws of nature, produce disease and death, as a beacon to others to avoid splitting on the same rock.

The prevalence of the plague in London till the year 1666 operated in a proper manner on the conduct of our ancestors; and the removal of nuisances, the construction of drains, the widening of the streets, and the giving more room and air to the houses, had the effect of eradicating completely this dreadful disorder, and of adding greatly to the health and happiness of the inhabitants.

In the history of every epidemic it has almost invariably been observed that the lower classes of people, whose food was poor and insufficient, and who lived crowded together in small and dirty houses, were the principal victims. In what other manner can Nature point out to us that, if we increase too fast for the means of subsistence, so as to render it necessary for a considerable part of the society to live in this miserable manner, we have offended against one of her laws? This law she has declared exactly in the same manner as she declares that intemperance in eating and drinking will be followed by ill health; and that, however grateful it may be to us at the moment to indulge this propensity to excess, such indulgence will ultimately produce unhappiness. It is as much a law of nature that repletion is bad for the human frame, as that eating and drinking, unattended with this consequence, are good for it.

An implicit obedience to the impulses of our natural passions would lead us into the wildest and most fatal extravagancies; and yet we have the strongest reasons for believing that all these passions are so necessary to our being, that they could not be generally weakened or diminished, without injuring our happiness. The most powerful and universal of all our desires is the desire of food, and of those things, such as clothing, houses, &c. which are immediately necessary to relieve us from the pains of hunger and cold. It is acknowledged by all, that these desires put in motion the greatest part of that activity, from which the multiplied improvements and advantages of civilized life are derived; and that the pursuit of these objects, and the gratification of these desires, form the principal happiness of the larger half of mankind, civilized or uncivilized, and are indispensably necessary to the more refined enjoyments of the other half. We are all conscious of the inestimable benefits that we derive from these desires, when directed in a certain manner; but we are equally conscious of the evils resulting from them, when not directed in this manner; so much so, that society has taken upon itself to punish most severely what it considers as an irregular gratification of them. And yet the desires in both cases are equally natural, and, abstractedly considered, equally virtuous. The act of the hungry man who satisfies his appetite by taking a loaf from the shelf of another, is in no respect to be distinguished from the act of him who does the same thing with a loaf of his own, but by its consequences. From the consideration of these consequences, we feel the most perfect conviction, that, if people were not prevented from gratifying their natural desires with the loaves in the possession of others, the number of loaves would universally diminish. This experience is the foundation of the laws relating to property, and of the distinctions of virtue and vice, in the gratification of desires otherwise perfectly the same.

If the pleasure arising from the gratification of these propensities were universally diminished in vividness, violations of property would become less frequent; but this advantage would be greatly overbalanced by the narrowing of the sources of enjoyment. The diminution in the quantity of all those productions, which contribute to human gratification, would be much greater in proportion than the diminution of thefts; and the loss of general happiness on the one side would be beyond comparison greater than the gain of happiness on the other. When we contemplate the constant and severe toils of the greatest part of mankind, it is impossible not to be forcibly impressed with the reflection, that the sources of human happiness would be most cruelly diminished, if the prospect of a good meal, a warm house, and a comfortable fireside in the evening, were not incitements sufficiently vivid, to give interest and cheerfulness to the labours and privations of the day.

After the desire of food, the most powerful and general of our desires is the passion between the sexes, taken in an enlarged sense. Of the happiness spread over human life by this passion very few are unconscious. Virtuous love, exalted by friendship, seems to be that sort of mixture of sensual and intellectual enjoyment, particularly suited to the nature of man, and most powerfully calculated to awaken the sympathies of the soul, and produce the most exquisite gratifications. Perhaps there is scarcely a man, who has once experienced the genuine delight of virtuous love, however great his intellectual pleasures may have been, who does not look back to that period as the sunny spot in his whole life, where his imagination loves most to bask, which he recollects and contemplates with the fondest regret, and which he would wish to live over again.

It has been said by Mr. Godwin, in order to shew the evident inferiority of the pleasures of sense, "Strip the commerce of the sexes of all its attendant circumstances, and it would be generally despised." He might as well say to a man who admires trees, strip them of their spreading branches and lovely foliage, and what beauty can you see in a bare pole? But it was the tree with the branches and foliage, and not without them, that excited admiration. It is "the symmetry of person, the vivacity, the voluptuous softness of temper, the affectionate kindness of feeling, the imagination, and the wit"[74] of a woman, which excites the passion of love, and not the mere distinction of her being a female.

It is a very great mistake to suppose that the passion between the sexes only operates and influences human conduct, when the immediate gratification of it is in contemplation. The formation and steady pursuit of some particular plan of life has been justly considered as one of the most permanent sources of happiness; but I am inclined to believe, that there are not many of these plans formed, which are not connected in a considerable degree with the prospect of the gratification of this passion, and with the support of children arising from it. The evening meal, the warm house, and the comfortable fireside, would lose half their interest, if we were to exclude the idea of some object of affection, with whom they were to be shared.

We have also great reason to believe that the passion between the sexes has the most powerful tendency to soften and meliorate the human character, and keep it more alive to all the kindlier emotions of benevolence and pity. Observations on savage life have generally tended to prove that nations, in which this passion appeared to be less vivid, were distinguished by a ferocious and malignant spirit, and particularly by tyranny and cruelty to the sex. If, indeed, this bond of conjugal affection were considerably weakened, it seems probable, either that the man would make use of his superior physical strength, and turn his wife into a slave, as among the generality of savages; or, at best, that every little inequality of temper, which must necessarily occur between two persons, would produce a total alienation of affection; and this could hardly take place, without a diminution of parental fondness and care, which would have the most fatal effect on the happiness of society.

It may be further remarked, and observations on the human character in different countries warrant us in the conclusion, that the passion is stronger, and its general effects in producing gentleness, kindness, and suavity of manners, much more powerful, where obstacles are thrown in the way of very early and universal gratification. In some of the southern countries, where every impulse may be almost immediately indulged, the passion sinks into mere animal desire, is soon weakened and almost extinguished by excess, and its influence on the character is extremely confined. But, in European countries, where, though the women are not secluded, yet manners have imposed considerable restraints on this gratification, the passion not only rises in force, but in the universality and beneficial tendency of its effects; and has often the greatest influence in the formation and improvement of the character, where it is the least gratified.

Considering then the passion between the sexes in all its bearings and relations, and including the endearing engagement of parent and child resulting from it, few will be disposed to deny that it is one of the principal ingredients of human happiness. Yet experience teaches us that much evil flows from the irregular gratification of it; and though the evil be of little weight in the scale, when compared with the good, yet its absolute

---

[74]Political Justice vol. i. b. i. c. v. p. 72. 8vo.

quantity cannot be inconsiderable, on account of the strength and universality of the passion. It is evident, however, from the general conduct of all governments in their distribution of punishments, that the evil resulting from this cause is not so great and so immediately dangerous to society, as the irregular gratification of the desire of property; but placing this evil in the most formidable point of view, we should evidently purchase a diminution of it at a very high price, by the extinction or diminution of the passion which causes it; a change, which would probably convert human life either into a cold and cheerless blank, or a scene of savage and merciless ferocity.

A careful attention to the remote as well as immediate effect of all the human passions, and all the general laws of nature, leads us strongly to the conclusion, that, under the present constitution of things, few or none of them will admit of being greatly diminished, without narrowing the sources of good, more powerfully than the sources of evil. And the reason seems to be obvious. They are, in fact, the materials of all our pleasures, as well as of all our pains; of all our happiness, as well as of all our misery; of all our virtues, as well as of all our vices. It must therefore be regulation and direction that are wanted, not diminution or extinction.

It is justly observed by Paley, that "Human passions are either necessary to human welfare, or capable of being made, and in a great majority of instances are in fact made, conducive to its happiness. These passions are strong and general; and perhaps would not answer their purpose, unless they were so. But strength and generality, when it is expedient that particular circumstances should be respected, become, if left to themselves, excess and misdirection. From which excess and misdirection the vices of mankind (the causes no doubt of much misery) appear to spring. This account, while it shews us the principle of vice, shews us at the same time the province of reason and self-government.[75]

Our virtue, therefore, as reasonable beings, evidently consists in educing from the general materials, which the Creator has placed under our guidance, the greatest sum of human happiness; and as natural impulses are abstractedly considered good, and only to be distinguished by their consequences, a strict attention to these consequences, and the regulation of our conduct conformably to them, must be considered as our principal duty.

The fecundity of the human species is, in some respects, a distinct consideration from the passion between the sexes, as it evidently depends more upon the power of women in bearing children, than upon the strength and weakness of this passion. It is a law however exactly similar in its great features to all the other laws of nature. It is strong and general, and apparently would not admit of any very considerable diminution, without being inadequate to its object; the evils arising from it are incidental to those necessary qualities of strength and generality; and these evils are capable of being very greatly mitigated and rendered comparatively light by human energy and virtue. We cannot but conceive that it is an object of the Creator, that the earth should be replenished; and it appears to me clear, that this could not be effected without a tendency in population to increase faster than food; and

[75]Natural Theology, c. xxvi. p. 547.

as, with the present law of increase, the peopling of the earth does not proceed very rapidly, we have undoubtedly some reason to believe, that this law is not too powerful for its apparent object. The desire of the means of subsistence would be comparatively confined in its effects, and would fail of producing that general activity so necessary to the improvement of the human faculties, were it not for the strong and universal effort of population to increase with greater rapidity than its supplies. If these two tendencies were exactly balanced, I do not see what motive there would be sufficiently strong to overcome the acknowledged indolence of man, and make him proceed in the cultivation of the soil. The population of any large territory, however fertile, would be as likely to stop at five hundred, or five thousand, as at five millions, or fifty millions. Such a balance therefore would clearly defeat one great purpose of creation; and if the question be merely a question of degree, a question of a little more or a little less strength, we may fairly distrust our competence to judge of the precise quantity necessary to answer the object with the smallest sum of incidental evil. In the present state of things we appear to have under our guidance a great power, capable of peopling a desert region in a small number of years; and yet, under other circumstances, capable of being confined by human energy and virtue to any limits however narrow, at the expense of a small comparative quantity of evil. The analogy of all the other laws of nature would be completely violated, if in this instance alone there were no provision for accidental failures, no resources against the vices of mankind, or the partial mischeifs resulting from other general laws. To effect the apparent object without any attendant evil, it is evident that a perpetual change in the law of increase would be necessary, varying with the varying circumstances of each country. But instead of this, it is not only more consonant to the analogy of the other parts of nature, but we have reason to think that it is more conducive to the formation and improvement of the human mind, that the law should be uniform, and the evils incidental to it, under certain circumstances, left to be mitigated or removed by man himself. His duties in this case vary with his situation; he is thus kept more alive to the consequences of his actions; and his faculties have evidently greater play and opportunity of improvement, than if the evil were removed by a perpetual change of the law according to circumstances.

Even if from passions too easily subdued, or the facility of illicit intercourse, a state of celibacy were a matter of indifference, and not a state of some privation, the end of nature in the peopling of the earth would be apparently liable to be defeated. It is of the very utmost importance to the happiness of mankind, that population should not increase too fast; but it does not appear, that the object to be accomplished would admit of any considerable diminution in the desire of marriage. It is clearly the duty of each individual not to marry till he has a prospect of supporting his children; but it is at the same time to be wished that he should retain undiminished his desire of marriage, in order that he may exert himself to realize this prospect, and be stimulated to make provision for the support of greater numbers.

It is evidently therefore regulation and direction which are required with regard to the principle of population, not diminution or alteration. And if moral restraint be the only virtuous mode of avoiding the incidental evils arising from this principle, our obligation to practise it will evidently rest exactly upon the same foundation as our obligation to practise any of the other virtues.

Whatever indulgence we may be disposed to allow to occasional failures in the discharge of a duty of acknowledged difficulty, yet of the strict line of duty we cannot doubt. Our obligation not to marry till we have a fair prospect of being able to support our children will appear to deserve the attention of the moralist, if it can be proved that an attention to this obligation is of most powerful effect in the prevention of misery; and that, if it were the general custom to follow the first impulse of nature, and marry at the age of puberty, the universal prevalence of every known virtue in the greatest conceivable degree, would fail of rescuing society from the most wretched and desperate state of want, and all the diseases and famines which usually accompany it.

4.2.2 Book IV, Chapter II: Of the Effects Which Would Result to Society from the Prevalence of Moral Restraint.

One of the principal reasons which have prevented an assent to the doctrine of the constant tendency of population to increase beyond the means of subsistence, is a great unwillingness to believe that the Deity would by the laws of nature bring beings into existence, which by the laws of nature could not be supported in that existence. But if, in addition to that general activity and direction of our industry put in motion by these laws, we further consider that the incidental evils arising from them are constantly directing our attention to the proper check to population, moral restraint; and if it appear that, by a strict obedience to the duties pointed out to us by the light of nature and reason, and confirmed and sanctioned by revelation, these evils may be avoided, the objection will, I trust, be removed, and all apparent imputation on the goodness of the Deity be done away.

The heathen moralists never represented happiness as attainable on earth, but through the medium of virtue; and among their virtues prudence ranked in the first class, and by some was even considered as including every other. The Christian religion places our present as well as future happiness in the exercise of those virtues which tend to fit us for a state of superior enjoyment; and the subjection of the passions to the guidance of reason, which, if not the whole, is a principal branch of prudence, is in consequence most particularly inculcated.

If, for the sake of illustration, we might be permitted to draw a picture of society, in which each individual endeavoured to attain happiness by the strict fulfilment of those duties, which the most enlightened of the ancient philosophers deduced from the laws of nature, and which have been directly taught, and received such powerful sanctions in the moral code of Christianity, it would present a very different scene from that which we now contemplate. Every act, which was prompted by the desire of immediate gratification, but which threatened an ultimate overbalance of pain, would be considered as a breach of duty; and consequently no man, whose earnings were only sufficient to maintain two children, would put himself in a situation in which he might have to maintain four or five, however he might be prompted to it by the passion of love. This prudential restraint, if it were generally adopted, by narrowing the supply of labour in the market, would, in the natural course of things, soon raise its price. The period of delayed gratification would be passed in saving the earnings which were above the wants of a single man, and in acquiring habits of sobriety, industry and economy, which would enable him in a few years to enter into the matrimonial contract without fear of its consequences. The operation of the preventive check in this way, by constantly keeping the population within the limits of the food, though constantly following its increase, would give a real value to the rise of wages and the sums saved by labourers before marriage, very different from those forced advances in the price of labour or arbitrary parochial donations, which, in proportion to their magnitude and extensiveness, must of necessity be followed by a proportional advance in the price of provisions. As the wages of labour would thus be sufficient to maintain with decency a large family, and as every married couple would set out with a sum for contingencies, all abject poverty would be removed from society; or would at least be confined to a very few, who had fallen into misfortunes, against which no prudence or foresight could provide.

The interval between the age of puberty and the period at which each individual might venture on marriage must, according to the supposition, be passed in strict chastity; because the law of chastity cannot be violated without producing evil. The effect of any thing like a promiscuous intercourse, which prevents the birth of children, is evidently to weaken the best affections of the heart, and in a very marked manner to degrade the female character. And any other intercourse would, without improper arts, loving as many children into the society as marriage, with a much greater probability of their becoming a burden to it.

These considerations shew that the virtue of chastity is not, as some have supposed, a forced produce of artificial society; but that it has the most real and solid foundation in nature and reason; being apparently the only virtuous means of avoiding the vice and misery which result so often from the principle of population.

In such a society as we have been supposing, it might be necessary for some of both sexes to pass many of the early years of life in the single state; and if this were general, there would certainly be room for a much greater number to marry afterwards, so that fewer, upon the whole, would be condemned to pass their lives in celibacy. If the custom of not marrying early prevailed generally, and if violations of chastity were equally dishonourable in both sexes, a more familiar and friendly intercourse between them might take place without danger. Two young people might converse together intimately without its being immediately supposed that they either intended marriage or intrigue; and a much better opportunity would thus be given to both sexes of finding out kindred dispositions, and of forming those strong and lasting attachments, without which the married state is generally more productive of misery than of happiness. The earlier years of life would not be spent without love, though without the full gratification of it. The passion, instead of being extinguished, as it now too frequently is, by early sensuality, would only be repressed for a time, that it might afterwards burn with a brighter, purer, and steadier flame; and the happiness of the married state, instead of only affording the means of immediate indulgence, would be looked forward to as the prize of industry and virtue, and the reward of a genuine and constant attachment.[76]

The passion of love is a powerful stimulus in the formation of character, and often prompts to the most noble and generous exertions; but this is only when the affections are centred in one object, and generally when full gratification is delayed by difficulties.[77] The heart is perhaps never so much disposed to virtuous conduct, and certainly at no time is the virtue of chastity so little difficult to men, as when under the influence of such a passion. Late marriages taking place in this way would be very different from those of the same name at present, where the union is too frequently prompted solely by interested views, and the parties meet, not unfrequently, with exhausted constitutions, and generally with exhausted affections. The late marriages at present are indeed principally confined to the men; of whom there are few, however advanced in life, who, if they determine to marry, do not fix their choice on a young wife. A young woman without fortune, when she has passed her twenty-fifth year, begins to fear, and with reason, that she may lead a life of celibacy; and with a heart capable of forming a strong attachment, feels, as each year creeps on, her hopes of finding an object on which to rest her affections gradually diminishing, and the uneasiness of her situation aggravated by the silly and unjust prejudices of the world. If the general age of marriage among women were later, the period of youth and hope would be prolonged, and fewer would be ultimately disappointed.

That a change of this kind would be a most decided advantage to the more virtuous half of society, we cannot for a moment doubt. However impatiently the privation might be borne by the men, it would be supported by the women readily and cheerfully; and if they could look forward with just confidence to marriage at twenty-seven or twenty-eight, I fully believe, that, if the matter were left to their free choice, they would clearly prefer waiting till this period, to the being involved in all the cares of a large family at twenty-five. The most eligible age of marriage however could not be fixed; but must depend entirely on

[76]Dr. Currie, in his interesting observations on the character and condition of the Scotch peasantry, prefixed to his Life of Burns, remarks, with a just knowledge of human nature, that, "in appreciating the happiness and virtue of a community, there is perhaps no single criterion on which so much dependence may be placed as the state of the intercourse between the sexes. Where this displays ardour of attachment, accompanied by purity of conduct, the character and the influence of women rise, our imperfect nature mounts in the scale of moral excellence; and from the source of this single affection, a stream of felicity descends, which branches into a thousand rivulets that enrich and adorn the field of life. Where the attachment between the sexes sinks into an appetite, the heritage of our species is comparatively poor, and man approaches to the condition of the brutes that perish." Vol. i. p. 18.

[77]Dr. Currie observes, that the Scottish peasant in the course of his passion often exerts a spirit of adventure, of which a Spanish cavalier need not be ashamed." Burns' Works, vol. i. p. 16. It is not to be doubted that this kind of romantic passion, which, Dr. C. says, characterizes the attachment of the humblest people of Scotland, and which has been greatly fostered hp the elevation of mind given to them by a superior education, has had a most powerful and most beneficial influence on the national character.

circumstances and situation. There is no period of human life, at which nature more strongly prompts to an union of the sexes, than from seventeen or eighteen to twenty. In every society above that state of depression, which almost excludes reason and foresight, these early tendencies must necessarily be restrained; and if, in the actual state of things, such a restraint on the impulses of nature be found unavoidable, at what time can we be consistently released from it, but at that period, whatever it may be, when, in the existing circumstances of the society, a fair prospect presents itself of maintaining a family?

The difficulty of moral restraint will perhaps be objected to this doctrine. To him who does not acknowledge the authority of the Christian religion, I have only to say that, after the most careful investigation, this virtue appears to be absolutely necessary, in order to avoid certain evils which would otherwise result from the general laws of nature. According to his own principles, it is his duty to pursue the greatest good consistent with these laws; and not to fail in this important end, and produce an overbalance of misery by a partial obedience to some of the dictates of nature, while he neglects others. The path of virtue, though it be the only path which leads to permanent happiness, has always been represented by the heathen moralists as of difficult ascent.

To the Christian I would say that the Scriptures most clearly and precisely point it out to us as our duty, to restrain our passions within the bounds of reason; and it is a palpable disobedience of this law to indulge our desires in such a manner as reason tells us will unavoidably end in misery. The Christian cannot consider the difficulty of moral restraint as any argument against its being his duty; since, in almost every page of the sacred writings, man is described as encompassed on all aides by temptations which it is extremely difficult to resist; and though no duties are enjoined, which do not contribute to his happiness on earth as well as in a future state, yet an undeviating obedience is never represented as an easy task.

There is in general so strong a tendency to love in early youth, that it is extremely difficult at this period to distinguish a genuine from a transient passion. If the earlier years of life were passed by both sexes in moral restraint, from the greater facility that this would give to the meeting of kindred dispositions, it might even admit of a doubt, whether more happy marriages would not take place, and consequently more pleasure from the passion of love, than in a state such as that of America, the circumstances of which allow of a very early union of the sexes. But if we compare the intercourse of the sexes in such a society as I have been supposing, with that which now exists in Europe, taken under all its circumstances, it may safely be asserted, that, independently of the load of misery which would be removed, the sum of pleasurable sensations from the passion of love would be increased in a very great degree.

If we could suppose such a system general, the accession of happiness to society in its internal economy would scarcely be greater than in its external relations. It might fairly be expected that war, that great pest of the human race, would, under such circumstances, soon cease to extend its ravages so widely and so frequently as it does at present.

One of its first causes and most powerful impulses was undoubtedly an insufficiency of room and food; and greatly as the circumstances of mankind have changed since it first began, the same cause still continues to operate and to produce, though in a smaller degree, the same effects. The ambition of princes would want instruments of destruction, if the distresses of the lower classes of people did not drive them under their standards. A recruiting serjeant always prays for a bad harvest and a want of employment, or, in other words, a redundant population.

In the earlier ages of the world, when war was the great business of mankind, and the drains of population from this cause were, beyond comparison, greater than in modern times, the legislators and statesmen of each country, adverting principally to the means of offence and defence, encouraged an increase of people in every possible way, fixed a stigma on barrenness and celibacy, and honoured marriage. The popular religions followed these prevailing opinions. In many countries the prolific power of nature was the object of solemn worship. In the religion of Mahomet, which was established by the sword, and the promulgation of which in consequence could not be unaccompanied by an extraordinary destruction of its followers, the procreation of children to glorify the Creator was laid down as one of the principal duties of man; and he, who had the most numerous offspring, was considered as having best answered the end of his creation. The prevalence of such moral sentiments had naturally a great effect in encouraging marriage; and the rapid procreation, which followed, was partly the effect and partly the cause of incessant war. The vacancies occasioned by former desolations made room for the rearing of fresh supplies; and the overflowing rapidity, with which these supplies followed, constantly furnished fresh incitements and fresh instruments for renewed hostilities. Under the influence of such moral sentiments, it is difficult to conceive how the fury of incessant war should ever abate.

It is a pleasing confirmation of the truth and divinity of the Christian religion, and of its being adapted to a more improved state of human society, that it places our duties respecting marriage and the procreation of children in a different light from that in which they were before beheld.

Without entering minutely into the subject, which would evidently lead too far, I think it will be admitted, that, if we apply the spirit of St. Paul's declarations respecting marriage to the present state of society and the known constitution of our nature, the natural inference seems to be, that, when marriage does not interfere with higher duties, it is right; when it does, it is wrong. According to the genuine principles of moral science, "The method of coming at the will of God from the light of nature is, to inquire into the tendency of the action to promote or diminish the general happiness."[78] There are perhaps few actions that tend so directly to diminish the general happiness, as to marry without the means of supporting children. He who commits this act, therefore, clearly offends against the will of God; and having become a burden on the society in which he lives, and plunged himself and family into a situation, in which virtuous habits are preserved with more difficulty than in any other, he appears to have violated his duty to his neighbours and to himself, and thus to have listened to the voice of passion in opposition to his higher obligations.

[78] Paley's Moral Philosophy, vol. i. b. ii, c. iv. p. 65.

In a society, such as I have supposed, all the members of which endeavour to attain happiness by obedience to the moral code derived from the light of nature, and enforced by strong sanctions in revealed religion, it is evident that no such marriages could take place; and the prevention of a redundant population, in this way, would remove one of the principal encouragements to offensive war; and at the same time tend powerfully to eradicate those two fatal political disorders, internal tyranny and internal tumult, which mutually produce each other.

Indisposed to a war of offence, in a war of defence, such a society would be strong as a rock of adamant. Where every family possessed the necessaries of live in plenty, and a decent portion of its comforts and conveniences, there could not exist that hope of change, or at best that melancholy and disheartening indifference to it, which sometimes prompts the lower classes of people to say, "Let what will come, we cannot be worse off than we are now." Every heart and hand will be united to repel an invader, when each individual felt the value of the solid advantages which he enjoyed, and a prospect of change presented only a prospect of being deprived of them.

As it appears therefore, that it is in the power of each individual to avoid all the evil consequences to himself and society resulting from the principle of population, by the practice of a virtue clearly dictated to him by the light of nature, and expressly enjoined in revealed religion; and as we have reason to think, that the exercise of this virtue to a certain degree would tend rather to increase than diminish individual happiness; we can have no reason to impeach the justice of the Deity, because his general laws make this virtue necessary, and punish our offences against it by the evils attendant upon vice, and the pains that accompany the various forms of premature death. A really virtuous society, such as I have supposed, would avoid these evils. It is the apparent object of the Creator to deter us from vice by the pains which accompany it, and to lead us to virtue by the happiness that it produces. This object appears to our conceptions to be worthy of a benevolent Creator. The laws of nature respecting population tend to promote this object. No imputation, therefore, on the benevolence of the Deity, can be founded on these laws, which is not equally applicable to any of the evils necessarily incidental to all imperfect state of existence.

4.2.3 Book IV, Chapter III: Of the Only Effectual Mode of Improving the Condition of the Poor.

He who publishes a moral code, or system of duties, however firmly he may be convinced of the strong obligation on each individual strictly to conform to it, has never the folly to imagine that it will be universally or even generally practised. But this is no valid objection against the publication of the code. If it were, the same objection would always have applied; we should be totally without general rules; and to the vices of mankind arising from temptation would be added a much longer list, than the have at present, of vices from ignorance.

Judging merely from the light of nature, if we feel convinced of the misery arising from a redundant population on the one hand, and of the evils and unhappiness, particularly to the female sex, arising from promiscuous intercourse, on the other, I do not see how it is possible for any person who acknowledges the principle of utility, as the great criterion of moral rules, to escape the conclusion, that moral restraint, or the abstaining from marriage till we are in a condition to support a family, with a perfectly moral conduct during that period, is the strict line of duty; and when revelation is taken into the question, this duty undoubtedly receives very powerful confirmation. At the same time I believe that few of my readers can be less sanguine than I am in their expectations of any sudden and great change in the general conduct of men on this subject: and the chief reason why in the last chapter I allowed myself to suppose the universal prevalence of this virtue was, that I might endeavour to remove any imputation on the goodness of the Deity, by shewing, that the evils arising from the principle of population were exactly of the same nature as the generality of other evils which excite fewer complaints; that they were increased by human ignorance and indolence, and diminished by human knowledge and virtue; and on the supposition that each individual strictly fulfilled his duty, would be almost totally removed; and this without any general diminution of those sources of pleasure, arising from the regulated indulgence of the passions, which have been justly considered as the principal ingredients of human happiness.

If it will answer any purpose of illustration, I see no harm in drawing the picture of a society, in which each individual is supposed strictly to fulfil his duties; nor does a writer appear to be justly liable to the imputation of being visionary, unless he make such universal or general obedience necessary to the practical utility of his system, and to that degree of moderate and partial improvement, which is all that can rationally be expected from the most complete knowledge of our duties.

But in this respect there is an essential difference between that improved state of society, which I have supposed in the last chapter, and most of the other speculations on this subject. The improvement there supposed, if we ever should make approaches towards it, is to be effected in the way in which we have been in the habit of seeing all the greatest improvements effected, by a direct application to the interest and happiness of each individual. It is not required of us to act from motives to which we are unaccustomed; to pursue a general good, which we may not distinctly comprehend, or the effect of which may be weakened by distance and diffusion. The happiness of the whole is to be the result of the happiness of individuals, and to begin first with them. No cooperation is required. Every step tells. He who performs his duty faithfully will reap the full fruits of it, whatever may be the number of others who fail. This duty is intelligible to the humblest capacity. It is merely, that he is not to bring beings into the world, for whom he cannot find the means of support. When once this subject is cleared from the obscurity thrown over it by parochial laws and private benevolence, every man must feel the strongest conviction of such an obligation. If he cannot support his children, they must starve; and if he marry in the face of a fair probability that he shall not be able to support his children, he is guilty of all the evils, which he thus brings upon himself, his wife and his offspring. It is clearly his interest, and will tend greatly to promote his happiness, to defer marrying, till by industry and economy he is in a capacity to support the children that he may reasonably expect from his marriage; and as he cannot in the mean time gratify his passions without violating an express command of God, and running a great risk of injuring himself; or some of his fellow-creatures, considerations of his own interest and happiness will dictate to him the strong obligation to a moral conduct while he remains unmarried.

However powerful may be the impulses of passion, they are generally in some degree modified by reason. And it does not seem entirely visionary to suppose that, if the true and permanent cause of poverty were clearly explained and forcibly brought home to each man's bosom, it would have some, and perhaps not an inconsiderable influence on his conduct; at least the experiment has never yet been fairly tried. Almost every thing, that has been hitherto done for the poor, has tended, as if with solicitous care, to throw a veil of obscurity over this subject, and to hide from them the true cause of their poverty. When the wages of labour are hardly sufficient to maintain two children, a man marries, and has five or six; he of course finds himself miserably distressed. He accuses the insufficiency of the price of labour to maintain a family. He accuses his parish for their tardy and sparing fulfilment of their obligation to assist him. He accuses the avarice of the rich, who suffer him to want what they can so well spare. He accuses the partial and unjust institutions of society, which have awarded him an inadequate share of the produce of the earth. He accuses perhaps the dispensations of Providence, which have assigned to him a place in society so beset with unavoidable distress and dependence. In searching for objects of accusation, he never adverts to the quarter from which his misfortunes originate. The last person that he would think of accusing is himself, on whom in fact the principal blame lies, except so far as he has been deceived by the higher classes of society. He may perhaps wish that he had not married, because he now feels the inconveniences of it; but it never enters into his head that he can have done any thing wrong. He has always been told, that to raise up subjects for his king and country is a very meritorious act. He has done this, and yet is suffering for it; and it cannot but strike him as most extremely unjust and cruel in his king and country, to allow him thus to suffer, in return for giving them what they are continually declaring that they particularly want.

Till these erroneous ideas have been corrected, and the language of nature and reason has been generally heard on the subject of population, instead of the language of error and prejudice, it cannot be said, that any fair experiment has been made with the understandings of the common people; and we cannot justly accuse them of improvidence and want of industry, till they act as they do now, after it has been brought home to their comprehensions, that they are themselves the cause of their own poverty; that the means of redress are in their own hands, and in the hands of no other persons whatever; that the society in which they live and the government which presides over it, are without any *direct* power in this respect; and that however ardently they may desire to relieve them, and whatever attempts they may make to do so, they are really and truly unable to execute what they benevolently wish but unjustly promise; that, when the wages of labour will not maintain a family, it is an incontrovertible sign that their king and country do not want more subjects, or at least that they cannot support them; that, if they marry in this case, so far from fulfilling a duty to society, they are throwing an useless burden on it, at the same time that they are plunging themselves into distress; and that they are acting directly contrary to the will of God, and bringing down upon themselves various diseases, which might all, or the greater part, have been avoided, if they had attended to the repeated admonitions which he gives by the general laws of nature to every being capable of reason.

Paley, in his Moral Philosophy, observes, that in "countries in which subsistence is become scarce, it behoves the state to watch over the public morals with increased solicitude; for nothing but the instinct of nature, under the restraint of chastity, will induce men to undertake the labour, or consent to the sacrifice of personal liberty and indulgence, which the support of a family in such circumstances requires."[79] That it is always the duty of a state to use every exertion likely to be effectual in discouraging vice and promoting virtue, and that no temporary circumstances ought to cause any relaxation in these exertions, is certainly true. The means therefore proposed are always good; but the particular end in view in this case appears to be absolutely criminal. We wish to force people into marriage, when from the acknowledged scarcity of subsistence they will have little chance of being able to support their children. We might as well force people into the water who are unable to swim. In both cases we rashly tempt Providence. Nor have we more reason to believe that a miracle will be worked to save us from the misery and mortality resulting from our conduct in the one case than in the other.

The object of those who really wish to better the condition of the lower classes of society, must be to raise the relative proportion between the price of labour and the price of provisions, so as to enable the labourer to command a larger share of the necessaries and comforts of life. We have hitherto principally attempted to attain this end by encouraging the married poor, and consequently increasing the number of labourers, and overstocking the market with a commodity which we still say that we wish to be dear. It would seem to have required no great spirit of divination to foretel the certain failure of such a plan of proceeding. There is nothing however like experience. It has been tried in many different countries, and for many hundred years, and the success has always been answerable to the nature of the scheme. It is really time now to try something else.

When it was found that oxygen, or pure vital air, would not cure consumptions as was expected, but rather aggravated their symptoms, trial was made of an air of the most opposite kind. I wish we had acted with the same philosophical spirit in our attempts to cure the disease of poverty; and having found that the pouring in of fresh supplies of labour only tended to aggravate the symptoms, had tried what would be the effect of withholding a little these supplies.

In all old and fully-peopled states it is from this method, and this along, that we can rationally expect any essential and permanent melioration in the condition of the labouring classes of the people.

[79] Vol. ii. c. xi. p. 352.

In an endeavour to raise the proportion of the quantity of provisions to the number of consumers in any country, our attention would naturally be first directed to the increasing of the absolute quantity of provisions; but finding that, as fast as we did this, the number of consumers more than kept pace with it, and that with all our exertions we were still as far as ever behind, we should be convinced, that our efforts directed only in this way would never succeed. It would appear to be setting the tortoise to catch the hare. Finding, therefore, that from the laws of nature we could not proportion the food to the population, our next attempt should naturally be, to proportion the population to the food. If we can persuade the hare to go to sleep, the tortoise may have some chance of overtaking her.

We are not however to relax our efforts in increasing the quantity of provisions, but to combine another effort with it; that of keeping the population, when once it has been overtaken, at such a distance behind, as to effect the relative proportion which we desire; and thus unite the two grind *desiderata,* a great actual population, and a state of society, in which abject poverty and dependence are comparatively but little known; two objects which are far from being incompatible.

If we be really serious in what appears to be the object of such general research, the mode of essentially and permanently bettering the condition of the poor, we must explain to them the true nature of their situation, and shew them, that the withholding of the supplies of labour is the only possible way of really raising its price, and that they themselves, being the possessors of this commodity, have alone the power to do this.

I cannot but consider this mode of diminishing poverty as so perfectly clear in theory, and so invariably confirmed by the analogy of every other commodity which is brought to market, that nothing but its being shewn to be calculated to produce greater evils than it proposes to remedy, can justify us in not making the attempt to put it into execution.

4.2.4 Book IV, Chapter IV: Objections to This Mode Considered.

One objection which perhaps will be made to this plan is, that from which alone it derives its value—a market rather understocked with labour. This must undoubtedly take place in a certain degree; but by no means in such a degree as to affect the wealth and prosperity of the country. But putting this subject of a market understocked with labour in the must unfavourable point of vices, if the rich will not submit to a slight inconvenience necessarily attendant on the attainment of what they profess to desire, they cannot really be in earnest in their professions. Their benevolence to the poor must be either childish play or hypocrisy; it must be, either to amuse them selves or to pacify the minds of the common people with a mere show of attention to their wants. To wish to better the condition of the poor by enabling them to command a greater quantity of the necessaries and comforts of life, and then to complain of high wages, is the act of a silly boy who gives away his cake and then cries for it. A market overstocked with labour, and an ample remuneration to each labourer, are objects perfectly incompatible with each other. In the annals of the world they never existed together; and to couple them even in imagination betrays a gross ignorance of the simplest principles of political economy.

A second objection that may be made to this plan is, the diminution of population that it would cause. It is to be considered, however, that this diminution is merely relative; and when once this relative diminution has been effected, by keeping the population stationary, while the supply of food has increased, it might then start afresh, and continue increasing for ages, with the increase of food, maintaining always nearly the same relative proportion to it. I can easily conceive that this country, with a proper direction of the national industry, might, in the course of some centuries, contain two or three times its present population, and yet every man in the kingdom be much better fed and clothed than he is at present. While the springs of industry continue in vigour, and a sufficient part of that industry is directed to agriculture, we need be under no apprehensions of a deficient population; and nothing perhaps would tend so strongly to excite a spirit of industry and economy among the poor, as a thorough knowledge that their happiness must always depend principally upon themselves; and that, if they obey their passions in opposition to their reason, or be not industrious and frugal while they are single, to save a sum for the common contingencies of the married state, they must expect to suffer the natural evils which Providence has prepared for those who disobey its repealed admonitions.

A third objection which may be started to this plan, and the only one which appears to me to have any kind of plausibility, is that, by endeavouring to urge the duty of moral restraint on the poor, we may increase the quantity of vice relating to the sex.

I should be extremely sorry to say any thing which could either directly or remotely be construed unfavourably to the cause of virtue; but I certainly cannot think that the vices which relate to the sex are the only vices which are to be considered in a moral question; or that they are even the greatest and most degrading to the human character. They can rarely or never be committed without producing unhappiness somewhere or other, and therefore ought always to be strongly reprobated: but there are other vices, the effects of which are still more pernicious; and there are other situations which lead more certainly to moral offences than the refraining from marriage. Powerful as may be the temptations to a breach of chastity, I am inclined to think that they are impotent, in comparison of the temptations arising from continued distress. A large class of women, and many men, I have no doubt, pass a considerable part of their lives consistently with the laws of chastity; but I believe there will be found very few who pass through the ordeal of squalid and hopeless poverty, or even of long continued embarrassed circumstances, without a great moral degradation of character.

In the higher and middle classes of society, it is a melancholy and distressing sight to observe, not unfrequently, a man of a noble and ingenuous disposition, once feelingly alive to a sense of honour and integrity, gradually sinking under the pressure of circumstances, making his excuses at first with a blush of conscious shame, afraid of seeing the faces of his friends from whom he may have borrowed money, reduced to the meanest tricks and subterfuges to delay or avoid the payment of his just debts; till ultimately grown familiar with falsehood and at enmity with the world he loses all the grace and dignity of man.

To the general prevalence of indigence, and the extraordinary encouragements which we afford in this country to a total want of foresight and prudence among the common people,[80] is to be attributed a considerable part of those continual depredations on property, and other more atrocious crimes which drive us to the painful resource of such a number of executions.[81] According to Mr. Colquhoun, above twenty thousand miserable individuals of various classes rise up every morning without knowing how or by what means they are to be supported during the passing day, or where, in many instances, they are to lodge on the succeeding night.[82] It is by these unhappy persons that the principal depredations on the public are committed: and supposing but few of them to be married, and driven to these acts from the necessity of supporting their children; yet still it is probably true, that the too great frequency of marriage amongst the poorest classes of society is one of the principal causes of the temptations to these crimes. A considerable part of these unhappy wretches will probably be found to be the offspring of such marriages, educated in workhouses where every vice is propagated, or bred up at home in filth and rags, with an utter ignorance of every moral obligation.[83] A still greater part perhaps consists of persons, who, being unable for some time to get employment owing to the full supply of labour, have been urged to these extremities by their temporary wants; and, having thus lost their characters, are rejected even when their labour may be wanted, by the well-founded caution of civil society.[84]

---

[80]Mr. Colquhoun, speaking of the poor-laws, observes, that "In spite of all the ingenious arguments which have been used in favour of a system, admitted to be wisely conceived in its origin, the effects it has produced incontestably prove that, with respect to the mass of the poor, there is something radically wrong in the execution. If it were not so, it is impossible that there could exist in the metropolis such an inconceivable portion of human misery, amidst examples of munificence and benevolence unparalleled in any age or country," Police of Metropolis, c. xiii. p 359.

In the effects of the poor-laws I fully agree with Mr. Colquhoun; but I cannot agree with him in admitting that the system was well conceived in its origin. I attribute still more evil to the original ill conception, than to the subsequent ill execution.

[81]Mr. Colquhoun observes, that "Indigence in the present state of society may be considered as a principal cause of the increase of crimes." Police of Metropolis, c. xiii. p. 352.

[82]Police of Metropolis, c. xi. p. 313.

[83]Id. c. xi. xii. p. 355. 370.

[84]Police of the Metropolis, c. xiii. p. 353, et seq. In so large a town as London, which must necessarily encourage a prodigious influx of strangers from the country, there must be always a great many persons out of work; and it is probable, that some public institution for the relief of the casual poor, upon a plan similar to that proposed by Mr. Colquhoun (c. xiii. p. 371) would, under very judicious management, produce more good than evil. But for this purpose it would be absolutely necessary that, if work were provided by the institution, the sum that a man could earn by it should be less than the worst paid common labour; otherwise the claimants would rapidly increase, and the funds would soon be inadequate to their object. In the institution at Hamburgh, which appears to have been the most successful of any yet established, the nature of the work was such, that, though paid above the usual price, a person could not easily earn by it more than eighteen pence a week. It was the determined principle of the managers of the institution, to reduce the support which they gave lower than what

When indigence does not produce overt acts of vice, it palsies every virtue. Under the continued temptations to a breach of chastity, occasional failures may take place, and the moral sensibility in other respects not be very strikingly impaired; but the continued temptations which beset hopeless poverty, and the strong sense of injustice that generally accompanies it from an ignorance of its true cause, tend so powerfully to sour the disposition, to harden the heart, and deaden the moral sense, that, generally speaking, virtue takes her flight clear away from the tainted spot, and does not often return.

Even with respect to the vices which relate to the sex, marriage has been found to be by no means a complete remedy. Among the higher classes, our Doctors' Commons, and the lives that many married men are known to lead, sufficiently prove this; and the same kind of vice, though not so much heard of among the lower classes of people, is probably in all our great towns not much less frequent.

Add to this, that abject poverty, particularly when joined with idleness, is a state the most unfavourable to chastity that can well be conceived. The passion is as strong, or nearly so, as in other situations; and every restraint on it from personal respect, or a sense of morality, is generally removed. There is a degree of squalid poverty, in which, if a girl was brought up, I should say, that her being really modest at twenty was an absolute miracle. Those persons must have extraordinary minds indeed, and such as are not usually formed under similar circumstances, who can continue to respect themselves when no other person whatever respects them. If the children thus brought up were even to marry at twenty, it is probable, that they would have passed some years in vicious habits before that period.

If after all, however, these arguments should appear insufficient; if the reprobate the idea of endeavouring to encourage the virtue of moral restraint among the poor, from a fear of producing vice; and if we think, that to facilitate marriage by all possible means is a point of the first consequence to the morality and happiness of the people; let us act consistently, and before we proceed, endeavour to make ourselves acquainted with the mode by which alone we can effect our object.

---

any industrious man or woman in such circumstances could earn. (Account of the Management of the Poor in Hamburgh, by C. Voght, p. 18.) And it is to this principle that they attribute their success. It should be observed however, that neither the institution at Hamburgh, nor that planned by Count Rumford in Bavaria, has subsisted long enough for us to be able to pronounce on their permanent good effects. It will not admit of a doubt, that institutions for the relief of the poor, on their first establishment, remove a great quantity of distress. The only question is, whether, as succeeding generations arise, the increasing funds necessary for their support, and the increasing numbers that become dependent, are not greater evils than that which was to be remedied; and whether the country will not ultimately be left with as much mendicity as before, besides all the poverty and dependence accumulated in the public institutions. This seems to be nearly the case in England at present. It may be doubted whether we should have more beggars if we laid no poor-laws.

4.2.5 Book IV, Chapter V: Of the Consequences of Pursuing the Opposite Mode.

It is an evident truth that, whatever may be the rate of increase in the means of subsistence, the increase of population must be limited by it, at least after the food has once been divided into the smallest shares that will support life. All the children born, beyond what would be required to keep up the population to this level, must necessarily perish, unless room be made for them by the deaths of grown persons. It has appeared indeed clearly in the course of this work, that in all old states the marriages and births depend principally upon the deaths, and that there is no encouragement to early unions so powerful as a great mortality. To act consistently therefore, we should facilitate, instead of foolishly and vainly endeavouring to impede, the operations of nature in producing this mortality; and if we dread the too frequent visitation of the horrid form of famine, we should sedulously encourage the other forms of destruction, which we compel nature to use. Instead of recommending cleanliness to the poor, we should encourage contrary habits. In our towns we should make the streets narrower, crowd more people into the houses, and court the return of the plague. In the country, we should build our villages near stagnant pools, and particularly encourage settlements in all marshy and unwholesome situations.[85] But above all, we should reprobate specific remedies for ravaging diseases; and those benevolent, but much mistaken men, who have thought they were doing a service to mankind by projecting schemes for the total extirpation of particular disorders. If by these and similar means the annual mortality were increased from 1 in 36 or 40, to 1 in 18 or 20, we might probably every one of us marry at the age of puberty, and yet few be absolutely starved.

If, however, we all marry at this age, and yet still continue our exertions to impede the operations of nature, we may rest assured that all our efforts will be vain. Nature will not, nor cannot, be defeated in her purposes. The necessary mortality must come, in some form or other; and the extirpation of one disease will only be the signal for the birth of another perhaps more fatal. We cannot lower the waters of misery by pressing them down in different places, which must necessarily make them rise somewhere else; the only way in which we can hope to effect our purpose, is by drawing them off. To this course nature is constantly directing our attention by the chastisements which await a contrary conduct. These chastisements are more or less severe, in proportion to the degree in which her

[85] Necker, speaking of the proportion of the births in France, makes use of a new and instructive expression on this subject, though he hardly seems to be sufficiently aware of it himself. He says, "Le nombre des naissances est à celui des habitans de un à vingt-trois et vingt-quatre daps les lieux *contrariés par la nature, ou par des circonstances morales; ce même* rapport, dans la plus grande partie de la France, est de un à 25, 25½, 8 26." Administ. des Finances, tom. i. c. ix. p. 254. 12mo. It appears, therefore, that we have nothing more to do, than to settle people in marshy situations, and oppress them by a bad government, in order to attain what politicians have hitherto considered as so desirable—a great proportion of marriages and a great proportion of births.

admonitions produce their intended effect. In this country at present these admonitions are by no means entirely neglected. The preventive check to population prevails to a considerable degree, and her chastisements are in consequence moderate: but if we were all to marry at the age of puberty, they would be severe indeed. Political evils would probably be added to physical. A people goaded by constant distress, and visited by frequent returns of famine, could not be kept down but by a cruel despotism. We should approach to the state of the people in Egypt or Abyssinia; and I would ask, whether in that case it is probable, that we should be more virtuous?

Physicians have long remarked the great changes which take place in diseases; and that, while some appear to yield to the efforts of human care and skill, others seem to become in proportion more malignant and fatal. Dr. William Heberden published, not long since, some valuable observations on this subject deduced from the London bills of mortality. In his preface, speaking of these bills, he says, "the gradual changes they exhibit in particular diseases correspond to the alterations, which in time are known to take place in the channels through which the great stream of mortality is constantly flowing."[86] In the body of his work, afterwards, speaking of some particular diseases, he observes with that candour which always distinguishes true science; "It is not easy to give a satisfactory reason for all the changes which may be observed to take place in the history of diseases. Nor is it any disgrace to physicians, if their causes are often so gradual in their operation, or so subtle, as to elude investigation."[87]

I hope I shall not be accused of presumption, in venturing to suggest that, under certain circumstances, such changes must take place; and perhaps without any alteration in those proximate causes, which are usually looked to on these occasions. If this should appear to be true, it will not seem extraordinary that the most skilful and scientific physicians, whose business it is principally to investigate proximate causes, should sometimes search for these causes in vain.

In the country which keeps its population at a certain standard, if the average number of marriages and births be given, it is evident that the average number of deaths will also be given; and, to use Dr. Heberden's metaphor, the channels, through which the great stream of mortality is constantly flowing, will always convey off a given quantity. Now if we stop up any of these channels, it is perfectly clear, that the stream of mortality must run with greater force through some of the other channels; that is, if we eradicate some diseases, others will become proportionally more fatal. In this case the only distinguishable cause is the damming up a necessary outlet of mortality.[88] Nature, in the attainment of her great purposes, seems always to seize upon the weakest part. If this part be made strong by human skill, she seizes upon the next weakest part, and so on in succession; not like a

[86] Observations on the Increase and Decrease of different Diseases. Preface, p. 5. 4to. 1801.

[87] Id. p. 43, 4to. 1801.

[88] The way in which it operates is probably by increasing poverty, in consequence of a supply of labour too rapid for the demand.

capricious deity, with an intention to sport with our sufferings, and constantly to defeat our labours; but, like a kind, though sometimes severe instructor, with the intention of teaching us to make all parts strong, and to chase vise and misery from the earth. In avoiding one fault we are too apt to run into some other; but we always find Nature faithful to her great object, at every false step we commit, ready to admonish us of our errors, by the infliction of some physical or moral evil. If the prevalence of the preventive check to population in a sufficient degree were to remove many of those diseases, which now afflict us, yet be accompanied by a considerable increase of the vice of promiscuous intercourse, it is probable that the disorders and unhappiness, the physical and moral evils arising from this vice, would increase in strength and degree; and, admonishing us severely of our error, would point to the only line of conduct approved by nature, reason and religion, abstinence from marriage till we can support our children, and chastity till that period arrives.

In the case just stated, in which the population and the number of marriages are supposed to be fixed, the necessity of a change in the mortality of some diseases, from the diminution or extinction of others, is capable of mathematical demonstration. The only obscurity which can possibly involve this subject, arises from taking into consideration the effect that might be produced by a diminution of mortality in increasing the population, or in decreasing the number of marriages. That the removal of any of the particular causes of mortality can have no further effect upon population than the means of subsistence will allow, and that it has no certain and necessary influence on these means of subsistence, are facts of which the reader must be already convinced. Of its operation in tending to prevent marriage, by diminishing the demand for fresh supplies of children, I have no doubt; and there is reason to think, that it had this effect in no inconsiderable degree on the extinction of the plague, which had so long and so dreadfully ravaged this country. Dr. Heberden draws a striking picture of the favourable change observed in the health of the people of England since this period; and justly attributes it to the improvements which have gradually taken place, not only in London but in all great towns; and in the manner of living throughout the kingdom, particularly with respect to cleanliness and ventilation.[89] But these causes would not have produced the effect observed, if they had not been accompanied by an increase of the preventive check; and probably the spirit of cleanliness, and better mode of living, which then began to prevail, by spreading more generally a decent and useful pride, principally contributed to this increase. The diminution in the number of marriages, however, was not sufficient to make up for the great decrease of mortality, from the extinction of the plague, and the striking reduction of the deaths in the dysentery.[90] While these and some other disorders became almost evanescent, consumption, palsy, apoplexy, gout, lunacy, and the small-pox, became more mortal.[91] The widening of these drains was necessary to carry off the population which still remained

[89] Observations on Increase and Decrease of Diseases, p.35.

[90] Id. p. 34.

[91] Id. p. 35, et seq.

redundant, notwithstanding the increased operation of the preventive check, and the part which was annually disposed of and enabled to subsist by the increase of agriculture.

Dr. Haygarth, in the Sketch of his benevolent plan for the extermination of the casual small-pox, draws a frightful picture of the mortality which has been occasioned by this distemper, attributes to it the slow progress of population, and makes some curious calculations on the favourable effects which would be produced in this respect by its extermination.[92] His conclusions, however, I fear, would not follow from his premises. I am far from doubting that millions and millions of human beings have been destroyed by the small-pox. But were its devastations, as Dr. Haygarth supposes, many thousand degrees greater than the plague,[93] I should still doubt whether the average population of the earth had been diminished by them. The small-pox is certainly one of the channels, and a very broad one, which nature has opened for the last thousand years, to keep down the population to the level of the means of subsistence; but had this been closed, others would have become wider, or new ones would have been formed. In ancient times the mortality from war and the plague was incomparably greater than in modern. On the gradual diminution of this stream of mortality, the generation and almost universal prevalence of the small-pox is a great and striking instance of one of those changes in the channels of mortality, which ought to awaken our attention and animate us to patient and persevering investigation. For my own part I feel not the slightest doubt, that, if the introduction of the cow-pox should extirpate the small-pox, and yet the number of marriages continue the same, we shall find a very perceptible difference in the increased mortality of some other diseases. Nothing could prevent this effect but a sudden start in our agriculture; and if this should take place, it will not be so much owing to the number of children saved from death by the cow-pox inoculation, as to the alarms occasioned among the people of property by the late scarcities,[94] and to the increased gains of farmers, which have been so absurdly reprobated. I am strongly however inclined to believe that the number of marriages will not, in this case, remain the same; but that the gradual light, which may be expected to be thrown on this interesting topic of human inquiry, will teach us how to make the extinction of a mortal disorder a real blessing to us, a real improvement in the general health and happiness of the society.

If, on contemplating the increase of vice which might contingently follow an attempt to inculcate the duty of moral restraint, and the increase of misery that must necessarily follow the attempts to encourage marriage and population, we come to the conclusion, not to interfere in any respect, but to leave every man to his own free choice, and responsible only to God for the evil which he does in either way; this is all I contend for; I would on no account do more; but I contend, that at present we are very far from doing this.

---

[92] Vol. i. part ii. sect. v, and vi.

[93] Id. s. viii. p. 164.

[94] The scarce harvests of 1799 and 1800. The start here alluded to, certainly took place from 1801 to 1814, and provision was really made for the diminished mortality.

Among the lower classes of society, where the point is of the greatest importance, the poor-laws afford a direct, constant and systematical encouragement to marriage, by removing from each individual that heavy responsibility, which he would incur by the laws of nature, for bringing beings into the world which he could not support. Our private benevolence has the same direction as the poor-laws, and almost invariably tends to encourage marriage, and to equalize as much as possible the circumstances of married and single men.

Among the higher classes of people, the superior distinctions which married women receive, and the marked inattentions to which single women of advanced age are exposed, enable many men, who are agreeable neither in mind nor person, and are besides in the wane of life, to choose a partner among the young and fair, instead of being confined, as nature seems to dictate, to persons of nearly their own age and accomplishments. It is scarcely to be doubted, that the fear of being an old maid, and of that silly and unjust ridicule, which folly sometimes attaches to this came, drives many women into the marriage union with men whom they dislike, or at best to whom they are perfectly indifferent. Such marriages must to every delicate mind appear little better than legal propositions; and they often burden the earth with unnecessary children, without compensating for it by an accession of happiness and virtue to the parties themselves.

Throughout all the ranks of society the prevailing opinions respecting the duty and obligation of marriage cannot but have a very powerful influence. The man who thinks that, in going out of the world without leaving representatives behind him, he shall have failed in an important duty to society, will be disposed to force rather than to repress his inclinations on this subject; and when his reason represents to him the difficulties attending a family, he will endeavour not to attend to these suggestions, will still determine to venture, and will hope that, in the discharge of what he conceives to be his duty, he shall not be deserted by Providence.

In a civilized country, such as England, where a taste for the decencies and comforts of life prevails among a very large class of people, it is not possible that the encouragements to marriage from positive institutions and prevailing opinions should entirely obscure the light of nature and reason on this subject; but still they contribute to make it comparatively weak and indistinct. And till this obscurity is removed, and the poor are undeceived with respect to the principal cause of their poverty, and taught to know, that their happiness or misery must depend chiefly upon themselves, it cannot be said that, with regard to the great question of marriage, we leave every man to his own free and fair choice.

4.2.6 Book IV, Chapter VI: Effects of the Knowledge of the Principal Cause of Poverty on Civil Liberty.

It may appear, perhaps, that a doctrine, which attributes the greatest part of the sufferings of the lower classes of society exclusively to themselves, is unfavourable to the cause of liberty, as affording a tempting opportunity to governments of oppressing their subjects at pleasure, and laying the whole blame on the laws of nature and the imprudence of the poor. We are not, however, to trust to first appearances; and I am strongly disposed to believe that those who will be at the pains to consider this subject deeply will be convinced, that nothing would so powerfully contribute to the advancement of rational freedom, as a thorough knowledge generally circulated of the principal cause of poverty; and that the ignorance of this cause, and the natural consequences of this ignorance, form, at present, one of the chief obstacles to its progress.

The pressure of distress on the lower classes of people, together with the habit of attributing this distress to their rulers, appears to me to be the rock of defence, the castle, the guardian spirit of despotism. It affords to the tyrant the fatal and unanswerable plea of necessity. It is the reason why every free government tends constantly to destruction; and that its appointed guardians become daily less jealous of the encroachments of power. It is the reason why so many noble efforts in the cause of freedom have failed; and why almost every revolution, after long and painful sacrifices, has terminated in a military despotism. While any dissatisfied man of talents has power to persuade the lower classes of people that all their poverty and distress arise solely from the iniquity of the government, though, perhaps, the greatest part of what they suffer is unconnected with this cause, it is evident that the seeds of fresh discontents and fresh revolutions are continually sowing. When an established government has been destroyed, finding that their poverty is not removed, their resentment naturally falls upon the successors to power; and when these have been immolated without producing the desired effect, other sacrifices are called for, and so on without end. Are we to be surprised that, under such circumstances, the majority of well-disposed people, finding that a government with proper restrictions is unable to support itself against the revolutionary spirit, and weary and exhausted with perpetual change to which they can see no end, should give up the struggle in despair, and throw themselves into the arms of the first power which can afford them protection against the horrors of anarchy?

A mob, which is generally the growth of a redundant population goaded by resentment for real sufferings, but totally ignorant of the quarter from which they originate, is of all monsters the most fatal to freedom. It fosters a prevailing tyranny, and engenders one where it was not; and though, in its dreadful fits of resentment, it appears occasionally to devour its unsightly offspring; yet no sooner is the horrid deed committed, than, however unwilling it may be to propagate such a breed, it immediately groans with a new birth.

Of the tendency of mobs to produce tyranny we may not, perhaps, be long without an example in this country. As a friend to freedom, and naturally an enemy to large standing armies, it is with extreme reluctance that I am compelled to acknowledge that, had it not been for the great organized force in the country, the distresses of the people during the late scarcities,[95] encouraged by the extreme ignorance and folly of many among the higher classes, might have driven them to commit the most dreadful outrages, and ultimately to

[95] 1800 and 1801.

involve the country in all the horrors of famine. Should such periods often recur, (a recurrence which we have too much reason to apprehend from the present state of the country,) the prospect which opens to our view is melancholy in the extreme. The English constitution will be seen hastening with rapid strides to the *Euthanasia* foretold by Hume, unless its progress be interrupted by some popular commotion; and this alternative presents a picture still more appalling to the imagination. If political discontents were blended with the cries of hunger, and a revolution were to take place by the instrumentality of a mob clamouring for want of food, the consequences would be unceasing change and unceasing carnage, the bloody career of which nothing but the establishment of some complete despotism could arrest.

We can scarcely believe that the appointed guardians of British liberty should quietly have acquiesced in those gradual encroachments of power which have taken place of late years, but from the apprehension of these still more dreadful evils. Great as has been the influence of corruption, I cannot yet think so meanly of the country gentlemen of England, as to believe that they would thus have given up a part of their birthright of liberty, if they had not been actuated by a real and genuine fear that it was then in greater danger from the people than from the crown. They appeared to surrender themselves to government, on condition of being protected from the mob; but they never would have made this melancholy and disheartening surrender, if such a mob had not existed either in reality or in imagination. That the fears on this subject were artfully exaggerated and increased beyond the limits of just apprehension, is undeniable; but I think it is also undeniable that the frequent declamations which were heard against the unjust institutions of society, and the delusive arguments on equality which were circulated among the lower classes, gave us just reason to suppose that, if the *vox populi* had been allowed to speak, it would have appeared to be the voice of error and absurdity, instead of the *vox Dei.*

To say that our conduct is not to be regulated by circumstances, is to betray an ignorance of the most solid and incontrovertible principles of morality. Though the admission of this principle may sometimes afford a cloak to changes of opinion that do not result from the purest motives; yet the admission of a contrary principle would be productive of infinitely worse consequences. The phrase of "existing circumstances" has, I believe, not unfrequently created a smile in the English House of Commons; but the smile should have been reserved for the application of the phrase, and not have been excited by the phrase itself. A very frequent repetition of it has indeed, of itself, rather a suspicious air; and its application should always be watched with the most jealous and anxious attention; but no man ought to be judged *in limine* for saying, that existing circumstances had obliged him to alter his opinions and conduct. The country gentlemen were, perhaps, too easily convinced that existing circumstances called upon them to give up some of the most valuable privileges of Englishmen; but as far as they were really convinced of this obligation, they acted consistently with the clearest rule of morality.

The degree of power to be given to the civil government, and the measure of our submission to it, must be determined by general expediency; and in judging of this expediency every circumstance is to be taken into consideration; particularly the state of public opinion, and the degree of ignorance and delusion prevailing among the common people. The patriot who might be called upon by the love of his country to join with heart and hand in a rising of the people for some specific attainable object of reform, if he knew that they were enlightened respecting their own situation, and would stop short when they had attained their demand, would be called upon by the same motive to submit to very great oppression rather than give the slightest countenance to a popular tumult, the members of which, at least the greater number of them, were persuaded that the destruction of the Parliament, the Lord Mayor, and the monopolizers, would make bread cheap, and that a revolution would enable them all to support their families. In this case it is more the ignorance and delusion of the lower classes of the people that occasions the oppression, than the actual disposition of the government to tyranny.

That there is, however, in all power a constant tendency to encroach is an incontrovertible truth, and cannot be too strongly inculcated. The checks, which are necessary to secure the liberty of the subject, will always in some degree embarrass and delay the operations of the executive government. The members of this government feeling these inconveniences, while they are exerting themselves, as they conceive, in the service of their country, and conscious, perhaps, of no ill intention towards the people, will naturally be disposed, on every occasion, to demand the suspension or abolition of these checks; but if once the convenience of ministers be put in competition with the liberties of the people, and we get into a habit of relying on fair assurances and personal character, instead of examining, with the most scrupulous and jealous care, the merits of each particular case, there is an end of British freedom. If we once admit the principle, that the government must know better with regard to the quantity of power which it wants, than we can possibly do with our limited means of information, and that, therefore, it is our duty to surrender up our private judgments, we may just as well at the same time surrender up the whole of our constitution. Government is a quarter in which liberty is not, nor cannot be, very faithfully preserved. If we are wanting to ourselves, and inattentive to our great interests in this respect, it is the height of folly and unreasonableness to expect that government will attend to them for us. Should the British constitution ultimately lapse into a despotism, as has been prophesied, I shall think that the country gentlemen of England will have much more to answer for than the ministers.

To do the country gentlemen justice, however, I should readily acknowledge that in the partial desertion of their posts as guardians of British freedom, which has already taken place, they have been actuated more by fear than corruption. And the principal reason of this fear was, I conceive, the ignorance and delusions of the common people, and the prospective horrors which were contemplated, if in such a state of mind they should by any revolutionary movement obtain an ascendant.

The circulation of Paine's Rights of Man, it is supposed, has done great mischief among the lower and middling classes of people in this country. This is probably true; but not because man is without rights, or that these rights ought not to be known; but because Mr. Paine has fallen into some fundamental errors respecting the principles of government, and in many important points has shewn himself totally unacquainted with the structure of society, and the different moral effects to be expected from the physical difference between this country and America. Mobs of the same description as those collections of people known by this name in Europe could not exist in America. The number of people without property is there, from the physical state of the country, comparatively small; and therefore the civil power, which is to protect property, cannot require the same degree of strength. Mr. Paine very justly observes, that whatever the apparent cause of any riots may be, the real one is always want of happiness; but when he goes on to say, it shews that something is wrong in the system of government, that injures the felicity by which society is to be preserved, he falls into the common error of attributing all want of happiness to government. It is evident that this want of happiness might have existed, and from ignorance might have been the principal cause of the riots; and yet be almost wholly unconnected with any of the proceedings of government. The redundant population of an old state furnishes materials of unhappiness, unknown to such a state as that of America; and if an attempt were to be made to remedy this unhappiness by distributing the produce of the taxes to the poorer classes of society, according to the plan proposed by Mr. Paine, the evil would be aggravated a hundred fold, and in a very short time no sum that the society could possibly raise would be adequate to the proposed object.

Nothing would so effectually counteract the mischiefs occasioned by Mr. Paine's Rights of Man, as a general knowledge of the real rights of man. What these rights are it is not my business at present to explain; but there is one right which man has generally been thought to possess, which I am confident he neither does nor can possess—a right to subsistence when his labour will not fairly purchase it. Our laws indeed say that he has this right, and bind the society to furnish employment and food to those who cannot get them in the regular market; but in so doing they attempt to reverse the laws of nature; and it is in consequence to be expected, not only that they should fail in their object, but that the poor, who were intended to be benefited, should suffer most cruelly from the inhuman deceit thus practised upon them.

The Abbé Raynal has said, that "Avant toutes les loix sociales l'homme avoit le droit de subsister."[96] He might with just as much propriety have said that, before the institution of social laws, every man had a right to live a hundred years. Undoubtedly he had then, and has still, a good right to live a hundred years, nay a thousand *if he can,* without interfering with the right of others to live; but the affair in both cases is principally an affair of power not of right. Social laws very greatly increase this power, by enabling a much greater number to subsist than could subsist without them, and so far very greatly enlarge *le droit de subsister;* but, neither before nor after the institution of social laws, could an unlimited number subsist; and before as well as since, he who ceased to have the power ceased to have the right.

If the great truths on these subjects were more generally circulated, and the lower classes of people could be convinced that by the laws of nature, independently of any particular

[96]Raynal Hist. des Indes, vol. x. s. x. p. 322, 8vo.

institutions, except the great one of property, which is absolutely necessary in order to attain any considerable produce, no person has any claim of *right* on society for subsistence, if his labour will not purchase it, the greatest part of the mischievous declamation on the unjust institutions of society would fall powerless to the ground. The poor are by no means inclined to be visionary. Their distresses are always real, though they are not attributed to the real causes. If these causes were properly explained to them, and they were taught to know what part of their present distress was attributable to government, and what part to causes totally unconnected with it, discontent and irritation among the lower classes of people would shew themselves much less frequently than at present; and when they did shew themselves, would be much less to be dreaded. The efforts of turbulent and discontented men in the middle classes of society might safely be disregarded, if the poor were so far enlightened respecting the real nature of their situation, as to be aware that by aiding them in their schemes of renovation, they would probably be promoting the ambitious views of others, without in any respect benefiting themselves. The country gentlemen and men of property in England might securely return to a wholesome jealousy of the encroachments of power; and instead of daily sacrificing the liberties of the subject on the altar of public safety, might, without any just apprehension from the people, not only tread back their late steps, but firmly insist upon those gradual reforms, which the lapse of time and the storms of the political world have rendered necessary, to prevent the gradual destruction of the British constitution.

All improvements in governments must necessarily originate with persons of some education; and these will of course be found among the people of property. Whatever may be said of a few, it is impossible to suppose that the great mass of the people of property should be really interested in the abuses of government. They merely submit to them from the fear that an endeavour to remove them might be productive of greater evils. Could we but take away this fear, reform and improvement would proceed with as much facility as the removal of nuisances, or the paving and lighting of the streets. In human life we are continually called upon to submit to a lesser evil in order to avoid a greater; and it is the part of a wise man to do this readily and cheerfully; but no wise man will submit to any evil, if he can get rid of it without danger. Remove all apprehension from the tyranny or folly of the people, and the tyranny of government could not stand a moment. It would then appear in its proper deformity, without palliation, without pretext, without protector. Naturally feeble in itself, when it was once stripped naked, and deprived of the support of public opinion and of the great plea of necessity, it would fall without a struggle. Its few interested defenders would hide their heads abashed, and would be ashamed any longer to support a cause, for which no human ingenuity could invent a plausible argument.

The most successful supporters of tyranny are without doubt those general declaimers, who attribute the distresses of the poor, and almost all the evils to which society is subject, to human institutions and the iniquity of governments. The falsity of these accusations, and the dreadful consequences that would result from their being generally admitted and acted upon, make it absolutely necessary that they should at all events be resisted; not only on account of the immediate revolutionary horrors to be expected from a movement of the people acting under such impressions, (a consideration which must at all times have very great weight); but also on account of the extreme probability that such a revolution would terminate in a much worse despotism than that which it had destroyed. On these grounds a genuine friend of freedom, a zealous advocate for the real rights of man, might be found among the defenders of a considerable degree of tyranny. A cause bad in itself might be supported by the good and the virtuous, merely because that which was opposed to it was much worse; and because it was absolutely necessary at the moment to make a choice between the two. Whatever therefore may be the intention of those indiscriminate accusations against governments, their real effect undoubtedly is, to add a weight of talents and principles to the prevailing power, which it never would have received otherwise.

It is a truth, which I trust has been sufficiently proved in the course of this work, that under a government constructed upon the best and purest principles, and executed by men of the highest talents and integrity, the most squalid poverty and wretchedness might universally prevail from an inattention to the prudential check to population. And as this cause of unhappiness has hitherto been so little understood, that the efforts of society have always tended rather to aggravate than to lessen it, we have the strongest reasons for supposing that, in all the governments with which we are acquainted, a great part of the misery to be observed among the lower classes of the people arises from this cause.

The inference therefore which Mr. Paine and others have drawn against governments from the unhappiness of the people, is palpably unfair; and before we give a sanction to such accusations, it is a debt we owe to truth and justice, to ascertain how much of this unhappiness arises from the principle of population, and how much is fairly to be attributed to government. When this distinction has been properly made, and all the vague, indefinite and false accusations removed, government would remain, as it ought to be, clearly responsible for the rest; and the amount of this would still be such as to make the responsibility very considerable. Though government has but little power in the direct and immediate relief of poverty, yet its indirect influence on the prosperity of its subjects is striking and incontestable. And the reason is, that though it is comparatively impotent in its efforts to make the food of a country keep pace with an unrestricted increase of population, yet its influence is great in giving the best direction to those checks, which in some form or other must necessarily take palace. It has clearly appeared in the former part of this work, that the most despotic and worst-governed countries, however low they might be in actual population, were uniformly the most populous in proportion to their means of subsistence; and the necessary effect of this state of things must of course be very low wages. In such countries the checks to population arise more from the sickness and mortality consequent on poverty, than from the prudence and foresight which restrain the frequency and universality of early marriages. The checks are more of the positive and less of the preventive kind.

The first grand requisite to the growth of prudential habits is the perfect security of property; and the next perhaps is that respectability and importance, which are given to the lower classes by equal laws, and the possession of some influence in the framing of them. The more excellent therefore is the government, the more does it tend to generate that prudence and elevation of sentiment, by which alone in the present state of our being poverty can be avoided.

It has been sometimes asserted that the only reason why it is advantageous that the people should have some share in the government, is that a representation of the people tends best to secure the framing of good and equal laws; but that, if the same object could be attained under a despotism, the same advantage would accrue to the community. If however the representative system, by securing to the lower classes of society a more equal and liberal mode of treatment from their superiors, gives to each individual a greater personal respectability, and a greater fear of personal degradation; it is evident that it will powerfully co-operate with the security of property in animating the exertions of industry, and in generating habits of prudence; and thus more powerfully tend to increase the riches and prosperity of the lower classes of the community, than if the same laws had existed under a despotism.

But though the tendency of a free constitution and a good government to diminish poverty be certain; yet their effect in this way must necessarily be indirect and slow, and very different from the direct and immediate relief, which the lower classes of people are too frequently in the habit of looking forward to as the consequence of a revolution. This habit of expecting too much, and the irritation occasioned by disappointment, continually give a wrong direction to their efforts in favour of liberty, and constantly tend to defeat the accomplishment of those gradual reforms in government, and that slow melioration of the condition of the lower classes of society, which are really attainable. It is of the very highest importance therefore, to know distinctly what government cannot do, as well as what it can. If I were called upon to name the cause, which, in my conception, had more than any other contributed to the very slow progress of freedom, so disheartening to every liberal mind, I should say that it was the confusion that had existed respecting the causes of the unhappiness and discontents which prevail in society; and the advantage which governments had been able to take, and indeed had been compelled to take, of this confusion, to confirm and strengthen their power. I cannot help thinking therefore, that a knowledge generally circulated, that the principal cause of want and unhappiness is only indirectly connected with government, and totally beyond its power directly to remove, and that it depends upon the conduct of the poor themselves, would, instead of giving any advantage to governments, give a great additional weight to the popular side of the question, by removing the dangers with which from ignorance it is at present accompanied, and thus tend, in a very powerful manner, to promote the cause of rational freedom.

4.2.7 Book IV, Chapter VII: Continuation of the Same Subject.[97]

The reasonings of the foregoing chapter have been strikingly confirmed by the events of the last two or three years. Perhaps there never was a period when more erroneous views were formed by the lower classes of society of the effects to be expected from reforms in the

---

[97] Written in 1817.

government, when these erroneous views were more immediately founded on a total misapprehension of the principal cause of poverty, and when they more directly led to results unfavourable to liberty.

One of the main causes of complaint against the government has been, that a considerable number of labourers, who are both able and willing to work, are wholly out of employment, and unable consequently to command the necessaries of life. That this state of things is one of the most afflicting events that can occur in civilized life, that it is a natural and pardonable cause of discontent among the lower classes of society, and that every effort should be made by the higher classes to mitigate it, consistently with a proper care not to render it permanent, no man of humanity can doubt. But that such a state of things may occur in the best-conducted and most economical government that ever existed is as certain, as that governments have not the power of commanding with effect the resources of a country to be progressive, when they are naturally stationary or declining.

It will be allowed that periods of prosperity may occur in any well-governed state, during which an extraordinary stimulus may be given to its wealth and population, which cannot in its nature be permanent. If, for instance, new channels of trade are opened, new colonies are possessed, new inventions take place in machinery, and new and great improvements are made in agriculture, it is quite obvious that while the markets at home and abroad will readily take off at advantageous prices the increasing produce, there must be a rapid increase of capital, and an unusual stimulus given to the population. On the other hand, if subsequently these channels of trade are either closed by accident, or contracted by foreign competition; if colonies are lost, or the same produce is supplied from other quarters; if the markets, either from glut or competition, cease to extend with the extension of the new machinery; and if the improvements in agriculture from any cause whatever cease to be progressive, it is as obvious that, just at the time when the stimulus to population has produced its greatest effect, the means of employing and supporting this population may, in the natural course of things, and without any fault whatever in the government, become deficient. This failure must unavoidably produce great distress among the labouring classes of society; but it is quite clear that no inference can be drawn from this distress, that a radical change is required in the government; and the attempt to accomplish such a change might only aggravate the evil.

It has been supposed in this case, that the government has in no respect by its conduct contributed to the pressure in question, a supposition which in practice perhaps will rarely be borne out by the fact. It is unquestionably in the power of a government to produce great distress by war and taxation, and it requires some skill to distinguish the distress which is the natural result of these causes, from that which is occasioned in the way just described. In our own case unquestionably both descriptions of causes have combined, but the former in a greater degree than the latter. War and taxation, as far as they operate directly and simply, tend to destroy or retard the progress of capital, produce, and population; but during the late war these checks to prosperity have been much more than overbalanced by a combination of circumstances which has given an extraordinary stimulus to production. That for this overbalance of advantages the country cannot be considered as much indebted to the government, is most certain. The government during the last twenty-five years has shewn no very great love either of peace or liberty; and no particular economy in the use of the national resources. It has proceeded in a very straight-forward manner to spend great sums in war, and to raise them by very heavy taxes. It has no doubt done its part towards the dilapidation of the national resources. But still the broad fact must stare every impartial observer in the face, that at the end of the war in 1814 the national resources were not dilapidated; and that not only were the wealth and population of the country considerably greater than they were at the commencement of the war, but that they had increased in the interval at a more rapid rate than was ever experienced before.

Perhaps this may justly be considered as one of the most extraordinary facts in history; and it certainly follows from it, that the sufferings of the country since the peace have not been occasioned so much by the usual and most natural effects to be expected from war and taxation, as by the sudden ceasing of an extraordinary stimulus to production, the distresses consequent upon which, though increased no doubt by the weight of taxation, do not essentially arise from it, and are not directly, therefore, and immediately, to be relieved by its removal.

That the labouring classes of society should not be fully aware that the main causes of their distress are, to a certain extent and for a certain time, irremediable, is natural enough; and that they should listen much more readily and willingly to those who confidently promise immediate relief, rather than to those who can only tell them unpalatable truths, is by no means surprising. But it must be allowed that full advantage has been taken by the popular orators and writers of a crisis which has given them so much power. Partly from ignorance, and partly from design, every thing that could tend to enlighten the labouring classes as to the real nature of their situation, and encourage them to bear an unavoidable pressure with patience, has been either sedulously kept out of their view, or clamorously reprobated; and every thing that could tend to deceive them, to aggravate and encourage their discontents, and to raise unreasonable and extravagant expectations as to the relief to be expected from reform, has been as sedulously brought forward. If under these circumstances the reforms proposed had been accomplished, it is impossible that the people should not have been most cruelly disappointed; and under a system of universal suffrage and annual parliaments, a general disappointment of the people would probably lead to every sort of experiment in government, till the career of change was stopped by a military despotism. The warmest friends of genuine liberty might justly feel alarmed at such a prospect. To a cause conducted upon such principles, and likely to be attended with such results, they could not of course, consistently with their duty, lend any assistance. And if with great difficulty, and against the sense of the great mass of petitioners, they were to effect a more moderate and more really useful reform, they could not but feel certain that the unavoidable disappointment of the people would be attributed to the half-measures which had been pursued; and that they would be either forced to proceed to more radical changes, or submit to a total loss of their influence and popularity by stopping short while the distresses of the people were unrelieved, their discontents unallayed, and the great *panacea* on which they had built their sanguine expectations untried.

These considerations have naturally paralyzed the exertions of the best friends of liberty; and those salutary reforms which are acknowledged to be necessary in order to repair the breaches of time, and improve the fabric of our constitution, are thus rendered much more difficult, and consequently much less probable.

But not only have the false expectations and extravagant demands suggested by the leaders of the people given an easy victory to government over every proposition for reform, whether violent or moderate, but they have furnished the most fatal instruments of offensive attack against the constitution itself. They are naturally calculated to excite some alarm, and to check moderate reform; but alarm, when once excited, seldom knows where to stop, and the causes of it are particularly liable to be exaggerated. There is reason to believe that it has been under the influence of exaggerated statements, and of inferences drawn by exaggerated fears from these statements, that acts unfavourable to liberty have been passed without an adequate necessity. But the power of creating these exaggerated fears, and of passing these acts, has been unquestionably furnished by the extravagant expectations of the people. And it must be allowed that the present times furnish a very striking illustration of the doctrine, that an ignorance of the principal cause of poverty is peculiarly unfavourable, and that a knowledge of it must be peculiarly favourable, to the cause of civil liberty.

4.2.8 Book IV, Chapter VIII: Plan of the Gradual Abolition of the Poor Laws Proposed.

If the principles in the preceding chapters should stand the test of examination, and we should ever feel the obligation of endeavouring to act upon them, the next inquiry would be, in what way we ought practically to proceed. The first grand obstacle which presents itself in this country is the system of the poor-laws, which has been justly stated to be an evil, in comparison of which the national debt, with all its magnitude of terror, is of little moment.[98] The rapidity with which the poor's rates have increased of late years presents us indeed with the prospect of such an extraordinary proportion of paupers in the society, as would seem to be incredible in a nation flourishing in arts, agriculture and commerce, and with a government which has generally been allowed to be the best that has hitherto stood the test of experience.[99]

Greatly as we may be shocked at such a prospect, and ardently as we may wish to remove it, the evil is now so deeply seated, and the relief given by the poor-laws so widely extended, that no man of humanity could venture to propose their immediate abolition. To

[98] Reports of the Society for bettering the Condition of the Poor, vol. iii. p. 21.

[99] If the poor's rates continue increasing as rapidly as they have done on the average of the last ten years, how melancholy are our future prospects! The system of the poor-laws bus been justly stated by the French to be *la plaie politique de l'Angleterre la plus devorante.* (Comité de Mendicité.)

mitigate their effects however and stop their future increase, to which, if left to continue upon their present plan, we can see no probable termination, it has been proposed to fix the whole sum to be raised at its present rate, or any other that might be determined upon; and to make a law, that on no account this sum should be exceeded. The objection to this plan is, that a very large sum would be still to be raised, and a great number of people to be supported; the consequence of which would be, that the poor would not be easily able to distinguish the alteration that had been made. Each individual would think that he had as good a right to be supported when he was in want as any other person; and those who unfortunately chanced to be in distress, when the fixed sum had been collected, would think themselves particularly ill used on being excluded from all assistance, while so many others were enjoying this advantage. If the sum collected were divided among all that were in want, however their numbers might increase, though such a plan would not be so unfair with regard to those who became dependent after the sum had been fixed, it would undoubtedly be very hard upon those who had been in the habit of receiving a more liberal supply, and had done nothing to justify its being taken from them: and in both cases it would certainly be unjust in the society to undertake the support of the poor, and yet, if their numbers increased, to feed them so sparingly, that they must necessarily die of hunger and disease.

I have reflected much on the subject of the poor-laws, and hope therefore that I shall be excused in venturing to suggest a mode of their gradual abolition, to which I confess that at present I can see no material objection. Of this indeed I feel nearly convinced, that, should we ever become so fully sensible of the wide-spreading tyranny, dependence, indolence and unhappiness which they create, as seriously to make an effort to abolish them, we shall be compelled by a sense of justice to adopt the principle, if not the plan, which I shall mention. It seems impossible to get rid of so extensive a system of support, consistently with humanity, without applying ourselves directly to its vital principle, and endeavouring to counteract that deeply-seated cause which occasions the rapid growth of all such establishments, and invariably renders them inadequate to their object.

As a previous step even to any considerable alteration in the present system, which would contract or stop the increase of the relief to be given, it appears to me that we are bound in justice and honour formally to disclaim the *right* of the poor to support.

To this end, I should propose a regulation to be made, declaring, that no child born from any marriage, taking place after the expiration of a year from the date of the law, and no illegitimate child born two years from the same date, should ever be entitled to parish assistance. And to give a more general knowledge of this law, and to enforce it more strongly on the minds of the lower classes of people, the clergyman of each parish should, after the publication of banns, read a short address, stating the strong obligation on every man to support his own children; the impropriety, and even immorality, of marrying without a prospect of being able to do this; the evils which had resulted to the poor themselves from the attempt which had been made to assist by public institutions in a duty which ought to be exclusively appropriated to parents; and the absolute necessity which had at length appeared of abandoning all such institutions, on account of their producing effects totally opposite to those which were intended.

This would operate as a fair, distinct and precise notice, which no man could well mistake; and, without pressing hard on any particular individuals, would at once throw off the rising generation from that miserable and helpless dependence upon the government and the rich, the moral as well as physical consequences of which are almost incalculable.

After the public notice which I have proposed had been given, and the system of poor-laws had ceased with regard to the rising generation, if any man chose to marry, without a prospect of being able to support a family, he should have the most perfect liberty so to do. Though to marry, in this case, is, in my opinion, clearly an immoral act, yet it is not one which society can justly take upon itself to prevent or punish; because the punishment provided for it by the laws of nature falls directly and most severely upon the individual who commits the act, and through him, only more remotely and feebly, on the society. When nature will govern and punish for us, it is a very miserable ambition to wish to snatch the rod from her hands, and draw upon ourselves the odium of executioner. To the punishment therefore of nature he should be left, the punishment of want. He has erred in the face of a most clear and precise warning, and can have no just reason to complain of any person but himself when he feels the consequences of his error. All parish assistance should be denied him; and he should be left to the uncertain support of private charity. He should be taught to know, that the laws of nature, which are the laws of God, had doomed him and his family to suffer for disobeying their repeated admonitions; that he had no claim of *right* on society for the smallest portion of food, beyond that which his labour would fairly purchase; and that if he and his family were saved from feeling the natural consequences of his imprudence, he would owe it to the pity of some kind benefactor, to whom, therefore, he ought to be bound by the strongest ties of gratitude.

If this system were pursued, we need be under no apprehensions that the number of persons in extreme want would be beyond the power and the will of the benevolent to supply. The sphere for the exercise of private charity would, probably, not be greater than it is at present; and the principal difficulty would be, to restrain the hand of benevolence from assisting those in, distress in so indiscriminate a manner as to encourage indolence and want of foresight in others.

With regard to illegitimate children, after the proper notice had been given, they should not be allowed to have any claim to parish assistance, but be left entirely to the support of private charity. If the parents desert their child, they ought to be made answerable for the crime. The infant is, comparatively speaking, of little value to the society, as others will immediately supply its place. Its principal value is on account of its being the object of one of the most delightful passions in human nature—parental affection. But if this value be disregarded by those who are alone in a capacity to feel it, the society cannot be called upon to put itself in their place; and has no further business in its protection than to punish the crime of desertion or intentional ill treatment in the persons whose duty it is to provide for it.

At present the child is taken under the protection of the parish,[100] and generally dies, at least in London, within the first year. The loss to the society is the same; but the crime is diluted by the number of people concerned, and the death passes as a visitation of Providence, instead of being considered as the necessary consequence of the conduct of its parents, for which they ought to be held responsible to God and to society.

The desertion of both parents, however, is not so common as the desertion of one. When a servant or labouring man has an illegitimate child, his running away is perfectly a matter of course; and it is by no means uncommon for a man who has a wife and large family to withdraw into a distant county, and leave them to the parish; indeed I once heard a hard-working good sort of man propose to do this, as the best mode of providing for a wife and six children.[101] If the simple fact of these frequent desertions were related in some countries, a strange inference would be drawn against the English character; but the wonder would cease when our public institutions were explained.

By the laws of nature, a child is confided directly and exclusively to the protection of its parents. By the laws of nature, the mother of a child is confided almost as strongly and exclusively to the man who is the father of it. If these ties were suffered to remain in the state in which nature has left them, and the man were convinced that the woman and the child depended solely upon him for support, I scarcely believe that there are ten men breathing so atrocious as to desert them. But our laws, in opposition to the laws of nature, say, that if the parents forsake their child, other persons will undertake to support it; or, if the man forsake the woman, that she shall still meet with protection elsewhere; that is, we take all possible pains to weaken and render null the ties of nature, and then say that men are unnatural. But the fact is, that the society itself, in its body politic, is the unnatural character, for framing laws that thus counteract the laws of nature, and give premiums to the violation of the best and most honourable feelings of the human heart.

It is a common thing in most parishes, when the father of an illegitimate child can be seized, to endeavour to frighten him into marriage by the terrors of a jail; but such a proceeding cannot surely be too strongly reprobated. In the first place, it is a most shallow policy in the parish officers; for, if they succeed, the effect upon the present system will generally be, that of having three or four children to provide for, instead of one. And in the next place it is difficult to conceive a more gross and scandalous profanation of a religious ceremony. Those who believe that the character of a woman is restored by such a forced engagement, or that the moral worth of the man is enhanced by affirming a lie before God, have, I confess, very different ideas of delicacy and morality from those which I have been taught to consider as just. If a man deceive a woman into a connexion with him under a

[100] I fully agree with Sir F. M. Eden, in thinking that the constant public support which deserted children receive is the cause of their very great numbers in the two most opulent countries of Europe, France and England. State of the Poor, vol. i. p. 339.

[101] "That many of the poorer classes of the community avail themselves of the liberality of the law, and leave their wives and children on the parish, the reader will find abundant proof in the subsequent part of this work." Sir F. M. Eden on the State of the Poor, vol. i. p. 339.

promise of marriage, he has undoubtedly been guilty of a most atrocious act, and there are few crimes which merit a more severe punishment; but the last that I should choose is that which will oblige him to affirm another falsehood, which will probably render the woman that he is to be joined to miserable, and will burden the society with a family of paupers.

The obligation on every man to support his children, whether legitimate or illegitimate, is so clear and strong, that it would be just to arm society with any power to enforce it, which would be likely to answer the purpose. But I am inclined to believe that no exercise of the civil power, however rigorous, would be half so effectual as a knowledge generally circulated, that children were in future to depend solely for support upon their parents, and would be left only to casual charity if they were deserted.

It may appear to be hard that a mother and her children, who have been guilty of no particular crime themselves, should suffer for the ill conduct of the father; but this is one of the invariable laws of nature; and, knowing this, we should think twice upon the subject, and be very sure of the ground on which we go, before we presume *systematically* to counteract it.

I have often heard the goodness of the Deity impeached on account of that part of the Decalogue in which he declares that he will visit the sins of the father upon the children; but the objection has not perhaps been sufficiently considered. Without a most complete and fundamental change in the whole constitution of human nature; without making man an angel, or at least something totally different from what he is at present; it seems absolutely necessary that such a law should prevail. Would it not require a perpetual miracle, which is perhaps a contradiction in terms, to prevent children from being affected in their moral and civil condition by the conduct of their parents? What man is there, that has been brought up by his parents, who is not at the present moment enjoying something from their virtues, or suffering something from their vices; who, in his moral character, has not been elevated in some degree by their prudence, their justice, their benevolence, their temperance, or depressed by the contraries; who, in his civil condition, has not been raised by their reputation, their foresight, their industry, their good fortune, or lowered by their want of character, their imprudence, their indolence, and their adversity? And how much does a knowledge of this transmission of blessings contribute to excite and invigorate virtuous exertion? Proceeding upon this certainty, how ardent and incessant are the efforts of parents to give their children a good education, and to provide for their future situation in the world! If a man could neglect or desert his wife and children without their suffering any injury, how many individuals there are, who, not being very fond of their wives, or being tired of the shackles of matrimony, would withdraw from household cares and difficulties, and resume their liberty and independence as single men! But the consideration that children may suffer for the faults of their parents has a strong hold even upon vice; and many who are in such a state of mind, as to disregard the consequence of their habitual course of life, as far as relates to themselves, are yet greatly anxious that their children should not suffer from their vices and follies. In the moral government of the world, it seems evidently necessary, that the sins of the fathers should be visited upon the children; and if in our overweening vanity we imagine, that we can govern a private society better by endeavouring *systematically* to counteract this law, I am inclined to believe, that we shall find ourselves very greatly mistaken.

If the plan which I have proposed were adopted, the poor's rates in a few years would begin very rapidly to decrease, and in no great length of time would be completely extinguished; and yet, as far as it appears to me at present, no individual would be either deceived or injured, and consequently no person could have a just right to complain.

The abolition of the poor-laws however is not of itself sufficient; and the obvious answer to those who lay too much stress on this system is, to desire them to look at the state of the poor in some other countries where such laws do not prevail, and to compare it with their condition in England. But this comparison, it must be acknowledged, is in many respects unfair; and would by no means decide the question of the utility or inutility of such a system. England possesses very great natural and political advantages, in which perhaps the countries, that we should in this case compare with her, would be found to be palpably deficient. The nature of her soil and climate is such, that those almost universal failures in the crops of grain, which are known in some countries, never occur in England. Her insular situation and extended commerce are peculiarly favourable for importation. Her numerous manufactures employ nearly all the hands that are not engaged in agriculture, and afford the means of a regular distribution of the annual produce of the land and labour to the whole of her inhabitants. But, above all, throughout a very large class of people, a decided taste for the conveniences and comforts of life, a strong desire of bettering their condition (that master-spring of public prosperity,) and, in consequence, a most laudable spirit of industry and foresight, are observed to prevail. These dispositions, so contrary to the hopeless indolence remarked in despotic countries, are generated by the constitution of the English government, and the excellence of its laws, which secure to every individual the produce of his industry. When, therefore, on a comparison with other countries, England appears to have the advantage in the state of her poor, the superiority is entirely to be attributed to these favourable circumstances, and not to the poor-laws. A woman with one bad feature, may greatly excel in beauty some other, who may have this individual feature tolerably good; but it would be rather strange to assert, in consequence, that the superior beauty of the former was occasioned by this particular deformity. The poor-laws have constantly tended to counteract the natural and acquired advantages of this country. Fortunately these disadvantages have been so considerable, that though weakened they could not be overcome; and to these advantages, together with the checks to marriage, which the laws themselves create, it is owing that England has been able to bear up so long against this pernicious system. Probably there is not any other country in the world, except perhaps Holland before the revolution, which could have acted upon it so completely for the same period of time, without utter ruin.

It has been proposed by some to establish poor-laws in Ireland; but from the depressed state of the common people, there is little reason to doubt, that, on the establishment of such laws, the whole of the landed property would very soon be absorbed, or the system be given up in despair.

In Sweden, from the dearths which are not unfrequent, owing to the general failure of crops in an unpropitious climate and the impossibility of great importations in a poor country, an attempt to establish a system of parochial relief such as that in England (if it were not speedily abandoned from the physical impossibility of executing it) would level the property of the kingdom from one end to the other, and convulse the social system in such a manner, as absolutely to prevent it from recovering its former state on the return of plenty.

Even in France, with all her advantages of situation and climate, the tendency to population is so great, and the want of foresight among the lower classes of the people so remarkable, that if poor-laws were established, the landed property would soon sink under the burden, and the wretchedness of the people at the same time be increased. On these considerations the committee *de Mendicité,* at the beginning of the revolution, very properly and judiciously rejected the establishment of such a system, which had been proposed.

The exception of Holland, if it were an exception, would arise from very particular circumstances—her extensive foreign trade and her numerous colonial emigrations, compared with the smallness of her territory, together with the extreme unhealthiness of a great part of the country, which occasions a much greater average mortality than is common in other states. These, I conceive, were the unobserved causes which principally contributed to render Holland so famous for the management of her poor, and able to employ and support all who applied for relief.

No part of Germany is sufficiently rich to support an extensive system of parochial relief; but I am inclined to think, that from the absence of it the lower classes of the people, in some parts of Germany, are in a better situation than those of the same class in England. In Switzerland, for the same reason, their condition, before the late troubles, was perhaps universally superior: and in a journey through the duchies of Holstein and Sleswick belonging to Denmark, the houses of the lower classes of people appeared to me to be neater and better, and in general there were fewer indications of poverty and wretchedness among them than among the same ranks in this country.

Even in Norway, notwithstanding the disadvantage of a severe and uncertain climate, from the little that I saw in a few weeks' residence in the country, and the information that I could collect from others, I am inclined to think that the poor are, on the average, better off than in England. Their houses and clothing are often superior; and though they have no white bread, they have much more meat, fish and milk than our labourers; and I particularly remarked, that the farmers' boys were much stouter and healthier looking lads than those of the same description in England. This degree of happiness, superior to what could be expected from the soil and climate, arises almost exclusively from the degree in which the preventive check to population operates. The establishment of a system of poor-laws, which would destroy this check, would at once sink the lower classes of the people into a state of the most miserable poverty and wretchedness; would diminish their industry, and consequently the produce of the land and labour of the country; would weaken the resources of ingenuity in times of scarcity; and ultimately involve the country in all the horrors of continual famines.

If, as in Ireland, Spain, and many countries of the more southern climates, the people are in so degraded a state, as to propagate their species without regard to consequences, it matters little whether they have poor-laws or not. Misery in all its various forms must be the predominant check to their increase. Poor-laws, indeed, will always tend to aggravate the evil, by diminishing the general resources of the country; and in such a state of things can exist only for a very short time; but with or without them, no stretch of human ingenuity and exertion can rescue the people from the most extreme poverty and wretchedness.

4.2.9 Book IV, Chapter IX: Of the Modes of Correcting the Prevailing Opinions on Population.

It is not enough to abolish all the positive institutions which encourage population; but we must endeavour, at the same time, to correct the prevailing opinions which have the same, or perhaps even a more powerful effect. This must necessarily be a work of time; and can only be done by circulating juster notions on these subjects in writings and conversation; and by endeavouring to impress as strongly as possible on the public mind, that it is not the duty of man simply to propagate his species, but to propagate virtue and happiness; and that, if he has not a tolerably fair prospect of doing this, he is by no means called upon to leave descendants.

Among the higher ranks of society, we have not much reason to apprehend the too great frequency of marriage. Though the circulation of juster notions on this subject might, even in this part of the community, do much good, and prevent many unhappy marriages; yet whether we make particular exertions for this purpose or not, we may rest assured that the degree of proper pride and spirit of independence almost invariably connected with education and a certain rank in life will secure the operation of the prudential check to marriage to a considerable extent. All that the society can reasonably require of its members is, that they should not have families without being able to support them. This may be fairly enjoined as a positive duty. Every restraint beyond this must be considered as a matter of choice and taste; but from what we already know of the habits which prevail among the higher ranks of life, we have reason to think that little more is wanted to attain the object required, than to award a greater degree of respect and of personal liberty to single women, and to place them nearer upon a level with married women;—a change, which, independently of any particular purpose in view, the plainest principles of equity seem to demand.

If, among the higher classes of society, the object of securing the operation of the prudential check to marriage to a sufficient degree appear to be attainable without much difficulty, the obvious mode of proceeding with the lower classes of society, where the point is of the principal importance, is to endeavour to infuse into them a portion of that knowledge and foresight, which so much facilitates the attainment of this object in the educated part of the community.

The fairest chance of accomplishing this end would probably be by the establishment of a system of parochial education upon a plan similar to that proposed by Adam Smith.[102] In addition to the usual subjects of instruction, and those which he has mentioned, I should be disposed to lay considerable stress on the frequent explanation of the real state of the lower classes of society, as affected by the principle of population, and their consequent dependence on themselves for the chief part of their happiness or misery. It would be by no means necessary or proper in these explanations to underrate, in the smallest degree, the desirableness of marriage. It should always be represented as, what it really is, a state peculiarly suited to the nature of man, and calculated greatly to advance his happiness and remove the temptations to vice; but, like property or any other desirable object, its advantages should be shewn to be unattainable, except under certain conditions. And a strong conviction in a young man of the great desirableness of marriage, with a conviction at the same time that the power of supporting a family was the only condition which would enable him really to enjoy its blessings, would be the most effectual motive imaginable to industry and sobriety before marriage, and would powerfully urge him to save that superfluity of income which single labourers necessarily possess, for the accomplishment of a rational and desirable object, instead of dissipating it, as is now usually done, in idleness and vice.

If in the course of time a few of the simplest principles of political economy could be added to the instructions given in these schools, the benefit to society would be almost incalculable.[103] In some conversations with labouring men, during the late

[102]Wealth of Nations, vol. iii. b. v. c. i. p. 187.

[103]Adam Smith proposes, that the elementary parts of geometry and mechanics should be taught in these parish schools; and I cannot help thinking, that the common principles by which markets are regulated might be made sufficiently clear, to be of considerable use. It is certainly a subject that, as it interests the lower classes of people very nearly, would be likely to attract their attention. At the same time it must be confessed, that it is impossible to be in any degree sanguine on this point, recollecting how very ignorant in general the educated part of the community is of these principles. If, however, political economy cannot be taught to the common people, I really think that it ought to form a branch of university education. Scotland has set us an example in this respect, which we ought not to be so slow to imitate. It is of the utmost importance, that the gentlemen of the country, and particularly the clergy, should not from ignorance aggravate the evils of scarcity, every time that it unfortunately occurs. During the late dearths, half of the gentlemen and clergymen in the kingdom richly deserved to have been prosecuted for sedition. After inflaming the minds of the common people against the farmers and corn dealers, by the manner in which they talked of them or preached about them, it was but a feeble antidote to the poison which they had infused, coldly to observe that, however the poor might be oppressed or cheated, it was their duty to keep the peace. It was little better than Antony's repeated declaration, that the conspirators were all honourable men; which did not save either their houses or their persons from the attacks of the mob. Political economy is perhaps the only science, of which it may be said that the ignorance of it is not merely a deprivation of good, but produces great positive evil.

[1825.] This note was written in 1803; and it is particularly gratifying to me, at the end of the year 1825, to see that what I stated as so desirable twenty-two years ago, seems to be now on the eve of its

scarcities,[104] I confess that I was to the last degree disheartened, at observing their inveterate prejudices on the subject of grain; and I felt very strongly the almost absolute incompatibility of a government really free with such a degree of ignorance. The delusions are of such a nature, that, if acted upon, they must at all events be repressed by force; and it is extremely difficult to give such a power to the government as will be sufficient at all times for this purpose, without the risk of its being employed improperly, and endangering the liberty of the subject.

We have lavished immense sums on the poor, which we have every reason to think have constantly tended to aggravate their misery. But in their education and in the circulation of those important political truths that most nearly concern them, which are perhaps the only means in our power of really raising their condition, and of making them happier men and more peaceable subjects, the have been miserably deficient. It is surely a great national disgrace, that the education of the lower classes of people in England should be left merely to a few Sunday schools, supported by a subscription from individuals, who can give to the course of instruction in them any kind of bias which they please. And even the improvement of Sunday schools (for, objectionable as they are in some points of view, and imperfect in all, I cannot but consider them as an improvement) is of very late date.[105]

The arguments which have been urged against instructing the people appear to me to be not only illiberal, but to the last degree feeble; and they ought, on the contrary, to be extremely forcible, and to be supported by the most obvious and striking necessity, to warrant us in withholding the means of raising the condition of the lower classes of people, when they are, in our power. Those who will not listen to any answer to these arguments drawn from theory, cannot, I think, refuse the testimony of experience; and I would ask, whether the advantage of superior instruction which the lower classes of people in Scotland are known to possess, has appeared to have any tendency towards creating a spirit of tumult and discontent amongst them. And yet, from the natural inferiority of its soil and climate, the pressure of want is more constant, and the dearths are not only more frequent, but more dreadful than in England. In the case of Scotland, the knowledge circulated among the common people, though not sufficient essentially to better their condition by increasing, in an adequate degree, their habits of prudence and foresight, has yet the effect of making them bear with patience the evils which they suffer, from being aware of the folly and inefficacy of turbulence. The quiet and peaceable habits of the instructed Scotch peasant,

accomplishment. The increasing attention which in the interval has been paid generally to the science of political economy; the lectures which have been given at Cambridge, London, and Liverpool; the chair which has lately been established at Oxford; the projected University in the Metropolis; and, above all, the Mechanics Institution, open the fairest prospect that, within a moderate period of time, the fundamental principles of political economy will, to a very useful extent, be known to the higher, middle, and a most important portion of the working classes of society in England.

[104] 1800 and 1801.

[105] Written in 1803.

compared with the turbulent disposition of the ignorant Irishman, ought not to be without effect upon every impartial reasoner.

The principal argument which I have heard advanced against a system of national education in England is, that the common people would be put in a capacity to read such works as those of Paine, and that the consequences would probably be fatal to government. But on this subject I agree most cordially with Adam Smith[106] in thinking, that an instructed and well-informed people would be much less likely to be led away by inflammatory writings, and much better able to detect the false declamation of interested and ambitious demagogues, than an ignorant people. One or two readers in a parish are sufficient to circulate any quantity of sedition; and if these be gained to the democratic side, they will probably have the power of doing much more mischief, by selecting the passages best suited to their hearers, and choosing the moments when their oratory is likely to have the most effect, than if each individual in the parish had been in a capacity to read and judge of the whole work himself; and at the same time to read and judge of the opposing arguments, which we may suppose would also reach him.

But in addition to this, a double weight would undoubtedly be added to the observation of Adam Smith, if these schools were made the means of instructing the people in the real nature of their situation; if they were taught, what is really true, that without an increase of their own industry and prudence no change of government could essentially better their condition; that, though they might get rid of some particular grievance, yet in the great point of supporting their families they would be but little, or perhaps not at all benefited; that a revolution would not alter in their favour the proportion of the supply of labour to the demand, or the quantity of food to the number of the consumers; and that if the supply of labour were greater than the demand, and the demand for food greater than the supply, they might suffer the utmost severity of want, under the freest, the most perfect, and best executed government, that the human imagination could conceive.

A knowledge of these truths so obviously tends to promote peace and quietness, to weaken the effect of inflammatory writings and to prevent all unreasonable and ill-directed opposition to the constituted authorities, that those who would still object to the instruction of the people may fairly be suspected of a wish to encourage their ignorance, as a pretext for tyranny, and an opportunity of increasing the power and the influence of the executive government.

Besides explaining the real situation of the lower classes of society, as depending principally upon themselves for their happiness or misery, the parochial schools would, by early instruction and the judicious distribution of rewards, have the fairest chance of training up the rising generation in habits of sobriety, industry, independence and prudence, and in a proper discharge of their religious duties; which would raise them from their present degraded state, and approximate them, in some degree, to the middle classes of society, whose habits, generally speaking, are certainly superior.

[106]Wealth of Nations, vol. iii. b. v. c. i. p. 193.

In most countries, among the lower classes of people, there appears to be something like a standard of wretchedness, a point below which they will not continue to marry and propagate their species. This standard is different in different countries, and is formed by various concurring circumstances of soil, climate, government, degree of knowledge, and civilization, &c. The principal circumstances which contribute to raise it are liberty, security of property, the diffusion of knowledge, and a taste for the conveniences and the comforts of life. Those which contribute principally to lower it are despotism and ignorance.

In an attempt to better the condition of the labouring classes of society, our object should be to raise this standard as high as possible, by cultivating a spirit of independence, a decent pride, and a taste for cleanliness and comfort. The effect of a good government in increasing the prudential habits and personal respectability of the lower classes of society has already been insisted on; but certainly this effect will always be incomplete without a good system of education; and, indeed, it may be said that no government can approach to perfection, that does not provide for the instruction of the people. The benefits derived from education are among those, which may be enjoyed without restriction of numbers; and, as it is in the power of governments to confer these benefits, it is undoubtedly their duty to do it.

4.2.10 Book IV, Chapter X: Of the Direction of Our Charity.

An important and interesting inquiry yet remains, relating to the mode of directing our private charity, so as not to interfere with the great object in view, of meliorating the condition of the labouring classes of people, by preventing the population from pressing too hard against the limits of the means of subsistence.

The emotion which prompts us to relieve our fellow-creatures in distress is, like all our other natural passions, general, and, in some degree, indiscriminate and blind. Our feelings of compassion may be worked up to a higher pitch by a well-wrought scene in a play, or a fictitious tale in a novel, than by almost any events in real life: and if among ten petitioners we were to listen only to the first impulses of our feelings without making further inquiries, we should undoubtedly give our assistance to the best actor of the party. It is evident, therefore, that the impulse of benevolence, like the impulses of love, of anger, of ambition, the desire of eating and drinking, or any other of our natural propensities, must be regulated by experience, and frequently brought to the test of utility, or it will defeat its intended purpose.

The apparent object of the passion between the sexes is the continuation of the species, and the formation of such an intimate union of views and interests between two persons as will best promote their happiness, and, at the name time, secure the proper degree of attention to the helplessness of infancy and the education of the rising generation; but if every man were to obey at all times the impulses of nature in the gratification of this passion, without regard to consequences, the principal part of these important objects would not be attained, and even the continuation of the species might be defeated by a promiscuous intercourse.

The apparent end of the impulse of benevolence is, to draw the whole human race together, but more particularly that part of it which is of our own nation and kindred, in the bonds of brotherly love; and by giving men an interest in the happiness and misery of their fellow-creatures, to prompt them, as they have power, to mitigate the partial evils arising from general laws, and thus to increase the sum of human happiness; but if our benevolence be indiscriminate, and the degree of apparent distress be made the sole measure of our liberality, it is evident that it will be exercised almost exclusively upon common beggars, while modest unobtrusive merit, struggling with unavoidable difficulties, yet still maintaining some slight appearances of decency and cleanliness, will be totally neglected. We shall raise the worthless above the worthy; we shall encourage indolence and check industry; and in the most marked manner subtract from the sum of human happiness.

Our experience has, indeed, informed us that the impulse of benevolence is not so strong as the passion between the sexes and that, generally speaking, there is much less danger to be apprehended from the indulgence of the former than of the latter; but independently of this experience and of the moral codes founded upon it, we should be as much justified in a general indulgence of the former passion as in following indiscriminately every impulse of our benevolence. They are both natural passions, excited by their appropriate objects, and to the gratification of which we are prompted by the pleasurable sensations which accompany them. As animals, or till we know their consequences, our only business is to follow these dictates of nature; but, as reasonable beings, we are under the strongest obligations to attend to their consequences; and if they be evil to ourselves or others, we may justly consider it as an indication, that such a mode of indulging these passions is not suited to our state or conformable to the will of God. As moral agents, therefore, it is clearly our duty to restrain their indulgence in these particular directions; and by thus carefully examining the consequences of our natural passions, and frequently bringing them to the test of utility, gradually to acquire a habit of gratifying them only in that way, which, being unattended with evil, will clearly add to the sum of human happiness, and fulfil the apparent purpose of the Creator.

Though utility, therefore, can never be the immediate excitement to the gratification of any passion, it is the test by which alone we can know, independently of the revealed will of God, whether it ought or ought not to be indulged; and is therefore the surest criterion of moral rules which can be collected from the light of nature. All the moral codes, which have inculcated the subjection of the passions to reason, have been, as I conceive, really built upon this foundation, whether the promulgators of them were aware of it or not.

I remind the reader of these truths, in order to apply them to the habitual direction of our charity; and if we keep the criterion of utility constantly in view, we may find ample room for the exercise of our benevolence, without interfering with the great purpose which we have to accomplish.

One of the most valuable parts of charity is its effect upon the giver. It is more blessed to give than to receive. Supposing it to be allowed that the exercise of our benevolence in acts of charity is not, upon the whole, really beneficial to the poor; yet we could never sanction any endeavour to extinguish an impulse, the proper gratification of which has so evident a tendency to purify and exalt the human mind. But it is particularly satisfactory and pleasing to find that the mode of exercising our charity, which, when brought to the test of utility, will appear to be most beneficial to the poor, is precisely that, which will have the best and most improving effect on the mind of the donor.

The quality of charity, like that of mercy, is not strained; It droppeth as the gentle rain from Heav'n

Upon the earth beneath,"

The immense sums distributed to the poor in this country by the parochial laws are improperly called charity. They want its most distinguishing attribute; and, as might be expected from an attempt to force that which loses its essence the moment it ceases to be voluntary, their effects upon those from whom they are collected are as prejudicial as on those to whom they are distributed. On the side of the receivers of this miscalled charity, instead of real relief, we find accumulated distress and more extended poverty; on the side of the givers, instead of pleasurable sensations, unceasing discontent and irritation.

In the great charitable institutions supported by voluntary contributions, some of which are certainly of a prejudicial tendency, the subscriptions, I am inclined to fear, are sometimes given grudgingly, and rather because they are expected by the world from certain stations and certain fortunes, than because they are prompted by motives of genuine benevolence; and as the greater part of the subscribers do not interest themselves in the management of the funds or in the fate of the particular objects relieved, it is not to be expected that this kind of charity should have any strikingly beneficial influence on the minds of the majority who exercise it.

Even in the relief of common beggars, we shall find that we are often as much influenced by the desire of getting rid of the importunities of a disgusting object, as by the pleasure of relieving it. We wish that it had not fallen in our way, rather than rejoice in the opportunity given us of assisting a fellow-creature. We feel a painful emotion at the sight of so much apparent misery; but the pittance we give does not relieve it. We know that it is totally inadequate to produce any essential effect. We know besides, that we shall be addressed in the same manner at the corner of the next street; and we know that we are liable to the grossest impositions. We hurry therefore sometimes by such objects, and shut our ears to their importunate demands. We give no more than we can help giving without doing actual violence to our feelings. Our charity is in some degree forced; and, like forced charity, it leaves no satisfactory impression on the mind, and cannot therefore have any very beneficial and improving effect on the heart and affections.

But it is far otherwise with that voluntary and active charity, which makes itself acquainted with the objects which it relieves; which seems to feel, and to be proud of the bond that unites the rich with the poor; which enters into their hourses, informs itself not only of their wants, but of their habits and dispositions; checks the hopes of clamorous and obtrusive poverty, with no other recommendation but rags; and encourages, with adequate relief, the silent and retiring sufferer, labouring under unmerited difficulties. This mode of exercising our clarity presents a very different picture from that of any other; and its contrast with the common mode of parish relief cannot be better described than in the words of Mr. Townsend, in the conclusion of his admirable dissertation on the Poor-Laws. "Nothing in nature can be more disgusting than a parish pay-table, attendant upon which, in the same objects of misery, are too often found combined, snuff, gin, rags, vermin, insolence, and abusive language; nor in nature can any thing be more beautiful than the mild complacency of benevolence hastening to the humble cottage to relieve the wants of industry and virtue, to feed the hungry, to clothe the naked, and to soothe the sorrows of the widow with her tender orphans; nothing can be more pleasing, unless it be their sparkling eyes, their bursting tears, and their uplifted hands, the artless expressions of unfeigned gratitude for unexpected favours. Such scenes will frequently occur, whenever men shall have power to dispose of their own property."

I conceive it to be almost impossible that any person could be much engaged in such scenes without daily making advances in virtue. No exercise of our affections can have a more evident tendency to purify and exalt the human mind. It is almost exclusively this species of charity that blesseth him that gives; and, in a general view, it is almost exclusively this species of charity which blesseth him that takes; at least it may be asserted that there are but few other modes of exercising our charity, in which large sums can be distributed, without a greater chance of producing evil than good.

The discretionary power of giving or withholding relief; which is, to a certain extent, vested in parish officers and justices, is of a very different nature, and will have a very different effect, from the discrimination which may be exercised by voluntary charity. Every man in this country, under certain circumstances, is entitled by law to parish assistance; and unless his disqualification is clearly proved, has a right to complain if it be withheld. The inquiries necessary to settle this point, and the extent of the relief to be granted, too often produce evasion and lying on the part of the petitioner, and afford an opening to partiality and oppression in the overseer. If the proposed relief be given, it is of course received with unthankfulness; and, if it be denied, the party generally thinks himself severely aggrieved, and feels resentment and indignation at his treatment.

In the distribution of voluntary charity nothing of this kind can take place. The person who receives it is made the proper subject of the pleasurable sensation of gratitude; and those who do not receive it cannot possibly conceive themselves in the slightest degree injured. Every man has a right to do what he will with his own, and cannot, in justice, be called upon to render a reason why he gives in the one case, and abstains from it in the other. This kind of despotic power, essential to voluntary charity, gives the greatest facility to the selection of worthy objects of relief, without being accompanied by any ill consequences; and has further a most beneficial effect from the degree of uncertainty which must necessarily be attached to it. It is in the highest degree important to the general happiness of the poor, that no man should look to charity as a fund on which he may confidently depend. He should be taught that his own exertions, his own industry and foresight, are his only just ground of dependence; that if these fail, assistance in his distresses can only be the subject of rational hope; and that even the foundation of this hope will depend in a considerable degree on his own good conduct and the consciousness that he has not involved himself in these difficulties by his indolence or imprudence.

That in the distribution of our charity we are under a strong moral obligation to inculcate this lesson on the poor by a proper discrimination, is a truth of which I cannot feel a doubt. If all could be completely relieved, and poverty banished from the country, even at the expense of three-fourths of the fortunes of the rich, I would be the last person to say a single syllable against relieving all, and making the degree of distress alone the measure of our bounty. But as experience has proved, I believe, without a single exception, that poverty and misery have always increased in proportion to the quantity of indiscriminate charity, are we not bound to infer, reasoning as we usually do from the laws of nature, that it is an intimation that such a mode of distribution is not the proper office of benevolence?

The laws of nature say, with St. Paul, "If a man will not work, neither shall he eat." They also say that he is not rashly to trust to Providence. They appear indeed to be constant and uniform for the express purpose of telling him what he is to trust to, and that, if he marry without a reasonable prospect of supporting a family, he must expect to suffer want. These intimations appear from the constitution of human nature to be absolutely necessary, and to have a strikingly beneficial tendency. If in the direction either of our public or our private charity we say that though a man will not work, yet he shall eat; and though he marry without being able to support a family, yet his family shall be supported; it is evident that we do not merely endeavour to mitigate the partial evils arising from general laws, but *regularly* and *systematically* to counteract the obviously beneficial effects of these general laws themselves. And we cannot easily conceive, that the Deity should implant any passion in the human breast for such a purpose.

In the great course of human events, the best-founded expectations will sometimes be disappointed; and industry, prudence, and virtue not only fail of their just reward, but be involved in unmerited calamities. Those who are thus suffering in spite of the best-directed endeavours to avoid it, and from causes which they could not be expected to foresee, are the genuine objects of charity. In relieving these, we exercise the appropriate office of benevolence, that of mitigating the partial evils arising from general laws; and in this direction of our charity therefore we need not apprehend any ill consequences. Such objects ought to be relieved, according to our means, liberally and adequately, even though the worthless were in much more severe distress.

When indeed this first claim on our benevolence was satisfied, we might then turn our attention to the idle and improvident; but the interests of human happiness most clearly require, that the relief which we afford them should not be abundant. We may perhaps take upon ourselves, with great caution, to mitigate the punishments which they are suffering from the laws of nature, but on no account to remove them entirely. They are deservedly at the bottom in the scale of society; and if we raise them from this situation, we not only palpably defeat the end of benevolence, but commit a most glaring injustice to those who are above them. They should on no account be enabled to command so much of the necessaries of life as can be obtained by the wages of common labour.

It is evident that these reasonings do not apply to those cases of urgent distress arising from disastrous accidents, unconnected with habits indolence and improvidence. If a man break a leg or an arm, we are not to stop to inquire into his moral character before we lend him our assistance; but in this case we are perfectly consistent, and the touchstone of utility completely justifies our conduct. By affording the most indiscriminate assistance in this way, we are in little danger of encouraging people to break their arms and legs. According to the touchstone of utility, the high approbation which Christ gave to the conduct of the good Samaritan, who followed the immediate impulse of his benevolence in relieving a stranger in the urgent distress of an accident, does not, in the smallest degree, contradict the expression of St. Paul, "If a man will not work, neither shall he eat."

We are not, however, in any case, to lose a present opportunity of doing good, from the mere supposition that we may meet possibly with a worthier object. In all doubtful cases it may safely be laid down as our duty to follow the natural impulse of our benevolence; but when, in fulfilling our obligations as reasonable beings to attend to the consequences of our actions, we have, from our own experience and that of others, drawn the conclusion that the exercise of our benevolence in one mode is prejudicial, and in another is beneficial in its effects; we are certainly bound, as moral agents, to check our natural propensities in the one direction, and to encourage them and acquire the habits of exercising them in the other.

4.2.11 Book IV, Chapter XI: Different Plans of Improving the Condition of the Poor Considered.

In the distribution of our charity, or in any efforts which we may make to better the condition of the lower classes of society, there is another point relating to the main argument of this work, to which we must be particularly attentive. We must on no account do any thing which tends directly to encourage marriage, or to remove, in any regular and systematic manner, that inequality of circumstances which ought always to exist between the single man and the man with a family. The writers who have best understood the principle of population appear to me all to have fallen into very important errors on this point.

Sir James Steuart, who was fully aware of what he calls vicious procreation, and of the misery that attends a redundant population, recommends, notwithstanding, the general establishment of foundling hospitals; the taking of children under certain circumstances from their parents, and supporting them at the expense of the state; and particularly laments the inequality of condition between the married and single man, so ill proportioned to their respective wants.[107] He forgets, in these instances, that if, without the encouragement to multiplication of foundling hospitals, or of public support for the children of some married persons, and under the discouragement of great pecuniary disadvantages on the side of the

[107] Political Economy, vol. i. b. i. c. xiii.

married man, population be still redundant, which is evinced by the inability of the poor to maintain all their children; it is a clear proof that the funds destined for the maintenance of labour cannot properly support a greater population; and that, if further encouragements to multiplication be given, and discouragements removed, the result must be, an increase somewhere or other of that vicious procreation, which he so justly reprobates.

Mr. Townsend, who in his Dissertation on the Poor Laws has treated this subject with great skill and perspicuity, appears to me to conclude with a proposal, which violates the principles on which he had reasoned so well. He wishes to make the benefit clubs, or friendly societies, which are now voluntarily established in many parishes, compulsory and universal; and proposes as a regulation that an unmarried man should pay a fourth part of his wages, and a married man with four children not more than a thirtieth part.[108]

I must first remark that the moment these subscriptions are made compulsory, they will necessarily operate exactly like a direct tax upon labour, which, as Adam Smith justly states, will always be paid, and in a more expensive manner, by the consumer. The landed interest therefore would receive little relief from this plan, but would pay the same sum as at present, only in the advanced price of labour and of commodities, instead of in the parish rates. A compulsory subscription of this kind would have almost all the bad effects of the present system of relief, and, though altered in name, would still possess the essential spirit of the poor-laws.

Dean Tucker, in some remarks on a plan of the same kind, proposed by Mr. Pew, observed that, after much talk and reflection on the subject, he had come to the conclusion, that they must be voluntary associations, and not compulsory assemblies. A voluntary subscription is like a tax upon a luxury, and does not necessarily raise the price of labour.

It should be recollected, also, that in a voluntary association of a small extent, over which each individual member can exercise a superintendence, it is highly probable that the original agreements will all be strictly fulfilled, or, if they be not, every man may at least have the redress of withdrawing himself from the club. But in an universal compulsory subscription, which must necessarily become a national concern, there would be no security whatever for the fulfilment of the original agreements; and when the funds failed, which they certainly would do, when all the idle and dissolute were included, instead of some of the most industrious and provident, as at present, a larger subscription would probably be demanded, and no man would have the right to refuse it. The evil would thus go on increasing as the poor-rates do now. If indeed the assistance given were always specific, and on no account to be increased, as in the present voluntary associations, this would certainly be a striking advantage; but the same advantage might be completely attained by a similar distribution of the sums collected by the parish rates. On the whole, therefore, it appears to me that, if the friendly societies were made universal and compulsory, it would be merely a different mode of collecting parish rates; and any particular mode of distribution might be as well adopted upon one system as upon the other.

[108] Dissertation on the Poor-Laws, p. 89, 2d edit. 1787.

With regard to the proposal of making single men pay a fourth part of their earnings weekly, and married men with families only a thirtieth part, it would evidently operate as a heavy fine upon bachelors, and a high bounty upon children; and is therefore directly adverse to the general spirit in which Mr. Townsend's excellent dissertation is written. Before he introduces this proposal, he lays it down as a general principle, that no system for the relief of the poor can be good, which does not regulate population by the demand for labour;[109] but this proposal clearly tends to encourage population without any reference to the demand for labour, and punishes a young man for his prudence in refraining from marriage, at a time, perhaps, when this demand may be so small, that the wages of labour are totally inadequate to the support of a family. I should be averse to any compulsory system whatever for the poor; but certainly if single men were compelled to pay a contribution for the future contingencies of the married state, they ought in justice to receive a benefit proportioned to the period of their privation; and the man who had contributed a fourth of his earnings for merely one year, ought not to be put upon a level with him who had contributed this proportion for ten years.

Mr. Arthur Young, in most of his works, appears clearly to understand the principle of population, and is fully aware of the evils which must necessarily result from an increase of people beyond the demand for labour, and the means of comfortable subsistence. In his Tour through France he has particularly laboured this point, and shewn most forcibly the misery which results in that country from the excess of population occasioned by the too great division of property. Such an increase he justly calls merely a multiplication of wretchedness. "Couples marry and procreate on the idea, not the reality, of a maintenance; they increase beyond the demand of towns and manufactures; and the consequence is, distress, and numbers dying of diseases arising from insufficient nourishment."[110]

In another place he quotes a very sensible passage from the Report of the Committee of Mendicity, which, alluding to the evils of overpopulation, concludes thus, "If faudroit enfin nécessairement que le prix de travail baissât par la plus grande concurrence de travailleurs, d'orésulteroit une indigence complette pour ceux qui ne trouveroient pas de travail, et une subsistence incomplette pour ceux mêmes auxquels il ne seroit pas réfusé." And in remarking upon this passage, he observes, "France itself affords an irrefragable proof of the truth of these sentiments; for I am clearly of opinion, from the observations I made in every province of the kingdom, that her population is so much beyond the proportion of her industry and labour, that she would be much more powerful and infinitely more flourishing, if she had five or six millions less of inhabitants. From her too great population she presents in every quarter such spectacles of wretchedness, as are absolutely inconsistent with that degree of national felicity which she was capable of attaining, even under the old government. A traveller much less attentive than I was to objects of this kind must see at every turn most unequivocal signs of distress. That these should exist, no one can wonder, who

[109]p. 84.

[110]Travels in France, vol. i. c. xii. p. 408.

considers the price of labour and of provisions, and the misery into which a small rise in the price of wheat throws the lower classes.[111]

"If you would see," he says, "a district with as little distress in it as is consistent with the political system of the old government of France, you must assuredly go where there are no little proprietors at all. You must visit the great farms in Beauce, Picardy, part of Normandy and Artois, and there you will find no more population than what is regularly employed, and regularly paid; and if in such districts you should, contrary to this rule, meet with much distress, it is twenty to one but that it is in a parish, which has some commons which tempt the poor to have cattle—to have property—and in consequence misery. When you are engaged in this political tour, finish it by seeing England, and I will shew you a set of peasants well clothed, well nourished, tolerably drunken from superfluity, well lodged, and at their ease; and yet amongst them, not one in a thousand has either land or cattle."[112] A little farther on, alluding to encouragements to marriage, he says of France, "The predominant evil of the kingdom is the having so great a population, that she can neither employ nor feed it; why then encourage marriage? Would you breed more people, because you have more already than you know what to do with? You have so great a competition for food, that your people are starving or in misery; and you would encourage the production of more, to increase that competition. It may almost be questioned, whether the contrary policy ought not to be embraced; whether difficulties should not be laid on the marriage of those who cannot make it appear that they have the prospect of maintaining the children that shall be the fruit of it? But why encourage marriages, which are sure to take place in all situations in which they ought to take place? There is no instance to be found of plenty of regular employment being first established, where marriages have not followed in a proportionate degree. The policy therefore, at best, is useless, and may be pernicious."

After having once so clearly understood the principle of population, as to express these and many other sentiments on the subject, equally just and important, it is not a little surprising to find Mr. Young, in a pamphlet, entitled *The Question of Scarcity plainly stated, and Remedies considered (published in* 1800,) observing, that "the means which would of all others perhaps tend most surely to prevent future scarcities so oppressive to the poor as the present, would be to secure to every country labourer in the kingdom, that has three children and upwards, half an acre of land for potatoes; and grass enough to feed one or two cows.[113] *** If each had his ample potatoe-ground and a cow, the price of wheat would be of little more consequence to them than it is to their brethren in Ireland."

"Every one admits the system to be good, but the question is how to enforce it."

I was by no means aware that the excellence of the system had been so generally admitted. For myself I strongly protest against being included in the general term of *every*

[111] Travels in France, vol. i. c. xvii. p. 469.

[112] Travels in France, vol. i. c. xvii. p. 471.

[113] P. 77.

one, as I should consider the adoption of this system as the most cruel and fatal blow to the happiness of the lower classes of people in this country that they had ever received.

Mr. Young, however, goes on to say, that "The magnitude of the object should make us disregard any difficulties, but such as are insuperable: none such would probably occur, if something like the following means were resorted to.

"I. Where there are common pastures, to give to a labouring man having children, a right to demand an allotment proportioned to the family, to be set out by the parish officers, &c., *** and a cow bought. Such labourer to have both for life, paying 40*s.* a year till the price of the cow, &c. was reimbursed: at his death to go to the labourer having the most numerous family, for life, paying shillings a week to the widow of his predecessor."

"II. Labourers thus demanding allotments by reason of their families to have land assigned and cows bought, till the proportion so allotted amounts to one of the extent of the common."

"III. In parishes where there are no commons, and the quality of the land adequate, every cottager having children, to whose cottage there is not within a given time land sufficient for a cow, and half an acre of potatoes, assigned at a fair average rent, subject to appeal to the sessions, to have a right to demand shillings per week of the parish for every child, till such land be assigned; leaving to landlords and tenants the means of doing it. Cows to be found by the parish under an annual reimbursement."[114]

"The great object is, by means of milk and potatoes, to take the mass of the country poor front the consumption of wheat, and to give them substitutes equally wholesome and nourishing, and as independent of scarcities, natural and artificial, as the providence of the Almighty will admit."[115]

Would not this plan operate, in the most direct manner, as an encouragement to marriage and a bounty on children, which Mr. Young has with so much justice reprobated in his travels in France? and does he seriously think that it would be an eligible thing to feed the mass of the people in this country on milk and potatoes, and make them as independent of the price of corn and demand for labour as their brethren in Ireland?

The specific cause of the poverty and misery of the lower classes of people in France and Ireland is, that from the extreme subdivision of property in the one country, and the facility of obtaining a cabin and potatoes in the other, a population is brought into existence, which is not demanded by the quantity of capital and employment in the country; and the consequence of which must therefore necessarily be, as is very justly expressed in the Report of the Committee of Mendicity before mentioned, to lower in general the price of labour by too great competition; from which must result complete indigence to those who cannot find employment, and an incomplete subsistence even to those who can.

---

[114]P. 78.

[115]P. 79.

The obvious tendency of Mr. Young's plan is, by encouraging marriage and furnishing a cheap food, independent of the price of corn, and of course of the demand for labour, to place the lower classes of people exactly in this situation.

It may perhaps be said, that our poor-laws at present regularly encourage marriage and children, by distributing relief in proportion to the size of families; and that this plan, which is proposed as a substitute, would merely do the same thing in a less objectionable manner. But surely, in endeavouring to get rid of the evil of the poor-laws, we ought not to retain their most pernicious quality; and Mr. Young must know as well as I do, that the principal reason why poor-laws have invariably been found ineffectual in the relief of the poor is, that they tend to encourage a population, which is not regulated by the demand for labour. Mr. Young himself, indeed, expressly takes notice of this effect in England, and observes that, notwithstanding the unrivalled prosperity of her manufactures, "population is sometimes too active, as we see clearly by the dangerous increase of poor's rates in country villages."[116]

But the fact is, that Mr. Young's plan would be incomparably more powerful in encouraging a population beyond the demand for labour, than our present poor-laws. A laudable repugnance to the receiving of parish relief, arising partly from a spirit of independence not yet extinct, and partly from the disagreeable mode in which the relief is given, undoubtedly deters many from marrying with a certainty of falling on the parish; and the proportion of births and marriages to the whole population, which has before been noticed, clearly proves that the poor-laws do not encourage marriage so much as might be expected from theory. But the case would be very different, if, when a labourer had an early marriage in contemplation, the terrific forms of workhouses and parish officers, which might disturb his resolution, were to be exchanged for the fascinating visions of land and cows. If the love of property, as Mr. Young has repeatedly said, will make a man do much, it would be rather strange if it would not make him marry; an action to which, it appears from experience, that he is by no means disinclined.

The population, which would be thus called into being, would be supported by the extended cultivation of potatoes, and would of course go on without any reference to the demand for labour. In the present state of things, notwithstanding the flourishing condition of our manufactures and the numerous chocks to our population, there is no practical problem so difficult, as to find employment for the poor; but this difficulty would evidently be aggravated a hundred fold, under the circumstances here supposed.

In Ireland, or in any other country, where the common food is potatoes, and every man who wishes to marry may obtain a piece of ground sufficient, when planted with this root, to support a family, prizes may be given till the treasury is exhausted for essays on the best means of employing the poor; but till some stop to the progress of population naturally

[116]Travels in France, vol. i. c. xvii. p. 470.

arising from this state of things takes place, the object in view is really a physical impossibility.[117]

Mr. Young has intimated, that, if the people were fed upon milk and potatoes, they would be more independent of scarcities than at present; but why this should be the case I really cannot comprehend. Undoubtedly people who live upon potatoes will not be much affected by a scarcity of wheat; but is there any contradiction in the supposition of a failure in the crops of potatoes? I believe it is generally understood that they are more liable to suffer damage during the winter than grain. From the much greater quantity of food yielded by a given piece of land when planted with potatoes, than under any other kind of cultivation, it would naturally happen that, for some time after the introduction of this root as the general food of the labouring classes of people, a greater quantity would be grown than was demanded, and they would live in plenty. Mr. Young, in his Travels through France, observes, that, "In districts which contain immense quantities of waste land of a certain degree of fertility, as in the roots of the Pyrenees, belonging to communities ready to sell them, economy and industry, animated with the views of settling and marrying, flourish greatly; in such neighbourhoods something like an American increase takes place, and if the land be cheap little distress is found. But as procreation goes on rapidly under such circumstances, the least check to subsistence is attended with great misery; as wastes becoming dearer, or the best portions being sold, or difficulties arising in the acquisition; all which circumstances I met with in those mountains. The moment that any impediment happens, the distress of such a people will be proportioned to the activity and vigour, which had animated population."[118]

This description will apply exactly to what would take place in this country, on the distribution of small portions of land to the common people, and the introduction of potatoes as their general food. For a time the change might appear beneficial, and of course the idea of property would make it, at first, highly acceptable to the poor; but, as Mr. Young in another place says, "You presently arrive at the limit, beyond which the earth, cultivate it as you please, will feed no more mouths; yet those simple manners, which instigate to marriage still continue; what then is the consequence, but the most dreadful misery imaginable?"[119]

When the commons were all divided, and difficulties began to occur in procuring potatoe-grounds, the habit of early marriages, which had been introduced, would occasion the most complicated distress; and when, from the increasing population, and diminishing

---

[117] Mr. Crumpe's Prize Essay on the best means of finding employment for the people is an excellent treatise, and contains most valuable information; but till the capital of the country is better proportioned to its population, it is perfectly chimerical to expect success in any project of the kind. I am also strongly disposed to believe that the indolent and turbulent habits of the lower Irish can never be corrected, while its potatoe system enables them to increase so much beyond the regular demand for labour.

[118] Travels in France, vol. i. c. xvii. p. 409.

[119] Travels in France, vol. i. c. xvii. p. 409.

sources of subsistence, the average growth of potatoes was not more than the average consumption, a scarcity of potatoes would be, in every respect, as probable as a scarcity of wheat at present; and, when it did arrive, it would be beyond all comparison more dreadful.

When the common people of a country live principally upon the dearest grain, as they do in England on wheat, they have great resources in a scarcity; and barley, oats, rice, cheap soups and potatoes, all present themselves as less expensive, yet at the same time wholesome means of nourishment; but when their habitual food is the lowest in this scale, they appear to be absolutely without resource, except in the bark of trees, like the poor Swedes; and a great portion of them must necessarily be starved.

The wages of labour will always be regulated mainly by the proportion of the supply of labour to the demand. And as, upon the potatoe system, a supply more than adequate to the demand would very soon take place, and this supply might be continued at a very cheap rate, on account of the cheapness of the food which would furnish it, the common price of labour would soon be regulated principally by the price of potatoes instead of the price of wheat, as at present; and the rags and wretched cabins of Ireland would follow of course.

When the demand for labour occasionally exceeds the supply, and wages are regulated by the price of the dearest grain, they will generally be such as to yield something besides mere food, and the common people may be able to obtain decent houses and decent clothing. If the contrast between the state of the French and English labourers, which Mr. Young has drawn, be in any degree near the truth, the advantage on the side of England has been occasioned precisely and exclusively by these two circumstances; and if, by the adoption of milk and potatoes as the general food of the common people, these circumstances were totally altered, so as to make the supply of labour constantly in a great excess above the demand for it, and regulate wages by the price of the cheapest food, the advantage would be immediately lost, and no efforts of benevolence could prevent the most general and abject poverty.

Upon the same principle it would by no means be eligible that the cheap soups of Count Rumford should be adopted as the general food of the common people. They are excellent inventions for the public institutions, and as occasional resources; but if they were once universally adopted by the poor, it would be impossible to prevent the price of labour from being regulated by them; and the labourer, though at first he might have more to spare for other expenses, besides food, would ultimately have much less to spare than before.

The desirable thing, with a view to the happiness of the common people, seems to be, that their habitual food should be dear, and their wages regulated by it; but that, in a scarcity, or other occasional distress, the cheaper food should be readily and cheerfully adopted.[120] With a view of rendering this transition easier, and at the same time of making an useful distinction between those who are dependent on parish relief and those who are

[120] It is certainly to be wished that every cottage in England should have a garden to it well stocked with vegetables. A little variety of food is in every point of view highly useful. Potatoes are undoubtedly a most valuable assistance, though I should be very sorry ever to see them the principal dependence of our labourers.

not, I should think that one plan, which Mr. Young proposes, would be extremely eligible. This is, "to pass an act prohibiting relief, so far as subsistence is concerned, in any other manner than by potatoes, rice and soup; not merely as a measure of the moment, but permanently."[121] I do not think that this plan would necessarily introduce these articles as the common food of the lower classes; and if it merely made the transition to them in periods of distress easier, and at the same time drew a more marked line than at present between dependence and independence, it would have a very beneficial effect.

As it is acknowledged that the introduction of milk and potatoes, or of cheap soups, as the general food of the lower classes of people, would lower the price of labour, perhaps some cold politician might propose to adopt the system, with a view of underselling foreigners in the markets of Europe. I should not envy the feelings which could suggest such a proposal. I really cannot conceive any thing much more detestable than the idea of knowingly condemning the labourers of this country to the rags and wretched cabins of Ireland, for the purpose of selling a few more broad cloths and calicoes.[122] The wealth and power of nations are, after all, only desirable as they contribute to happiness. In this point of view, I should be very far from undervaluing them, considering them, in general, as absolutely necessary means to attain the end; but if any particular case should occur, where they appear to be in direct opposition to each other, we cannot rationally doubt which ought to be preferred.

---

[121] Question of Scarcity, &c. p. 80. This might be done, at least with regard to workhouses. In assisting the poor at their own homes, it might be subject to some practical difficulties.

[122] In this observation I have not the least idea of alluding to Mr. Young, who, I firmly believe, ardently wishes to improve the condition of the lower classes of people; though I do not think that his plan would effect the object in view. He either did not see those consequences which I apprehend from it; or he has a better opinion of the happiness of the common people in Ireland than I have. In his his Irish Tour he seemed much struck with the plenty of potatoes which they possessed, and the absence of all apprehension of want. Had he travelled in 1800 and 1801, his impressions would by all accounts have been very different. From the facility which has hitherto prevailed in Ireland of procuring potatoe-grounds, scarcities have certainly been rare, and all the effects of the system have not yet been felt, though certainly enough to make it appear very far from desirable.

Mr. Young has since pursued his idea more in detail, in a pamphlet entitled, *An Inquiry into the Propriety of applying Wastes to the better Maintenance and Support of the Poor.* But the impression on my mind is still the same; and it appears to be calculated to assimilate the condition of the labourers of this country to that of the lower classes of the Irish. Mr. Young seems, in a most unaccountable manner, to have forgotten all his general principles on this subject. He has treated the question of a provision for the poor, as if it was merely, How to provide in the cheapest and best manner for a *given number* of people. If this had been the sole question, it would never have taken so many hundred years to resolve. But the real question is, How to provide for those who are in want, in such a manner as to present a continual accumulation of their numbers? and it will readily occur to the reader, that a plan of giving them land and cows cannot promise much success in this respect. If, after all the commons had been divided, the poor-laws were still to continue in force, no good reason can be assigned why the rates should not in a few years be as high us they are at present, independently of all that had been expended in the purchase of land and stock.

Fortunately, however, even on the narrowest political principles, the adoption of such a system would not answer. It has always been observed that those, who work chiefly on their own property, work very indolently and unwillingly when employed for others; and it must necessarily happen, when, from the general adoption of a very cheap food, the population of a country increases considerably beyond the demand for labour, that habits of idleness and turbulence will be generated, most peculiarly unfavourable to a flourishing state of manufactures. In spite of the cheapness of labour in Ireland, there are few manufactures which can be prepared in that country for foreign sale so cheap as in England: and this is in a great measure owing to the want of those industrious habits which can only be produced by regular employment.

4.2.12 Book IV, Chapter XII: Continuation of the Same Subject.[123]

The increasing portion of the society which has of late years become either wholly or partially dependent upon parish assistance, together with the increasing burden of the poor's rates on the landed property, has for some time been working a gradual change in the public opinion respecting the benefits resulting to the labouring classes of society, and to society in general, from a legal provision for the poor. But the distress which has followed the peace of 1814, and the great and sudden pressure which it has occasioned on the parish rates, have accelerated this change in a very marked manner. More just and enlightened views on the subject are daily gaining ground; the difficulties attending a legal provision for the poor are better understood, and more generally acknowledged; and opinions are now seen in print, and heard in conversation, which twenty years ago would almost have been considered as treason to the interests of the state.

This change of public opinion, stimulated by the severe pressure of the moment, has directed an unusual portion of attention to the subject of the poor-laws; and as it is acknowledged that the present system has essentially failed, various plans have been proposed either as substitutes or improvements. It may be useful to inquire shortly how far the plans which have already been published are calculated to accomplish the ends which they propose. It is generally thought that some measure of importance will be the result of the present state of public opinion. To the permanent success of any such measure, it is absolutely necessary that it should apply itself in some degree to the real source of the difficulty. Yet there is reason to fear, that, notwithstanding the present improved knowledge on the subject, this point may be too much overlooked.

Among the plans which appear to have excited a considerable degree of the public attention, is one of Mr. Owen. I have already adverted to some views of Mr. Owen in a chapter on Systems of Equality, and spoken of his experience with the respect which is justly due to it. If the question were merely how to accommodate, support and train, in the

---

[123]Written in 1817.

best manner, societies of 1200 people, there are perhaps few persons more entitled to attention than Mr. Owen: but in the plan which he has proposed, he seems totally to have overlooked the nature of the problem to be solved. This problem is, *How to provide for those who are in want, in such a manner as to prevent a continual increase of their numbers, and of the proportion which they bear to the whole society.* And it must be allowed that Mr. Owen's plan not only does not make the slightest approach towards accomplishing this object, but seems to be peculiarly calculated to effect an object exactly the reverse of it, that is, to increase and multiply the number of paupers.

If the establishments which he recommends could really be conducted according to his apparent intentions, the order of nature and the lessons of providence would indeed be in the most marked manner reversed; and the idle and profligate would be placed in a situation which might justly be the envy of the industrious and virtuous. The labourer or manufacturer who is now ill lodged and ill clothed, and obliged to work twelve hours a day to maintain his family, could have no motive to continue his exertions, if the reward for slackening them, and seeking parish assistance, was good lodging, good clothing, the maintenance and education of all his children, and the exchange of twelve hours hard work in an unwholesome manufactory for four or five hours of easy agricultural labour on a pleasant farm. Under these temptations, the numbers yearly falling into the new establishments from the labouring and manufacturing classes, together with the rapid increase by procreation of the societies themselves, would very soon render the first purchases of land utterly incompetent to their support. More land must then be purchased and fresh settlements made; and if the higher classes of society were bound to proceed in the system according to its apparent spirit and intention, there cannot be a doubt that the whole nation would shortly become a nation of paupers with a community of goods.

Such a result might not perhaps be alarming to Mr. Owen. It is just possible indeed that he may have had this result in contemplation when he proposed this plan, and have thought that it was the best mode of quietly introducing that community of goods which he believes is necessary to complete the virtue and happiness of society. But to those who totally dissent from him as to the effects to be expected from a community of goods; to those who are convinced that even his favourite doctrine, that a man can be trained to produce more than he consumes, which is no doubt true at present, may easily cease to be true, when cultivation is pushed beyond the bounds prescribed to it by private property;[124] the approaches towards a system of this kind will be considered as approaches towards a system of universal indolence, poverty, and wretchedness.

Upon the supposition, then, that Mr. Owen's plan could be effectively executed, and that the various pauper societies scattered over the country could at first be made to realize his most sanguine wishes, such might be expected to be their termination in a moderately short time, from the natural and necessary action of the principle of population.

[124] See vol. ii, c. x. b. iii. p. 154.

But it is probable that the other grand objection to all systems of common propriety would even at the very outset confound the experience of Mr. Owen, and destroy the happiness to which he looks forward. In the society at the Lanerk Mills, two powerful stimulants to industry and good conduct are in action, which would be totally wanting in the societies proposed. At Lanerk, the whole of every man's earnings is his own; and his power of maintaining himself, his wife and children, in decency and comfort, will be in exact proportion to his industry, sobriety, and economy. At Lanerk, also, if any workman be perseveringly indolent and negligent, if he get drunk and spoil his work, or if in any way he conduct himself essentially ill, he not only naturally suffers by the diminution of his earnings, but he may at any time be turned off, and the society be relieved from the influence and example of a profligate and dangerous member. On the other hand, in the pauper establishments proposed in the present plan, the industry, sobriety, and good conduct of each individual, would be very feebly indeed connected with his power of maintaining himself and family comfortably; and in the case of persevering idleness and misconduct, instead of the simple and effective remedy of dismission, recourse must be had to a system of direct punishment of some kind or other, determined and enforced by authority, which is always painful and distressing, and generally inefficient.

I confess it appears to me that the most successful experience in such an establishment as that of Lanerk, furnishes no ground whatever to say what could be done towards the improvement of society in an establishment where the produce of all the labour employed would go to a common stock, and dismissal, from the very nature and object of the institution, would be impossible, if under such disadvantages the proper management of these establishments were within the limits of possibility, what judgment, what firmness, what patience, would be required for the purpose! But where are such qualities to be found in sufficient abundance to manage one or two millions of people?

On the whole, then, it may be concluded, that Mr. Owen's plan would have to encounter obstacles that really appear to be insuperable, even at its first outset; and that if these could by any possible means be overcome, and the most complete success attained, the system would, without some most unnatural and unjust laws to prevent the progress of population, lead to a state of universal poverty and distress, in which, though all the rich might be made poor, none of the poor could be made rich,—not even so rich as a common labourer at present.

The plan for bettering the condition of the labouring classes of the community, published by Mr. Curwen, is professedly a slight sketch: but principles, not details, are what it is our present object to consider; and the principles on which he would proceed are declared with sufficient distinctness, when he states the great objects of his design to be,

1. Meliorating the present wretched condition of the lower orders of the people.
2. Equalizing by a new tax the present poor's rates, which *must* be raised for their relief.
3. And giving to all those who may think proper to place themselves under its protection, a voice in the local management and distribution of the fund destined for their support.

The first proposition is, of course, or ought to he, the object of every plan proposed. And the two last may be considered as the modes by which it is intended to accomplish it.

But it is obvious that these two propositions, though they may be both desirable on other accounts, not only do not really touch, but do not even propose to touch, the great problem. We wish to check the increase and diminish the proportion of paupers, in order to give greater wealth, happiness, and independence to the mass of the labouring classes. But the equalization of the poor's rates, simply considered, would have a very strong tendency to increase rather than to diminish the number of the dependent poor. At present the parochial rates fall so very heavily upon one particular species of property, that the persons, whose business it is to allow them, have in general a very strong interest indeed to keep them low; but if they fell equally on all sorts of property, and particularly if they were collected from large districts, or from counties, the local distributors would have comparatively but very feeble motives to reduce them, and they might be expected to increase with great rapidity.

It may be readily allowed, however, that the peculiar weight with which the poor's rates press upon land is essentially unfair. It is particularly hard upon some country parishes, where the births greatly exceed the deaths, owing to the constant emigrations which are taking place to towns and manufactories, that, under any circumstances, a great portion of these emigrants should be returned upon them when old, disabled, or out of work. Such parishes may be totally without the power of furnishing either work or support for all the persons born within their precincts. In fact, the same number would not have been born in them unless these emigrations had taken place. And it is certainly hard, therefore, that parishes so circumstanced should be obliged to receive and maintain all who may return to them in distress. Yet, in the present state of the country, the most pressing evil is not the weight upon the land, but the increasing proportion of paupers. And, as the equalization of the rates would certainly have a tendency to increase this proportion, I should be sorry to see such a measure introduced, even if it were easily practicable, unless accompanied by some very strong and decisive limitations to the continued increase of the rates so equalized.

The other proposition of Mr. Curwen will, in like manner, be found to afford no security against the increase of pauperism. We know perfectly well that the funds of the friendly societies, as they are at present constituted, though managed by the contributors themselves, are seldom distributed with the economy necessary to their permanent efficiency; and in the national societies proposed, as a considerable part of the fund would be derived from the poor's rates, there is certainly reason to expect that every question which could be influenced by the contributors would be determined on principles still more indulgent, and less economical.

On this account it may well be doubted, whether it would ever be advisable to mix any public money, derived from assessments, with the subscriptions of the labouring classes. The probable result would be, that in the case of any failure in the funds of such societies, arising from erroneous calculations and too liberal allowances, it would be expected that the whole of the deficiency should be made up by the assessments. And any rules which might have been made to limit the amount applied in this way would probably be but a feeble barrier against claims founded on a plan brought forward by the higher classes of society.

Another strong objection to this sort of union of parochial and private contributions is, that from the first the members of such societies could not justly feel themselves independent. If one half or one third of the fund were to be subscribed from the parish, they would stand upon a very different footing from the members of the present benefit-clubs. While so considerable a part of the allowances to which they might be entitled in sickness or in age would really come from the poor's rates, they would be apt to consider the plan as what, in many respects, it really would be,—only a different mode of raising the rates. If the system were to become general, the contributions of the labouring classes would have nearly the effects of a tax on labour, and such a tax has been generally considered as more unfavourable to industry and production than most other taxes.

The best part of Mr. Curwen's plan is that which proposes to give a credit to each contributor in proportion to the amount of his contributions, and to make his allowance in sickness, and his annuity in old age, dependent upon this amount; but this object could easily be accomplished without the objectionable accompaniments. It is also very properly observed, that "want of employment must furnish no claims on the society; for, if this excuse were to be admitted, it would most probably be attended with the most pernicious consequences." Yet it is at the same time rather rashly intimated, that employment must be found for all who are able to work; and in another place it is observed, that timely assistance would be afforded by these societies, without degradation, on all temporary occasions of suspended labour.

On the whole, when it is considered that a large and probably increasing amount of poor's rates would be subscribed to these societies; that on this account their members could hardly be considered as independent of parish assistance; and that the usual poor's rates would still remain to be applied as they are now, without any proposed limitations, there is little hope that Mr. Curwen's plan would be successful in diminishing the whole amount of the rates, and the proportion of dependent poor.

There are two errors respecting the management of the poor, into which the public seem inclined to fall at the present moment. The first is a disposition to attach too much importance to the effects of subscriptions from the poor themselves, without sufficient attention to the mode in which they are distributed. But the mode of distribution is much the more important point of the two; and if this be radically bad, it is of little consequence in what manner the subscriptions are raised, whether from the poor themselves, or from any other quarter. If the labouring classes were universally to contribute what might at first appear a very ample proportion of their earnings, for their own support in sickness and in old age, when out of work, and when the family consisted of more than two children; it is quite certain that the funds would become deficient. Such a mode of distribution implies a power of supporting a rapidly increasing and unlimited population on a limited territory, and must therefore terminate in aggravated poverty. Our present friendly societies or benefit-clubs aim at only limited objects, which are susceptible of calculation; yet many have failed, and many more it is understood are likely to fail from the insufficiency of their funds. If any society were to attempt to give much more extensive assistance to its members; if it were to endeavour to imitate what is partially effected by the poor-laws, or to accomplish those objects which Condorcet thought were within the power of proper calculations; the failure of its funds, however large at first, and from whatever sources derived, would be absolutely inevitable. In short, it cannot be too often or too strongly impressed upon the public, especially when any question for the improvement of the condition of the poor is in agitation, that no application of knowledge and ingenuity to this subject, no efforts either of the poor or of the rich, or both, in the form of contributions, or in any other way, can possibly place the labouring classes of society in such a state as to enable them to marry generally at the same age in an old and fully peopled country as they may do with perfect safety and advantage in a new one.

The other error towards which the public seems to incline at present is that of laying too much stress upon the *employment* of the poor. It seems to be thought that one of the principal causes of the failure of our present system is the not having properly executed that part of the 43d of Elizabeth which enjoins the purchase of materials to set the poor to work. It is certainly desirable, on many accounts, to employ the poor, when it is practicable, though it will always be extremely difficult to make people work actively who are without the usual and most natural motives to such exertions; and a system of coercion involves the necessity of placing great power in the hands of persons very likely to abuse it. Still however it is probable that the poor might be employed more than they have hitherto been, in a way to be advantageous to their habits and morals, without being prejudicial in other respects. But we should fall into the grossest error if we were to imagine that any essential part of the evils of the poor-laws, or of the difficulties under which we are at present labouring, has arisen from not employing the poor; or if we were to suppose that any possible scheme for giving work to all who are out of employment can ever in any degree apply to the source of these evils and difficulties, so as to prevent their recurrence. In no conceivable case can the forced employment of the poor, though managed in the most judicious manner, have any direct tendency to proportion more accurately the supply of labour to the natural demand for it. And without great care and caution it is obvious that it may have a pernicious effect of an opposite kind. When, for instance, from deficient demand or deficient capital, labour has a strong tendency to fall, if we keep it up to its usual price by creating an artificial demand by public subscriptions or advances from the government, we evidently prevent the population of the country from adjusting itself gradually to its diminished resources, and act much in the same manner as those who would prevent the price of corn from rising in a scarcity, which must necessarily terminate in increased distress.

Without then meaning to object to all plans for employing the poor, some of which, at certain times and with proper restrictions, may be useful as temporary measures, it is of great importance, in order to prevent ineffectual efforts and continued disappointments, to be fully aware that the permanent remedy which we are seeking cannot possibly come from this quarter.

It may indeed be affirmed with the most perfect confidence that there is only one class of causes from which any approaches towards a remedy can be rationally expected; and that consists of whatever has a tendency to increase the prudence and foresight of the labouring classes. This is the touchstone to which every plan proposed for the improvement of the condition of the poor should be applied. If the plan be such as to co-operate with the lessons of Nature and Providence, and to encourage and promote habits of prudence and foresight, essential and permanent benefit may be expected from it; if it has no tendency of this kind, it may possibly still be good as a temporary measure, and on other accounts, but we may be quite certain that it does not apply to the source of the specific evil for which we are seeking a remedy.

Of all the plans which have yet been proposed for the assistance of the labouring classes, the saving-banks, as far as they go, appear to me much the best, and the most likely, if they should become general, to effect a permanent improvement in the condition of the lower classes of society. By giving to each individual the full and entire benefit of his own industry and prudence, they are calculated greatly to strengthen the lessons of Nature and Providence; and a young man, who had been saving from fourteen or fifteen with a view to marriage at four or five and twenty, or perhaps much earlier, would probably be induced to wait two or three years longer if the times were unfavourable; if corn were high; if wages were low; or if the sum he had saved had been found by experience not to be sufficient to furnish a tolerable security against want. A habit of saving a portion of present earnings for future contingencies can scarcely be supposed to exist without general habits of prudence and foresight; and if the opportunity furnished by provident banks to individuals, of reaping the full benefit of saving, should render the practice general, it might rationally be expected that, under the varying resources of the country, the population would be adjusted to the actual demand for labour, at the expense of less pain and less poverty; and the remedy thus appears, so far as it goes, to apply to the very root of the evil.

The great object of saving-banks, however, is to prevent want and dependence by enabling the poor to provide against contingencies themselves. And in a natural state of society, such institutions, with the aid of private charity well directed, would probably be all the means necessary to produce the best practicable effects. In the present state of things in this country the case is essentially different. With so very large a body of poor habitually dependent upon public funds, the institution of saving-banks cannot be considered in the light of substitutes for the poor's rates. The problem how to support those who are in want in such a manner as not continually to increase the proportion which they bear to the whole society will still remain to be solved. But if any plan should be adopted either of gradually abolishing or gradually reducing and fixing the amount of the poor's rates, saving-banks would essentially assist it; at the same time that they would receive a most powerful aid in return.

In the actual state of things, they have been established at a period likely to be particularly unfavourable to them—a period of very general distress, and of the most extensive parochial assistance; and the success which has attended them, even under these disadvantages, seems clearly to shew, that in a period of prosperity and good wages, combined with a prospect of diminished parochial assistance, they might spread very extensively, and have a considerable effect on the general habits of a people.

With a view to give them greater encouragement at the present moment, an act has been passed allowing persons to receive parish assistance at the discretion of the justices, although they may have funds of their own under a certain amount in a saving-bank. But this is probably a shortsighted policy. It is sacrificing the principle for which saving-banks are established, to obtain an advantage which, on this very account, will be comparatively of little value. We wish to teach the labouring classes to rely more upon their own exertions and resources, as the only way of really improving their condition; yet we reward their saving by making them still dependent upon that very species of assistance which it is our object that they should avoid. The progress of saving-banks under such a regulation will be but an equivocal and uncertain symptom of good; whereas without such a regulation every step would tell, every fresh deposit would prove the growth of a desire to become independent of parish assistance; and both the great extension of the friendly societies and the success of the saving-banks in proportion to the time they have been established, clearly shew that much progress might be expected in these institutions under favourable circumstances, without resorting to a measure which is evidently calculated to sacrifice the end to the means.

With regard to the plans which have been talked of for reducing and limiting the poor's rates, they are certainly of a kind to apply to the root of the evil; but they would be obviously unjust without a formal retraction of the *right* of the poor to support; and for many years they would unquestionably be much more harsh in their operation than the plan of abolition which I have ventured to propose in a preceding chapter. At the same tune, if it be thought that this country cannot entirely get rid of a system which has been so long interwoven in its frame, a limitation of the amount of the poor's rates, or rather of their proportion to the wealth and population of the country which would be more rational and just, accompanied with a very full and fair notice of the nature of the change to be made, might be productive of essential benefit, and do much towards improving the habits and happiness of the poor.

4.2.13 Book IV, Chapter XIII: Of the Necessity of General Principles on This Subject.

Temporarily missing material.

of giving assistance are so often directly opposite to the general and permanent effects.

It has been observed in particular districts, where cottagers are possessed of small pieces of land, and are in the habit of keeping cows, that during the late scarcities some of them were able to support themselves without parish assistance, and others with comparatively little.[125]

According to the partial view in which this subject has been always contemplated, a general inference has been drawn from such instances, that, if we could place all our labourers in a similar situation, they would all be equally comfortable, and equally independent of the parish. This is an inference, however, that by no means follows. The advantage, which cottagers who at present keep cows enjoy, arises in a great measure from its being peculiar, and would be considerably diminished if it were made general.

A farmer or gentleman has, we will suppose, a certain number of cottages on his farm. Being a liberal man, and liking to see all the people about him comfortable, he may join a piece of land to each cottage sufficient to keep one or two cows, and give besides high wages. His labourers will of course live in plenty, and be able to rear up large families; but his farm may not require many hands; and though he may choose to pay those which he employs well, he will not probably wish to have more labourers on his land than his work requires. He does not therefore build more houses; and the children of the labourers whom he employs must evidently emigrate, and settle in other countries. While such a system continues peculiar to certain families or certain districts, the emigrants would easily be able to find work in other places; and it cannot be doubted that the individual labourers employed on these farms are in an enviable situation, and such as we might naturally wish was the lot of all our labourers. But it is perfectly clear that such a system could not, in the nature of things, possess the same advantages, if it were made general; because there would then be no countries to which the children could emigrate with the same prospect of finding work. Population would evidently increase beyond the demand of towns and manufactories, and the price of labour would universally fall.

It should be observed also, that one of the reasons why the labourers who at present keep cows are so comfortable, is, that they are able to make considerable profit of the milk which they do not use themselves; an advantage which would evidently be very much diminished, if the system were universal. And though they were certainly able to struggle through the late scarcities with less assistance than their neighbours, as might naturally be expected, from their having other resources besides the article which in those individual years was scarce; yet if the system were universal, there can be no reason assigned why they would not be subject to suffer as much from a scarcity of grass and a mortality among cows;[126] as our common labourers do now from a scarcity of wheat. We should be extremely cautious

[125] See an Inquiry into the state of Cottagers in the Counties of Lincoln and Rutland, by Robert Gourlay. Annals of Agriculture, vol. xxxvii. p. 514.

[126] At present the loss of a cow, which must now and then happen, is generally remedied by a petition and subscription; and as the event is considered as a most serious misfortune to a labourer, these petitions are for the most part attended to; but if the cow system were universal, losses would occur so frequently, that they could not possibly be repaired in the same way, and families would be continually dropping from comparative plenty into want.

therefore of trusting to such appearances, and of drawing a general inference from this kind of partial experience.

The main principle, on which the society for increasing the comforts and bettering the condition of the poor professes to proceed, is excellent. To give effect to that master-spring of industry, the desire of bettering our condition[127] is the true mode of improving the state of the lower classes; and we may safely agree with Sir Thomas Bernard, in one of his able prefaces, that whatever encourages and promotes habits of industry, prudence, foresight, virtue and cleanliness, among the poor, is beneficial to them and to the country; and whatever removes or diminishes the incitements to any of these qualities is detrimental to the state, and pernicious to the individual.[128]

Sir Thomas Bernard indeed himself seems in general to be fully aware of the difficulties, which the society has to contend with in the accomplishment of its object. But still it appears to be in some danger of falling into the error before alluded to, of drawing general inferences from insufficient experience. Without adverting to the plans respecting cheaper foods and parish shops, recommended by individuals, the beneficial effects of which depend entirely upon their being peculiar to certain families or certain parishes, and would be lost if they were general, by lowering the wages of labour; I shall only notice one observation of a more comprehensive nature, which occurs in the preface to the second volume of the Reports. It is there remarked that the experience of the society seemed to warrant the conclusion, that the best mode of relieving the poor was, by assisting them at their own homes, and placing out their children as soon as possible in different employments, apprenticeships, &c. I really believe that this is the best, and it is certainly the most agreeable, mode in which occasional and discriminate assistance can be given. But it is evident that it must be done with caution, and cannot be adopted as a general principle, and made the foundation of universal practice. It is open exactly to the same objection as the cow system, which has just been noticed, and that part of the act of the 43d of Elizabeth, which directs the overseers to employ and provide for the children of the poor. A particular parish, where all the children, as soon as they were of a proper age, were taken from their parents, and placed out in proper situations, might be very comfortable; but if the system were general, and the poor saw that all their children would be thus provided for, every employment would presently be overstocked with hands, and the consequences need not be again repeated.

Nothing can be more clear than that it is within the power of money, and of the exertions of the rich, adequately to relieve a particular family, a particular parish, and even a particular district. But it will be equally clear, if we reflect a moment on the subject, that it is totally out of their power to relieve the whole country in the same way; at least without providing a regular vent for the overflowing numbers in emigration, or without the

[127] Preface to vol. ii. of the Reports.

[128] Preface to vol. iii. of the Reports.

prevalence of a particular virtue among the poor, which the distribution of this assistance tends obviously to discourage.

Even industry itself is, in this respect, not very different from money. A man who possesses a certain portion of it, above what is usually possessed by his neighbours, will, in the actual state of things, be almost sure of a competent livelihood; but if all his neighbours were to become at once as industrious as himself, the absolute portion of industry which he before possessed would no longer be a security against want. Hume fell into a great error, when he asserted that "almost all the moral as well as natural evils of human life arise from idleness;" and for the cure of these ills required only that the whole species should possess naturally an equal diligence with that, which many individuals are able to attain by habit and reflection.[129] It is evident that this given degree of industry possessed by the whole species, if not combined with another virtue of which he takes no notice, would totally fail of rescuing society from want and misery, and would scarcely remove a single moral or physical evil of all those to which he alludes.

I am aware of an objection, which will, with great appearance of justice, be urged against the general scope of these reasonings. It will be said that to argue thus is at once to object to every mode of assisting the poor, as it is impossible, in the nature of things, to assist people individually, without altering their relative situation in society, and proportionally depressing others; and that as those who have families are the persons naturally most subject to distress, and as we are certainly not called upon to assist those who do not want our aid, we must necessarily, if we act at all, relieve those who have children, and thus encourage marriage and population.

I have already observed, however, and I here repeat it again, that the general principles on these subjects ought not to be pushed too far, though they should always be kept in view; and that many cases may occur, in which the good resulting from the relief of the present distress may more than overbalance the evil to be apprehended from the remote consequence.

All relief in instances of distress, not arising from idle and improvident habits, clearly comes under this description; and in general it may be observed, that it is only that kind of *systematic* and *certain* relief, on which the poor can confidently depend, whatever may be their conduct, that violates general principles in such a manner as to make it clear that the general consequence is worse than the particular evil.

Independently of this discriminate and occasional assistance, the beneficial effects of which I have fully allowed in a preceding chapter, I have before endeavoured to shew, that much might be expected from a better and more general system of education. Every thing that can be done in this way has indeed a very peculiar value; because education is one of those advantages, which not only all may share without interfering with each other, but the raising of one person may actually contribute to the raising of others. If, for instance, a man by education acquires that decent kind of pride and those juster habits of thinking, which

[129] Dialogues on Natural Religion, part xi. p. 212.

will prevent him from burdening society with a family of children which he cannot support, his conduct, as far as an individual instance can go, tends evidently to improve the condition of his fellow-labourers; and a contrary conduct from ignorance would tend as evidently to depress it.

I cannot help thinking also, that something might be done towards bettering the situation of the poor by a general improvement of their cottages, if care were taken, at the same time, not to make them so large as to allow of two families settling in them; and not to increase their number faster than the demand for labour required. One of the most salutary and least pernicious checks to the frequency of early marriages in this country is the difficulty of procuring a cottage, and the laudable habits which prompt a labourer rather to defer his marriage some years in the expectation of a vacancy, than to content himself with a wretched mud cabin, like those in Ireland.[130]

Even the cow system, upon a more confined plan, might not to be open to objection. With any view of making if a substitute for the poor-laws, and of giving labourers a right to demand land and cows in proportion to their families; or of taking the common people from the consumption of wheat, and feeding them on milk and potatoes; it appears to me, I confess, truly preposterous: but if it were so ordered as merely to provide a comfortable situation for the better and more industrious labourers, and to supply at the same time a very important want among the poor in general, that of milk for their children; I think that it would be extremely beneficial, and might be made a very powerful incitement to habits of industry, economy and prudence. With this view, however, it is evident that only a certain portion of labourers in each parish could be included in the plan; that good conduct, and not mere distress, should have the most valid claim to preference; that too much attention should not be paid to the number of children; and that universally, those who had saved money enough for the purchase of a cow, should be preferred to those who required to be furnished with one by the parish.[131]

One should undoubtedly be extremely unwilling not to make as much use as possible of that known stimulus to industry and economy, the desire of, and attachment to, property: but it should be recollected that the good effects of this stimulus shew themselves principally when this property is to be procured or preserved by personal exertions; and

[130] Perhaps, however, this is not often left to his choice, on account of the fear which every parish has of increasing its poor. There are many ways by which our poor-laws operate in counteracting their first obvious tendency to increase population, and this is one of them. I have little doubt that it is almost exclusively owing to these counteracting causes, that we have been able to persevere in this system so long, and that the condition of the poor has not been so much injured by it as might have been expected.

[131] The act of Elizabeth, which prohibited the building of cottages, unless four acres of land were annexed to than, is probably impracticable in a manufacturing country like England; but, upon this principle, certainly the greatest part of the poor might possess land; because the difficulty of procuring such cottages would always operate as a powerful check to their increase. The effect of such a plan would be very different from that of Mr. Young.

that they are by no means so general under other circumstances. If any idle man with a family could demand and obtain a cow and some land, I should expect to see both very often neglected.

It has been observed that those cottagers, who keep cows, are more industrious and more regular in their conduct, than those who do not. This is probably true, and what might naturally be expected; but the inference that the way to make all people industrious is to give them cows, may by no means be quite so certain. Most of those who keep cows at present have purchased them with the fruits of their own industry. It is therefore more just to say that their industry has given them a cow, than that a cow has given them their industry; though I would by no means be understood to imply that the sudden possession of property never generates industrious habits.

The practical good effects, which have been already experienced from cottagers keeping cows,[132] arise in fact from the system being nearly such as the confined plan which I have mentioned. In the districts where cottagers of this description most abound, they do not bear a very large proportion to the population of the whole parish; they consist in general of the better sort of labourers, who have been able to purchase their own cows; and the peculiar comforts of their situation arise as much from the relative as the positive advantages which they possess.

From observing therefore their industry and comforts, we should be very cautious of inferring that we could give the same industry and comforts to all the lower classes of people, by giving them the same possessions. There is nothing, that has given rise to such a cloud of errors, as a confusion between relative and positive, and between cause and effect.

It may be said, however, that any plan of generally improving the cottages of the poor, or of enabling more of them to keep cows, would evidently give them the power of rearing a greater number of children, and, by thus encouraging population, violate the principles which I have endeavoured to establish. But if I have been successful in making the reader comprehend the principal bent of this work, he will be aware that the precise reason why I think that more children ought not to be born than the country can support is, that the greatest possible number of those that are born may be supported. We cannot, in the nature of things, assist the poor in any way, without enabling them to rear up to manhood a greater number of their children. But this is, of all other things, the most desirable, both with regard to individuals and the public. Every loss of a child from the consequences of poverty must evidently be preceded and accompanied by great misery to individuals; and in a public view, every child that dies under ten years of age is a loss to the nation of all that had been expended in its subsistence till that period. Consequently, in every point of view, a decrease of mortality at all ages is what we ought to aim at. We cannot however effect this object, without first crowding the population in some degree by making more children grow up to manhood; but we shall do no harm in this respect, if, at the same time, we can impress these

[132] Inquiry into the State of Cottagers in Counties of Lincoln and Rutland, by Robert Gourlay. Annals of Agriculture, vol. xxxvii. p. 514.

children with the idea, that, to possess the same advantages as their parents, they must defer marriage till they have a fair prospect of being able to maintain a family. And it must be candidly confessed that, if we cannot do this, all our former efforts will have been thrown away. It is not in the nature of things that any permanent and general improvement in the condition of the poor can be effected without an increase in the preventive check; and unless this take place, either with or without our efforts, every thing that is done for the poor must be temporary and partial: a diminution of mortality at present will be balanced by an increased mortality in future; and the improvement of their condition in one place will proportionally depress it in another. This is a truth so important, and so little understood, that it can scarcely be too often insisted on.

Paley, in a chapter on population, provision, &c., in his Moral Philosophy, observes, that the condition most favourable to the population of a country, and at the same time to its general happiness, is, "that of a laborious frugal people ministering to the demands of an opulent luxurious nation."[133] Such a form of society has not, it must be confessed, an inviting aspect. Nothing but the conviction of its being absolutely necessary could reconcile us to the idea of ten millions of people condemned to incessant toil, and to the privation of every thing but absolute necessaries, in order to minister to the excessive luxuries of the other million. But the fact is, that such a form of society is by no means necessary. It is by no means necessary that the rich should be excessively luxurious, in order to support the manufactures of a country; or that the poor should be deprived of all luxuries, in order to make them sufficiently numerous. The best, and in every point of view the most advantageous manufactures in this country, are those which are consumed by the great body of the people. The manufactures which are confined exclusively to the rich are not only trivial, on account of the comparative smallness of their quantity, but are further liable to the great disadvantage of producing much occasional misery among those employed in them, from changes of fashion. It is the diffusion of luxury therefore among the mass of the people, and not an excess of it in a few, that seems to be most advantageous, both with regard to national wealth and national happiness; and what Paley considers as the true evil and proper danger of luxury, I should be disposed to consider as its true good and peculiar advantage. If, indeed, it be allowed that in every society, not in the state of a new colony, some powerful check to population must prevail; and if it be observed that a taste for the comforts and conveniences of life will prevent people from marrying, under the certainty of being deprived of these advantages; it must be allowed that we can hardly expect to find any check to marriage so little prejudicial to the happiness and virtue of society as the general prevalence of such a taste; and consequently, that the extension of luxury in this

[133] Vol. ii. c. xi. p. 359. From a passage in Paley's Natural Theology, I am inclined to think that subsequent reflection induced him to modify some of his former ideas on the subject of population. He states most justly (ch. xxv. p. 539.) that mankind will in every country breed up to a certain point of distress. If this be allowed, that country will evidently be the happiest, where the degree of distress at this point is the least; and consequently, if the diffusion of luxury, by producing the check sooner, tend to diminish this degree of distress, it is certainly desirable.

sense of the term is particularly desirable, and one of the best means of raising that standard of wretchedness alluded to in a former chapter.

It has been generally found that the middle parts of society are most favourable to virtuous and industrious habits, and to the growth of all kinds of talents. But it is evident that all cannot be in the middle. Superior and inferior parts are in the nature of things absolutely necessary; and not only necessary, but strikingly beneficial. If no man could hope to rise, or fear to fall in society; if industry did not bring with it its reward, and indolence its punishment; we could not expect to see that animated activity in bettering our condition, which now forms the master-spring of public prosperity. But in contemplating the different states of Europe, we observe a very considerable difference in the relative proportions of the superior, the middle and the inferior parts; and from the effect of these differences it seems probable, that our best-grounded expectations of an increase in the happiness of the mass of human society are founded in the prospect of an increase in the relative proportions of the middle parts. And if the lower classes of people had acquired the habit of proportioning the supplies of labour to a stationary or even decreasing demand, without an increase of misery and mortality, as at present, we might even venture to indulge a hope that at some future period the processes for abridging human labour, the progress of which has of late years been so rapid, might ultimately supply all the wants of the most wealthy society with less personal effort than at present; and if they did not diminish the severity of individual exertion, might, at least, diminish the number of those employed in severe toil. If the lowest classes of society were thus diminished, and the middle classes increased, each labourer might indulge a more rational hope of rising by diligence and exertion into a better station; the rewards of industry and virtue would be increased in number; the lottery of human society would appear to consist of fewer blanks and more prizes; and the sum of social happiness would be evidently augmented.

To indulge, however, in any distant views of this kind, unaccompanied by the evils usually attendant on a stationary or decreasing demand for labour, we must suppose the general prevalence of such prudential habits among the poor, as would prevent them from marrying, when the actual price of labour, joined to what they might have saved in their single state, would not give them the prospect of being able to support a wife and five or six children without assistance. And undoubtedly such a degree of prudential restraint would produce a very striking melioration in the condition of the lower classes of people.

It may be said, perhaps, that even this degree of prudence might not always avail, as when a man marries he cannot tell what number of children he shall have, and many have more than six. This is certainly true; and in this case I do not think that any evil would result from making a certain allowance to every child above this number; not with a view of rewarding a man for his large family, but merely of relieving him from a species of distress which it would be unreasonable in us to expect that he should calculate upon. And with this view, the relief should be merely such as to place him exactly in the same situation as if he had had six children. Montesquieu disapproves of an edict of Lewis the Fourteenth, which give certain pensions to those who had ten and twelve children, as being of no use in encouraging population.[134] For the very reason that he disapproves of it, I should think that some law of the kind might be adopted without danger, and might relieve particular individuals from a very pressing and unlooked for distress, without operating in any respect as an encouragement to marriage.

If at some future period any approach should be made towards the more general prevalence of prudential habits with respect to marriage among the poor, from which alone any permanent and general improvement of their condition can arise; I do not think that the narrowest politician need be alarmed at it, from the fear of its occasioning such an advance in the price of labour, as will enable our commercial competitors to undersell us in foreign markets. There are four circumstances that might be expected to accompany it, which would probably either prevent, or fully counterbalance any effect of this kind. These are, 1st, the more equable and lower price of provisions, from the demand being less frequently above the supply. 2dly, the removal of that heavy burden on agriculture, and that great addition to the present wages of labour, the poor's rates. 3dly, the national saving of a great part of that sum, which is expended without return in the support of those children who die prematurely from the consequences of poverty. And, lastly, the more general prevalence of economical and industrious habits, particularly among unmarried men, which would prevent that indolence, drunkenness and waste of labour, which at present are too frequently a consequence of high wages.

4.2.14 Book IV, Chapter XIV: Of Our Rational Expectations Respecting the Future Improvement of Society.

In taking a general and concluding view of our rational expectations respecting the mitigation of the evils arising from the principle of population, it may be observed that though the increase of population in a geometrical ratio be incontrovertible, and the period of doubling, when unchecked, has been uniformly stated in this work rather below than above the truth; yet there are some natural consequences of the progress of society and civilization, which necessarily repress its full effects. These are, more particularly, great towns and manufactures, in which we can scarcely hope, and certainly not expect, to see any very material change. It is undoubtedly our duty, and in every point of view highly desirable, to make towns and manufacturing employments as little injurious as possible to the duration of human life; but, after all our efforts, it is probable that they will always remain less healthy than country situations and country employments; and consequently, operating as positive checks, will diminish in some degree the necessity of the preventive check.

In every old state, it is observed that a considerable number of grown-up people remain for a time unmarried. The duty of practising the common and acknowledged rules of

[134] Esprit des Loix, liv. xxiii. c. xxvii.

morality during this period has never been controverted in theory, however it may have been opposed in practice. This branch of the duty of moral restraint has scarcely been touched by the reasonings of this work. It rests on the same foundation as before, neither stronger nor weaker. And knowing how incompletely this duty has hitherto been fulfilled, it would certainly be visionary to expect that in future it would be completely fulfilled.

The part which has been affected by the reasonings of this work is not therefore that which relates to our conduct during the period of celibacy, but to the duty of extending this period till we have a prospect of being able to maintain our children. And it is by no means visionary to indulge a hope of some favourable change in this respect; because it is found by experience that the prevalence of this kind of prudential restraint is extremely different in different countries, and in the same countries at different periods.

It cannot be doubted that throughout Europe in general, and most particularly in the northern states, a decided change has taken place in the operation of prudential restraint, since the prevalence of those warlike and enterprising habits which destroyed so many people. In later times the gradual diminution and almost total extinction of the plagues, which so frequently visited Europe in the seventeenth and beginnings of the eighteenth centuries, produced a change of the same kind. And in this country, it is not to be doubted that the proportion of marriages has become smaller since the improvement of our towns, the less frequent returns of epidemics, and the adoption of habits of greater cleanliness. During the late scarcities it appears that the number of marriages diminished;[135] and the same motives which prevented many people from marrying during such a period, would operate precisely in the same way, if, in future, the additional number of children reared to manhood from the introduction of the cow-pox, were to be such as to crowd all employments, lower the price of labour, and make it more difficult to support a family.

Universally, the practice of mankind on the subject of marriage has been much superior to their theories; and however frequent may have been the declamations on the duty of entering into this state, and the advantage of early unions to prevent vice, each individual has practically found it necessary to consider of the means of supporting a family, before he ventured to take so important a step. That great *vis medicatrix reipublicæ,* the desire of bettering our condition, and the fear of making it worse, has been constantly in action, and has been constantly directing people into the right road, in spite of all the declamations which tended to lead them aside. Owing to this powerful spring of health in every state, which is nothing more than an inference from the general course of the laws of nature, irresistibly forced on each man's attention, the prudential check to marriage has increased in Europe; and it cannot be unreasonable to conclude that it will still make further advances. If this take place without any marked and decided increase of a vicious intercourse with the sex, the happiness of society will evidently be promoted by it; and with regard to the danger of such increase, it is consolatory to remark that those countries in Europe, where marriages are the latest or least frequent, are by no means particularly distinguished by vices of this

[135] 1800 and 1801.

kind. It has appeared, that Norway, Switzerland, England, and Scotland, are above all the rest in the prevalence of the preventive check; and though I do not mean to insist particularly on the virtuous habits of these countries, yet I think that no person would select them as the countries most marked for profligacy of manners. Indeed, from the little that I know of the continent, I should have been inclined to select them as most distinguished for contrary habits, and as rather above than below their neighbours in the chastity of their women, and consequently in the virtuous habits of their men. Experience therefore seems to teach us that it is possible for moral and physical causes to counteract the effects that might at first be expected from an increase of the check to marriage; but allowing all the weight to these effects which is in any degree probable, it may be safely asserted, that the diminution of the vices arising from indigence would fully counterbalance them; and that all the advantages of diminished mortality and superior comforts, which would certainly result from an increase of the preventive check, may be placed entirely on the side of the gains to the cause of happiness and virtue.

It is less the object of the present work to propose new plans of improving society, than to inculcate the necessity of resting contented with that mode of improvement which already in has part been acted upon as dictated by the course of nature, and of not obstructing the advances which would otherwise be made in this way.

It would be undoubtedly highly advantageous that all our positive institutions, and the whole tenour of our conduct to the poor, should be such as actively to co-operate with that lesson of prudence inculcated by the common course of human events; and if we take upon ourselves sometimes to mitigate the natural punishments of imprudence, that we could balance it by increasing the rewards of an opposite conduct. But much would be done, if merely the institutions which directly tend to encourage marriage were gradually changed, and we ceased to circulate opinions and inculcate doctrines, which positively counteract the lessons of nature.

The limited good, which it is sometimes in our power to effect, is often lost by attempting too much, and by making the adoption of some particular plan essentially necessary even to a partial degree of success. In the practical application of the reasonings of this work, I hope that I have avoided this error. I wish to press on the recollection of the reader, that, though I may have given some new views of old facts, and may have indulged in the contemplation of a considerable degree of *possible* improvement, that I might not shut out that prime cheerer hope; yet in my expectations of probable improvement and in suggesting the means of accomplishing it, I have been very cautious. The gradual abolition of the poor-laws has already often been proposed, in consequence of the practical evils which have been found to flow from them, and the danger of their becoming a weight absolutely intolerable on the landed property of the kingdom. The establishment of a more extensive system of national education has neither the advantage of novelty with some, nor its disadvantages with others, to recommend it. The practical good effects of education have long been experienced in Scotland; and almost every person who has been placed in a situation to judge, has given his testimony that education appears to have a considerable effect in the prevention of crimes,[136] and the promotion of industry, morality and regular conduct. Yet these are the only plans which have been offered; and though the adoption of them in the modes suggested would very powerfully contribute to forward the object of this work and better the condition of the poor; yet if nothing be done in this way, I shall not absolutely despair of some partial good resulting from the general effects of the reasoning.

If the principles which I have endeavoured to establish be false, I most sincerely hope to see them completely refuted; but if they be true, the subject is so important, and interests the question of human happiness so nearly, that it is impossible they should not in time be more fully known and more generally circulated, whether any particular efforts be made for the purpose or not.

Among the higher and middle classes of society, the effect of this knowledge will, I hope, be to direct without relaxing their efforts in bettering the condition of the poor; to shew them what they can and what they cannot do; and that, although much may be done by advice and instruction, by encouraging habits of prudence and cleanliness, by discriminate charity, and by any mode of bettering the present condition of the poor which is followed by an increase of the preventive check; yet that, without this last effect, all the former efforts would be futile; and that, in any old and well-peopled state, to assist the poor in such a manner as to enable them to marry as early as they please, and rear up large families, is a physical impossibility. This knowledge, by tending to prevent the rich from destroying the good effects of their own exertions, and wasting their efforts in a direction where success is unattainable, would confine their attention to the proper objects, and thus enable them to do more good.

Among the poor themselves, its effects would be still more important. That the principal and most permanent cause of poverty has little or no *direct* relation to forms of government, or the unequal division of property; and that, as the rich do not in reality possess the *power* of finding employment and maintenance for the poor, the poor cannot, in the nature of things, possess the *right* to demand them; are important truths flowing from the principle of population, which, when, properly explained, would by no means be above the most ordinary comprehensions. And it is evident that every man in the lower classes of society, who became acquainted with these truths, would be disposed to bear the distresses in which he might be involved with more patience; would feel less discontent and irritation at the government and the higher classes of society, on account of his poverty; would be on all occasions less disposed to insubordination and turbulence; and if he received assistance, either from any public institution or from the hand of private charity, he would receive it with more thankfulness, and more justly appreciate its value.

---

[136] Mr. Howard found fewer prisoners in Switzerland and Scotland than in other countries, which is attributed to a more regular education among the lower classes of the Swiss and the Scotch. During the number of years which the late Mr. Fielding presided at Bow-street, only six Scotchmen were brought before him. He used to say, that of the persons committed the greater part were Irish. Preface to vol. iii. of the Reports of the Society for bettering the Condition of the Poor, p. 32.

If these truths were by degrees more generally known, (which in the course of time does not seem to be improbable from the natural effects of the mutual interchange of opinions,) the lower classes of people, as a body, would become more peaceable and orderly, would be less inclined to tumultuous proceedings in seasons of scarcity, and would at all times be less influenced by inflammatory and seditious publications, from knowing how little the price of labour and the means of supporting a family depend upon a revolution. The mere knowledge of these truths, even if they did not operate sufficiently to produce any marked change in the prudential habits of the poor with regard to marriage, would still have a most beneficial effect on their conduct in a political light; and undoubtedly, one of the most valuable of these effects would be the power, that would result to the higher and middle classes of society, of gradually improving their governments,[137] without the apprehension of those revolutionary excesses, the fear of which, at present, threatens to deprive Europe even of that degree of liberty, which she had before experienced to be practicable, and the salutary effects of which she had long enjoyed.

From a review of the state of society in former periods, compared with the present, I should certainly say that the evils resulting from the principle of population have rather diminished than increased, even under the disadvantage of an almost total ignorance of the real cause. And if we can indulge the hope that this ignorance will be gradually dissipated, it does not seem unreasonable to expect that they will be still further diminished. The increase of absolute population, which will of course take place, will evidently tend but little to weaken this expectation, as every thing depends upon the relative proportion between population and food, and not on the absolute number of people. In the former part of this work it appeared that the countries, which possessed the fewest people, often suffered the most from the effects of the principle of population; and it can scarcely be doubted that, taking Europe throughout, fewer famines and fewer diseases arising from want have prevailed in the last century than in those which preceded it.

On the whole, therefore, though our future prospects respecting the mitigation of the evils arising from the principle of population may not be so bright as we could wish, yet they are far from being entirely disheartening, and by no means preclude that gradual and progressive improvement in human satiety, which, before the late wild speculations on this subject, was the object of rational expectation. To the laws of property and marriage, and to the apparently narrow principle of self-interest which prompts each individual to exert himself in bettering his condition, we are indebted for all the noblest exertions of human genius, for every thing that distinguishes the civilized from the savage state. A strict inquiry

[137] I cannot believe that the removal of all unjust grounds of discontent against constituted authorities would render the people torpid and indifferent to advantages, which are really attainable. The blessings of civil liberty are so great that they surely cannot need the aid of false colouring to make them desirable. I should be sorry to think that the lower classes of people could never by animated to assert their rights but by means of such illusory promises, as will generally make the remedy of resistance much worse than the disease which it was intended to cure.

Appendix I.

into the principle of population obliges us to conclude that we shall never be able to throw down the ladder, by which we have risen to this eminence; but it by no means proves, that we may not rise higher by the same means. The structure of society, in its great features, will probably always remain unchanged. We have every reason to believe that it will always consist of a class of proprietors and a class of labourers; but the condition of each, and the proportion which they bear to each other, may be so altered, as greatly to improve the harmony and beauty of the whole. It would indeed be a melancholy reflection that, while the views of physical science are daily enlarging, so as scarcely to be bounded by the most distant horizon, the science of moral and political philosophy should be confined within such narrow limits, or at best be so feeble in its influence, as to be unable to counteract the obstacles to human happiness arising from a single cause. But however formidable these obstacles may have appeared in some parts of this work, it is hoped that the general result of the inquiry is such, as not to make us give up the improvement of human society in despair. The partial good which seems to be attainable is worthy of all our exertions; is sufficient to direct our efforts, and animate our prospects. And although we cannot expect that the virtue and happiness of mankind will keep pace with the brilliant career of physical discovery; yet, if we are not wanting to ourselves, we may confidently indulge the hope that, to no unimportant extent, they will be influenced by its progress and will partake in its success.

4.2.15 Appendix I,: 1807

In the preface to the second edition of this Essay, I expressed a hope that the detailed manner in which I had treated the subject and pursued it to its consequences, though it might open the door to many objections, and expose me to much severity of criticism, might be subservient to the important end of bringing a subject so nearly connected with the happiness of society into more general notice. Conformably to the same views I should always have felt willing to enter into the discussion of any serious objections that were made to my principles or conclusions, to abandon those which appeared to be false, and to throw further lights, if I could, on those which appeared to be true. But though the work has excited a degree of public attention much greater than I should have presumed to expect, yet very little has been written to controvert it; and of that little, the greatest part is so full of illiberal declamation, and so entirely destitute of argument, as to be evidently beneath notice. What I have to say therefore at present, will be directed rather more to the objections which have been urged in conversation, than to those which have appeared in print. My object is to correct some of the misrepresentations which have gone abroad respecting two or three of the most important points of the Essay; and I should feel greatly obliged to those who have not had leisure to read the whole work, if they would cast their eyes over the few following pages, that they may not, from the partial and incorrect statements which they have heard, mistake the import of some of my opinions, and attribute to me others which I have never held.

The first grand objection that has been made to my principles is, that they contradict the original command of the Creator, to increase and multiply and replenish the earth. But those who have urged this objection have certainly either not read the work, or have directed their attention solely to a few detached passages, and have been unable to seize the bent and spirit of the whole. I am fully of opinion, that it is the duty of man to obey this command of his Creator; nor is there, in my recollection, a single passage in the work, which, taken with the context, can, to any reader of intelligence, warrant the contrary inference.

Every express command given to man by his Creator is given in subordination to those great and uniform laws of nature, which he had previously established; and we are forbidden both by reason and religion to expect that these laws will be changed in order to enable us to execute more readily any particular precept. It is undoubtedly true that, if man were enabled miraculously to live without food, the earth would be very rapidly replenished: but as we have not the slightest ground of hope that such a miracle will be worked for this purpose, it becomes our positive duty as reasonable creatures, and with a view of executing the commands of our Creator, to inquire into the laws which he has established for the multiplication of the species. And when we find, not only from the speculative contemplation of those laws, but from the far more powerful and imperious suggestions of our senses, that man cannot live without food, it is a folly exactly of the same *kind* to attempt to obey the will of our Creator by increasing population without reference to the means of its support, as to attempt to obtain an abundant crop of corn by sowing it on the way-side and in hedges, where it cannot receive its proper nourishment. Which is it, I would ask, that best seconds the benevolent intentions of the Creator in covering the earth with esculent vegetables, he who with care and foresight duly ploughs and prepares a piece of ground, and sows no more seed than he expects will grow up to maturity, or he who scatters a profusion of seed indifferently over the land, without reference to the soil on which it falls, or any previous preparation for its reception?

It is an utter misconception of any argument to infer that I am an enemy to population. I am only am enemy to vice and misery, and consequently to that unfavourable proportion between population and food, which produces these evils. But this unfavourable proportion has no necessary connection with the quantity of absolute population which a country may contain. On the contrary, it is more frequently found in countries which are very thinly peopled, than in those which are populous.

The bent of my argument on the subject of population may be illustrated by the instance of a pasture farm. If a young grazier were told to stock his land well, as on his stock would depend his profits and the ultimate success of his undertaking, he would certainly have been told nothing but what was strictly true: and he would have to accuse himself, not his advisers, if, in pursuance of these instructions, he were to push the breeding of his cattle till they became lean and half-starved. His instructor, when he talked of the advantages of a large stock, meant undoubtedly stock in proper condition, and not such a stock as, though it might be numerically greater, was in value much less. The expression of stocking a farm well does not refer to particular numbers, but merely to that proportion which is best adapted to the farm, whether it be a poor or a rich one, whether it will carry fifty head of cattle or five hundred. It is undoubtedly extremely desirable that it should carry the greater number, and every effort should be made to effect this object: but surely that farmer could not be considered as an enemy to a large quantity of stock, who should insist upon the folly and impropriety of attempting to breed such a quantity, before the land was put into a condition to bear it.

The arguments which I have need respecting the increase of population are exactly of the same nature as these just mentioned. I believe that it is the intention of the Creator that the earth should be replenished; but certainly with a healthy, virtuous and happy population, not an unhealthy, vicious and miserable one. And if, in endeavouring to obey the command to increase and multiply, we people it only with beings of this latter description and suffer accordingly, we have no right to impeach the justice of the command, but our irrational mode of executing it.

In the desirableness of a great and efficient population, I do not differ from the warmest advocates of increase. I am perfectly ready to acknowledge with the writers of old that it is not extent of territory, but extent of population that measures the power of states. It is only as to the mode of obtaining a vigorous and efficient population that I differ from them; and in thus differing I conceive myself entirely borne out by experience, that great test of all human speculations.

It appears from the undoubted testimony of registers, that a large proportion of marriages and births is by no means necessarily connected with a rapid increase of population, but is often found in countries where it is either stationary or increasing very slowly. The population of such countries is not only comparatively inefficient from the general poverty and misery of the inhabitants, but invariably contains a much larger proportion of persons in those stages of life, in which they are unable to contribute their share to the resources or the defence of the state.

This is most strikingly illustrated in an instance which I have quoted from M. Muret, in a chaplet on Switzerland, where it appeared, that in proportion to the same population, the Lyonais produced 16 births, the Pays de Vaud 11, and a particular parish in the Alps only 8; but that at the age of 20 these three very different numbers were all reduced to the same. In the Lyonais nearly half of the population was under the age of puberty, in the Pays de Vaud one-third, and in the parish of the Alps only non-fourth. The inference from such facts is unavoidable, and of the highest importance to society.

The poorer of a country to increase its resources or defend its possessions must depend principally upon its efficient population, upon that part of the population which is of an age to be employed effectually in agriculture, commerce or war; but it appears with an evidence little short of demonstration, that in a country, the resources of which do not naturally call for a larger proportion of births, such an increase, so far from tending to increase this efficient population, would tend materially to diminish it. It would undoubtedly, at first, increase the number of souls in proportion to the means of subsistence, and therefore cruelly increase the pressure of want; but the numbers of persons rising annually to the age of puberty might not be so great as before, a larger part of the produce would be distributed without return to children who would never reach manhood, and the additional population, instead of giving additional strength to the country, would essentially lessen this strength, and operate as a constant obstacle to the creation of new resources.

We are a little dazzled at present by the population and power of France, and it is known that she has always had a large proportion of births: but if any reliance can be placed on what are considered as the best authorities an this subject, it is quite certain that the advantages which she enjoys do not arise from any thing peculiar in the structure of her population, but solely from the great absolute quantity of it, derived from her immense extent of fertile territory.

Necker, speaking of the population of France, says, that it is so comprised, that a million of individuals present neither the same force in war, nor the same capacity for labour, as an equal number in a country where the people are less oppressed and fewer die in infancy. And the view which Arthur Young has given of the state of the lower classes of the people at the time he travelled in France, which was just at the commencement of the revolution, leads directly to the same conclusion. According to the *Statistique Générale et Particulière de la France,* lately published, the proportion of the population under twenty is almost 9/62; in England, if increasing no faster than France, it would probably not be much more than 7/26. Consequently out of a population of ten millions England would have a million more of persons above twenty than France, and would upon this supposition have at least three or four hundred thousand more males of a military age. If our population were of the same description as that of France, it must be increased numerically by none than a million and a half, in order to enable us to produce from England and Wales the same number of persons above the age of twenty as at present; and if we had only an increase of a million, our efficient strength in agriculture, commerce and war, would be in the most decided manner diminished, while at the same time the distresses of the lower classes would be dreadfully increased. Can any rational man say that an additional population of this description would be desirable, either in a moral or political view? And yet this is the kind of population which invariably results from direct encouragements to marriage, or from the want of that personal respectability which is occasioned by ignorance and despotism.

It may perhaps be true that France fills her armies with greater facility and less interruption to the usual labours of her inhabitants than England; and it must be acknowledged that poverty and want of employment are powerful aids to a recruiting serjeant; but it would not be a very humane project to keep our people always in want, for the sake of enlisting them cheaper; nor would it be a very politic project to diminish our wealth and strength with the same economical view. We cannot attain incompatible objects. If we possess the advantage of being able to keep nearly all our people constantly employed, either in agriculture or commerce, we cannot expect to retain the opposite advantage of their being always at leisure, and willing to enlist for a very small sum. But we may rest perfectly assured that while we have the efficient population, we shall never want men to fill our armies, if we propose to them adequate motives.

In many parts of the Essay I have dwelt much on the advantage of rearing the requisite population of any country from the smallest number of births. I have stated expressly, that a decrease of mortality at all ages is what we ought chiefly to aim at; and as the best criterion of happiness and good government, instead of the largeness of the proportion of births, which was the usual mode of judging, I have proposed the smallness of the proportion dying under the age of puberty. Conscious that I had never intentionally deviated from these principles, I might well be rather surprised to hear that I had been considered by some as an enemy to the introduction of the vaccine inoculation, which is calculated to attain the very end which I have uniformly considered as so desirable. I have indeed intimated what I still continue most firmly to believe, that if the resources of the country would not permanently admit of a greatly accelerated rate of increase in the population (and whether they would or not must certainly depend upon other causes besides the number of lives saved by the vaccine inoculation,) one of two things would happen, either an increased mortality of some other diseases, or a diminution in the proportion of births. But I have expressed my conviction that the latter effect would take place; and therefore consistently with the opinions which I have always maintained, I ought to be, and am, one of the warmest friends to the introduction of the cow-pox. In making every exertion which I think likely to be effectual, to increase the comforts and diminish the mortality among the poor, I act in the most exact conformity to my principles. Whether those are equally consistent who profess to have the same object in view, and yet measure the happiness of nations by the large proportion of marriages and births, is a point which they would do well to consider.

It has been said by some, that the natural checks to population will always be sufficient to keep it within bounds, without resorting to any other aids; and one ingenious writer has remarked that I have not deduced a single original fact from real observation, to prove the inefficiency of the checks which already prevail. These remarks are correctly true and are truisms exactly of the same kind as the assertion that men cannot live without food. For, undoubtedly as long as this continues to be a law of his nature, what are here called the natural checks cannot possibly fail of being effectual. Besides the curious truism that these assertions involve, they proceed upon the very strange supposition, that the *ultimate* object of my work is to check population; as if any thing could be more desirable than the most rapid increase of population, unaccompanied by vice and misery. But of course my ultimate object is to diminish vice and misery, and any checks to population, which may have been suggested, are solely as means to accomplish this end. To a rational being, the prudential check to population ought to be considered as equally natural with the check from poverty and premature mortality which these gentlemen seem to think so entirely sufficient and satisfactory; and it will readily occur to the intelligent reader, that one class of checks may be substituted for another, not only without essentially diminishing the population of a country, but even under a constantly progressive increase of it.

On the possibility of increasing very considerably the effective population of this country, I have expressed myself in some parts of my work more sanguinely, perhaps, than experience would warrant. I have said, that in the course of some centuries it might contain two or three times as many inhabitants as at present and yet every person be both better fed and better clothed. And in the comparison of the increase of population and food at the beginning of the Essay, that the argument might not seem to depend upon a difference of opinion respecting facts, I have allowed the produce of the earth to be unlimited, which is certainly going too far. It is not a little curious therefore, that it should still continue to be urged against me as an argument, that this country might contain two or three times as many inhabitants; and it is still more curious, that some persons, who have allowed the different ratios of increase on which all my principal conclusions are founded, have still asserted that no difficulty or distress could arise from population, till the productions of the earth could not be further increased. I doubt whether a stronger instance could readily be produced of the total absence of the power of reasoning, than this assertion, after such a concession, affords. It involves a greater absurdity than the saying, that because a farm can by proper management be made to carry an additional stock of four head of cattle every year, that therefore no difficulty or inconvenience would arise if an additional forty were placed in it yearly.

The power of the earth to produce subsistence is certainly not unlimited, but it is strictly speaking indefinite; that is, its limits are not defined, and the time will probably never arrive when we shall be able to say, that no further labour or ingenuity of man could make further additions to it. But the power of obtaining an additional quantity of food from the earth by proper management, and in a certain time, has the most remote relation imaginable to the power of keeping pace with an unrestricted increase of population. The knowledge and industry, which would enable the natives of New Holland to make the best use of the natural resources of their country, must, without an absolute miracle, come to them gradually and slowly; and even then, as it has amply appeared, would be perfectly ineffectual as to the grand object; but the passions which prompt to the increase of population are always in full vigour, and are ready to produce their full effect even in a state of the most helpless ignorance and barbarism. It will be readily allowed, that the reason why New Holland, in proportion to its natural powers, is not so populous as China, is the want of those human institutions which protect property and encourage industry; but the misery and vice which prevail almost equally in both countries, from the tendency of population to increase faster than the means of subsistence, form a distinct consideration, and miss from a distinct cause. They arise from the incomplete discipline of the human passions; and no person with the slightest knowledge of mankind has ever had the hardihood to affirm that human institutions could completely discipline all the human passions. But I have already treated this subject so fully in, the course of the work, that I am ashamed to add any thing further here.

The next grand objection which has been urged against me, is my denial of the *right* of the poor to support.

Those who would maintain this objection with any degree of consistency, are bound to shew, that the different ratios of increase with respect to population and food, which I attempted to establish at the beginning of the Essay, are fundamentally erroneous; since on the supposition of then being true, the conclusion is inevitable. If it appear, as it must appear on these ratios being allowed, that it is not possible for the industry of man to produce on a limited territory sufficient food for all that would be born, if every person were to marry at the time when he was first prompted to it by inclination, it follows irresistably, that all cannot have a *right* to support. Let us for a moment suppose an equal division of property in any country. If under these circumstances one half of the society were by prudential habits so to regulate their increase, that it exactly kept pace with their increasing cultivation, it is evident that the individuals of this portion of society would always remain as rich as at first. If the other half during the same time married at the age of puberty, when they would probably feel most inclined to it, it is evident that they would soon become wretchedly poor. But upon what plea of justice or equity could this second half of the society claim a right, in virtue of their poverty, to any of the possessions of the first half? This poverty had arisen entirely from their own ignorance or imprudence; and it would be perfectly clear, from the manner in which it had come upon them, that if their plea were admitted, and they were not suffered to feel the particular evils resulting from their conduct the whole society would shortly be involved in the same degree of wretchedness. Any voluntary and temporary assistance, which might be given as a measure of charity by the richer members of the society to the others, while they were learning to make a better use of the lessons of nature, would be quite a distinct consideration, and without doubt most properly applied; but nothing like a claim of *right* to support can possibly be maintained, till we deny the premises; till we affirm that the American increase of population is a miracle, and does not arise from the greater facility of obtaining the means of subsistence.

In fact, whatever we may say in our declamations on this subject, almost the whole of our *conduct* is founded on the non-existence of this right. If the poor had really a claim of *right* to support, I do not think that any man could justify his wearing broad cloth, or eating as much meat as he likes for dinner; and those who assert this right, and yet are rolling in their carriages, living every day luxuriously, and keeping even their horses on food of which their fellow-creatures are in want, mist be allowed to act with the greatest inconsistency. Taking an individual instance with out reference to consequences, it appears to me that Mr. Godwin's argument is irresistible. Can it be pretended for a moment that a part of the mutton which I expect to eat to-day would not be much more beneficially employed on some hard-working labourer, who has not perhaps tasted animal food for the last week, or on some poor family who cannot command sufficient food of any kind fully to satisfy the cravings of hunger? If these instances were not of a nature to multiply in proportion as such wants were indiscriminately gratified, the gratification of them as it would be practicable, would be highly beneficial; and in this case I should not have the smallest hesitation in most fully allowing the right. But as it appears clearly, both from theory and experience, that, if the claim were allowed, it would soon increase beyond the *possibility* of satisfying it; and that the practical attempt to do so would involve the human race in the most wretched and universal poverty; it follows necessarily that our conduct, which denies the right, is more suited to the present state of our being, than our declamations which allow it.

The great Author of nature, indeed, with that wisdom which is apparent in all his works, has not left this conclusion to the cold and speculative consideration of general consequences. By making the passion of self-love beyond comparison stronger then the passion of benevolence, he has at once impelled us to that line of conduct, which is essential to the preservation of the human race. If all that might be born could be adequately supplied, we cannot doubt, that he would have made the desire of giving to others as ardent as that of supplying ourselves. But since, under the present constitution of things, this is not so, he has enjoined every man to pursue, as his primary object, his own safety and happiness, and the safety and happiness of those immediately connected with him; and it is highly instructive to observe that, in proportion as the sphere contracts and the power of giving effectual assistance increases, the desire increases at the same time. In the case of children, who have certainly a claim of *right* to the support and protection of their parents, we generally find parental affection nearly as strong as self-love: and except in a few anomalous cases, the last morsel will be divided into equal shares.

By this wise provision the most ignorant are led to promote the general happiness, an end which they would have totally failed to attain, if the moving principle of their conduct had been benevolence. Benevolence indeed, as the great and constant source of action, would require the most perfect knowledge of causes and effects, and therefore can only be the attribute of the Deity. In a being so shortsighted as man, it would lead into the grossest errors, and soon transform the fair and cultivated soil of civilized society into a dreary scene of want and confusion.

But though benevolence cannot in the present state of our being be the great moving principle of human actions, yet, as the kind corrector of the evils arising from the other stronger passion, it is essential to human happiness; it is the balm and consolation and grace of human life, the source of our noblest efforts in the cause of virtue, and of our purest and most refined pleasures. Conformably to that system of general laws, according to which the Supreme Being appears with very few exceptions to act, a passim so strong and general as self-love could not prevail without producing much partial evil: and to prevent this passion from degenerating into the odious vice of selfishness, to make us sympathize in the pains and pleasures of our fellow-creatures, and feel the same *kind* of interest in their happiness and misery as in our own, though diminished in degree; to prompt us often to put ourselves in their place, that we may understand their wants, acknowledge their rights and do them good as we have opportunity; and to remind us continually, that even the passion which urges us to procure plenty for ourselves was not implanted in us for our own exclusive advantage, but as the means of procuring the greatest plenty for all; these appear to be the objects and offices of benevolence. In every situation of life there is ample room for the exercise of this virtue; and as each individual rises in society, as he advances in knowledge and excellence, as his power of doing good to others becomes greater, and the necessary attention to his own wants less, it will naturally come in for an increasing share among his constant motives of action. In situations of high trust and influence it ought to have a very large share, and in all public institutions it should be the great moving principle. Though we have often reason to fear that our benevolence may not take the most beneficial direction, we need never apprehend that there will be too much of it in society. The foundations of that passion on which our preservation depends, are fixed so deeply in our nature, that no reasonings or addresses to our feelings can essentially disturb it. It is just therefore and proper that all the positive precepts should be on the side of the weaker impulse; and we may safely endeavour to increase and extend its influence as much as we are able, if at the same time we are constantly on the watch, to prevent the evil which may arise from its misapplication.

The law, which in this country entitles the poor to relief, is undoubtedly different from a full acknowledgment of the natural right; and from this difference, and the many counteracting causes that arise from the mode of its execution, it will not of course be attended with the same consequences. But still it is an approximation to a full acknowledgment, and as such appears to produce much evil, both with regard to the habits and the temper of the poor. I have in consequence ventured to suggest a plan of gradual abolition, which, as might be expected, has not met with universal approbation. I can readily understand any objections that may be made to it on the plea, that, the right having been once acknowledged in this country, the revocation of it might at first excite discontents; and I should therefore most fully concur in the propriety of proceeding with the greatest caution, and of using all possible means of preventing any sudden shock to the opinions of the poor. But I have never been able to comprehend the grounds of the further assertion, which I have sometimes heard made, that if the poor were really convinced that they had no claim of right to relief, they would in general be more inclined to be discontented and seditious. On these occasions, the only way I have of judging is to put myself in imagination in the place of the poor man, and consider how I should feel in his situation. If I were told that the rich, by the laws of nature and the laws of the land, were bound to support me, I could not, in the first place, feel much obligation for such support; and, in the next place, if I were given any food of an inferior kind, and could not see the absolute necessity of the change, which would probably be the case, I should think that I had good reason to complain. I should feel, that the laws had been violated to my injury, and that I had been unjustly deprived of my right. Under these circumstances, though I might be deterred by the fear of an armed force from committing any overt acts of resistance, yet I should consider myself as perfectly justified in so doing, if this fear were removed; and the injury, which I believed that I had suffered, might produce the most unfavourable effects on my general dispositions towards the higher classes of society. I cannot indeed conceive any thing more irritating to the human feelings, than to experience that degree of distress, which, in spite of all our poor-laws and benevolence, is not unfrequently felt in this country; and yet to believe that these sufferings were not brought upon me either by my own faults, or by the operation of those general laws which, like the tempest, the blight or the pestilence, are continually falling hard on particular individuals, while others entirely escape, but were occasioned solely by the avarice and injustice of the higher classes of society.

On the contrary, if I firmly believed that by the laws of nature, which are the laws of God, I had no claim of *right* to support, I should, in the first place, feel myself more strongly bound to a life of industry and frugality; but if want, notwithstanding, came upon me, I should consider it in the light of sickness, as an evil incidental to my present state of being, and which, if I could not avoid, it was my duty to bear with fortitude and resignation. I should know from past experience, that the best title I could have to the assistance of the benevolent would be, the not having brought myself into distress by my own idleness or extravagance. What I received would have the best effect on my feelings towards the higher classes. Even if it were much inferior to what I had been accustomed to, it would still, instead of an injury, be on obligation; and conscious that I had no claim of *right,* nothing but the dread of absolute famine, which might overcome all other considerations, could palliate the guilt of resistance.

I cannot help believing that, if the poor in this country were convinced that they had no claim of *right* to support, and yet in scarcities and all cases of urgent distress were liberally relieved, which I think they would be; the bond, which unites the rich with the poor, would be drawn much closer than at present; and the lower classes of society, as they would have less real reason for imitation and discontent, would be much less subject to these uneasy sensations.

Among those who have objected to my declaration, that the poor have no claim of *right* to support, is Mr. Young, who, with a harshness not quite becoming a candid inquirer after truth, has called my proposal for the gradual abolition of the poor-laws a horrible plan, and asserted that the execution of it would be a most iniquitous proceeding. Let this plan however be compared for a moment with that which he himself and others have proposed, of fixing the sum of the poor's rates, which on no account is to be increased. Under such a law, if the distresses of the poor mere to be aggravated tenfold, either by the increase of numbers or the recurrence of a scarcity, the same sum would invariably be appropriated to their relief. If the statute which gives the poor a right to support were to remain unexpunged, we should add to the cruelty of starving them the injustice of still *professing* to relieve them. If this statute were expunged or altered, we should virtually deny the right of the poor to support, and only retain the absurdity of saying, that they had a right to a certain sum; an absurdity on which Mr. Young justly comments with much severity in the case of France. In both cases the hardships which they would suffer would be much more severe, and would come upon them in a much more unprepared state, than upon the plan proposed in the Essay.

According to this plan, all that are already married, and even all that are engaged to marry during the course of the year, and all their children, would be relieved as usual; and only those who marry subsequently, and who of course may be supposed to have made better provision for contingencies, would be out of the pale of relief.

Any plan for the abolition of the poor-laws must presuppose a general acknowledgment that they are essentially wrong, and that it is necessary to tread back our steps. With this acknowledgment, whatever objections may be made to my plan, in the too frequently short-sighted views of policy, I have no fear of comparing it with any other that has yet been advanced, in point of justice and humanity; and of course the terms iniquitous and horrible "pass by me like the idle wind, which I regard not."

Mr. Young, it would appear, has now given up this plan. He has pleaded for the privilege of being inconsistent, and has given such reasons for it that I am disposed to acquiesce in them, provided he confines the exercise of this privilege to different publications, in the interval between which he may have collected new facts. But I still think it not quite allowable in the same publication: and yet it appears that in the very paper, in which he has so severely condemned my scheme, the same arguments, which he has used to reprobate it, are applicable with equal force against his own proposal, as there explained.

He allows that his plan can provide only for a certain number of families, and has nothing to do with the increase from them; but in allowing this, he allows that it does not reach the grand difficulty attending a provision for the poor. In this most essential point, after reprobating me for saying, that the poor have no claim of *right* to support, he is compelled to adopt the very same conclusion; and to own that "it might be prudent to consider the misery to which the progressive population might be subject, when there was not a sufficient demand for them in towns and manufactures, as an evil which it was absolutely and physically impossible to prevent." Now the sole reason why I say that the poor have no claim of *right* to support, is the physical impossibility of relieving this progressive population. Mr. Young expressly acknowledges this physical impossibility; yet with an inconsistency scarcely credible still declaims against my declaration.

The power, which the society may possess of relieving a certain portion of the poor, is a consideration perfectly distinct from the general question; and I am quite sure I have never said that it is not our duty to do all the good that is practicable. But this limited power of assisting individuals cannot possibly establish a general right. If the poor have really a natural right to support, and if nor present laws be only a confirmation of this right, it ought certainly to extend unimpaired to all who are in distress, to the increase from the cottagers as well as to the cottagers themselves; and it would be a palpable injustice in the society, to adopt Mr. Young's plan, and purchase from the present generation the disfranchisement of their posterity.

Mr. Young objects very strongly to that passage of the Essay, in which I observe that a man, who plunges himself into poverty, and dependence by marrying without any prospect of bring able to maintain his family, has more reason to accuse himself than the price of labour, the parish, the avarice of the rich, the institutions of society, and the dispensations of Providence; except as far as he has been deceived by those who ought to have instructed him. In answer to this, Mr. Young says that the poor fellow is justified in every one of these complaints, that of Providence alone excepted; and that, seeing other cottagers living comfortably with three or four acres of land, he has cause to accuse institutions which deny him that which the rich could well spare, and which would give him all he wants. I would beg Mr. Young for a moment to consider how the matter would stand, if his own plan were completely executed. After all the commons had been divided as he has proposed, if a labourer had more than one son, in what respect would the second or third be in a different situation from the man that I have supposed? Mr. Young cannot possibly mean to say that, if he had the very natural desire of marrying at twenty, he would still have a right to complain that the society did not give him a house and three or four acres of land. He has indeed expressly denied this absurd consequence, though in so doing he has directly contradicted the declaration just quoted. The progressive population, he says, would, according to his system, be cut off from the influence of the poor-laws, and the encouragement to marry would remain exactly in that proportion less than at present. Under these circumstances, without land, without the prospect of parish relief, and with the price of labour only sufficient to maintain two children, can Mr. Young seriously think that the poor man, if he be really aware of his situation, does not do wrong in marrying, and ought not to accuse himself for following what Mr. Young calls the dictates of God, of nature and of revelation? Mr. Young cannot be unaware of the wretchedness that must inevitably follow a marriage under such circumstances. His plan makes no provision whatever for altering these circumstances. He must therefore totally disregard all the misery arising from excessive poverty; or, if he allows that these supernumerary members must necessarily wait, either till a cottage with land becomes vacant in the country, or that by emigrating to towns they can find the means of providing for a family, all the declamation, which he has urged with such pomp against deferring marriage in my system, would be equally applicable in his own. In fact, if Mr. Young's plan really attained the object, which it professes to have in view, that of bettering the condition of the poor; and did not defeat its intent by encouraging a too rapid multiplication, and consequently lowering the price of labour; it cannot be doubted that not only the supernumerary members just mentioned, but all the labouring poor, must wait longer before they could marry than they do at present.

The following proposition may be said to be capable of mathematical demonstration. In a country, the resources of which will not permanently admit of an increase of population more rapid than the existing rate, no improvement in the condition of the people, which would lead to diminish mortality, could *possibly* take place without being accompanied by a smaller proportion of births, supposing of course no particular increase of emigration. To a person who has considered the subject, there is no proposition in Euclid, which brings home to the mind a stronger conviction than this; and there is no truth so invariably confirmed by all the registers of births, deaths and marriages, that have ever been collected. In this country it has appeared that, according to the returns of the Population Act in 1801, the proportion of births to deaths was about 4 to 3. This proportion with a mortality of 1 in 40 would double the population in 83 years and a half; and as we cannot suppose that the country could admit of more than a quadrupled population in the next hundred and sixty-six years, we may safely say that its resources will not allow of a permanent rate of increase greater than that which was then taking place. But if this be granted, it follows as a direct conclusion, that if Mr. Young's plan, or any other, really succeeded in bettering the condition of the poor, and enabling them to rear more of their children, the vacancies in cottages in proportion to the number of expectants would happen slower than at present, and the age of marriage]must inevitably be later.

With regard to the expression of later marriages, it should always be recollected that it refers to no particular age, but is entirely comparative. The marriages in England are later than in France, the natural consequence of that prudence and respectability generated by a better government; and can we doubt that good has been the result? The marriages in this country now are later than they were before the revolution; and I feel firmly persuaded, that the increased healthiness observed of late years could not possibly have taken place without this accompanying circumstance. Two or three years in the average age of in marriage, by lengthening each generation, and tending, in a small degree, both to diminish the prolifickness of marriages, and the number of born living to be married, may make a considerable difference in the rate of increase, and be adequate to allow for a considerably diminished mortality. But I would on no account talk of any limits whatever. The only plain and intelligible measure with regard to marriage, is the having a fair prospect of being able to maintain a family. If the possession of one of Mr. Young's cottages would give the labourer this prospect, he would be quite right to marry; but if it did not, or if he could only obtain a rented house without land, and the wages of labour were only sufficient to maintain two children, does Mr. Young, who cuts him off from the influence of the poor-laws, presume to say, that he would still he right in marrying?

Mr. Young has asserted that I have made perfect chastity in the single state absolutely necessary to the success of my plan; but this surely is a misrepresentation. Perfect virtue is, indeed, necessary to enable man to avoid *all* the moral and physical evils which depend upon his own conduct; but who ever expected perfect virtue upon earth? I have said, what I conceive to be strictly true, that it is our duty to defer marriage till we can feed our children; and that it is also our duty not to indulge ourselves in vicious gratifications; but I have never said that I expected either, much less both, of these duties to be completely fulfilled. In this, and a number of other cases, it may happen that the violation of one of two duties will enable a man to perform the other with greater facility; but if they be really both duties, and both practicable, no power *on earth* can absolve a man from the guilt of violating either. This can only be done by that God, who can weigh the crime against the temptation, and will temper justice with mercy. The moralist is still bound to inculcate the practice of both duties; and each individual must be left to act under the temptations to which he is exposed, as his conscience shall dictate. Whatever I may have said in drawing a picture *professedly* visionary, for the sake of illustration; in the practical application of my principles I have taken man as he is, with all his imperfections on his head. And thus viewing him, and knowing that some checks to population must exist, I have not the slightest hesitation in saying, that the prudential check to marriage is better than premature mortality. And in this decision I feel myself completely justified by experience.

In every instance that can be traced, in which an improved government has given to its subjects a greater degree of foresight, industry, and personal dignity, these effects, under similar circumstances of increase, have invariably been accompanied by a diminished proportion of marriages. This is a proof that an increase of moral worth in the general character is not, at least, *incompatible* with an increase of temptations with respect to one particular vice; and the instances of Norway, Switzerland, England, and Scotland, adduced in the last chapter of the Essay, shew that, in comparing different countries together, a smaller proportion of marriages and births does not necessarily imply the greater prevalence even of this particular vice. This is surely quite enough for the legislator. He cannot estimate, with tolerable accuracy, the degree in which chastity in the single state prevails.

His general conclusions must be founded on general results, and these are clearly in his favour.

To much of Mr. Young's plan, as he has at present explained it, I should by no means object. The peculiar evil which I apprehended from it, that of taking the poor from the consumption of wheat, and feeding them on milk and potatoes, might certainly be avoided by a limitation of the number of cottages; and I entirely agree with him in thinking, that we should not be deterred from making 500,000 families more comfortable, because we cannot extend the same relief to all the rest. I have, indeed, myself ventured to recommend a general improvement of cottages, and even the cow system on a limited scale; and, perhaps with proper precautions, a certain portion of land might be given to a considerable body of the labouring classes.

If the law which entitles the poor to support were to be repealed, I should most highly approve of any plan which would tend to render such repeal more palatable on its first promulgation: and, in this view, some kind of compact with the poor might be very desirable. A plan of letting land to labourers, under certain conditions, has lately been tried in the parish of Long Newnton, in Gloucestershire; and the result, with a general proposal founded on it, has been submitted to the public by Mr. Estcourt. The present success has been very striking; but, in this, and every other case of the kind, we should always bear in mind, that no experiment respecting a provision for the poor can be said to be complete till succeeding generations have arisen. I doubt if ever there has been an instance of any thing like a liberal institution for the poor, which did not succeed on its first establishment, however it might have failed afterwards. But this consideration should by no means deter us from making such experiments, when present good is to be obtained by them, and a future overbalance of evil is not justly to be apprehended. It should only make us less rash in drawing our inferences.

With regard to the general question of the advantages to the lower classes of possessing land, it should be recollected that such possessions are by no means a novelty. Formerly this system prevailed in almost every country with which we are acquainted, and prevails at present in many countries, where the peasants are far from being remarkable for their comforts, but are, on the contrary, very poor, and particularly subject to scarcities. With respect to this latter evil, indeed, it is quite obvious that a peasantry which depends principally on its possessions in land, must be more exposed to it than one which depends on the general wages of labour. When a year of deficient crops occurs in a country of any extent and diversity of soil, it is always partial, and some districts are more affected than others. But when a bad crop of grass, corn, or potatoes, or a mortality among cattle, falls on a poor man, whose principal dependence is on two or three acres of land, he is in the most deplorable and helpless situation. He is comparatively without money to purchase supplies, and is not for a moment to be compared with the man who depends on the wages of labour, and who will, of course, be able to purchase that portion of the general crop, whatever it may be, to which his relative situation in the society entities him. In Sweden, where the farmers' labourers are paid principally in land, and often keep two or three cows, it is not uncommon for the peasants of one district to be almost starving, while their neighbours at a little distance are living in comparative plenty. It will be found indeed generally, that, in almost all the countries which are particularly subject to scarcities and famines, either the farms are very small, or the labourers are paid principally in land. China, Indostan, and the former state of the Highlands of Scotland, furnish some proofs among many others of the truth of this observation; and in reference to the small properties of France, Mr. Young himself, in his Tour, particularly notices the distress arising from the least failure of the crops; and observes, that such a deficiency, as in England posses almost without notice, in France is attended with dreadful calamities.

Should any plan, therefore, of assisting the poor by land be adopted in this country, it would be absolutely essential to its ultimate success, to prevent them from making it their principal dependence. And this might probably be done by attending strictly to the two following rules. Not to let the division of land be so great as to interrupt the cottager essentially in his usual labours; and always to stop in the farther distribution of land and cottages, when the price of labour, independently of any assistance from land, would not, at the average price of corn, maintain three, or, at least, two children. Could the matter be so ordered, that the labourer, in working for others, should still continue to earn the same real command over the necessaries of life that he did before, a very great accession of comfort and happiness might accrue to the poor from the possession of land, without any evil that I can foresee at present. But if these points were not attended to, I should certainly fear an approximation to the state of the poor in France, Sweden, and Ireland; nor do I think that any of the partial experiments that have yet taken place afford the slightest presumption to the contrary. The result of these experiments is, indeed, exactly such as one should have expected. Who could ever have doubted that, if, without lowering the price of labour, or taking the labourer off from his usual occupations, you could give him the produce of one or two acres of land and the benefit of a cow, you would decidedly raise his condition? But it by no means follows that he would retain this advantage, if the system were so extended, as to make the land his principal dependence, to lower the price of labour, and, in the language of Mr. Young, to take the poor from the consumption of wheat and feed them all milk and potatoes. It does not appear to me so marvellous as it does to Mr. Young that the very same system, which in Lincolnshire and Rutlandshire may produce now the most comfortable peasantry in the British dominions, should in the end, if extended without proper precautions, assimilate the condition of the labourers of this country to that of the lower classes of the Irish.

It is generally dangerous and impolitic in a government to take it upon itself to regulate the supply of any commodity in request; and probably the supply of labourers forms no exception to the general rule. I would on no account, therefore propose a positive law to regulate their increase; but as any assistance which the society might give them cannot, in the nature of doings, be unlimited, the line may fairly be drawn where we please; and with regard to the increase from this point, every thing would be left as before to individual exertion and individual speculation.

If any plan of this kind were adopted by the government I cannot help thinking that it might be made the means of giving the best kind of encouragement and reward to those who we employed in our defence. If the period of enlisting were only for a limited time, and at the expiration of that time every person who had conducted himself well were entitled to a house and a small portion of land, if a country labourer, and to a tenement in a town and a small pension, if an artificer (all inalienable), a very strong motive would be held out to young men, not only to enter into the service of their country, but to behave well in that service; and, in a short time, there would be such a martial population at home as the unfortunate state of Europe seems in a most peculiar manner to require. As it is only limited assistance that the society can possibly give, it seems in every respect fair and proper that, in regulating this limit, some important end should be attained.

If the poor-laws be allowed to remain exactly in their present state, we ought, at least, to be aware to what cause it is owing, that their effects have not been more pernicious than they are observed to be; that we may not complain of, or alter those parts, without which we should really not have the power of continuing them. The law which obliges each parish to maintain its own poor is open to many objections. It keeps the overseers and churchwardens continually on the watch to prevent new comers, and constantly in a state of dispute with other parishes. It thus prevents the free circulation of labour from place to place, and renders its price very unequal in different parts of the kingdom. It disposes all landlords rather to pull down than to build cottages on their estates; and this scarcity of habitations in the country, by driving more to the towns than would otherwise have gone, gives a relative discouragement to agriculture, and a relative encouragement to manufactures. These, it must be allowed, are no inconsiderable evils; but if the cause which occasions them were removed, evils of much greater magnitude would follow. I agree with Mr. Young in thinking that there is scarcely a parish in the kingdom, where, if more cottages were built, and let at any tolerably moderate rents they would not be immediately filled with new couples. I even agree with him in thinking that, in some places, this want of habitations operates too strongly in preventing marriage. But I have not the least doubt that, considered generally, its operation in the present state of things is most beneficial; and that it is almost exclusively owing to this cause that we have been able so long to continue the poor-laws. If any man could build a hovel by the road-side, or on the neighbouring waste, without molestation; and yet were secure that he and his family would always be supplied with work and food by the parish, if they were not readily to be obtained elsewhere; I do not believe that it would be long before the physical impossibility of executing the letter of the poor-laws would appear. It is of importance, therefore, to be aware that it is not because this or any other society has really the power of employing and supporting all that might be born, that we have been able to cntinue the present system; but because by the indirect operation of this system, not adverted to at the time of its establishment and frequently reprobated since, the number of births is always very greatly limited, and thus reduced within the pale of possible support.

The obvious tendency of the poor-laws is certainly to encourage marriage; but a closer attention to all their indirect us well as direct effects may make it a matter of doubt to what extent they really do this. They clearly tend, in their general operation, to discourage sobriety and economy, to encourage idleness and the desertion of children, and to put virtue and vice more on a level than they otherwise would be; but I will not presume to say positively that they greatly encourage population. It is certain that the proportion of births in this country compared with others in similar circumstances is very small; but this was to be expected from the superiority of the government, the more respectable state of the people, and the more general diffusion of a taste for cleanliness and conveniences. And it will readily occur to the reader, that owing to these causes, combined with the twofold operation of the poor-laws, it must be extremely difficult to ascertain, with any degree of precision, what has been their effect on population.

The only argument of a general nature against the Essay, which strikes me as having any considerable force, is the following. It is against the application of its principles, not the principles themselves, and has not, that I know of, been yet advanced in its present form. It may be said that, according to my own reasonings and the facts stated in my work, it appears that the diminished proportion of births, which I consider as absolutely necessary to the permanent improvement of the condition of the poor, invariably follows an improved government, and the greater degree of personal respectability which it gives to the lower classes of society. Consequently allowing the desirableness of the end, it is not necessary, in order to obtain it, to risk the promulgation of any new opinions which may alarm the prejudices of the poor, and the effect of which we cannot with certainly foresee; but we have only to proceed in improving our civil polity, conferring the benefits of education upon all, and removing every obstacle to the general extension of all those privileges and advantages which may be enjoyed in common; and we may be quite sure that the effect, to which I look forward, and which can alone render these advantages permanent, will follow.

I acknowledge the truth and force of this argument, and have only to observe, in answer to it, that it is difficult to conceive, that we should not proceed with more celerity and certainly towards the end in view, if the principal causes, which tend to promote or retard it, were generally known. In particular, I cannot help looking forward to a very decided improvement in the habits and temper of the lower classes, when their real situation has been clearly explained to them; and if this were done gradually and cautiously, and accompanied with proper moral and religions instructions, I should not expect any danger from it. I am always unwilling to believe, that the general dissemination of truth is prejudicial. Cases of this kind are undoubtedly conceivable; but they should be admitted with very great caution. If the general presumption in favour of the advantage of truth were once essentially shaken, all ardour in its cause would share the same fate; and the interests of knowledge and virtue most decidedly suffer. It is, besides, a species of arrogance not lightly to be encouraged, for any man to suppose that he has penetrated further into the laws of nature than the great Author of them intended, further than is consistent with the good of mankind.

Under these impressions I have freely given my opinions to the public. In the truth of the general principles of the Essay I confess that I feel such a confidence, that, till something has been advanced against them very different indeed from any thing that has hitherto appeared, I cannot help considering them as incontrovertible. With regard to the application of these principles, the case is certainly different; and as dangers of opposite kinds are to be guarded against, the subject will, of course, admit of much latitude of opinion. At all events, however, it must be allowed that, whatever may be our determination respecting the advantages or disadvantages of endeavouring to circulate the truths on this subject among the poor, it most be highly advantagous that they should be known to all those who have it in their power to influence the laws and institutions of society. That the body of an army should not in all cases know the particulars of their situation may possibly be desirable; but that the leaders should be in the same state of ignorance will hardly, I think, be contended.

If it be really true, that without a diminished proportion of births we cannot attain any *permanent* improvement in the health and happiness of the mass of the people, and cannot secure that description of population, which, by containing a larger share of adults, is best calculated to create fresh resources, and consequently to encourage a continued increase of efficient population; it is surely of the highest importance that this should be known, that, if we take no steps directly to promote this effect, we should not, under the influence of the former prejudices on this subject, endeavour to counteract it. And if it he thought unadviseable, to abolish the poor-laws, it cannot be doubted, that a knowledge of those general principles, which render them inefficient in their humane intentions, might be applied so far to modify them and regulate their execution, as to remove many of the evils with which they are accompanied, and make them less objectionable.

There is only one subject more which I shall notice, and that is rather a matter of feeling than of argument. Many persons, whose understandings are not so constituted that they can regulate their belief or disbelief by their likes or dislikes, have professed their perfect conviction of the truth of the general principles contained in the Essay; but, at the same time, have lamented this conviction, as throwing a darker shade over our vices of human nature, and tending particularly to narrow our prospects of future improvement. In these feelings I cannot agree with them. If, from a review of the past, I could not only believe that a fundamental and very extraordinary improvement in human society was possible, but feel a firm confidence that it would take place, I should undoubtedly be grieved to find, that I had overlooked some cause, the operation of which would at once blast my hopes. But if the contemplation of the past history of mankind, from which alone we can judge of the future, renders it almost impossible to feel such a confidence, I confess that I had much rather believe that some real and deeply-seated difficulty existed, the constant struggle with which was calculated to rouse the natural inactivity of man, to call forth his facilities, and invigorate and improve his mind; a species of difficulty, which it must be allowed is most eminently and peculiarly suited to a state of probation; than that nearly all the evils of life might with the most perfect facility be removed, but for the perverseness and wickedness of those who influence human institutions.

A person who held this latter opinion must necessarily live in a constant state of irritation and disappointment. The ardent expectations, with which he might begin life, would soon receive the most cruel check. The regular progress of society, under the most favourable circumstances, would to him appear slow and unsatisfactory; but instead even of this regular progress, his eye would be more frequently presented with retrograde movements, and the most disheartening reverses. The changes, to which he had looked forward with delight, would be found big with new and unlooked-for evils; and the characters, on which he had reposed the most confidence, would be seen frequently deserting his favourite cause, either from the lessons of experience or the temptations of wealth and power. In this state of constant disappointment, he would be but too apt to attribute every thing to the worst motives; he would be inclined to give up the cause of improvement in despair; and judging of the whole from a part, nothing but a peculiar goodness of heart and amiableness of disposition could preserve him from that sickly and disgusting misanthropy, which is but too frequently the end of such characters.

On the contrary, a person who held the other opinion, as he would set out with more moderate expectations, would of course be less liable to disappointment. A comparison of the best with the worst states of society, and the obvious inference from analogy, that the best were capable of further improvement, would constantly present to his mind a prospect sufficiently animating to warrant his most persevering exertions. But aware of the difficulties with which the subject was surrounded, knowing how often in the attempt to attain one object some other had been lost, and that, though society had made rapid advances in some directions, it had been comparatively stationary in others, he would be constantly prepared for failures. These failures, instead of creating despair, would only create knowledge; instead of checking his ardour, would give it a wiser and more successful direction; and, having founded his opinion of mankind on broad and general grounds, the disappointment of any particular views would not change this opinion; but even in declining age he would probably be found believing as firmly in the reality and general prevalence of virtue as in the existence and frequency of vice; and to the last, looking forward with a just confidence to those improvements in society, which the history of the past, in spite of all the reverses with which it is accompanied, seems clearly to warrant.

It may be true that, if ignorance is bliss, 'tis folly to be wise; but if ignorance be out bliss, as in the present instance; if all false views of society must not only impede decidedly the progress of improvement, but necessarily, terminate in the most bitter disappointments to the individuals who formed them; I shall always think that the feelings and prospects of those, who make the justest estimates of our future expectations, are the most consolatory; and that the characters of this description are happier themselves, at the same time that they are beyond comparison more likely to contribute to the improvement and happiness of society.

4.2.16 Appendix II,: 1817

Since the publication of the last edition of this Essay in 1807, two Works have appeared, the avowed objects of which are directly to oppose its principles and conclusions. These are the *Principles of Population and Production,* by Mr. Weyland; and *an Inquiry into the Principle of Population,* by Mr. James Grahame.

I would willingly leave the question as it at present stands to the judgment of the public, without any attempt on my part to influence it further by a more particular reply; but as I professed my readiness to enter into the discussion of any serious objections to my principles and conclusions, which were brought forward in a spirit of candour and truth; and as one at least of the publications above mentioned may be so characterised, and the other is by no means deficient in personal respect; I am induced shortly to notice them.

I should not however have thought it necessary to advert to Mr. Grahame's publication, which is a slight work without any very distinct object in view, if it did not afford some strange specimens of misrepresentation, which it may be useful to point out.

Mr. Grahame in his second chapter, speaking of the tendency exhibited by the law of human increase to a redundance of population, observes, that some philosophers have considered this tendency as a mark of the foresight of nature, which has thus provided a ready supply for the waste of life occasioned by human vices and passions; while "others, of whom Mr. Malthus is the leader, regards the vices and follies of human nature, and their various products, famine, disease and war as *benevolent remedies* by which nature has enabled human beings to correct the disorders that would arise from that redundance of population which the unrestrained operation of her laws would create."

These are the opinions imputed to me and the philosophers with whom I am associated. If the imputation were just, we have certainly on many accounts great reason to be ashamed of ourselves. For what are we made to say? In the first place, we are stated to assert that famine is a benevolent remedy for *want of food,* as redundance of population admits of no other interpretation than that of a people ill supplied with the means of subsistence, and consequently the benevolent remedy of famine here noticed can only apply to the disorders arising from scarcity of food.

Secondly; we are said to affirm that nature enables human beings by means of diseases to correct the disorders that would arise from a redundance of population;—that is, that mankind willingly and purposely create diseases, with a view to prevent those diseases which are the necessary consequence of a redundant population, and are not worse or more mortal than the means of prevention.

And thirdly, it is imputed to us generally, that we consider the vices and follies of mankind as benevolent remedies for the disorders arising from a redundant population; and it follows as a matter of course that these vices ought to be encouraged rather than reprobated.

It would not be easy to compress in so small a compass a greater quantity of absurdity, inconsistency, and unfounded assertion.

The two first imputations may perhaps be peculiar to Mr. Grahame; and protection from them may be found in their gross absurdity and inconsistency. With regard to the third, it must be allowed that it has not the merit of novelty. Although it is scarcely less absurd than the two others, and has been shewn to be an opinion no where to be found in the Essay, nor legitimately to be inferred from any part of it, it has been continually repeated in various quarters for fourteen years, and now appears in the pages of Mr. Grahame. For the last time I will now notice it; and should it still continue to be brought forward, I think I may be fairly excused from paying the slightest further attention either to the imputation itself, or to those who advance it.

If I had merely stated that the tendency of the human race to increase faster than the means of subsistence was kept to a level with these means by some or other of the forms of vice and misery, and that these evils were absolutely unavoidable, and incapable of being diminished by any human efforts; still I could not with any semblance of justice be accused of considering vice and misery as the remedies of these evils, instead of the very evils themselves. As well nearly might I be open to Mr. Grahame's imputations of considering the famine and disease necessarily arising from a scarcity of food as a benevolent remedy for the evils which this scarcity occasions.

But I have not so stated the proposition. I have not considered the evils of vice and misery arising from a redundant population as unavoidable, and incapable of being diminished. On the contrary, I have pointed out a mode by which these evils may be removed or mitigated by removing or mitigating their cause. I have endeavoured to shew that this may be done consistently with human virtue and happiness. I have never considered any possible increase of population as an evil, except as far as it might increase the proportion of vice and misery. Vice and misery, and these alone, are the evil, which it has been my great object to contend against. I have expressly proposed moral restraint as their rational and proper remedy; and whether the remedy be good or bad, adequate or inadequate, the proposal itself, and the stress which I have laid upon it, is an incontrovertible proof that I never can have considered vice and misery as themselves remedies.

But not only does the general tenour of my work, and the specific object of the latter part of it, clearly shew, that I do not consider vice and misery as remedies; but particular passages in various parts of it are so distinct on the subject, as not to admit of being misunderstood by the most perverse blindness.

It is therefore quite inconceivable that any writer with the slightest pretension to respectability should venture to bring forward such imputations; and it must be allowed to shew either such a degree of ignorance, or such a total want of candour, as utterly to disqualify him for the discussion of such subjects.

But Mr. Grahame's misrepresentations are not confined to the passage above referred to. In his introduction he observes that, in order to check a redundant population, the evils of which I consider as much nearer than Mr. Wallace, I "recommend immediate recourse to human efforts, to the restraints prescribed by Condorcet, for the correction or mitigation of the evil." This is an assertion entirely without foundation. I have never adverted to the check suggested by Condorcet without the most marked disapprobation. Indeed I should always particularly reprobate any artificial and unnatural modes of checking population, both on account of their immorality and their tendency to remove a necessary stimulus to industry. If it were possible for each married couple to limit by a wish the number of their children, there is certainly reason to fear that the indolence of the human race would be very greatly increased; and that neither the population of individual countries, nor of the whole earth, would ever reach its natural and proper extent. But the restraints which I have recommended are quite of a different character. They ae not only pointed out by reason and sanctioned by religion, but tend in the most marked manner to stimulate industry. It is not easy to conceive a more powerful encouragement to exertion and good conduct than the looking forward to marriage as a state peculiarly desirable: but only to be enjoyed in comfort, by the acquisition of habits of industry, economy, and prudence. And it is in this light that I have always wished to place it.

In speaking of the poor-laws in this country, and of their tendency (particularly as they have been lately administered) to eradicate all remaining spirit of independence among our peasantry, I observe that, "hard as it may appear in individual instances, dependent poverty ought to be held disgraceful;" by which of course I only mean that such a proper degree of pride as will induce a labouring man to make great exertions as in Scotland, in order to prevent himself or his nearest relations from falling upon the parish, is very desirable, with a view to the happiness of the lower classes of society. The interpretation which Mr. Grahame gives to this passage is, that the rich "are so to imbitter the pressure of indigence by the stings of contumely, that men may be driven by their pride to prefer even the refuge of despair to the condition of dependence!!"—a curious specimen of misrepresentation and exaggeration.

I have written a chapter expressly on the practical direction of our charity; and in detached passages elsewhere have paid a just tribute to the exalted virtue of benevolence. To those who have read these parts of my work, and have attended to the general tone and spirit of the whole, I willingly appeal, if they are but tolerably candid, against these charges of Mr. Grahame, which intimate that I would root out the virtues of charity and benevolence without regard to the exaltation which they bestow on the moral dignity of our nature; and that in my view the "rich are required only to harden their hearts against calamity, and to prevent the charitable visitings of their nature from keeping alive in them that virtue which is often the only moral link between them and their fellow-mortals." It is not indeed easy to suppose that Mr. Grahame can have read the chapter to which I allude, as both the letter and spirit of it contradict, in the most express and remarkable manner, the imputations conveyed in the above passages.

These are a few specimens of Mr. Grahame's misrepresentations, which might easily be multiplied; but on this subject I will only further remark that it shews no inconsiderable want of candour to continue attacking and dwelling upon passages which have ceased to form a part of the work controverted. And this Mr. Grahame has done in more instances than one, although he could hardly fail to know that he was combating expressions and passages which I have seen reason to alter or expunge.

I really should not have thought it worth while to entire these misrepresentations of Mr. Grahame, if, in spite of them, the style and tone of his publication had not appeared to me to be entitled to more respect than most of my opponents.

With regard to the substance and aim of Mr. Grahame's work, it seems to be intended to shew that emigration is the remedy provided by nature for a redundant population; and that if this remedy cannot be adequately applied, there is no other that can be proposed, which will not lead to consequences worse than the evil itself. These are two points which I have considered at length in the Essay; and it cannot be necessary to repeat any of the arguments here. Emigration, if it could be freely used, has been shewn to be a resource which could not be of long duration. It cannot therefore under any circumstances be considered as an adequate remedy. The latter position is a matter of opinion, and may rationally be held by any person who sees reason to think it well founded. It appears to me, I confess, that experience most decidedly contradicts it; but to those who think otherwise, there is nothing more to be said than that they are bound in consistency to acquiesce in the necessary consequences of their opinion. These consequences are, that the poverty and wretchedness arising from a redundant population, or, in other words, from very low wages and want of employment, are absolutely irremediable, and must be continually increasing as the population of the earth proceeds; and that all the efforts of legislative wisdom and private charity, though they may afford a wholesome and beneficial exercise of human virtue, and may occasionally alter the distribution and vary the pressure of human misery, can do absolutely nothing towards diminishing the general amount or checking the increasing weight of this pressure.

Mr. Weyland's work is of a much more elaborate description than that of Mr. Grahame. It has also a very definite object in view: and although, when he enters into the details of his subject, he is compelled entirely to agree with me respecting the checks which practically keep down population to the level of the means of subsistence, and has not in fact given a single reason for the slow progress of population in the advanced stages of society, that does not clearly and incontrovertibly come under the heads of moral restraint, vice, or misery; yet it must be allowed that he sets out with a bold and distinct denial of my premises, and finishes, as he ought to do from such a beginning, by drawing the most opposite conclusions.

After stating fairly my main propositions, and adverting to the conclusion which I have drawn from them, Mr. Weyland says, "Granting the premises, it is indeed obvious that this conclusion is undeniable."

I desire no other concession than this; and if my premises can be shewn to rest on unsolid foundations, I will most readily give up the inferences I have drawn from them.

To determine the point here at issue it cannot be necessary for me to repeat the proofs of these premises derived both from theory and experience, which have already so fully been brought forwards. It has been allowed that they have been stated with tolerable clearness; and it is known that many persons have considered them as unassailable, who still refuse to admit the consequences to which they appear to lead. All that ran be required therefore on the present occasion is to examine the validity of the objections to these premises brought forward by Mr. Weyland.

Mr. Weyland observes, "that the origin of what are conceived to be the mistakes and false reasonings, with respect to the principle of population, appears to be the assumption of a tendency to increase in the human species, the quickest that can be proved possible in any particular state of society, as that which is natural and theoretically possible in all; and the characterizing of every cause which tends to prevent such quickest possibly rate as checks to the natural and spontaneous tendency of population to increase; but as checks evidently insufficient to stem the progress of an overwhelming torrent. This seems as eligible a mode of reasoning, as if one were to assume the height of the Irish giant in the natural standard of the stature of men, and to cell every reason, which may be suggested as likely to prevent the generality of men from reaching it, checks to their growth."

Mr. Weyland has here most unhappily chosen his illustration, as it is in no respect applicable to the case. In order to illustrate the different rates at which population increases in different countries, by the different heights of men, the following comparison and inference would be much more to the purpose.

If in a particular country we observed that all the people had weights of different sizes upon their heads, and that invariably each individual was tall or short in proportion to the smallness or greatness of the pressure upon him; that every person was observed to grow when the weight he carried was either removed or diminished, and that the few among the whole people who were exempted from this burden, were very decidedly taller than the rest; would it not be quite justifiable to infer, that the weights which the people carried were the cause of their being in general so short; and that the height of those without weights might fairly be considered as the standard to which it might be expected that the great mass would arrive, if their growth were unrestricted?

For what is it in fact, which we really observe with regard to the different rates of increase in different countries? Do we not see that in almost every state to which we can direct our attention, the natural tendency to increase is repressed by the difficulty which the mass of the people find in procuring an ample portion of the necessaries of life, which shews itself more immediately in some or other of the forms of moral restraint, vice, and misery? Do we not see that invariably the rates of increase are fast or slow, according as the pressure of these checks is light or heavy; and that in consequence Spain increases at one rate, France at another, England at a third, Ireland at a fourth, parts of Russia at a fifth, parts of Spanish America at a sixth, and the United States of North America, at a seventh? Do we not see that, whenever the resources of any country increase so as to create a great demand for labour and give the lower classes of society a greater command over the necessaries of life, the population of such country, though it might before have been stationary or proceeding very slowly, begins immediately to make a start forwards? And do we not see that in those few countries, or districts of countries, where the pressure arising from the difficulty of procuring the necessaries and conveniences of life is almost entirely removed, and where in consequence the checks to early marriages are very few, and large families are maintained with perfect facility, the rate at which the population increases is always the greatest?

And when to these broad and glaring facts we add, that neither theory nor experience will justify us in believing either that the passion between the sexes, or the natural prolifickness of women, diminishes in the progress of society; when we further consider that the climate of the United States of America is not particularly healthy, and that the qualities which mainly distinguish it from other countries, are its rapid production and distribution of the means of subsistence;—is not the induction as legitimate and correct as possible, that the varying weight of the difficulties attending the maintenance of families, and the moral restraint, vice, and misery which these difficulties necessarily generate, are the causes of the varying rates of increase observable in different countries; and that, so far from having any reason to consider the American rate of increase as peculiar, unnatural, and gigantic, we are bound by every law of induction and analogy to conclude that these is scarcely a state in Europe where, if the marriages were as early, the means of maintaining large families as ample, and the employments of the labouring classes as healthy, the rate of increase would not be as rapid, and in some cases, I have no doubt, even more rapid than in the United States of America.

Another of Mr. Weyland's curious illustrations is the following:—He says that the *physical tendency* of a people in a commercial and manufacturing to double their number in twenty-five years is "as absolutely gone as the tendency of a bean to shoot up further into the air, after it has arrived at its full growth;" and that to assume such a *tendency* is to build a theory upon a mere shadow, "which, when brought to the test, is directly at variance with experience of the fact; and as unsafe to act upon, as would be that of a general who should assume the force of a musket-shot to be double its actual range, and then should calculate upon the death of all his enemies as soon as he had drawn up his own men for battle within this line of assumed efficiency."

Now I am not in the least aware who it is that has assumed the act *actual* range of the shot, or the actual progress of population in different countries, as very different from what it is observed to be; and therefore cannot see how the illustration, as brought forward by Mr. Weyland, applies, or how I can be said to resemble his miscalculating general. What I have really done is this (if he will allow me the use of his own metaphor): having observed that the range of musket-balls, projected from similar barrels and with the same quantity of powder of the same strength, was, under different circumstances, very different, I applied myself to consider what these circumstances were; and, having found that the range of each ball was greater or less in proportion to the smaller or greater number of the obstacles which it met with in its course, or the rarity or density of the medium through which it passed, I was led to infer that the variety of range observed was owing to these obstacles; and I consequently thought it a more correct and legitimate conclusion, and one more consonant both to theory and experience, to say that the *natural tendency* to a range of a certain extent, or the force impressed upon the ball, was always the same, and the actual range, whether long or short, only altered by external resistance; than to conclude that the different distances to which the balls reached must proceed from some mysterious change its the *natural tendency* of each bullet at different times, although no observable difference could be noticed either in the barrel or the charge.

I leave Mr. Weyland to determine which would be the conclusion of the natural philosopher, who was observing the different velocities and ranges of projectiles passing through resisting media of different densities; and I do not see why the moral and political philosopher should proceed upon principles so totally Opposite.

But the only arguments of Mr. Weyland against the *natural tendency* of the human race to increase faster than the means of subsistence, are a few of these illustrations which he has so unhappily applied, together with the acknowledged fact, that countries under different circumstances and in different stages of their progress, do really increase at very different rates.

Without dwelling therefore longer on such illustrations, it may be observed, with regard to the fact of the different rates of increase in different countries, that as long as it is a law of our nature that man cannot live without food, these different rates are as absolutely and strictly *necessary* as the differences in the power of producing food in countries more or less exhausted: and that to infer from these different rates of increase, as they are actually found to take place, that "population has a *natural tendency* to keep within the powers of the soil to afford it subsistence in every gradation through, which society passes," is just as rational as to infer that every man has a *natural tendency* to remain in prison who is necessarily confined to it by four strong walls; or that the pine of the crowded Norwegian forest has no *natural* tendency to shoot out lateral branches, because there is no room for their growth. And yet this is Mr. Weyland's first and grand proposition, on which the whole of his work turns!

But though Mr. Weyland has not proved, or approached towards proving, that the *natural* tendency of population to increase is not unlimited; though he has not advanced a single reason to make it appear probable that a thousand millions would not be doubled in twenty-five years just as easily as a thousand, if moral restraint, vice and misery, were equally removed in both cases; yet there is one part of his argument, which undoubtedly might, under certain circumstances, be true; and if true, though it would in no respect impeach the premises of the Essay, it would essentially affect some of its conclusions.

The argument may be stated shortly thus;—that the natural division of labour arising from a very advanced state of society, particularly in countries where the land is rich, and great improvements have taken place in agriculture, might throw so large a portion of the people into towns, and engage so many in unhealthy occupations, that the immediate checks to population might be too powerful to be overcome even by an abundance of food.

It is admitted that this is a possible case; and, foreseeing this possibility, I provided for it in the terms in which the second proposition of the Essay was enunciated.

The only practical question then worth attending to between me and Mr. Weyland is, whether cases of the kind above stated are to be considered in the light in which I have considered them in the Essay, as exceptions of very rare occurrence, or in the light in which Mr. Weyland has considered them, as a state of things naturally accompanying every stage in the progress of improvement. On either supposition, population would still be repressed by some or other of the forms of moral restraint, vice or misery; but the moral and political conclusions, in the actual state of almost all countries, would be essentially different. On the one supposition moral restraint would, except in a few cases of the rarest occurrence, be one of the most useful and necessary of virtues; and on the other, it would be one of the most useless and unnecessary.

This question can only be determined by an appeal to experience. Mr. Weyland is always ready to refer to the state of this country; and, in fact, may be said almost to have built his system upon the peculiar policy of a single state. But the reference in this case will entirely contradict his theory. He has brought forward some elaborate calculations to show the extreme difficulty with which the births of the country supply the demands of the towns and manufactories. In looking over them, the reader, without other information, would be disposed to feel considerable alarm at the prospect of depopulation impending over the country; or at least he would be convinced that we were within a hair's breadth of that formidable point of *non-reproduction,* at which, according to Mr. Weyland, the population *naturally* comes to a full stop before the means of subsistence cease to be progressive.

These calculations were certainly as applicable twenty years ago as they are now; and indeed they are chiefly founded on observations which were made at a greater distance of time than the period here noticed. But what has happened since? In spite of the enlargement of all our towns; in spite of the most rapid increase of manufactories, and of the proportion of people employed in them; in spite of the most extraordinary and unusual demands for the army and navy; in short, in spite of a state of things which, according to Mr. Weyland's theory, ought to have brought us long since to the point of *non-reproduction,* the population of the country has advanced at a rate more rapid than was ever known at any period of its history. During the ten years from 1800 to 1811, as I have mentioned in a former part of this work, the population of this country (even after making an allowance for the presumed deficiency of the returns in the first enumeration) increased at a rate which would double its numbers in fifty-five years.

This fact appears to me at once a full and complete refutation of the doctrine, that, as society advances, the increased indisposition to marriage and increased mortality in great towns and manufactories always overcome the principle of increase; and that, in the language of Mr. Weyland, "population, so far from having an inconvenient tendency uniformly to press against the means of subsistence, becomes by degrees very slow in overtaking those means."

With this acknowledged and glaring fact before him, and with the most striking evidences staring him in the face, that even, during this period of rapid increase, thousands both in the country and in towns were prevented from marrying so early as they would have done, if they had possessed sufficient means of supporting a family independently of parish relief, it is quite inconceivable how a man of sense could bewilder himself in such a maze of futile calculations, and come to a conclusion so diametrically opposite to experience.

The fact already noticed, as it applies to the most advanced stage of society known in Europe, and proves incontrovertibly that the actual checks to population, even in the most improved countries, arise principally from an insufficiency of subsistence, and soon yield to increased resources, notwithstanding the increase of towns and manufactories, may I think fairly be considered as quite decisive of the question at issue.

But in treating of so general and extensive a subject as the Principle of Population, it would surely not be just to take our examples and illustrations only from a single state. And in looking at the other countries Mr. Weyland's doctrine on population is, if possible, still more completely contradicted. Where, I would ask, are the great towns and manufactories in Switzerland, Norway and Sweden, which are to act as *the graves of mankind,* and to prevent the possibility of a redundant population? In Sweden the proportion of the people living in the country is to those who live in towns as 13 to 1; in England this proportion is about 2 to 1; and yet England increases much faster than Sweden. How is this to be reconciled with the doctrine that the progress of civilization and improvement is always accompanied by a correspondent abatement in the natural tendency of population to increase? Norway, Sweden and Switzerland have not on the whole been ill governed; but where are the necessary "anticipating alterations," which, according to Mr. Weyland, arise in every society as the powers of the soil diminish, and "render so many persons unwilling to marry, and so many more, who do marry, incapable of reproducing their own numbers, and of replacing the deficiency in the remainder." What is it that in these countries indisposes people to marry, but the absolute hopelessness of being able to support their families? What is it that renders many more who do marry incapable of reproducing their own numbers, but the diseases generated by excessive poverty—by all insufficient supply of the necessaries of life? Can any man of reflection look at these and many of the other countries of Europe, and then venture to state that there is no moral reason for repressing the inclination to early marriages: when it cannot be denied that the alternative of not repressing it must necessarily and unavoidably be premature mortality from excessive poverty? And is it possible to know that in few or none of the countries of Europe the wages of labour, determined in the common way by the supply and the demand, can support in health large families; and yet assert that population does not press against the means of subsistence, and that "the evils of a redundant population can never be necessarily felt by a country till it is actually peopled up to the fall capacity of its resources."

Mr. Weyland really appears to have dictated his book with his eyes blindfolded, and his ears stopped. I have a great respect for his character and intentions; but I must say that it has never been my fortune to meet with a theory so uniformly contradicted by experience. The very slightest glance at the different countries of Europe shews with a force amounting to demonstration, that to all practical purposes the *natural tendency* of population to increase may be considered us a given quantity; and that the actual increase is regulated by the varying resources of each country for the employment and maintenance of labour, in whatever stage of its progress it may be, whether it is agricultural or manufacturing, whether it has few or many towns. Of course this actual increase, or the actual limits of population, must always be far short of the utmost powers of the earth to produce food; first, because we can never rationally suppose that the human skill and industry actually exerted are directed in the best *possible* manner towards the production of food; and secondly, because, as I have stated more particularly in a former part of this work, the greatest production of food which the powers of the earth would admit cannot possibly take place under a system of private property. But this acknowledged truth obviously affects only the actual quantity of food, and the actual number of people, and has not the most distant relation to the question respecting the *natural tendency* of population to increase beyond the powers of the earth to produce food for it.

The observations already made are sufficient to shew that the four main propositions of Mr. Weyland, which depend upon the first, are quite unsupported by any appearances in the state of human society, as it is known to us in the countries with which we are acquainted. The last of these four propositions is the following:—"This tendency" (meaning the natural tendency of population to keep within the powers of the soil to afford it subsistence) "will have its complete operation so as constantly to maintain the people in comfort and plenty in proportion as religion, morality, rational liberty and security of person and property approach the attainment of a perfect influence."

In the morality here noticed, moral or prudential restraint from marriage is not included: and so understood, I have no hesitation in saying that this proposition appears to me more directly to contradict the observed laws of nature, than to assert that Norway might easily grow food for a thousand millions of inhabitants. I trust that I am disposed to attach as much importance to the effects of morality and religion on the happiness of society, even as Mr. Weyland; but among the moral duties, I certainly include a restraint upon the inclination to an early marriage when there is no reasonable prospect of maintenance for a family; and unless this species of virtuous self-denial be included in morality, I am quite at issue with Mr. Weyland; and so distinctly deny his proposition as to say that no degree of religion and morality, no degree of rational liberty and security of person and property, can, under the existing laws of nature, place the lower classes of society in a state of comfort and plenty.

With regard to Mr. Weyland's fifth and last proposition, I have already answered it in a note which I have added, in the present edition, to the last chapter of the third book, and will only observe here that an illustration to shew the precedence of population to food, which I believe was first brought forward by an anonymous writer, and appears so to have pleased Mr. Grahame as to induce him to repeat it twice, is one which I would willingly take to prove the very opposite doctrine to that which it was meant to support. The apprehension that an increasing population would starve unless a previous increase of food were procured for it, has been ridiculed by comparing it with the apprehension that increasing numbers would be obliged to go naked unless a previous increase of clothes should precede their births. Now however well or ill-founded maybe our apprehensions in the former case, they are certainly quite justifiable in the latter; at least society has always acted as if it thought so. In the course of the next twenty-four hours there will be about 800 children born in England and Wales; and I will venture to say that there are not ten out of the whole number that come at the expected time, for whom clothes are not prepared before their births. It is said to be dangerous to meddle with edged tools which we do not know how to handle; and it is equally dangerous to meddle with illustrations which we do not know how to apply, and which may tend to prove exactly the reverse of what we wish.

On Mr. Weyland's theory it will not be necessary further to enlarge. With regard to the practical conclusions which he has drawn from it in our own country, they are such as might be expected from the nature of the premises. If population, instead of leaving a tendency to press against the means of subsistence, becomes by degrees very slow in overtaking them, Mr. Weyland's inference that we ought to encourage the increase of the labouring classes by abundant parochial assistance to families, might perhaps be maintained. But if his premises be entirely wrong, while his conclusions are still acted upon, the consequence must be, a constantly increasing amount of unnecessary pauperism and dependence. Already above one-fourth of the population of England and Wales have been dependent upon parish relief; and if the system which Mr. Weyland recommends, and which has been so generally adopted in the midland counties, should extend itself over the whole kingdom, there is really no saying to what height the level of pauperism may rise. While the practice of making an allowance from the parish for every child above two is confined to the labourers in agriculture, whom Mr. Weyland considers as the breeders of the country, it is essentially unjust, as it lowers without compensation the wages of the manufacturer and artificer: and when it shall become just by including the whole of the working classes, what a dreadful picture does it present! what a scene of equality, indolence, rags and dependence, among one-half or three-fourths of the society! Under such a system to expect any essential benefit from *saving banks* or any other institutions to promote industry and economy is perfectly preposterous. When the wages of labour are reduced to the level to which this system foods, there will be neither power nor motive to save.

Mr. Weyland strangely attributes much of the wealth and prosperity of England to the cheap population which it raises by means of the poor-laws; and seems to think that, if labour had been allowed to settle at its natural rate, and all workmen had been paid in proportion to their skill and industry, whether with or without families, we should never have attained that commercial and manufacturing ascendancy by which we have been so eminently distinguished.

A practical refutation of so ill founded an opinion may be seen in the state of Scotland, which in proportion to its natural resources has certainly increased in agriculture, manufactures and commerce, during the lost fifty years, still more rapidly than England, although it may fairly be said to have been essentially without poor-laws.

It is not easy to determine what is the price of labour most favourable to the progress of wealth. It is certainly conceivable that it may be too high for the prosperity of foreign commerce. But I believe it is much more frequently too low; and I doubt if there has ever been an instance in any country of very great prosperity in foreign commerce, where the working classes have not had good money wages. It is impossible to sell very largely without being able to buy very largely; and no country can buy very largely in which the working classes are not in such a state as to be able to purchase foreign commodities.

But nothing tends to place the lower classes of society in this state so much as a demand for labour which is allowed to take its natural course, and which therefore pays the unmarried man and the man with a family at the same rate; and consequently gives at once to a very large mass of the working classes the power of purchasing foreign articles of consumption, and of paying taxes on luxuries to no inconsiderable extent. While, on the other hand, nothing would tend so effectually to destroy the power of the working classes of society to purchase either home manufactures or foreign articles of consumption, or to pay taxes on luxuries, as the practice of doling out to each member of a family an allowance, in the shape of wages and parish relief combined, just sufficient, or only a very little more than to furnish them with the mere food necessary for their maintenance.

To shew that, in looking forward to such an increased operation of prudential restraint as would greatly improve the condition of the poor, it is not necessary to suppose extravagant and impossible wages, as Mr. Weyland seems to think, I will refer to the proposition of a practical man on the subject of the price of labour; and certainly much would be done, if this proposition could be realized, though it must be effected in a very different way from that which he has proposed.

It has been recommended by Mr. Arthur Young so to adjust the wages of day-labour as to make them at all times equivalent to the purchase of a peck of wheat. This quantity, he says, was earned by country labourers during a considerable period of the last century, when the poor-rates were low, and not granted to assist in the maintenance of those who were able to work. And he goes on to observe that, "as the labourer would (in this case) receive 70 bushels of wheat for 47 weeks' labour, exclusive of five weeks for harvest; and as a family of six persons consumes in a year no more than 48 bushels; it is clear that such wages of labour would cut off every pretence of parochial assistance; and of necessity the conclusion would follow, that all right to it in men thus paid should be annihilated for ever."

An adjustment of this kind, either enforced by law, or used as a guide in the distribution of parish assistance, as suggested by Mr. Young, would be open to insuperable objections. At particular times it might be the means of converting a dearth into a famine. And in its general operation, and supposing no change of habits among the labouring classes, it would be tantamount to saying that, under all circumstances, whether the affairs of the country were prosperous or adverse; whether its resources in land were still great, or nearly exhausted; the population ought to increase exactly at the same rate,—a conclusion which involves an impossibility.

If, however, this adjustment, instead of being enforced by law, were produced by the increasing operation of the prudential check to marriage, the effect would be totally different, and in the highest degree beneficial to society. A gradual change in the habits of the labouring classes would then effect the necessary retardation in the rate of increase, and would proportion the supply of labour to the effective demand, as society continued to advance, not only without the pressure of a diminishing quantity of food, but under the enjoyment of an increased quantity of conveniences and comforts; and in the progress of cultivation and wealth, the condition of the lower classes of society would be in a State of constant improvement.

A peck of wheat a day cannot be considered in any light as excessive wages. In the early periods of cultivation, indeed, when corn is low in exchangeable value, much more is frequently earned; but in such a country as England, where the price of corn, compared with manufactures and foreign commodities, is high, it would do much towards placing the great mass of the labouring classes in a state of comparative comfort and independence; and it would be extremely desirable, with a view to the virtue and happiness of human society, that no land should be taken into cultivation that could not pay the labourers employed upon it to this amount.

With these wages as the average minimum, all those who were unmarried, or, being married, had small families, would be extremely well off; while those who had large families, though they would unquestionably be subjected sometimes to a severe pressure, would in general be able, by the sacrifice of conveniences and comforts, to support themselves without parish assistance. And not only would the amount and distribution of the wages of labour greatly increase the stimulus to industry and economy throughout all the working classes of the society, and place the great body of them in a very superior situation, but it would furnish them with the means of making an effectual demand for a great amount of foreign commodities and domestic manufactures, and thus, at the same time that it would promote individual and general happiness, would advance the mercantile and manufacturing prosperity of the country.

Mr. Weyland, however, finds it utterly impossible to reconcile the necessity of moral restraint either with the nature of man, or the plain dictates of religion on the subject of marriage. Whether the check to population, which he would substitute for it, is more consistent with the nature of a rational being, the precepts of revelation, and the benevolence of the Deity, must be left to the judgment of the reader. This check, it is already known, is no other than the unhealthiness and mortality of towns and manufactories. And though I have never felt any difficulty in reconciling to the goodness of the Deity the necessity of practising the virtue of moral restraint in a state allowed to be a state of discipline and trial; yet I confess that I could make no attempt to reason on the subject, if I were obliged to believe, with Mr. Weyland, that a large proportion of the human race was doomed by the inscrutable ordinations of Providence to a premature death in large towns.

If indeed such peculiar unhealthiness and mortality were the proper and natural check to the progress of population in the advanced stages of society, we should justly have reason to apprehend that, by improving the healthiness of our towns and manufactories, as we have done in England during the last twenty years, we might really defeat the designs of Providence. And though I have too much respect for Mr. Weyland to suppose that he would deprecate all attempts to diminish the mortality of towns, and render manufactories less destructive to the health of the children employed in them; yet certainly his principles lead to this conclusion, since his theory has been completely destroyed by those laudable efforts which have made the mortality of England—a country abounding in towns and manufactories, less than the mortality of Sweden—a country in a state almost purely agricultural.

It was my object in the two chapters on *Moral Restraint,* and its *Effects on Society,* to shew that the evils arising from the principle of population were exactly of the same nature as the evils arising from the excessive or irregular gratification of the human passions in general; and that from the existence of these evils we had no more reason to conclude that the principle of increase was too strong for the purpose intended by the Creator, than to infer, from the existence of the vices arising from the human passions, that these passions required diminution or extinction, instead of regulation and direction.

If this view of the subject be allowed to be correct, it will naturally follow that, notwithstanding the acknowledged evils occasioned by the principle of population, the advantages derived from it under the present constitution of things nary very greatly overbalance them.

A slight sketch of the nature of these advantages, as far as the main object of the Essay would allow, was given in the two chapters to which I have alluded; but the subject has lately been pursued with great ability in the Work of Mr. Sumner on the Records of the Creation; and I am happy to refer to it as containing a masterly developement and completion of views, of which only an intimation could be given in the Essay.

I fully agree with Mr. Sumner as to the beneficial effects which result from the principle of population, and feel entirely convinced that the natural tendency of the human race to increase faster than the possible increase of the means of subsistence could not be either destroyed or essentially diminished without diminishing that hope of rising and fear of falling in society, so necessary to the improvement of the human faculties and the advancement of human happiness. But with this conviction on my mind, I feel no wish to alter the view which I have given of the evils arising from the principle of population. These evils do not lose their name or nature because they are overbalanced by good; and to consider them in a different light on this account, and cease to call them evils, would be as irrational as the objecting to call the irregular indulgences of passion vicious, and to affirm that they lend to misery, because our passions are the main sources of human virtue and happiness.

I have always considered the principle of population as a law peculiarly suited to a state of discipline and trial. Indeed I believe that, in the whole range of the laws of nature with which we are acquainted, not one can be pointed out which in so remarkable a manner tends to strengthen and confirm this scriptural view of the state of man on earth. And as each individual has the power of avoiding the evil consequences to himself and society resulting from the principle of population by the practice of a virtue clearly dictated to him by the light of mature, and sanctioned by revealed religion, it must be allowed that the ways of God to man with regard to this great law of nature are completely vindicated.

I have, therefore, certainly felt surprise as well as regret that no inconsiderable part of the objections which have been made to the principles and conclusions of the Essay on Population has come from persons for whose moral and religious character I have so high a respect, that it would have been particularly gratifying to me to obtain their approbation and sanction. This effect has been attributed to some expressions used in the course of the work which have been thought too harsh, and not sufficiently indulgent to the weaknesses of human nature, and the feelings of Christian charity.

It is probable, that having found the bow bent too much one way, I was induced to bend it too much the other, in order to make it straight. But I shall always be quite ready to blot out any part of the work which is considered by a competent tribunal as having a tendency to prevent the bow from becoming finally straight, and to impede the progress of truth. In deference to this tribunal I have already expunged the passages which have been most objected to, and I have made some few further corrections of the same kind in the present edition. By these alterations I hope and believe that the work has been improved without impairing its principles. But I still trust that whether it is read with or without these alterations, every reader of candour must acknowledge that the practical design uppermost in the mind of the writer, with whatever want of judgment it may have been executed, is to improve the condition and increase the happiness of the lower classes of society.

4.2.17 Appendix III,: 1825

Since the last edition of this Work was published, an answer from Mr. Godwin has appeared; but the character of it, both as to matter and manner, is such, that I quite sure every candid and competent inquirer after truth will agree with me in thinking that it does not require a reply. To return abusive declamation in kind would be as unedifying to the reader as it would be disagreeable in me; and to argue seriously with one who denies the most glaring and best attested facts respecting the progress of America, Ireland, England, and other states, and brings forward Sweden, one of of the most barren and worst-supplied countries of Europe, as a specimen of what would be the natural increase of population under the greatest abundance of food, would evidently be quite vain with regard to the writer himself, and must be totally uncalled for by any of his readers whose authority could avail in the establishment of truth.

Correction to: Malthus: The Last Words. Comments and Reflections of the 6th and last Edition of Malthus's 'Principle of Population' (1826).

David Coleman

Correction to:
Chapter 1 in: David Coleman et al. (eds.), Overpopulation and the Limits to Growth–Our Concern, But Less of or Malthus, https://doi.org/10.1007/978-3-031-94503-8_1

The original version of the book was inadvertently published without a key line in the red graph. This has now been amended in the book and approved by the author.

The updated version of this chapter can be found at
https://doi.org/10.1007/978-3-031-94503-8_1

D. Coleman (ed.), *Thomas Robert Malthus – An Essay on the Principle of Population*,
Classic Texts in the Sciences,
https://doi.org/10.1007/978-3-031-94503-8_5

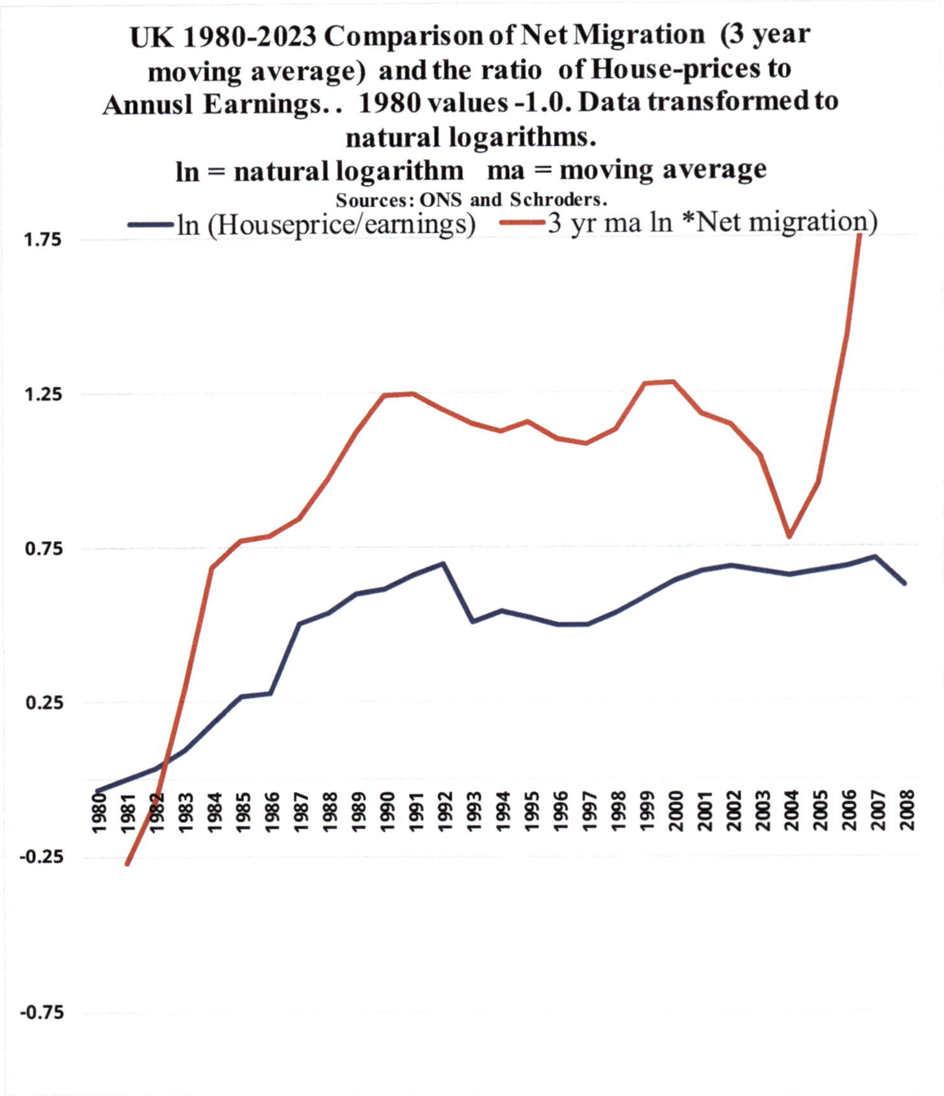

Fig. 1.5 Trend of house-prices and rents in relation to net migration

Index for Malthus Text

D. Coleman (ed.), *Thomas Robert Malthus – An Essay on the Principle of Population*, Classic Texts in the Sciences,
https://doi.org/10.1007/978-3-031-94503-8

Index

D. Coleman (ed.), *Thomas Robert Malthus – An Essay on the Principle of Population*, Classic Texts in the Sciences,
https://doi.org/10.1007/978-3-031-94503-8

MIX
Papier aus verantwortungsvollen Quellen
Paper from responsible sources
FSC® C105338

If you have any concerns about our products,
you can contact us on
ProductSafety@springernature.com

In case Publisher is established outside the EU,
the EU authorized representative is:
Springer Nature Customer Service Center GmbH
Europaplatz 3, 69115 Heidelberg, Germany

Printed by Libri Plureos GmbH
in Hamburg, Germany